ADVANCES IN
X-RAY ANALYSIS

Volume 15

ADVANCES IN X-RAY ANALYSIS

Volume 15

Edited by
Kurt F. J. Heinrich
National Bureau of Standards
Washington, D.C.

and

Charles S. Barrett, John B. Newkirk, and Clayton O. Ruud
Denver Research Institute
The University of Denver
Denver, Colorado

**Proceedings of the Twentieth Annual Conference on
Applications of X-Ray Analysis
Held August 11-13, 1971**

Sponsored by
University of Denver
Denver Research Institute
Metallurgy and Materials Science Division

℗

SPRINGER SCIENCE+BUSINESS MEDIA, LLC

Library of Congress Catalog Card Number 58-35928
ISBN 978-1-4613-9971-1   ISBN 978-1-4613-9969-8 (eBook)
DOI 10.1007/978-1-4613-9969-8

PREFACE

The application of solid-state detectors of high energy resolution to x-ray spectrometry, and the increasing use of computers in both measurement and data evaluation, are giving a new stimulus to x-ray techniques in analytical chemistry. The Twentieth Annual Denver X-ray Conference reflects this renewed interest in several ways.

The invited papers, grouped in Session I, review the characteristics of the detectors used in the measurement of x-rays. One paper is dedicated to the detection of single ions. Although such a subject may appear to be marginal to the purposes of the Denver Conference, we must recognize the affinity of techniques applied to similar purposes. Ion probe mass spectrometry is dedicated to tasks similar to those performed by x-ray spectrometry with the electron probe microanalyzer. Scientists and technologists will see these two techniques discussed in the same meetings.

The discussion of automation and programming is not limited to the two invited speakers, but extends to papers presented in more than one session. The matter of fluorescence analysis by isotope- and tube-excitation will also be of great interest to those concerned with the practical applications of x-ray techniques.

The communications contained in this volume, and the lively discussions which frequently followed the presentation of papers, attest to the vitality of the subjects which are the concern of the Annual Denver X-ray Conference.

Kurt F. J. Heinrich

As conference cochairman I would like to acknowledge the invited speakers and the session cochairmen. The topic of the invited speaker's paper or the title of the session at which the cochairman presided is listed below with the respective name and affiliation of each.

CHARACTERISTICS OF GAS-FLOW PROPORTIONAL COUNTERS. R. J. Liefeld, New Mexico State University, Las Cruces, New Mexico.

THE APPLICATION OF HIGH-RESOLUTION SOLID STATE DETECTORS TO X-RAY SPECTROMETRY--A REVIEW. R. L. Heath, Aerojet Nuclear Corporation, Idaho Falls, Idaho.

DETECTION OF SINGLE IONS BY PULSE COUNTING: APPLICATION TO ION MICROPROBE MASS ANALYZER. L. A. Dietz, General Electric Company, Schenectady, New York.

APPLICATION OF COMPUTERS IN ELECTRON PROBE AND X-RAY FLUORESCENCE ANALYSIS. S. D. Rasberry, National Bureau of Standards, Washington, D.C.

COMPUTER CONTROLLED X-RAY AND NEUTRON DIFFRACTION EXPERIMENTS. M. H. Mueller, Argonne National Laboratory, Argonne, Illinois.

SESSION I. SURVEYS OF DETECTION METHODS AND AUTOMATION SYSTEMS. K. F. J. Heinrich, National Bureau of Standards, Washington, D.C., and C. O. Ruud, University of Denver, Denver, Colorado.

SESSION II. AUTOMATION AND PROGRAMMING. M. H. Mueller, Argonne National Laboratory, Argonne, Illinois, and M. T. Hepworth, University of Denver, Denver, Colorado.

SESSION III. FLUORESCENCE ANALYSIS BY ISOTOPE- AND TUBE-EXCITATION. S. D. Rasberry, National Bureau of Standards, Washington, D.C., and J. C. Dempsey, U.S. Atomic Energy Commission, Washington, D.C.

SESSION IV. LINE BROADENING AND SPECTRAL PHENOMENA. R. L. Heath, Aerojet Nuclear Corporation, Idaho Falls, Idaho, and H. N. Barton, Dow Chemical Company, Golden, Colorado.

SESSION V. PROTON EXCITATION AND ADVANCED EQUIPMENT. R. E. Green, Jr., The Johns Hopkins University, Baltimore, Maryland, and C. E. Lundin, Denver Research Institute, Denver, Colorado.

SESSION VI. DIFFRACTION AND MATERIAL CHARACTERIZATION. R. J. Liefeld, New Mexico State University, Las Cruces, New Mexico, and A. Segmüller, IBM Thomas J. Watson Research Center, Yorktown Heights, New York.

I would like to thank the speakers, the session cochairmen, and the students and staff of the University of Denver's Research Institute for their invaluable contributions to the conference. Particularly, I would like to thank Mrs. Jeanne Cochran, Mr. Daniel Witkowsky, and Mr. Robert Clark for their organization and coordination of conference activities. Finally, the efforts of Dr. K. F. J. Heinrich in the arrangement of the invited speaker session and his enthusiastic participation in the conference are especially appreciated.

As with the previous three volumes of this series, the authors submitted their papers in final form for photoreproduction. The published discussions at the end of a few of the papers are there at the specific request of the speaker and the conferee whose question or comment appears.

Clayton O. Ruud

CONTENTS

Contents

THE APPLICATION OF HIGH-RESOLUTION SOLID STATE DETECTORS TO

X-RAY SPECTROMETRY - A REVIEW

R. L. Heath

Aerojet Nuclear Company, National Reactor Testing Station

Idaho Falls, Idaho

ABSTRACT

Developments during the past few years in solid-state radiation detectors and low-noise electronics employing field-effect transistors operated at cryogenic temperatures have resulted in the availability of high-resolution energy-dispersive spectrometers for a variety of applications in x-ray spectrometry. Using pulse-amplitude analysis techniques, these spectrometers make it possible to obtain multi-elemental analyses on a routine laboratory basis employing x-ray fluorescence techniques. The combination of these spectrometers with small, inexpensive on-line computer data systems makes it possible to obtain rapid on-line qualitative and quantitative analysis of samples in the laboratory and in special field applications. A general review of the present state of development in detectors, electronics and on-line data systems will be presented together with descriptions of applications of such equipment in the laboratory.

INTRODUCTION

In the past two or three years the use of energy-dispersive x-ray spectrometers based on lithium-ion drifted Si and Ge solid-state detectors has grown considerably. Although the experimental techniques, which were pioneered in nuclear physics laboratories, involve the use of many sophisticated techniques, the development of small digital computer systems has made it possible to develop laboratory systems which are capable of rapid qualitative and quantitative assay on a real-time basis. It is the purpose of this

1

review to present the state of development in this area and to
describe application of this equipment to routine laboratory
problems.

Recent developments in this field include improved energy
resolution (approaching 100 eV) due to the development of improved
low-noise preamplifiers which employ field-effect transistors
operated at cryogenic temperatures and optoelectronic and pulsed
feedback systems. This improvement in energy resolution makes it
possible to resolve adjacent elements down to carbon. To provide
the laboratory spectroscopist with a basis for comparison with
other techniques, a brief review of the principles and techniques
employed in photon spectrometry will be presented. This will
include a discussion of the elements of a modern pulse-amplitude
analysis system.

DETECTORS

The solid-state detector most frequently used in high-
resolution energy-dispersive systems at this time is the lithium-
ion drifted device. In 1960, Pell(1) first demonstrated that
lithium ions could be drifted through a single crystal of silicon
or germanium and successfully compensate for the presence of
acceptor impurities in semiconductor materials. Using this technique, lithium ions are used to create a P-I-N structure such as
is shown in Figure 1. These devices are fabricated by diffusing
lithium into the surface of an ingot of p-type Si or Ge, forming
an n-type region on one surface. A bias voltage is then applied to
the ingot at elevated temperature causing the lithium ions to drift
through the p-type material. During this process, an equilibrium
condition is established where lithium ions pair with atoms of the
doping material (gallium or indium in the case of p-type germanium)
creating a region with intrinsic properties. The term intrinsic
implies that charge carriers will have a very long lifetime with
all recombination processes reduced to a negligible level. Under
reverse bias, the charge carriers can then be collected from this
region, resulting in the production of a solid-state ionization
chamber with appreciable volume. Using the best available single
crystal germanium material, P-I-N structures have now been fabricated with intrinsic regions up to 2 centimeters in depth. Since
the mobility of lithium ions is not negligible at room temperature,
such devices, once fabricated, should be maintained at dry ice
temperature or below to preserve the structure.

Prior to any detailed discussion of the elements of a lithium-
ion drifted silicon x-ray spectrometer, let us first examine a few
examples of the performance of high-resolution systems. The development of low noise pulse amplifying systems which employ field-
effect transistors operating in the liquid nitrogen temperature

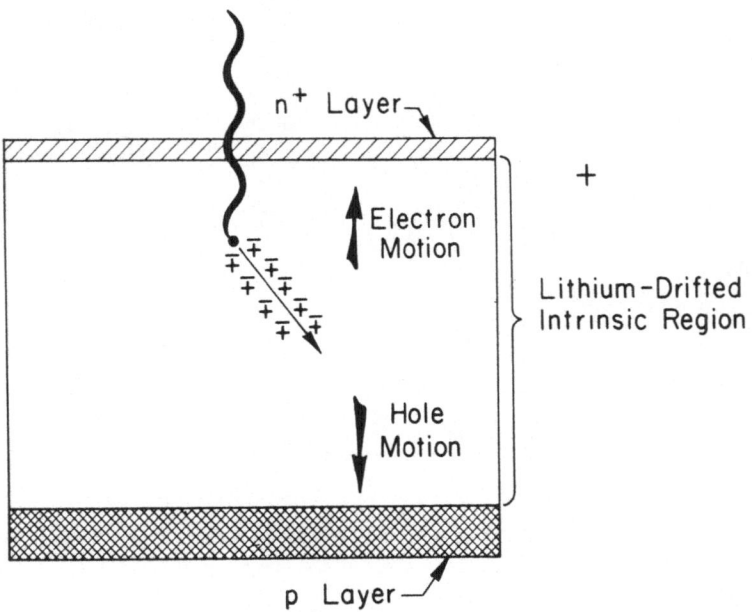

Fig. 1. Illustration of P-I-N planar structure of a lithium-drifted germanium solid-state detector.

range has made it possible to achieve energy resolution approaching the statistical limits imposed by the variance in the number of hole-electron pairs created in the sensitive volume of the detector. Figures 2 and 3 are pulse-height spectra obtained with systems using Si(Li) detectors and cryogenic FET preamplifiers. Figure 2 is a pulse-height spectrum obtained with a Si(Li) detector mounted in a cryostat with a cooled FET preamplifier. These data represent the low-energy photon spectrum emitted by the radioisotope ^{57}Co. This radioisotope decays by electron capture, giving rise to Fe x-rays and nuclear gamma rays with an energy of 14.4 keV. We observe that the K_α and K_β Fe x-rays are resolved. The measured resolution in the energy region (full width at half maximum) is 250 eV for this system. The noise contribution of the electronic system was 150 eV.

Other features of the spectrum indicate certain experimental problems which are important when analyzing x-ray spectra in detail. The small peak below the Fe x-ray results from events which occur very close to the surface of the detector with subsequent escape of the Si x-ray produced following a photoelectric interaction. Thus the energy loss in the detector is due only to the primary photoelectron. The two small peaks at the high energy end of the spectrum result from x-ray fluorescence in the indium-tin eutectic used to form an electrical contact on the surface of the detector. This can be eliminated by proper choice of contact materials.

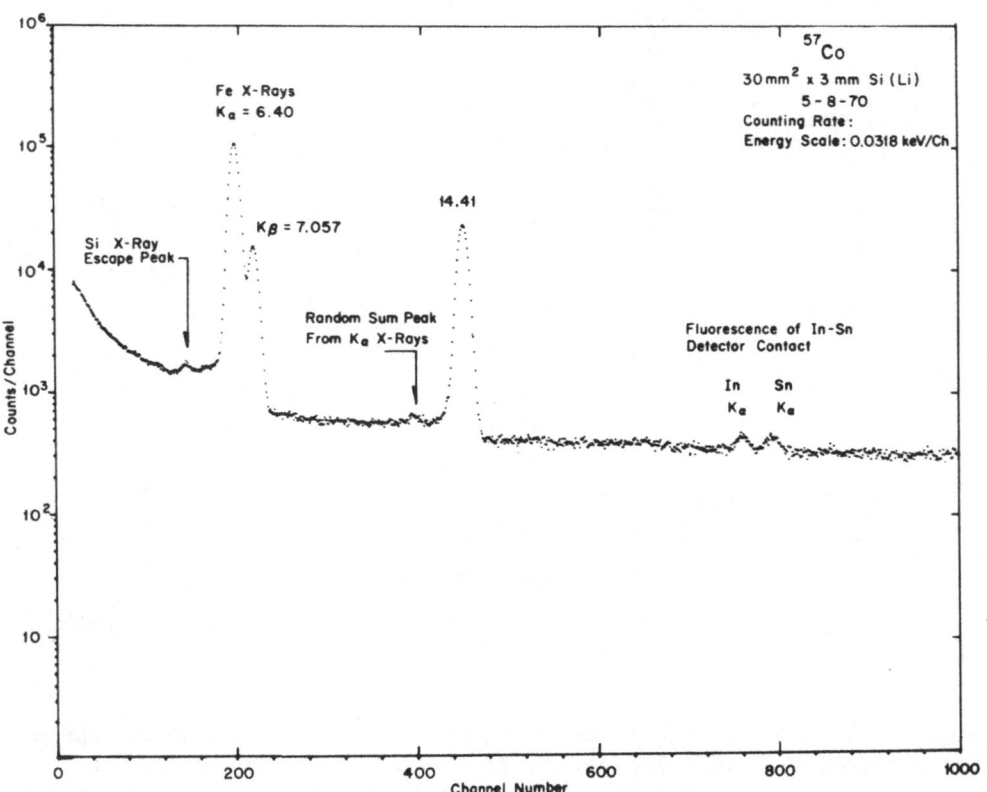

Fig. 2. Pulse-amplitude spectrum obtained using a ^{57}Co source and Si(Li) detector illustrating the low-energy photon spectrum obtained with a solid-state detector spectrometer. Spectrum shows Fe K_α and K_β x-rays and 14.4-keV gamma ray. Also illustrated are spurious effects due to Si x-ray escape from the surface of the detector and In and Sn x-rays produced from fluorescence in materials used in electrical contact on surface of solid-state detector.

While we are observing the x-ray spectra it should be pointed out that the continuum of pulses underlying the characteristic x-ray peaks is in part due to scattered radiation and partially due to events which occur in regions of reduced electric field within the detector, resulting in less than total charge collection. To reduce this effect, physical collimation of x-rays into the detector is advisable. Another solution is to use a guard-ring structure in the fabrication of x-ray detectors as suggested by Goulding(2).

Quite recently improvements in the noise performance of low-noise FET preamplifiers through the development of optoelectronic and pulsed feedback techniques have made it possible to achieve energy resolution approaching 100 eV. Figure 3 shows an ^{55}Fe

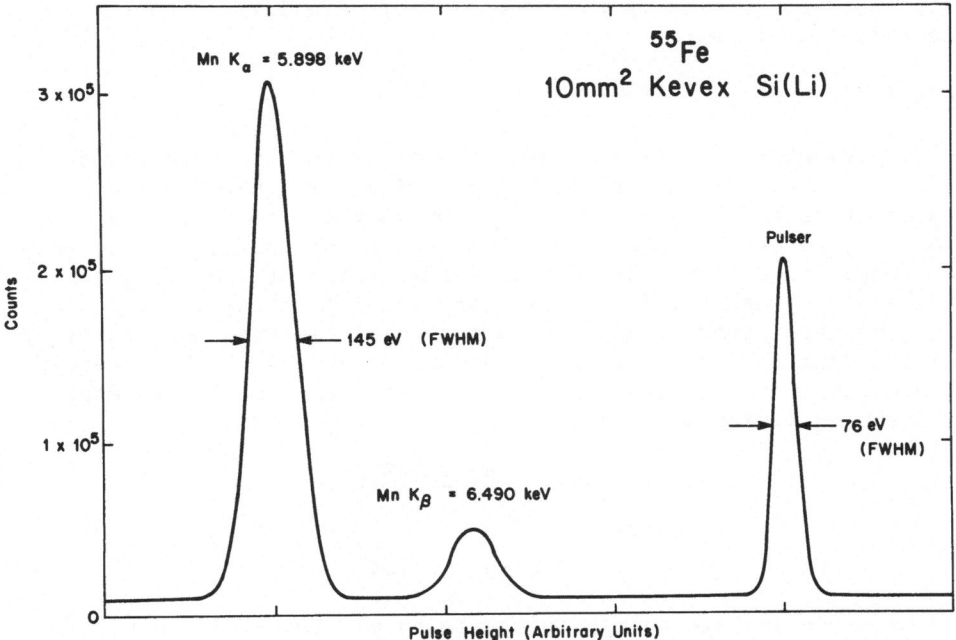

<u>Fig. 3</u>. Pulse-height spectrum of Mn x-rays emitted from a
^{55}Fe source, illustrating typical performance of a pulsed feedback
system with optical coupling.

spectrum obtained with a small Si(Li) detector using a pulsed
optical feedback system. As indicated the resolution of the Fe K_α
x-ray is 145 eV with electronic noise contribution of 76 eV as
indicated by the pulser line.

ENERGY RESOLUTION

Let us first review the factors influencing the energy
resolution of a solid-state detector system. The measured width
of a peak in a pulse-height spectrum resulting from interaction of
a primary photon by the photoelectric effect is used as a measure
of the resolution. This width is related to the statistical spread
in the number of electron-hole pairs created in the detector following an ionizing event. In addition to this fundamental limitation, the observed peak width also includes contributions from a
number of "noise sources" arising from other elements of the electronic system. These may be attributed to problems in the detector

and the characteristics of the preamplifier, amplifier, and the
pulse-height analyzer.

Detector

In germanium, on the average, 2.94 eV of energy is required
to produce an electron-hole pair (3.6 eV for silicon). Following
the production of a high-energy electron in the detector, energy
is lost either by the production of electron-hole pairs or processes
which result in optical photon collisions. Since the sharing of
this energy between these two competing modes is statistical in
nature, we must speak of an average number of electron-hole pairs
resulting from the loss of a given amount of energy in the detector.
It is customary to express the experimental resolution of a detector in the following manner:

$$\text{FWHM (keV)} = 2.355 \sqrt{FE\varepsilon}$$

where

      E = the energy of the electron in keV,

      ε = the average energy to create an electron-hole pair,

      F = the Fano factor, which is related to the yield or
          fractional amount of total energy absorbed which results
          in the production of electron-hole pairs.

Other sources of noise which can be attributed to a practical
detector configuration include (a) fluctuations in detector leakage
current, (b) charge collection problems resulting from trapping of
charge carriers or variation in charge collection time which will
produce a variation in pulse rise-time, and (c) shunt resistance
noise sources in the preamplifier input circuit.

It is of interest to determine a value for the Fano factor, F,
to obtain an estimate of the ultimate limit in energy resolution
to be expected from a detector. Figure 4 shows the results of
measurements made in our laboratory to obtain an experimental value
for the Fano factor in lithium-ion drifted silicon devices. This
figure indicates experimental values for the energy resolution of
a silicon spectrometer system as a function of energy in the x-ray
region. To obtain an estimate of the detector line width, the
contribution from all other factors can be assumed to be given by
the observed width of a mercury-switch tail pulse generator measured simultaneously with the x-ray source. The resulting line
width for the detector, after removal of the contribution from
system noise, is shown by the triangles. Also plotted for comparison are two solid curves representing the expected line width for
Fano factor values of 0.2 and 0.15 for silicon. The experimental
points are seen to agree quite well with a value of 0.15 for the

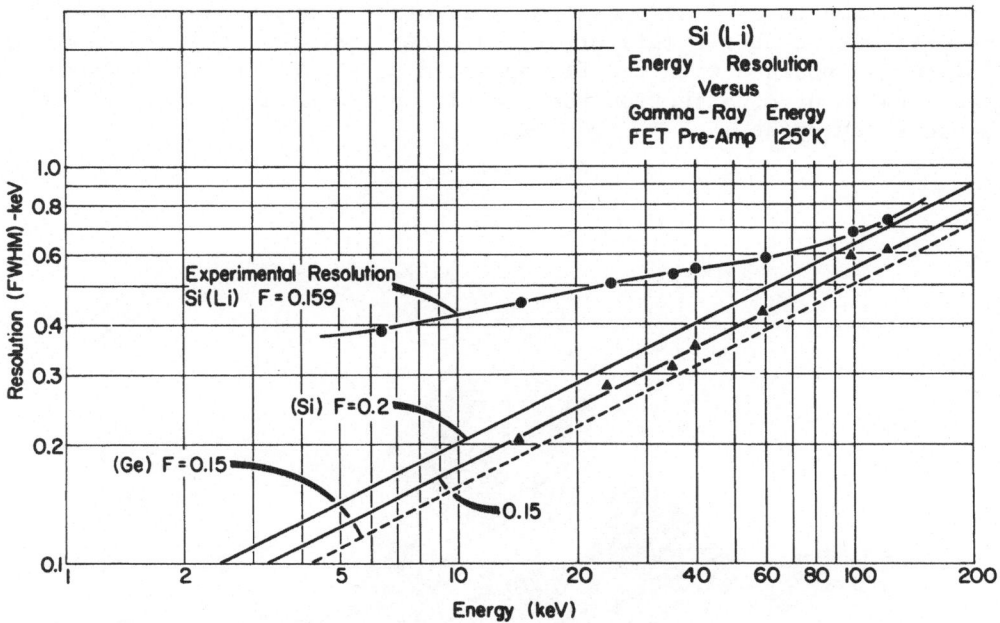

Fig. 4. Measured energy resolution vs. x-ray energy for
Si(Li), showing value for the Fano factor in silicon devices.

Fano factor in silicon. Also indicated for comparison by the
dashed line is the expected resolution for lithium-drifted
germanium detectors for F = 0.15.

From an examination of these data one may conclude that (1) in
the x-ray region, the experimental values for energy resolution are
limited largely by preamplifier noise, and (2) that improved resolution could be obtained by using germanium devices because of the
lower values for the energy required to create an electron-hole
pair in germanium.

Preamplifier

As an example of the electronic systems employed to achieve
the ultimate in energy resolution with these devices, let us examine the cryogenic units presently being used in our laboratory.

The preamplifier presently in use was developed at this
laboratory and uses 2N 3823 field-effect transistors (FET) in the

input stages to achieve excellent resolution characteristics when
coupled with either Si(Li) or Ge(Li) detectors.(3) Optimum resolution is obtained with the FET operated at 140°K. Figure 5 shows
the details of construction for a cooled preamplifier installed in
a vacuum cryostat.

Fig. 5. Photograph of assembly of FET preamplifier in cryostat
prior to mounting of detector.

The performance of lithium-ion drifted silicon devices is
quite sensitive to operating temperature. In view of the advantages of using silicon in the x-ray energy region, (1) availability
with thin entrance windows, and (2) lower sensitivity to high-energy
gamma rays, studies were conducted to determine an optimum operating
temperature for these devices for the detection of electrons and
low-energy photons. Figure 6 summarizes the results obtained. The
curves in the figure indicate energy resolution for three different
detectors for electrons and for 22-keV x-rays as a function of
detector temperature. The resolution for x-rays is relatively
independent of temperature below 200°K while resolution for electrons exhibits a relatively narrow minimum. This minimum appeared
at a different temperature for each detector tested. The difference

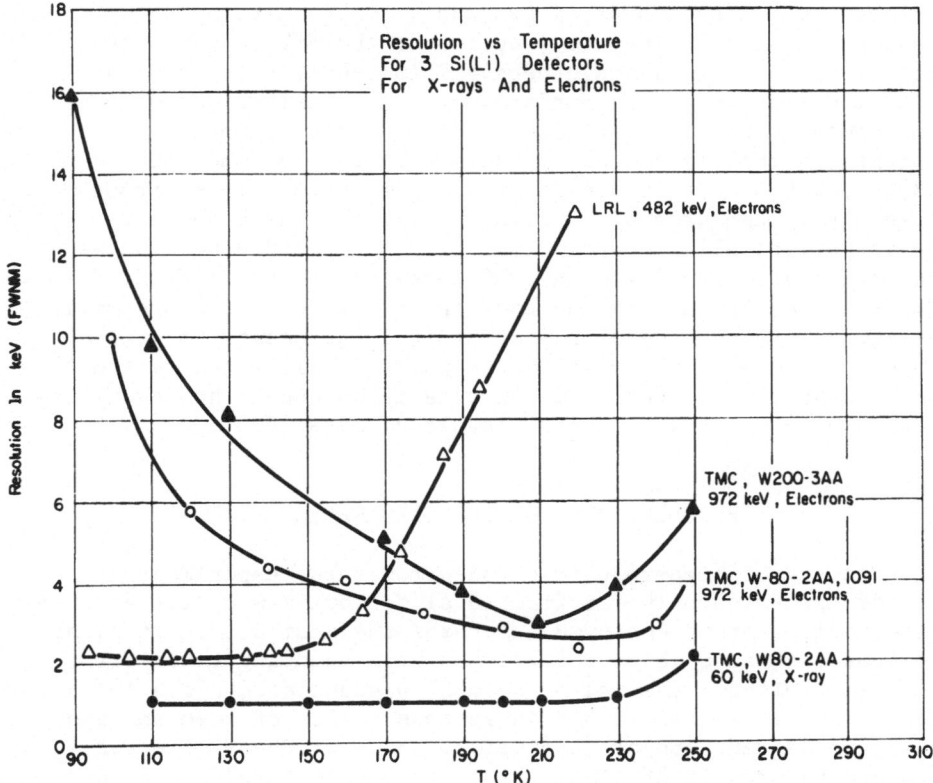

Fig. 6. A plot of resolution vs. operating temperature for several Si(Li) detectors. The variation in performance of the devices is thought to result from charge collection problems due to "trapping" at low temperatures.

in behavior for electron and x-ray detection seems to occur at all energies and is interpreted as the result of non-homogeneous charge collection problems. This suggests that there are still some difficulties associated with techniques used in the fabrication of detectors, or with the initial material from which they are made. This is born out by the fact that the inherent resolution to be expected from Si(Li) detectors seems to never quite be achieved in practice. Over the past two years a number of important improvements have been made in low-noise FET preamplifiers. The conventional charge-sensitive input stage employs a feedback loop consisting of a resistor and capacitor. Typically, low-noise preamplifiers employed feedback resistors having values in the thousands of Megohms. With selected components the limit on the noise performance of the input stage appeared to result in part from stray capacity and spurious noise arising in the feedback resistors. To overcome this limitation the optoelectronic feedback approach has been applied to achieve improved noise performance. The resistor

is replaced by a light coupling between a light-emitting diode and
the photo-sensitive drain-gate junction in the FET in circuits
developed by Goulding(4) and others. This resulted in reduced
capacity and improved noise performance. The limitations of this
system, however, resulted àt high counting rates, from non-
linearities in the light-coupling network. To overcome this, a
pulsed feedback scheme is presently being applied. This arrangement was first suggested by Kandiah(5). In this arrangement,
charge is periodically returned to the input by applying a pulse
through the light-coupling when a DC level sensor has indicated that
the DC level is approaching the non-linear region of the preamplifier stage. In this mode of operation it is necessary to gate off
the input to the ADC following charge pulse to allow the system to
return to a quiescent state. Using this technique it has been possible to achieve electronic noise levels of less than 100 eV.

PROBLEMS AT HIGH COUNTING RATES

     In the use of high-resolution spectrometers, experimental
conditions are frequently far from ideal for optimum performance of
the electronic systems employed. Perhaps the most difficult class
of problems are those which result when high counting rates are
encountered. Without the use of special precautions, operation at
high counting rates will result in serious shifts of zero and system gain, and degradation of resolution. Most of these observed
effects are the result of fluctuations in the zero reference baseline at the input of the analog-to-digital converter, produced by
random fluctuations. To provide some insight into these effects
and the nature of the specialized circuitry employed to correct
these problems, let us examine part (a) of Figure 7. To achieve
optimum signal-to-noise ratio in high-resolution systems, a monopolar pulse shape with RC equal integration and differentiation time
constants is generally employed. As indicated in Figure 7(a) the
output pulse shape from such a filter network will exhibit a more-
or-less Gaussian shape with equal rise and fall times. Following
the pulse there will be a negative undershoot with a long recovery
time constant (generally several hundred microseconds). At low
counting rates, the pulse will have returned to the original baseline and will have no influence on succeeding events. If, however,
at high rates, another pulse (indicated by the dotted line) occurs,
the negative tail of the preceding pulse will result in a reduced
amplitude measurement by the ADC. At reasonable rates (a few thousand pulses per second) the net effect of this will be an asymmetry
of peaks as indicated at the far right of the figure.

     A circuit technique, first proposed by Knowlin and
Blankenship(6), which has been termed "pole-zero cancellation", has
been developed which quite effectively reduces the undershoot from
a monopolar pulse. The net effect of pole-zero cancellation is

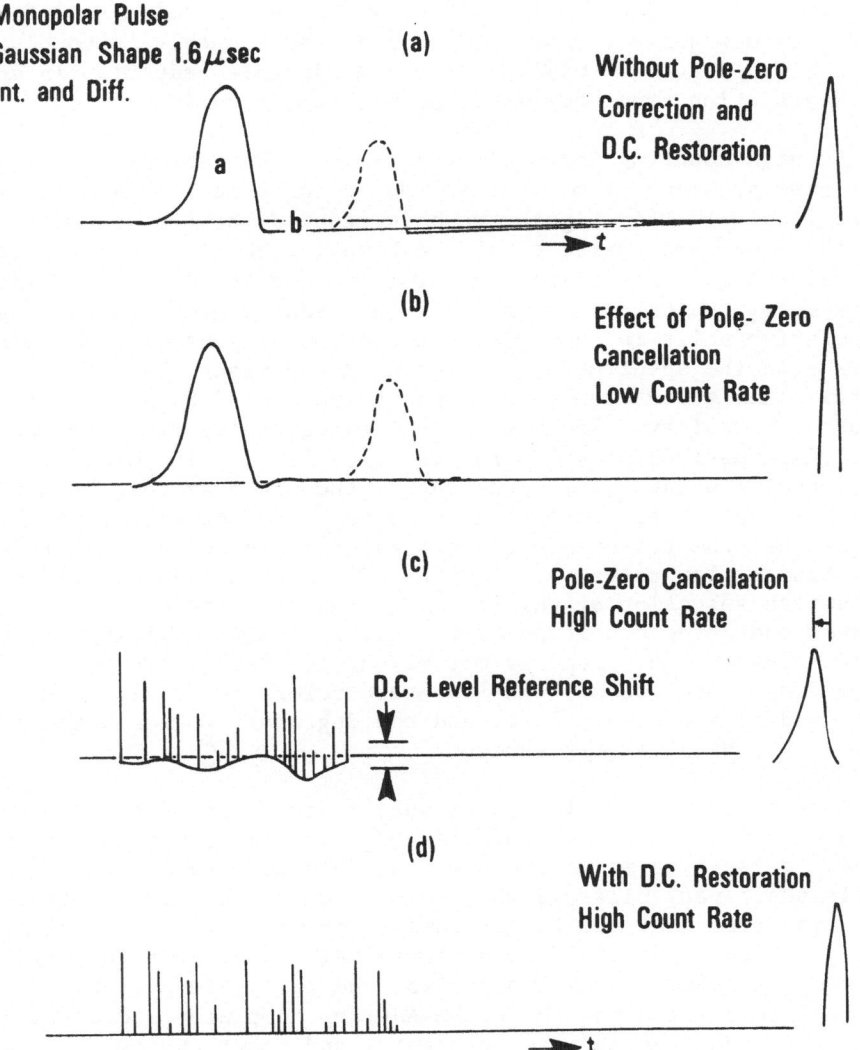

Fig. 7. Pictorial representation of problems which result in distortion of peaks in pulse-height spectra as a function of input signal rate: (a) system without pole-zero cancellation networks in amplifier at low rate, (b) system with pole-zero cancellation networks in amplifier at low input rate, (c) system at high input rates with no DC restoration and (d) system at high input rates with both pole-zero and DC restoration circuits optimized.

shown in Figure 7(b). Here we see that the undershoot returns quickly to the original baseline, thus pulses following will be unaffected. It should be pointed out that pole-zero cancellation must be accomplished on the complete preamplifier-amplifier system

to be effective. The use of such networks in low-noise amplifier
systems is now quite common, and if adjusted properly in combination
with a given detector, will provide considerable reduction in spectrum degradation from undershoot at moderate counting rates.

     At high counting rates (in excess of 5000 counts/sec), we
still have problems from fluctuations in the baseline as shown in
Figure 7(c). In this figure, we have reduced the time scale to
show the long-term character of the baseline shift. As a result of
residual charge on coupling capacitors within the AC coupled amplifier system, the baseline will vary in a random manner producing a
net negative shift in the zero reference and a general broadening
of peaks in the spectrum as indicated on the right side of the
figure. This effect can be reduced by the use of DC baseline-
restoring circuitry. The action of these circuits is to remove the
random fluctuations in the zero reference level. It should be
stated that the input time constant on the restorer must be optimized for best performance at high rates. The net effect of this
will be that the restorer will restore on noise and a small loss in
resolution at low rates will result. The most recent DC restorers
have switch selection of this time constant to permit selection of
the best operating conditions for a given experiment. The result
of restoring action, properly optimized, is illustrated in
Figure 7(d). Here we see that the zero reference baseline has been
maintained at a constant level and the degradation seen without the
restorer has been largely removed.

     At high input counting rates pulses are arriving closely
spaced in time at the input of the ADC. To simplify a discussion
of these effects it is most convenient to divide them into
two classes: peak pile-up, which occurs when two pulses arrive
within the time that the linear gate of the analog-to-digital converter is open, thus yielding a stored pulse amplitude which is
related to the sum of the two pulses; and tail pile-up, which
results from a situation where the ADC has completed a conversion,
stored the resulting address in memory, and reset the ADC. If any
deviation from the baseline as a result of a signal which was presented to the linear gate by the amplifier shortly before ADC reset
exists, and a second pulse arrives, the linear gate will then open
and the resultant amplitude will again be related to the sum of the
preceding pulse and the new event. A block diagram of a complete
spectrometer system which includes rejection circuitry for eliminating questionable events is shown in Figure 8.

Tail Pile-up Rejection

     Tail pile-up rejection is normally accomplished by including
in the ADC input circuitry an "inspection circuit" which requires
that the reference baseline be identically zero before a new pulse
can be accepted by the ADC. As shown in the figure, the shaping

DC Pulse Height Analysis System

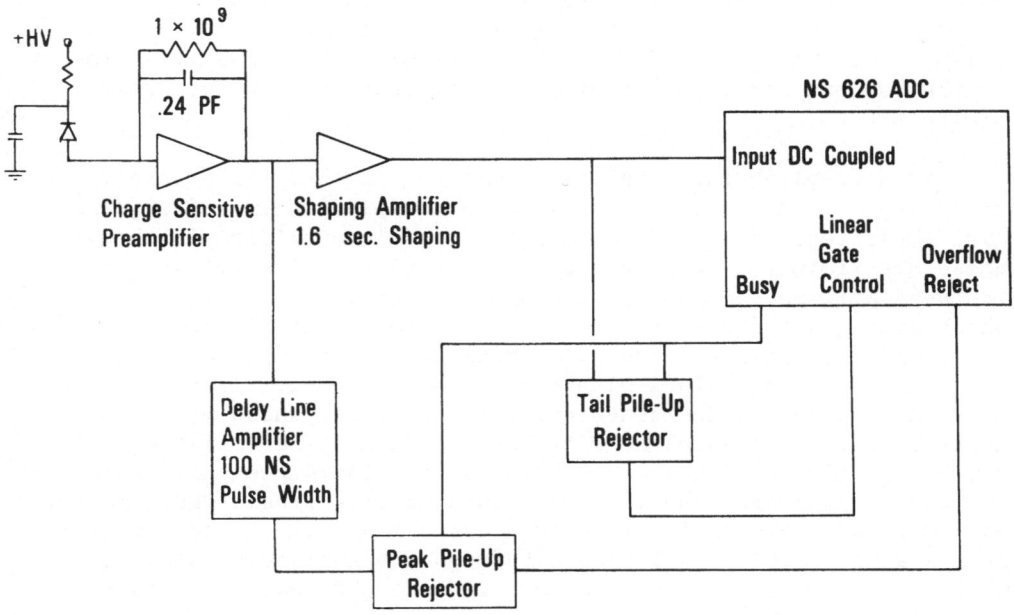

Fig. 8. Block diagram of completely DC coupled electronic
system designed for high-resolution spectrometry at high counting
rates. System employs both peak and tail pile-up rejection
circuitry.

amplifier output is fed into a sensitive baseline discriminator
whose threshold is preferably set in the middle of the baseline
noise band. The discriminator output and the ADC-busy signal feed
a gated bistable whose output controls the ADC linear gate. The
conditions that must be satisfied to open the linear gate are ADC-
not-busy, and baseline restored. The only condition that will close
the linear gate is the presence of the ADC-busy signal.

Peak Pile-up Rejection

Peak pile-up rejection is generally accomplished by special
external circuitry which incorporates fast differentiating networks
which inspect short intervals of time to determine whether or not
two pulses have occurred during the time an ADC conversion has been
requested. If this is the case, a reject signal will be generated
and the event rejected by the ADC. As indicated in Figure 8, we
are presently accomplishing this without delaying the signal pulse

train by utilizing the ADC reject circuitry. Occurrence of a
second pulse from the amplifier in a time period starting 50 nano-
seconds after the primary pulse rise and ending at the time the ADC
detects the peak amplitude of the primary pulse will cause a reject
pulse to be generated. The reject pulse will then inhibit storage
of the rundown capacitor information and cause an internal reset
of the ADC.

To accomplish this, pulses from the output of the preamplifier
are clipped to a 50 nanosecond width and amplified in a fast delay
line amplifier. The delayed output from this amplifier is then
noise discriminated such that the output of the discriminator in
the peak pile-up rejector produces pulses of uniform amplitude and
whose widths are equal to the noise baseline width of the delay-
line amplifier pulses. The discriminator output feeds an anti-
coincidence circuit along with the ADC-busy signal. If the ADC is
not busy, an anti-coincidence pulse is generated whose width is
equal to the noise baseline width of the delay-line amplifier output pulse. The trailing edge of the anti-coincidence pulse sets a
3-microsecond single shot. The outputs of this single-shot and the
anti-coincidence circuit, in turn, feed a fast coincidence network.
If a discriminator signal passes through the anti-coincidence gate
while the 3-microsecond single-shot is set, a reject pulse is then
generated.

System Performance

The performance of the DC-coupled amplifier system
incorporating tail pile-up rejection is illustrated in Figures 9
and 10. The system used to obtain these data employed a 3.5-cc
planar detector. The two figures indicate the shape of the 1.33-
MeV photopeak from a ^{60}Co source for input counting rates varying
from 2,000 pulses/sec to 75,000 pulses/sec. It is important in considering the degradation in resolution shown, that the system resolution is essentially intrinsic (i.e., limited almost entirely by
detector statistics). If the resolution of a system employed to
demonstrate high-rate performance does not meet this requirement,
the relative peak widths as a function of input counting rates may
be somewhat deceiving. Observing Figure 9, we see that the system
resolution at 2,000 c/sec is 1.92 keV (FWHM). Beneath each set of
peak shapes are listed the conditions (baseline inspection in or
out), input rate, resolution at both half-height and one-tenth of
the maximum peak amplitude, and the peak center location in channels.
Proceeding upward in counting rate, we note that essentially no
change in resolution or peak shape has occurred at a counting rate
of 25,000 pulses/sec. The peak width has increased by about 0.2
keV (10%). The peak location has essentially remained unchanged.
The difference of one channel represents a long-term drift in the
system since the data were taken on a different day from the two
preceding runs. At 50,000 pulses/sec, we see a further slight

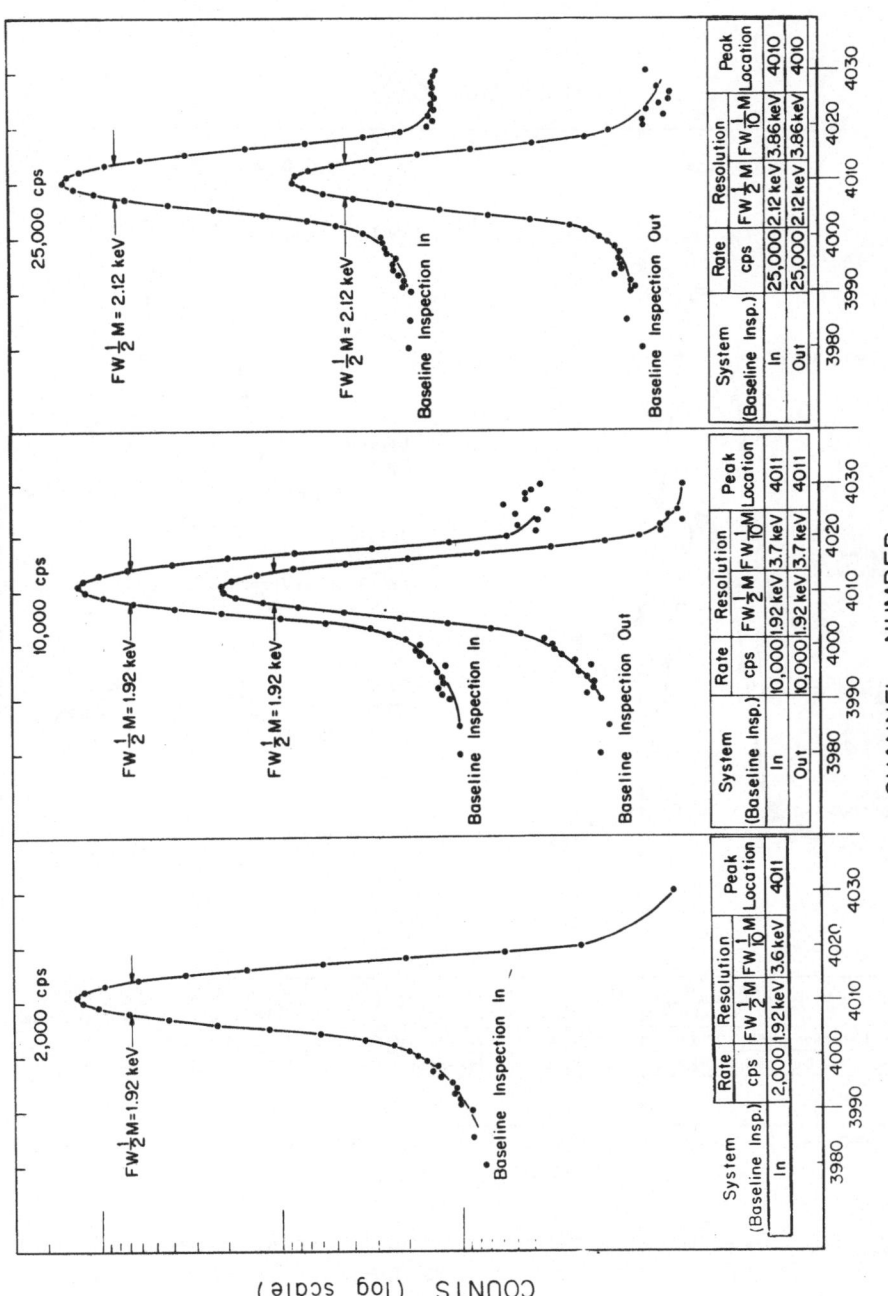

CHANNEL NUMBER

Fig. 9. Performance of DC-coupled system as a function of input counting rate showing the 1.33-MeV peak of ^{60}Co at input rates of 2,000, 10,000 and 25,000 pulses/sec.

1.33 MeV Co⁶⁰

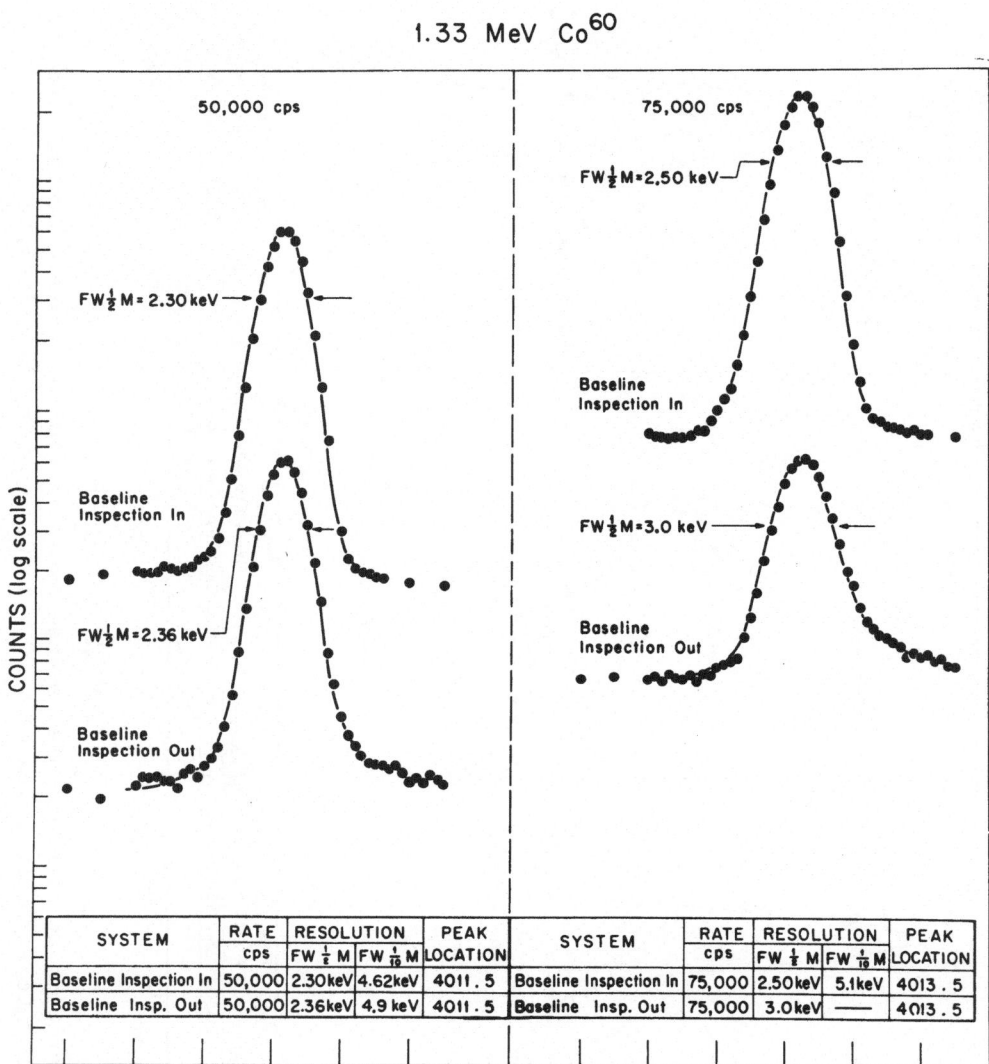

Fig. 10. System performance for ⁶⁰Co at input rates of 50,000 and 75,000 pulses/sec. For comparison, spectra are shown with and without tail pile-up rejection system in operation.

increase in peak width. It should be noted that at this rate, the
effect of baseline inspection is becoming significant. Without the
baseline inspection circuitry in operation, the peak is somewhat
asymmetric on the high-energy side and the resolution is somewhat
degraded from that which was measured with the tail pile-up rejection circuitry in operation. At 75,000 pulses/sec this effect is
even more dramatic. This is also attributed to long-term DC drift.
In short-term measurements we have seen no observable drift in zero
up to input rates of 100,000 pulses/sec. Observing the general
character of the spectra at high rates we see no asymmetry on the
low-energy side of the peaks, which would normally be the case even
with the best baseline DC restorer in an AC-coupled system. The
general rise in the continuum relative to the peak is attributed to
peak pile-up. This results from the fact that random coincidences
will remove pulses from the peak to distribute them over the
continuum.

SPECTROMETER SYSTEM

The elements of an energy-dispersive x-ray spectrometer are
shown in Figure 11. In this figure we have shown the block diagram
of a conventional laboratory pulse-height analyzer. As indicated,
it consists of an energy-sensitive detector, cryogenic FET preamplifier, linear pulse amplifier, and the pulse-height analyzer. The
pulse-height analyzer, here termed a "hard-wired" analyzer to differentiate it from a computer data system to perform the same task,
utilizes an analog-to-digital converter which normally divided a
0-10 volt pulse amplitude range into 256 to 8000 equal increments
or channels. The amplitude information is stored in a magnetic-
core memory for subsequent use. Laboratory pulse-height analyzers
normally provide an oscilloscope display to view data in final
stored form and a choice of readout devices which include magnetic
tape, digital printer, point plotter, or punched paper tape to provide for analysis of pulse-height data on large central computer
facilities. These instruments, which were highly developed for
energy spectrometry of gamma rays and charged particles by experimental nuclear physicists, have evolved into highly complex and
versatile laboratory tools. Pulse-height analyzers presently available incorporate a variety of functions to assist the user in the
acquisition and analysis of data. Succeeding experimental results
may be stored in different segments of magnetic core memory, displayed simultaneously, added or subtracted from one another, and
simple arithmetic operations performed to partially analyze the
pulse-height spectra stored in memory. While they are quite versatile, complete data analysis cannot be achieved without further
reduction, using either manual techniques or transfer of the data to
a central computer.

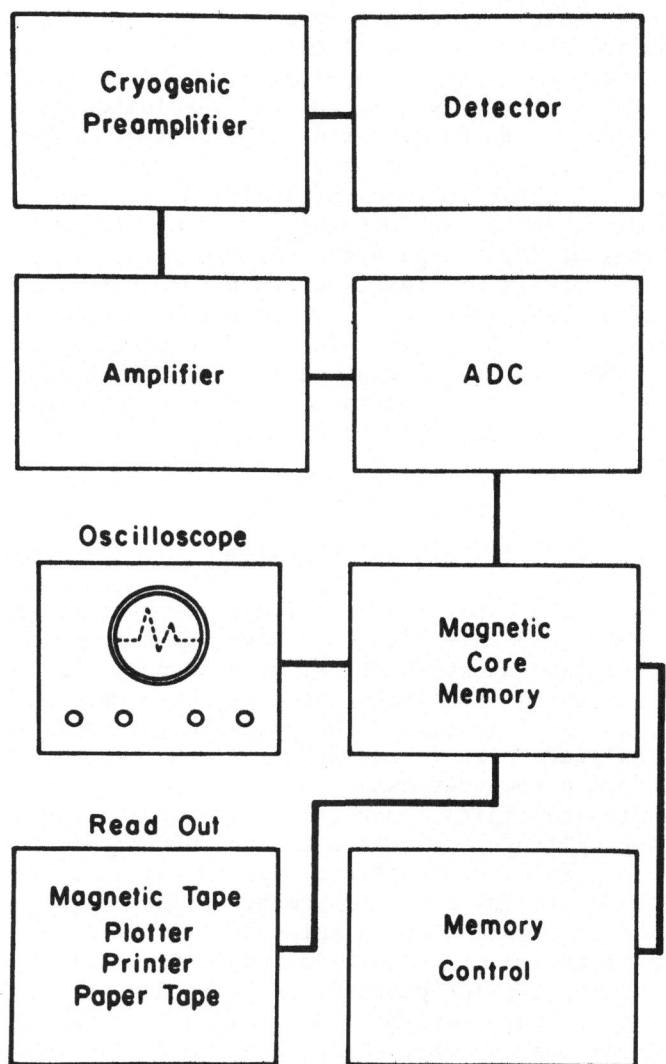

<u>Fig. 11</u>. Block diagram of conventional "hard-wired" pulse-height analyzer.

As described above, the Si(Li) spectrometer produces data in the form of a pulse-amplitude spectrum. A typical spectrum obtained from an x-ray fluorescence measurement is shown in Figure 12. This spectrum was obtained by irradiating a sample of Monazite (thorium-bearing) ore with gamma rays emitted from a radioactive source of ^{241}Am. The exciting radiation resulted in the production of x-rays characteristic of the major elements present in the ore sample. As indicated, the major constituents of the ore sample were thorium, and rare earth elements, characterized by the L-series

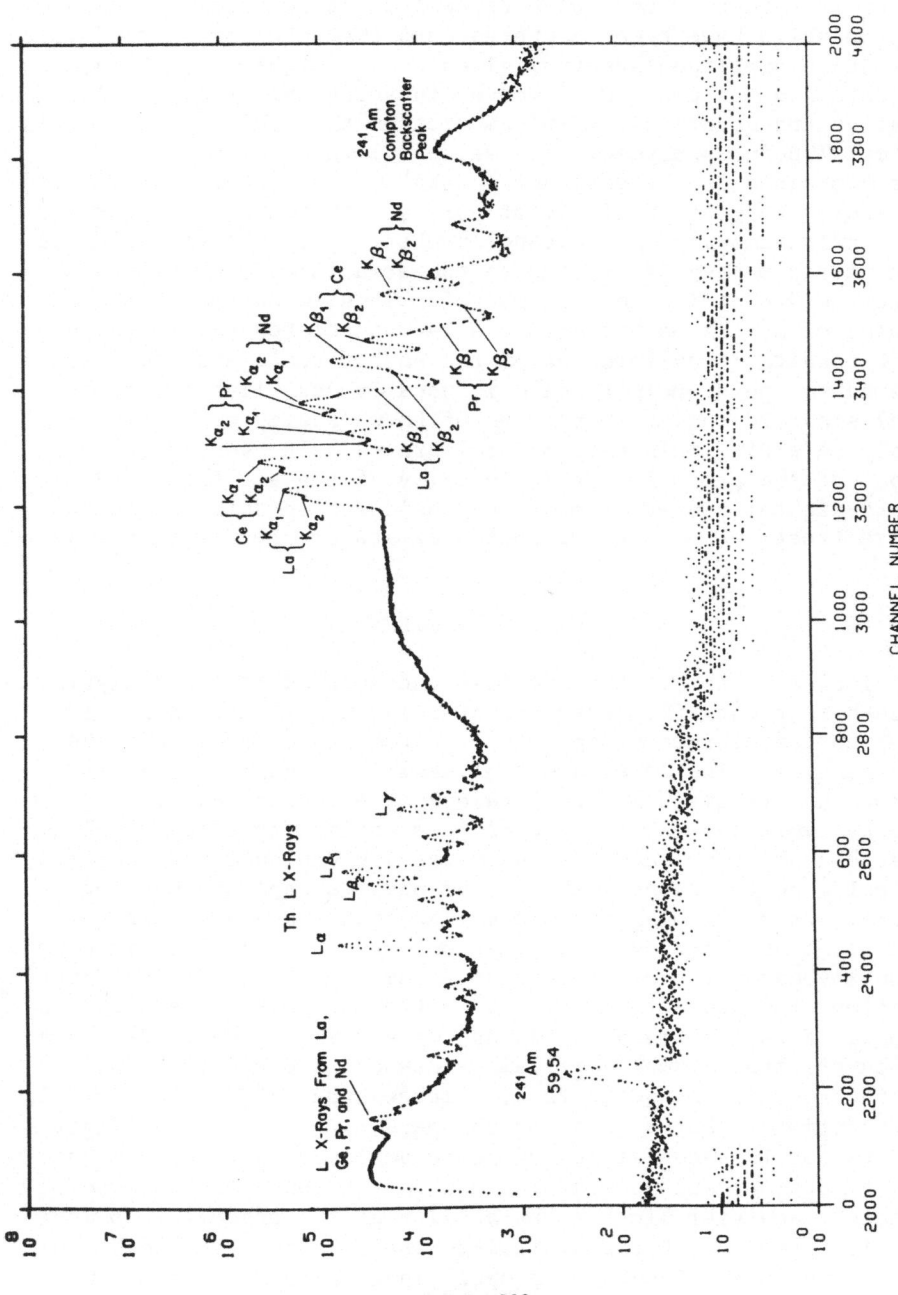

Fig. 12. Pulse-height spectrum of x-rays from Monazite ore sample excited with ^{241}Am isotopic source.

x-rays of thorium and the K-series x-rays of lanthanum, cerium,
praseodymium and neodymium. Also indicated in the spectrum are
two peaks at the upper end which represent photons from the exciting source which have been scattered into the detector. Examination
of this pulse-amplitude spectrum gives some insight into the nature
of the data and the problems involved in analyzing it to obtain
qualitative and quantitative information on the elements present in
the material being analyzed. The essential information in these
data is contained in the distinct "peaks" which appear superimposed
upon a slowly varying continuum of pulses. Photon energy is related to pulse amplitude or channel number, while the intensity of
a given photon energy is related to the area under each peak with
the background contribution subtracted. Thus, a determination of
the center of a peak on the abscissa may be related to energy, provided that suitable calibration of the energy scale has been made.
The energy vs. pulse-height scale is usually established with
internal standard radiation sources of known energy or by simultaneously injecting pulses from a calibrated pulse generator into
the input of the preamplifier. Linearity of the amplifier and
analog-to-digital converter must be considered in this procedure.
Extensive treatment of these problems exists in the literature.(7,8)

COMPUTER DATA SYSTEMS

     As indicated above, the problems encountered in the analysis
of pulse-height data from energy-dispersive spectrometers require
the application of rather sophisticated techniques to obtain the
desired results. The development of small, inexpensive digital
processors has resulted in the development of automated techniques
for the analysis and acquisition of pulse-height spectra. On-line
computer data systems have been successfully employed for the analysis of pulse-height spectra in nuclear physics applications for
some time. Of particular importance has been the development of
graphic techniques employing the use of large display oscilloscopes,
function keyboards and other peripheral devices to permit operator
interaction with the data to assist in the analysis procedure. As
an example, Figure 13 is a photograph of a small computer data system which utilizes a small digital processor (DEC PDP-8). This
system incorporates magnetic tape, a large-area CRT display oscilloscope, function keyboard, and control panel. In the use of digital
processors for the acquisition of pulse-amplitude data, the major
difference between such a system and a conventional "hard-wired"
pulse-height analyzer lies in the ability of the processor to modify
information provided by the digital-to-analog converter under control of a programmed sequence of operations which may assume any
form the user desires. It is this versatility which makes a digital
processor desirable for data acquisition. Figure 14 shows a block
diagram of a typical processor data acquisition and analysis system
suitable for application to x-ray fluorescence analysis. Comparing

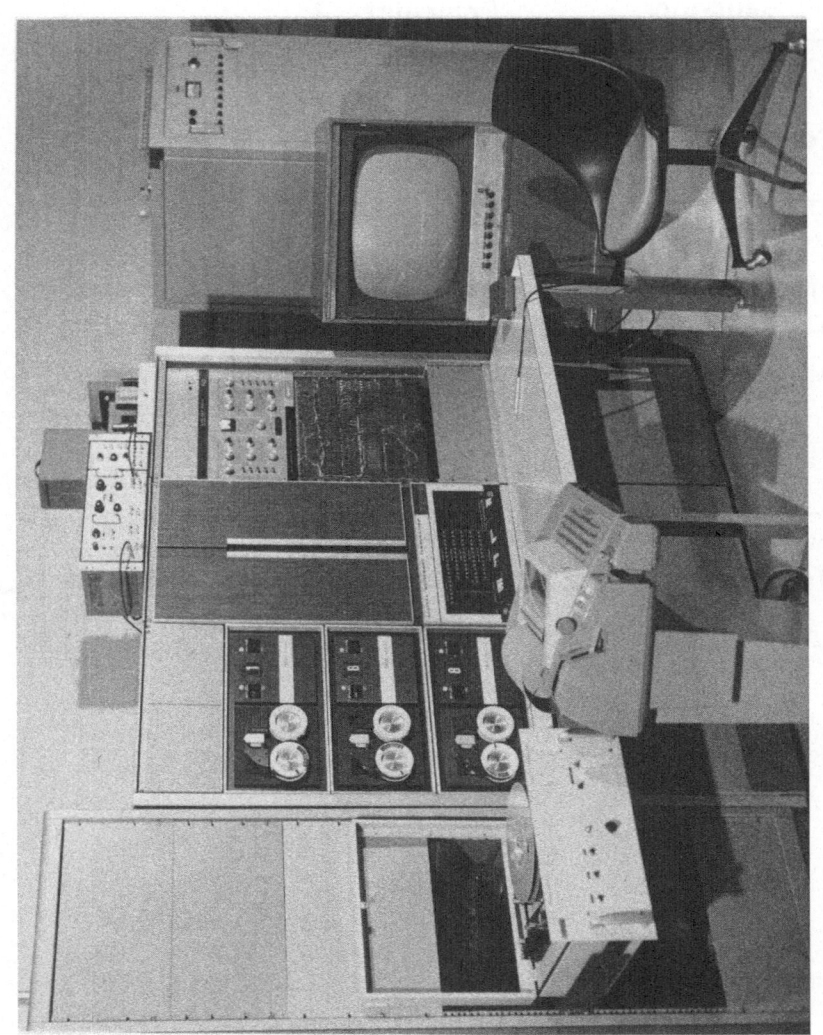

Fig. 13. Photograph of computer data system employing PDP-8 computer, magnetic tape
transports, function keyboard and CRT display oscilloscope for interactive analysis of pulse-
height data.

R. L. Heath

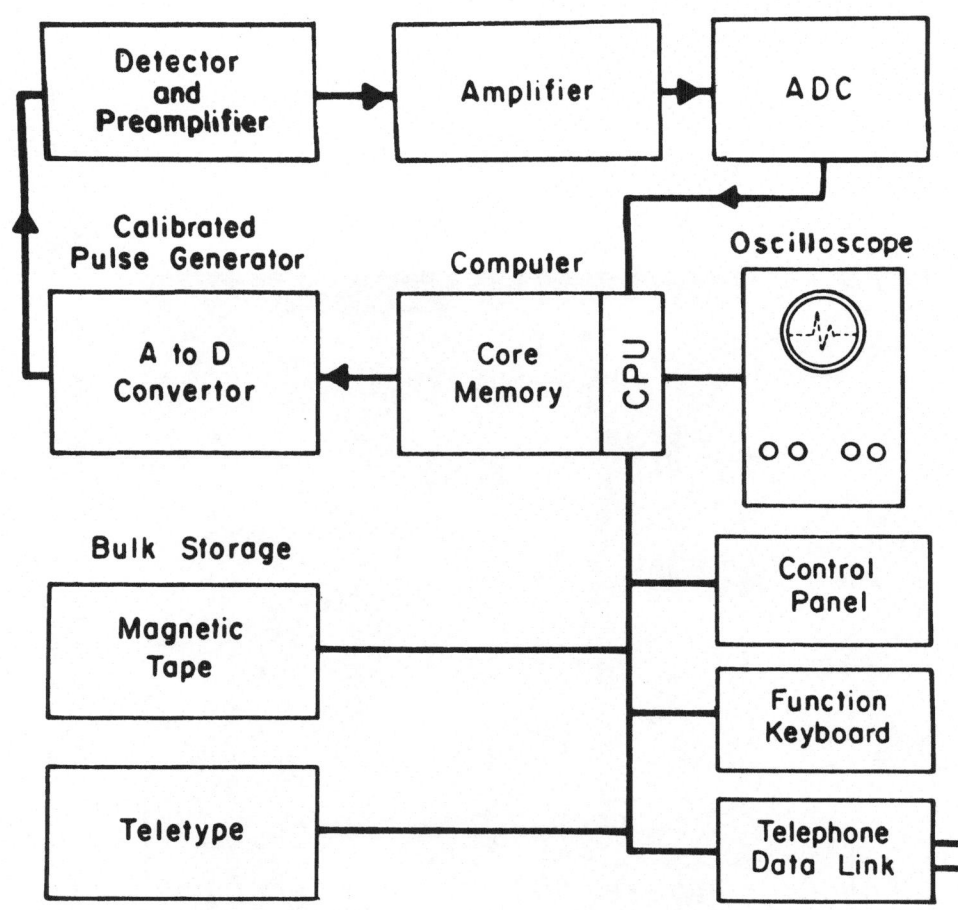

Fig. 14. Block diagram of typical computer-coupled pulse-height spectrometer for x-ray and gamma-ray spectrometry.

this system with the conventional pulse-height analyzer shown in Figure 11, we see that the major difference is the arithmetic unit of the processor and the peripheral devices which permit the operator to greatly expand his capability to perform operations upon the digital data residing in the central memory of the processor. In a conventional "hard-wired" pulse-height analyzer, analysis of a pulse from the detector produces a number corresponding to a particular address in the analyzer magnetic-core memory. This address is uniquely defined by the memory control circuitry and the procedure is then to initiate a process which adds the number "1" to the contents of that memory location. If this same number representing the amplitude of a given pulse is entered into the accumulator of a digital processor, this information can be modified at will. For example, it might be stored in a number of different locations for different purposes, modified to change the energy scale or to correct for non-linearities or instabilities in the electronic system, etc. The circuitry required to pass information to or from an external device through the accumulator or directly to the computer memory is termed an "interface". In general, information may be entered into a processor memory by two different modes:

(1) Direct memory access - in this mode, the address information is entered directly into dedicated memory locations much in the same fashion as in "hard-wired" analyzers. The transfer and add one is accomplished in one memory cycle on what is called a "cycle-stealing" basis. This implies that entry of information from the external source is made by interrupting the normal sequence of operation of the resident program in the machine for only one memory cycle, then the control of the machine is returned to the resident program sequence.

(2) Program interrupt - in this method of entry, the analog-to-digital converter signals the processor that it has completed the analysis of a pulse and requests that this information be handled by the processor. The normal program in operation at that time is then interrupted and a special program called to handle the information presented to the processor by the device which produced the "interrupt". This method of entry provides the ultimate in flexibility, at the expense of increased processing time.

Digital Control Panel

A significant difficulty encountered in the application of digital processors for on-line data acquisition and analysis results from the unfamiliarity of experimenters with such equipment. Normally, the only means for communication between user and machine is a teletypewriter keyboard. This means that normal setup and operation of a processor-controlled pulse-height analyzer must be accomplished by means of a very lengthy dialogue with the computer, using the keyboard. A concept which provides a logical alternative is to provide the user with a panel containing switches and knobs which may be used in exactly the same manner as with conventional "hard-wired" equipment. Figure 15 is a photograph of a digital control panel which is designed for interface to small digital processors. Such panels, which are nothing more than an array of rotary and push button switches, may be located remotely from the processor. Provided with a variable overlay, the switches may be made to represent any desired assortment of functions. The position of any switch is translated into a binary form in a register which is

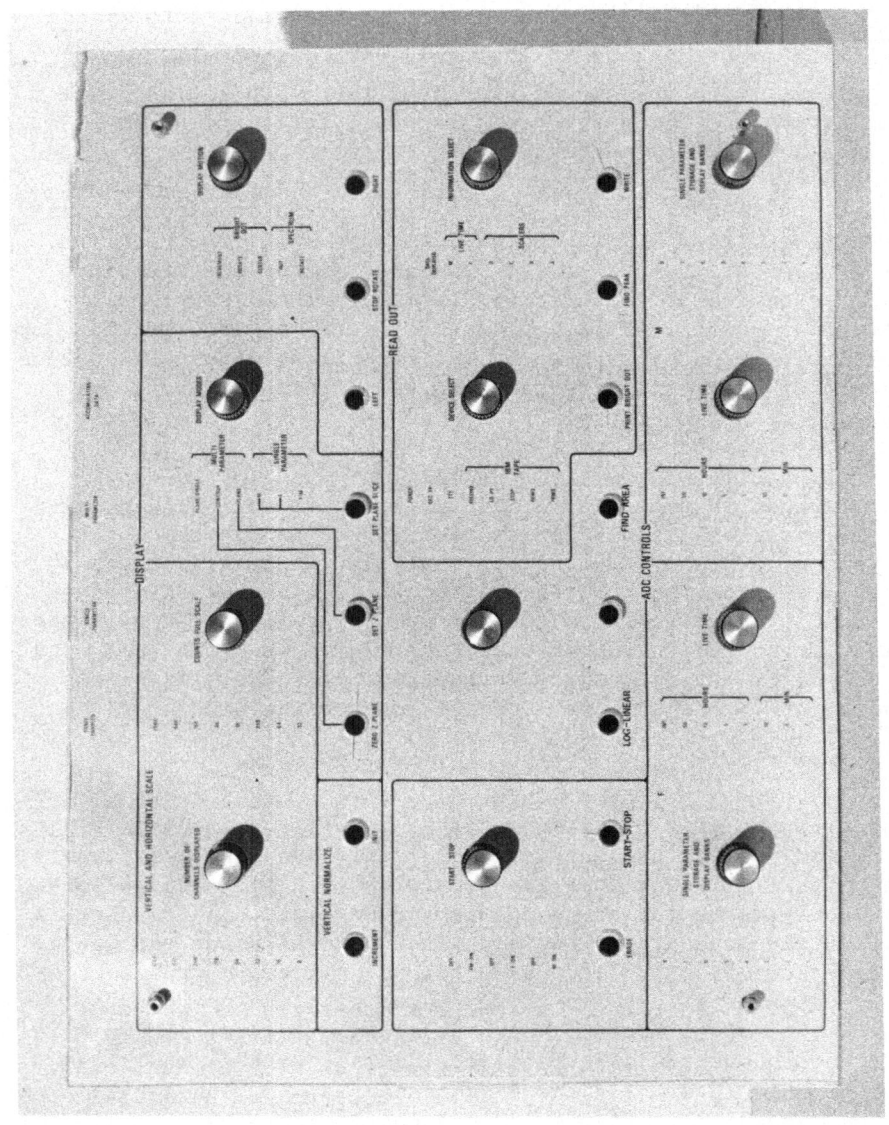

Fig. 15.   Photograph of digital control panel to provide convenient control and indication of experimental status of equipment for operator.

decoded by the processor to establish the selection of functions
desired. When a button is depressed or the position of a rotary
switch is changed, an interrupt is generated to the processor which
causes the binary register to be read and the new selection of
operations or functions initiated by calling special programs into
use by the processor. The advantages of such a panel are obvious.
The position of all switches on the panel gives a visual presentation of the operating status of the machine at any time, and changes
in function may be accomplished in a manner which is both convenient and familiar.

Graphical Display

A computer display system incorporates many devices to supplement the cathode-ray tube to achieve interaction between the user
and the program resident in the processor. The principal device
employed in systems of the type under discussion here is the function keyboard. This device, pictured at the lower left-hand corner
of the photograph in Figure 16, is simply a small keyboard which
operates in a manner similar to the digital control panel. The
function keyboard is also generally interfaced to the processor
through the interrupt structure. An interrupt "flag" is generated
whenever a key is depressed. This causes the computer to interrogate the keyboard to determine which key has been activated, then
using this information to implement any program desired by the
operator.

An example of how the function keyboard and the CRT are used
to achieve interaction between operator and the data is in the display of pulse-height data. Assume that a portion of the entire
spectrum (256 channels) of a 1024- or 4096-channel spectrum is displayed on the screen of the oscilloscope as shown in Figure 16. It
will be noted that the center channel in the display is intensified
to provide a reference. The operator, by depressing keys on the
function keyboard, may cause the spectrum to rotate past him from
left to right or from right to left, bringing onto the screen a
small portion of the entire spectrum for examination in detail.
During this process of rotation the intensified point remains fixed
in the center of the screen, so that the channel number or position
on the spectrum is changing. Using the combination of the rotation
and the fixed reference point to identify data points in the spectrum, it is possible to select information for performing a variety
of operations on the data. More insight into the use of interactive
graphics techniques will be provided in the discussion of data
processing presented below.

R. L. Heath

Fig. 16. Photograph of large-screen oscilloscope and function keyboard for analysis of data using interactive techniques.

DATA ANALYSIS

In the introduction, the basic nature of the data was discussed and general considerations for analysis presented. To properly assess the role of the on-line computer system applied to the analysis of x-ray fluorescence data, a more detailed discussion of the procedures required to obtain elemental quantitative results is required. As previously indicated, the basic information contained in a pulse-height spectrum is related to the energy and intensity of monoenergetic x-rays excited in the material under investigation. To obtain this information from the raw data it is necessary to perform the following operations:

(1)   locate the median positions of the symmetric "peaks" in the pulse-height distribution;

(2)   establish a channel-number vs. energy scale by suitable calibration techniques, taking into consideration the linearity of the electronic instrumentation;

(3)   determine the energy of the x-ray photons observed in the pulse-height spectrum from the information obtained in the first two steps;

(4)   determine the area (intensity) of the peaks in question with subtraction of the underlying continuum;

(5)   compare the calculated energies of the peaks with a list of energies of all the characteristic x-rays to identify the elements detected in the sample;

(6)   determine the concentration of each element present in the sample from the peak areas by suitable calibration techniques.

Since a detailed treatment of the techniques involved in this procedure is beyond the scope of this presentation, we will define the procedure in outline form to provide the reader with an understanding of the role of a digital processor in automating the analysis of such data.

Peak Location

As indicated in the experimental data, individual x rays or gamma rays appear as symmetrical peaks having finite width and height above the background. Due to the statistical nature of electronic noise, a peak may be represented by a Gaussian function having the form:

$$y(x) = y_o \, e^{-(x - x_o)^2/b_o}$$

where x_0 is the midpoint of the peak, b_0 is related to the width
of the peak (w_0) and y_0 is the peak maximum.

Using a linear least-squares procedure(9), data points in the
region of a peak may be fitted with a Gaussian distribution plus a
straight line to determine the most probably location of the peak
on the horizontal axis, and the peak area. This procedure is illustrated in Figure 17. In this figure are two cases. On the right-
hand side of the figure a single peak is shown, with all experimental quantities defined. A Gaussian function has been fit to the

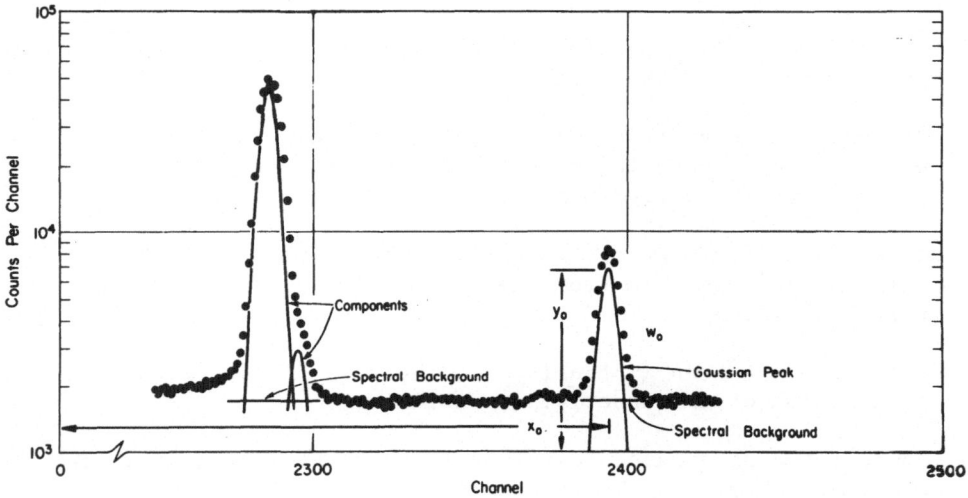

Fig. 17. Plot of typical pulse-height spectrum illustrating
technique for fitting data to determine peak position and area.

experimental points with a straight line background subtraction.
On the left-hand side we see a more complex situation. In this
case the peak consists of two unresolved components. The best
representation of this region of the spectrum would be two Gaussian
functions plus a background. To accomplish this fitting procedure
in a least-squares sense it is necessary to resort to what is termed
a non-linear regression technique. This technique, which is quite
sophisticated, has been described by Helmer et al.(10). Non-linear
regression fitting programs for the analysis of pulse-height spectra
require large blocks of computer memory and usually are restricted
to large computer installations. Techniques for location of peaks
in spectra to obtain peak position and area require the generation
of considerable input information. Although the process may be automated, it requires extensive and complex computer programs. Linear
least-squares peak fitting techniques have been implemented on small
on-line processors, using a CRT display and function keyboard for
input of the necessary information for the fitting procedure.

In order to implement completely automated data analysis of pulse-amplitude spectra it has been necessary to develop a procedure for locating peaks without operator intervention. Several approaches have been developed for this purpose. The simplest procedure in use locates peaks in a spectrum by examining the change in slope between data points, taking into account statistical variation in the data points.(11) In this technique, the reversal of direction of the slope between adjacent data points gives an indication of the presence of a peak. A second method in use makes use of the fact that higher-order derivatives of a Gaussian function provide a point of inflection through zero at the midpoint of the peak(12). The method which we have found to be most accurate and versatile for the location of peaks in pulse-height spectra utilizes a cross-correlation technique.(13) This procedure makes use of the fact that the resulting correlation function will indicate a positive correlation where similar functional forms appear in the data. In the case of pulse-height data, the resolution function (Gaussian) is correlated against the spectrum. The correlation spectrum which results will have peaks at almost the same channel position as the original spectrum and negative values at all other points. Figure 18 shows a Si(Li) x-ray spectrum and the corresponding correlation spectrum indicating the unique manner in

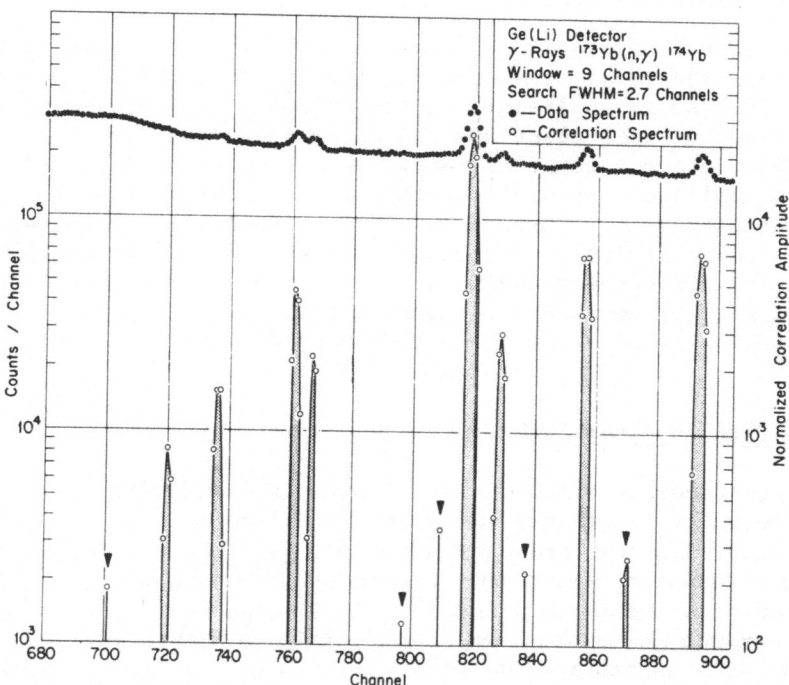

Fig. 18. Plot of Si(Li) x-ray spectrum showing result of cross-correlation of data with Gaussian function to locate position of peaks on pulse-height scale.

which this procedure selects and locates resolved peaks in the complex spectrum. Although the cross-correlation technique of locating peaks will not find unresolved peaks, it is most effective in locating significant peaks and can be efficiently programmed to automatically locate peaks with small on-line processors.

Peak Area Determination

Because the elemental concentration in a sample is related to the area in a given x-ray peak, it is necessary to determine the x-ray peak areas precisely. The most important consideration, assuming that a peak may be represented by a Gaussian function, is the determination of the baseline for background subtraction. The most successful approach to this problem, as illustrated by Figure 17, has been to select a region of the spectrum near a given peak which can provide a reasonable extrapolation under the peak to represent the background contribution. A straight line is fit to this region using a linear least-squares fit and then extrapolated under the peak to determine the background to subtract. Subsequent fit of the peak region with a Gaussian function then provides a value for the area under the peak.

Channel Number vs. Energy Scale

The energy calibration of automated systems is presently being achieved by injecting pulses from a calibrated pulse generator into the input of the preamplifier. In the computer-controlled systems the amplitude of this pulser may be determined under program control. Thus, using a computer-controlled pulse generator, pulses of two known amplitudes may be injected into the system simultaneously with detector pulses. Through the ADC interface, these pulses are tagged as pulser pulses and stored in a separate region of the computer memory. Subsequent analysis of the location of these two amplitude groups from the pulse generator provides a means of determining the zero and slope of the channel number vs. energy scale.

Composite Data Analysis Program

A block diagram of a complete program for analysis of pulse-height spectra is presented in Figure 19. After collection of a spectrum the following procedure is used: (1) the cross-correlation algorithm is used to search the spectrum with a Gaussian function to determine the location of all significant peaks, (2) the background contribution to each peak is determined, (3) a least-squares fit of the data points in the vicinity of each peak is performed to determine the centroid and area of each peak, (4) the information from the computer-controlled pulse generator is utilized to establish an energy vs. channel number scale and the energy of each peak is determined from the peak position, corrected for the pulse-height

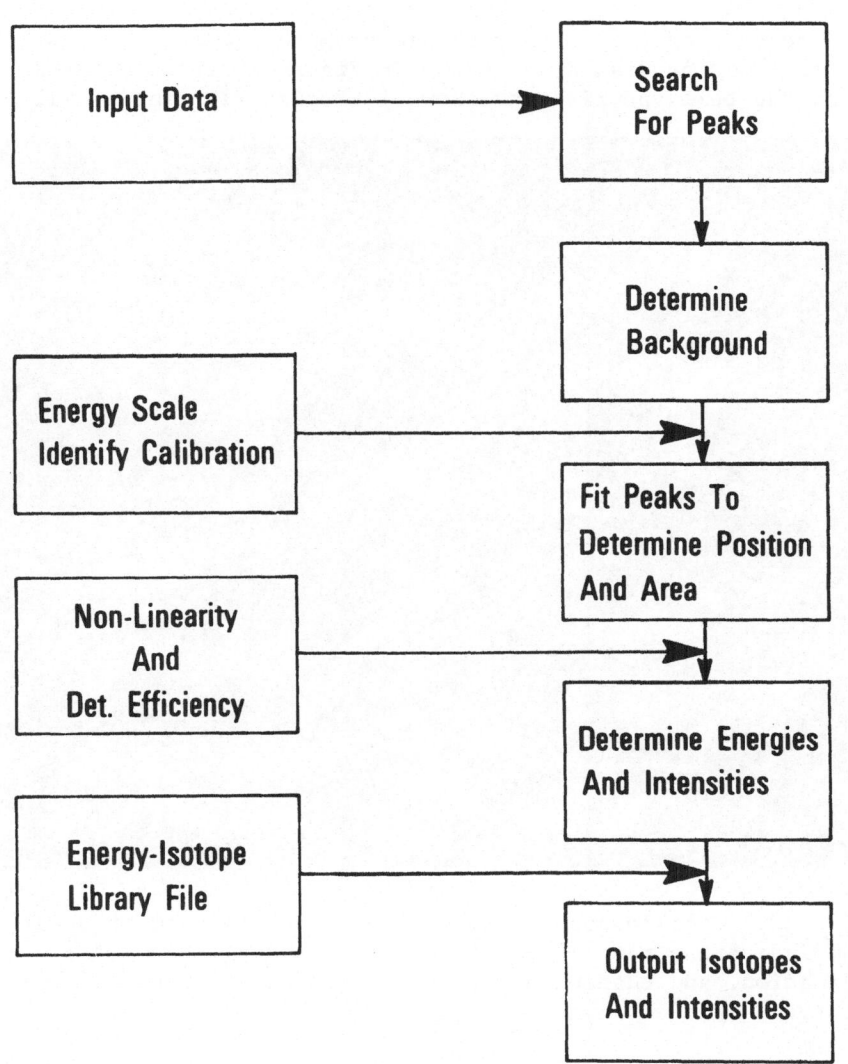

<u>Fig. 19</u>. Block diagram of data analysis program for automated analysis of pulse-height spectra obtained using Si(Li) and Ge(Li) spectrometers.

non-linearity, and the gain, and finally (5) the energies determined for each peak are compared with a reference list of known energies for all x-rays emitted by each element to provide elemental identification. The reference list of x-ray energies emitted by each element is listed in order of ascending energy on a magnetic tape and is read in segments into the computer for use with the elemental identification search routine.

The results of this data analysis program can be made available
on the oscilloscope display as shown in Figure 20 or as printed
output on the teletypewriter as shown in Figure 21. The oscillo-

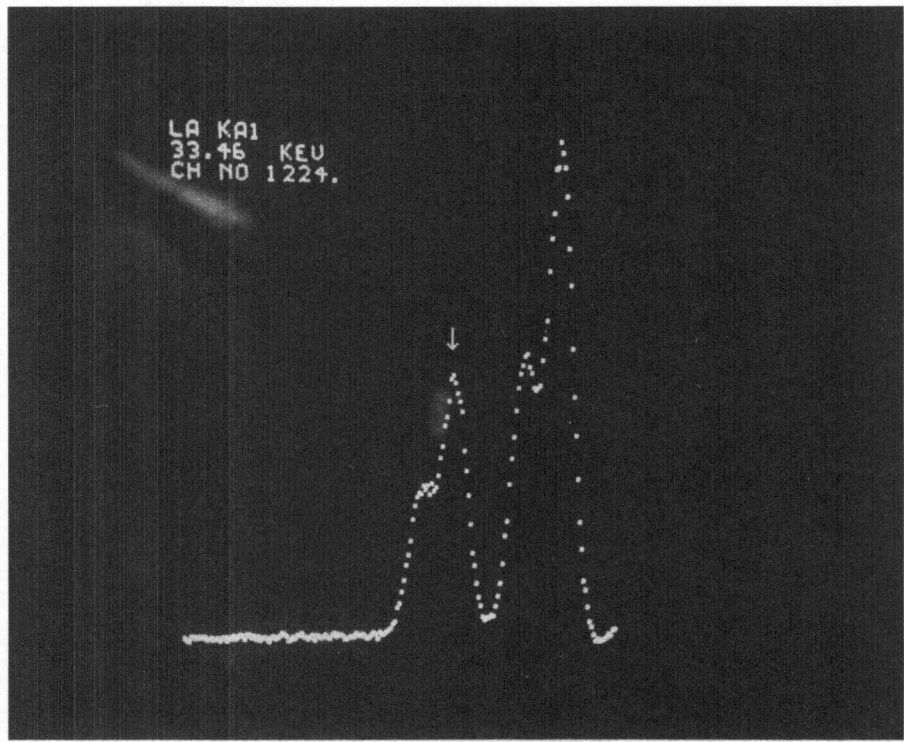

Fig. 20. Oscilloscope display of on-line analysis of x-ray
spectrum indicating identification of x-ray energy, elemental
identification, and channel number associated with peak indicated
by cursor.

scope display permits the operator to rotate the data displayed
on the oscilloscope, to interrogate this data as he chooses on a
peak-by-peak basis by the use of the function keyboard, and to display the output of the elemental analysis on the scope. In many
on-line process control applications, however, the program operator
wants to obtain the analysis automatically and as quickly as possible. In this case, an analysis appearing on the oscilloscope
screen in a format similar to the printed output is preferred.

Elemental Concentration

To obtain an absolute or even a relative analysis in x-ray
fluorescence spectroscopy it is necessary to calibrate the spectrometer with standard samples of known elemental composition. The

ENERGY	CENTER	FWHM	AREA	IDENTIFICATION
2.79	53.11	43.35	13736.05	
13.01	443.49	11.38	8578.30	TH LA1, BI LB1
14.97	518.36	10.65	1001.81	Y KA1, RB KB1
15.69	545.81	11.37	4141.73	TH LB2, ZR KA2, AC LB1
16.25	566.95	11.77	8102.06	TH LB1, AT LG1
18.08	637.02	28.55	1518.56	ZR KB2
18.99	671.93	18.87	3291.65	TH LG1
33.06	1208.95	12.66	18305.61	I KB2, LA KA2
33.46	1224.29	12.85	30329.20	LA KA1
34.32	1257.34	15.03	38701.77	CE KA2
34.74	1273.10	13.77	63284.88	CE KA1
36.03	1322.59	15.22	7308.43	PR KA1
36.86	1354.15	12.87	12005.27	ND KA2
37.38	1374.03	12.45	14705.93	ND KA1
39.25	1445.52	11.00	7824.44	CE KB1
40.21	1481.92	20.60	5646.09	CE KB2
42.25	1559.83	11.77	2726.69	ND KB1, GD KA2
59.62	2223.17	20.08	706.25	

Fig. 21. Teletypewriter printout of complete elemental analysis of Monazite sand using x-ray fluorescence technique.

standard can sometimes be another element already present or added to the sample to be analyzed (internal standard) or may be a separate set of standard samples whose matrix closely approximates the unknown samples. In either case the x-ray fluorescence spectrometer is calibrated by comparing the peak areas of the x-rays emitted from those elements to be analyzed in the unknown sample to the peak areas of the respective x-rays emitted from those elements of known concentration in the standard sample. If the elemental composition of the samples varies appreciably from sample to sample, it will be necessary to apply suitable corrections for the effect of matrix absorption. A thorough theoretical discussion of methods of correcting for matrix effects in non-dispersive x-ray fluorescence analysis is presented in ref. (14). These methods of matrix correction can be implemented on a small on-line processor to provide the user with a powerful analytical instrument for quantitative elemental analysis.

Applications

The number of applications of non-dispersive x-ray fluorescence spectroscopy is growing at an accelerating rate. Several excellent reviews of the field of non-dispersive x-ray fluorescence spectrometry have been recently published in the literature(15,16). These review articles indicate that biology, medicine, mining, metallurgy, geology, forensic science, public health and safety, archaeology, art, space exploration and industrial process and quality control are only some of the fields in which x-ray fluorescence analysis is presently being applied.

Obviously, the addition of the small on-line processor to non-dispersive x-ray fluorescence spectroscopy will result in an even larger number of applications and will make available a variety of analysis techniques presently not possible with off-line central processors.

ACKNOWLEDGMENTS

It is with great pleasure that the author acknowledges the contribution of many colleagues at the National Reactor Testing Station to the development of the techniques reported in this paper. In particular, a special thanks is extended to M. S. Cole, R. C. Davies, C. W. Richardson and E. W. Killian for the development of the computer data and graphics systems.

REFERENCES

1. E. M. Pell, J. Appl. Physics $\underline{31}$, 291 (1960).

2. F. S. Goulding <u>et al</u>., X-Ray Analysis, Vol. 15, 20th Annual Denver X-Ray Conference, Denver, Colorado, August 11-13, 1971.

3. K. F. Smith and J. E. Cline, IEEE Trans. on Nucl. Science, Vol. <u>NS-23</u>, No. 3, 468 (June 1966).

4. F. S. Goulding, J. Walton and D. Malone, Nucl. Instr. and Methods 71, 273 (1969).

5. A. Kandiah, "Semiconductor Nuclear Particle Detectors and Circuits", National Academy of Science Publication 15943, 495 (1969).

6. C. A. Knowlin and J. L. Blankenship, Rev. Sci. Instr. $\underline{36}$, 1830 (1965).

7. W. W. Black, Nucl. Instr. and Methods <u>53</u>, No. 2, 249 (1967).

8. R. L. Heath, in Modern Trends in Activation Analysis, National
    Bureau of Standards Special Publication <u>312</u>, Vol. II (1969).

9. R. L. Heath, Nucl. Instr. and Methods <u>43</u>, 209 (1966).

10. R. G. Helmer, R. L. Heath, M. Putnam and D. H. Gipson, Nucl.
    Instr. and Methods <u>57</u>, 46 (1969).

11. R. Gunnink and J. B. Niday, in Modern Trends in Activation
    Analysis (Proc. Int. Conf. Gaithersburg, Maryland, 1968:
    J. R. DeVoe, Editor) <u>2</u>, NBS, Washington, D.C., 1244 (1969).

12. H. P. Yule, Nucl. Instr. and Methods <u>54</u>, 61 (1967).

13. A. L. Connelly and W. W. Black, Nucl. Instr. and Methods <u>82</u>,
    141 (1970).

14. A. Lubecki, J. Radioanal. Chem. <u>2</u>, 3 (1969).

15. W. J. Campbell, Anal. Chem. <u>42</u>, 248B (1970).

16. F. J. Walter, IEEE Trans. on Nucl. Science, <u>NS-17</u>, No. 3, 196
    (June 1970).

DETECTION OF SINGLE IONS BY PULSE COUNTING:

APPLICATION TO ION MICROPROBE MASS ANALYZER

L. A. Dietz

General Electric Company, Knolls Atomic Power Laboratory*

Schenectady, New York 12301

ABSTRACT

The need to detect low intensity beams of resolved ions with keV energy is encountered frequently in mass spectrometry. It is most readily accomplished by using some form of pulse-counting detector in which ions are allowed to strike a target surface (conversion dynode) where each ion releases a pulse of low energy secondary electrons. The secondary electrons undergo further amplification in an electron multiplier or scintillator type of ion detector which is specially designed to detect single ions. Ion-to-electron conversion, amplification of the secondary electrons, efficiency of ion detection by pulse counting and dead-time corrections to observed random counting rates are all statistical processes which must be understood so that observed data can be corrected properly to obtain precise and accurate measurements. These processes are reviewed, along with some of the basic properties of ion-induced secondary electron emission from an amorphous Al_2O_3 thin film target. Ion detection in the ion microprobe is interpreted in light of these basic considerations, and several mass spectra are given for electron multiplier dynode surfaces.

INTRODUCTION

Pulse-counting detectors offer the ultimate sensitivity and accuracy that is attainable for recording the collision of an ener-

*Operated for the United States Atomic Energy Commission by the General Electric Company.

36

getic ion, atom, electron or photon with a target surface. The detection process begins at a conversion dynode where a primary particle produces secondary electrons or photons, or when a primary
photon produces a photoelectron. Charge amplification of the secondary electrons or photoelectrons may involve generation of secondary electron cascades between dynodes in vacuum or generation of
electron-hole pairs in a semiconductor. Renewed interest in detecting single ions is being generated by the advent of the ion
microprobe and by the extension of spark-source mass spectrometry
to the accurate assay of elemental concentrations below the ppm
level. Detection of single ions by pulse counting in electron
multipliers dates back to 1954 in our laboratory (1). It has long
been a standard technique in surface-ionization mass spectrometry.
This type of mass spectrometry is used for isotope-ratio analysis
of metallic elements.

More recently, ion sputtering has been applied to the detection and analysis of atoms in thin surface layers. This takes place
in the high-vacuum system of a mass spectrometer such as the ion
microprobe. The ion-sputtering processes at the ion source and at
the detector are shown schematically in Fig. 1. Here we see that
an energetic ion in the primary beam can dislodge negatively charged
or positively charged ions or neutral atoms. It also dislodges
electrons, molecular fragments and sometimes photons. The range of
the primary ions is of the order of 500 Å or less and the sputtered
particles are characteristic only of the first few atom layers of
the target surface or of a thin film of material deposited on the
surface. The ratio of charged to neutral species varies, but
typically is very low, i.e. about 1 ion in 100 - 1000 neutral atoms
ejected. The sputtering process is destructive. For a comprehensive review of the ion sputtering and secondary electron emission
processes see Carter and Colligon (2).

The mass analyzing system is shown schematically as a magnet.

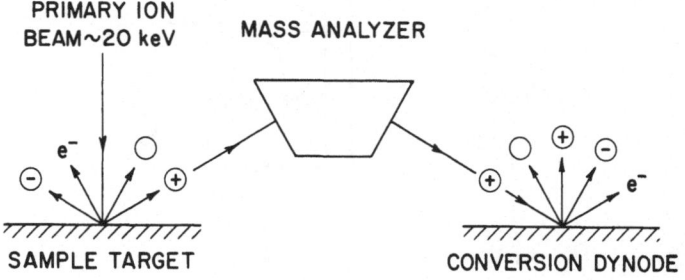

Fig. 1. Ion sputtering processes at source and detector.

The bombarding primary ion may be positively charged or negatively
charged and either positively charged or negatively charged ions
can be selected for mass separation and detection. For example,
we can choose a positively-charged sputtered ion, accelerate it to
1 or 2 keV, separate it from other ions of different mass by passing
it through a mass analyzing system and then allow it to strike the
conversion dynode of a suitable ion detector. At the detector the
sputtering process repeats, but with a much lower intensity than at
the source. Here the secondary electrons ejected by each single
ion impact are accelerated into an electron multiplier or scintillator detector for further amplification before they are pulse-
counted. The secondary electrons resulting from an ion impact are
released in a time less than 10^{-12} sec and the counting rates typically encountered do not exceed a few MHz. In a well-designed fast
pulse-counting system coincidence losses are low and can be calibrated accurately. Let us look at the detection process in more
detail and then come back to several examples of surface analysis
in the ion microprobe.

BASIC PROPERTIES OF SECONDARY ELECTRON EMISSION

We can gain considerable insight into the basic properties of
secondary electron emission and its relation to the precision and
accuracy of a measurement by: (a) measuring secondary electron
yield coefficients under varying experimental conditions, and (b)
by developing an appropriate statistical model to describe the
electron multiplication process in a pulse-counting detector.

Secondary Electron Yield

We have constructed a high-vacuum test apparatus to obtain
preliminary measurements of ion-induced secondary electron emission

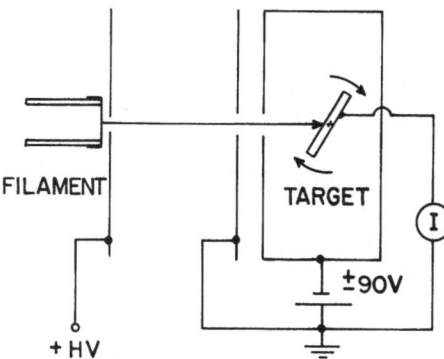

Fig. 2.   Secondary electron test apparatus.

from thin films of amorphous aluminum oxide (3). We use Al_2O_3 as
the surface of our conversion dynode. The apparatus is shown in
Fig. 2 and is housed in a bell-jar high vacuum system. The main
purpose of these measurements is to gain sufficient experience to
design and construct a more complex apparatus which includes mass
analysis of the primary ion beam.

Ions are emitted from a surface ionization filament and are
accelerated to energies up to 20 keV before they strike a flat target which is positioned several cm from the filament. The target
can be rotated about an axis at right angles to the path of the ion
beam. This arrangement allows one to measure secondary electron
yield as a function of ion mass, impact energy and angle of entry
into the target surface. Because there is no mass analysis in this
device, we are limited to such elements as Li, Ca, K, Ba and Cs
which can be made to emit pure ion beams by the surface ionization
technique. Primary ion current I_p is measured with the battery at
-90V to suppress secondary electrons and primary current I_p plus
secondary electron current I_e is measured at 90V. From these two
measurements the secondary electron yield γ is readily obtained as

$$\gamma = I_e/I_p = (I_e + I_p)/I_p - 1. \qquad (1)$$

An ion-converter dynode is prepared by evaporating 1000 Å or
more of pure aluminum onto a section of glass microscope slide to
which a thin-film electrical conductor has been attached. The glass
surface is smooth to perhaps 20 Å. Pressure during evaporation of
aluminum is in the low 10^{-9} Torr range. Aluminum surfaces are

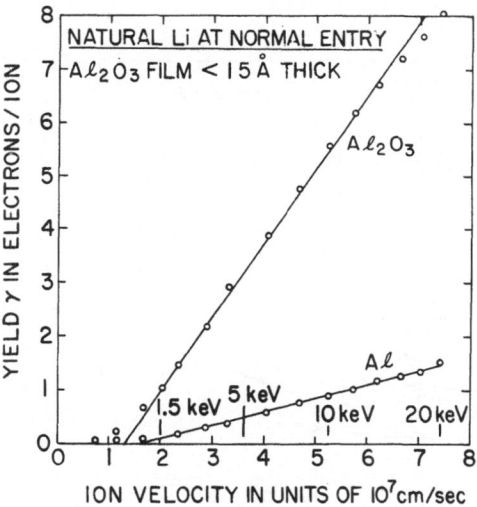

Fig. 3. Secondary electron yields from Al and Al_2O_3 surface
before and after anodizing.

oxidized to form the secondary electron emitting layer. This is
done by exposing the freshly deposited aluminum to pure oxygen for
a few hours or by chemically anodizing the surface to an accurately
known thickness of aluminum oxide. Oxide films formed by exposing
the aluminum surface to oxygen are of the order of 10 A or less in
thickness.

The shape of the secondary electron yield curve for Li, shown
in Fig. 3, also is typical for heavy monatomic ions. The results
indicate that secondary electron yield from a pure, freshly deposited aluminum surface before and after oxidation is approximately
a linear function of ion velocity. Furthermore, there is a defi-
nite threshold of velocity below which no secondary electrons are
released by the kinetic ejection mechanism. The theory of kinetic
ejection of secondary electrons by ion bombardment has been described by Parilis and Kishinevskii (4). We observe a velocity
threshold of about $(5 \pm 2) \times 10^6$ cm/sec for Ca-40$^+$, K-39$^+$, Ba-138$^+$,
and Cs-133$^+$. The velocity threshold for Li is somewhat higher.
Our results are in agreement with those of Schram (5) and his co-
workers. For a CuBe dynode in an electron multiplier they observed
a linear dependence of secondary electron yield with ion velocity
for all the noble gases and a velocity threshold of 5.5×10^6 cm/sec.
For lithium it should be noted that the velocity threshold corresponds to ~200 eV energy, whereas for a heavy ion such as uranium
it corresponds to ~3 keV. This means that a low-energy heavy ion
will produce fewer secondary electrons than a light ion of the same
energy. The net result is a low detection efficiency for the heavy
mass ion, unless the impact energy is increased to 10 keV or more.
The very low secondary electron yield for pure aluminum is typical
of that for pure metals in general.

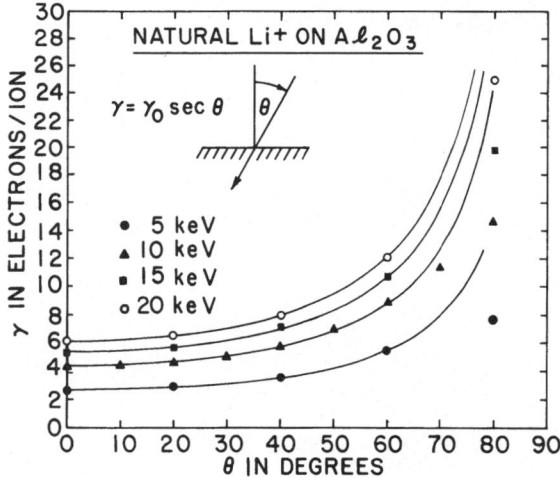

Fig. 4. Secondary electron yield as a function of impact
energy and angle of incidence. Lines are sec θ
and points are experimental data.

From simple considerations one expects the secondary electron yield at a given ion energy to be proportional to sec θ, where θ is the angle between the ion beam and the surface normal at the point of impact. Our results are shown in Fig. 4. We observe that this relation generally holds to about 60°, but between 60-80° may be somewhat lower than that predicted by the secant relationship. The secant relationship is well known for secondary electron emission from metallic surfaces and has been discussed by Kaminsky (6).

Escape Depth of Secondary Electrons

Another important parameter is the escape depth of internal secondary electrons formed below the surface of the ion-converter dynode.

To estimate this depth we successively anodized an aluminum thin film to 10, 20 and 30 Å of Al_2O_3 in an electrochemical cell. Secondary electron yield at normal incidence was measured after each anodization. Because the yield of the metallic substrate is 5-10 times lower than that of Al_2O_3, internal secondary electrons formed in the substrate, near the Al-Al_2O_3 interface, cannot contribute significantly to observed secondary electron yield. Davies and his co-workers (7,8) at Chalk River have measured the ranges of heavy ions in amorphous Al_2O_3. From their results we can infer that the range of 20 keV Li ions is much greater than 30 Å, therefore any increase in secondary electron yield with oxide thickness would indicate a mean escape depth greater than 10 Å. No such effect was observed and in fact the yield remained constant with increasing oxide thickness. From this we conclude that the attenuation length, or mean free path for escape from an amorphous Al_2O_3 film is less than 10 Å. This is somewhat disappointing, but is consistent with

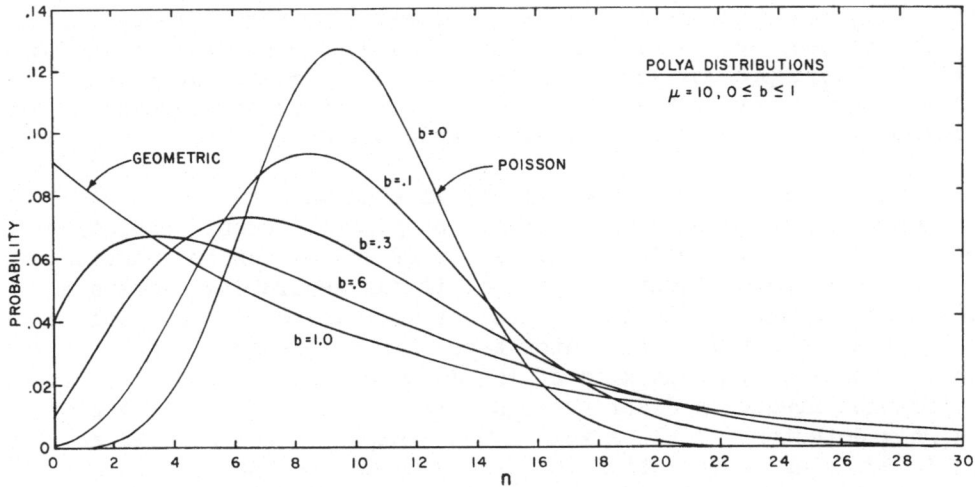

Fig. 5. Plots of several Polya frequency distributions.

a mean free path of 5 Å in Al_2O_3 reported by Collins and Davies (9) for hot electron transport in Al-Al_2O_3-Al tunnel cathodes.

Statistics of Secondary Electron Multiplication

As was mentioned earlier, the statistics of secondary electron emission are important in assessing the precision and accuracy with which small ion beams can be measured. We have applied a Polya statistical model to the secondary electron multiplication process in ion detectors used for pulse counting. Polya distributions can assume many shapes, as indicated in Fig. 5. The model has been published by Dietz (10), together with a description of the computer program used for calculations, so only its main features need be outlined here; with it one can predict the probability of occurence of 0, 1, 2, ..., n secondary electrons/ion impact. The Polya model is based on Taylor series expansions of generating functions. It describes the shape of an output pulse-height distribution of a detector and allows one to estimate detection efficiency for a wide range of pulse-counting detectors. Efficiency of a detector is defined as $1-P(0)$, where $P(0)$ is the probability of no secondary electrons being released in an ion impact. Shape is related to the statistical variance and magnitude of the secondary electron yield. The Polya distribution is a form of the negative binomial distribution and has the parameters μ and b. Mu is taken as the average secondary electron yield and b is a relative variance parameter that can take on values between 0 and 1. Relative variance is the sum of squared deviations from the mean, divided by the square of the mean. When b is zero, the frequency distribution is Poisson and when b is 1 the distribution is geometric or approximately exponential. Since the relative variance of a Polya distribution is $b + 1/\mu$ and that of a Poisson distribution is just $1/\mu$, it is seen that b is the additional relative variance over that of a Poisson distribution. Physically the b-parameter can be interpreted as a measure of uniformity of secondary electron yield from a dynode surface, assuming that the basic process from a given element of area follows the Poisson law. That is to say, when b is zero, secondary electron emission is uniform over an entire dynode surface.

Calculations become somewhat complicated when two or more secondary electron multiplying stages obeying different statistics are cascaded. Nevertheless, any number of stages can be cascaded and accurate computations can be made if the generating process at each stage is known. To illustrate the usefulness of the model we will apply it to the scintillator type of ion detector. The statistical processes in this kind of ion detector are described by Polya or binomial frequency distributions at each stage of amplification. The Daly detector (11) is a well-known example of the scintillator ion detector and a variation of it is used in the ion microprobe. A schematic diagram of a modern version of the Daly detector has

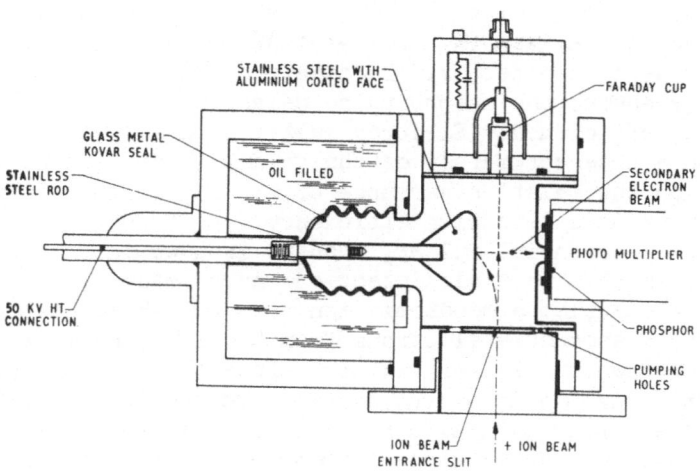

Fig. 6. Schematic diagram of Daly Detector.

been published by Young, Ridley and Daly (12) and is shown in
Fig. 6. In the ion-converter first stage an energetic, mass-analyzed primary ion releases a pulse of secondary electrons from a
highly polished thin-film surface which usually is Al_2O_3. These
secondary electrons are accelerated to about 40 keV and strike a
plastic scintillator, which is coupled optically to a photomultiplier tube. An output pulse-height distribution for N_2 ions
is shown in Fig. 7. A compound Polya distribution has been fitted
to it by the method described earlier (10). The pulse-height discriminator typically is set at a normalized pulse amplitude value of
0.2 to reject detector noise pulses below this level and accept all
signal pulses above this level. Photon generation in the scintillator second stage is commonly assumed to follow a Poisson distribution, but in a given detector may deviate somewhat from this, so
that the Polya distribution perhaps is a better choice. However,
when the gain is high it is difficult to distinguish the shape of a
Polya distribution from that of a Poisson distribution possessing
the same relative variance. Photon transfer to the photocathode,
photoelectron release and transfer to the first dynode of the
electron multiplying stages are all binomial processes which com-
bine into a single effective binomial process for the third stage.
Estimates for these processes are shown separately in Fig. 7. The
fourth stage represents the electron multiplying stages of the
photomultiplier. They also are described by Polya distributions,
as has been shown by Prescott (13). A good fit to the data is
obtained by adjusting the theoretical relative variance until it
equals the experimental relative variance. The Polya model fits
other kinds of ion detectors equally well, such as the electron
multiplier, which is a simpler detector.

Throughout our investigations of ion detection we have found that
the yield properties of the ion-converter dynode dominates the pulse
counting response of a detector. Secondary electron response of
the first stage shapes the output pulse-height distribution and
fixes detection efficiency. Electron multiplying stages following
the conversion dynode act merely as charge amplifiers and have
little effect on the relative variance of a pulse-height distribution or on intrinsic detection efficiency. An especially valuable
feature of the electron multiplier, when it is used in the pulse-
counting mode, is that particle detection can be made virtually
noiseless. For example, a monatomic ion with 20 keV impact energy
releases about 15 secondary electrons from the ion-converter dynode
in the KAPL electron multiplier (14,15). This allows the electronic
discrimination threshold to be set high enough to bias out practi-
cally all pulses caused by thermionic electron emission - even those
from the first stage. Under these favorable conditions background
counting rates can be reduced to less than 1 cpm and detection
efficiency is approximately 100%.

THE KAPL HIGH SPEED PULSE-COUNTING SYSTEM

Next we will describe our high speed electron multiplier pulse-
counting system and how it is calibrated. It is used on our surface ionization mass spectrometers, but the same considerations
apply to the ion microprobe. The design of a pulse-counting system
for mass spectrometer application centers around the fact that output pulses from an electron multiplier or scintillator ion detector

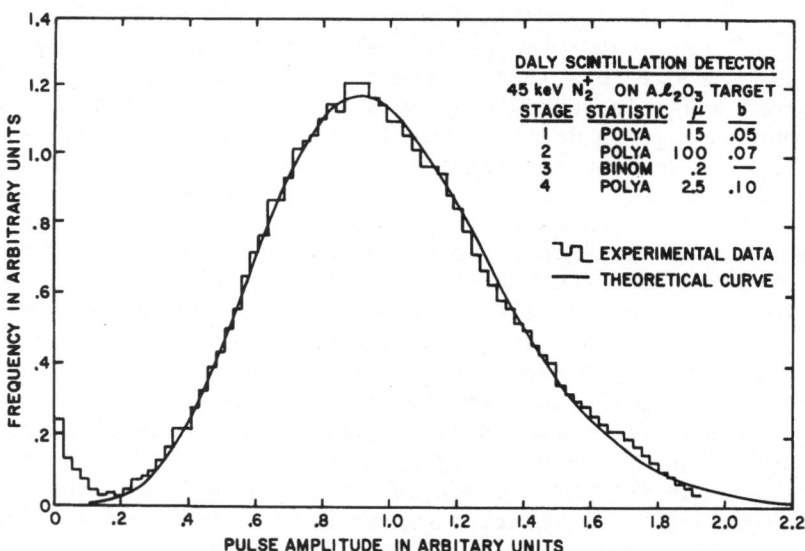

Fig. 7. Polya model fitted to output of Daly detector.
Average pulse amplitude is normalized to unity.

are randomly distributed both in amplitude and in time. The function of the discriminator/pulse amplifier is to transform random
pulse heights into standardized output pulses of constant amplitude
and shape, so that they can be counted accurately by a high speed
scaler. Our 100 MHz counting system is never used at anything
approaching a 100 MHz counting rate. Generally it is operated at
counting rates below 1 MHz. The main value of a high speed counting
system is that coincidence losses are minimized at counting rates
less than 1 MHz and can be calibrated accurately. This limitation
of counting systems is commonly misunderstood.

We begin with a photograph of our high gain electron multiplier,
which we construct ourselves. The electron multiplier operates in
a vacuum of 10^{-8} Torr. It is designed for high speed, high gain
and low noise. Maximum gain exceeds 5×10^9; operating gain is
1×10^8. The conversion dynode is flat and the ion beam is in-
clined at 60° to the surface normal in order to double the secondary
electron yield. The multiplier anode is coupled to a 50 ohm coaxial transmission line feedthrough which is tight to ultra-high
vacuum. Pulse rise time on the atmospheric pressure side of the
feedthrough is 1.8 nsec and pulse width at the base is less than
5 nsec. Background is about 0.01 cps and maximum count rate is
about 10^7 cps. Thus, the dynamic useful range of pulse-counting
rates is approximately 10^9:1. Intrinsic detection efficiency is
virtually unity for ions with 18-20 keV impact energy. A 100 MHz
solid state discriminator was specially designed by Sawada (16)
for use with the multiplier and later was improved by Hance and
Hanrahan (17). It is shown in Fig. 9. An important feature of
this circuit is the emitter-follower clipper first stage, which
amplifies small pulses and attenuates large pulses. Pulse height
discrimination is accomplished through use of a monostable tunnel
diode circuit. After further pulse amplification the output pulses

Fig. 8. Photograph of KAPL electron multiplier.

L. A. Dietz

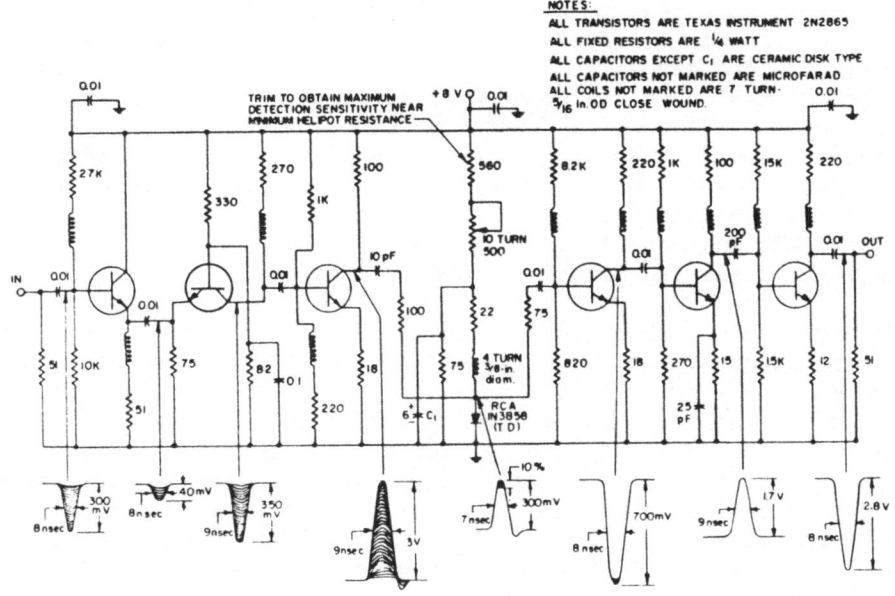

Fig. 9. Schematic diagram of KAPL discriminator.

are standardized to a width of 8 nsec FWHM and an amplitude of -2.8V. At the output the bandwidth of random amplitudes (Polya distribution) has been removed.

Of course, the randomness in time still remains. Coincidence losses are governed by the Poisson distribution. This process has been described in great detail by Rainwater and Wu (18) for pulse counting in radiochemistry. We have developed a simple and accurate technique to calibrate the dead time of an entire counting system. It entails measuring an isotope ratio at different counting rates and plotting the observed ratio as a function of counting rate. We have the analytical relationships in Eqs. (2) and (3),

$$N_1 = n_1/(1 - n_1\tau), \quad N_2 = n_2/(1 - n_2\tau), \tag{2}$$

$$R = N_1/N_2 = (R_0 - n_1\tau)/(1 - n_1\tau), \tag{3}$$

where R_0 equals n_1/n_2. Subscripts 1 and 2 refer to isotopic ion beams 1 and 2, respectively. The N's are true counting rates, the n's are observed counting rates and τ is the dead time of the entire counting system. It corresponds to the dead time of the slowest element in the system. For calibration purposes, the dead time equation is rearranged in linear from, as shown in Fig. 10. A weighted linear least squares analysis gives a dead time of 13.3 ± 1.4 nsec.

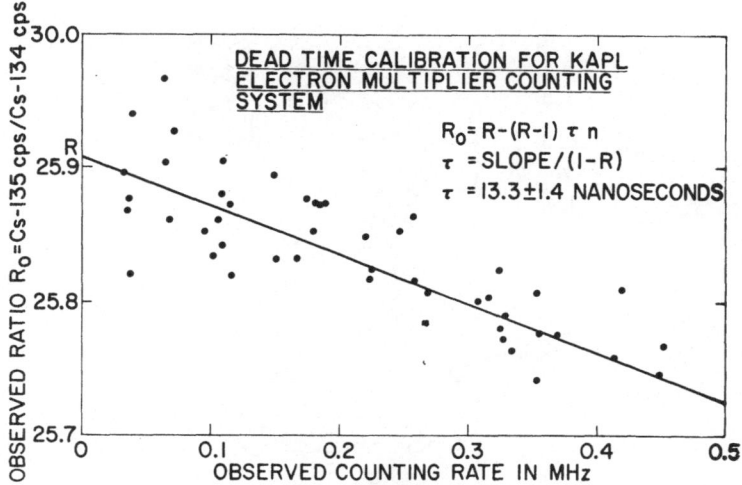

Fig. 10. Counting loss curve for dead time calibration.

The precision and accuracy of the pulse-counting method depends on small discriminations in the ion-forming and mass separation processes, but fundamentally is limited only by the statistics of the counting process. Like radio-chemical counting, the standard deviation precision of ion counting is equal to the square root of the number of counts. Using an electron multiplier, we have routinely achieved precisions of 0.05% (1 std. dev.) when using an internal standard technique (19) for measuring the half life of Cs-137.

ION MICROPROBE

Now let us look at ion detection in the ion microprobe and discuss briefly some results of surface analyses. The ion microprobe is truly an interdisciplinary analytical tool and will prove to be very useful in chemistry, physics, metallurgy and materials science. It is a hybrid spectrometer that combines the sample mounting and beam scanning techniques of the electron microprobe with the mass analyzing and ion-counting techniques used in mass spectrometry. From our viewpoint it is a new type of mass spectrometer.

We have the first production model of the Applied Research Laboratory ion microprobe. It was first described by Liebl (20) and is shown in Fig. 11. Primary ions are generated in a duoplasmatron discharge source, are accelerated up to 20 keV, and then are mass-analyzed to form a very pure primary beam, which is focused by condenser and objective lenses into a spot 2 microns in

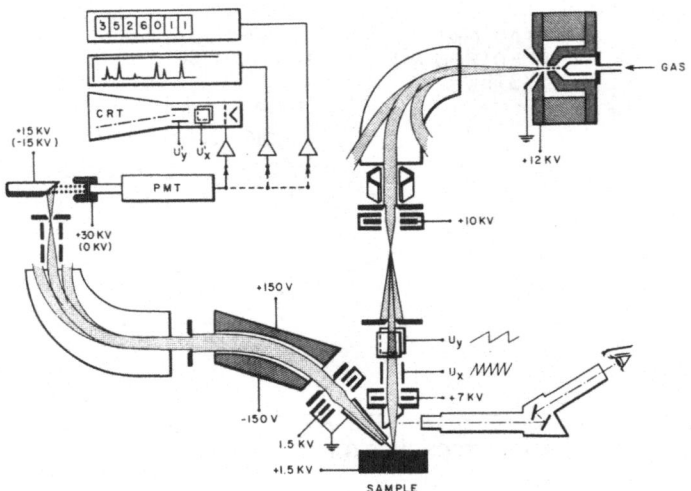

Fig. 11.   Schematic diagram of ARL ion microprobe.

diameter FWHM where it strikes the flat sample target. Secondary
ions formed by the sputtering process are accelerated through a
pickup electrode and pass through a double-focusing mass analyzer
before they strike the conversion dynode of a scintillator ion
detector. Detected ion current can be displayed visually on an
oscilloscope, on an x-y recorder, or on a high speed scaler. The
primary beam can be raster scanned across the target surface. A
very pure primary ion beam is essential to keep unwanted surface
reactions to a minimum.

Except for perhaps a few special cases, accurate calibration
techniques for the ion microprobe have yet to be developed to bring
its analytical accuracy within the 1% range. Nevertheless, there
is a growing body of knowledge about the sputtering process applied
to analytical measurements. Sputtered ions result from two-body
collisions of energetic primary ions with atoms in a target surface.
Ejected species are characteristic of only the first few atom layers
of the surface. Their intensity depends on the kind of bombarding
ion, its energy and the chemistry of the surface. The sputtered
ions tend to become neutralized as they leave the surface and the
number that survive depend on the electronic properties of the
surface. For example, for positively-charged sputtered ions the
survival rate is improved by forming strongly-bonded compounds on
the surface. This is done by bombarding it with a reactive gas,
such as oxygen. Formation of an oxide layer greatly reduces the
number of losely-bound electrons that might neutralize a sputtered
ion. According to Anderson (21), positive ion yields are inversely
proportional to the availability of surface electrons. Conversely,
the yield of negatively-charged sputtered ions is enhanced by
bombarding the surface with an electropositive element such as

cesium. Cesium reduces the work function of the surface and increases the availability of loosely-bound electrons. Thus, a
sputtered neutral atom with a large electron affinity can acquire
an additional electron and escape as a negatively-charged ion. It
is experimentally observed that a surface with a high secondary ion
yield also tends to have a high secondary electron yield.

Now let us look at some experimental results with the ion
microprobe. Anderson (22) has found that when a pure aluminum sample
is bombarded with a noble gas ion such as Ar-40[+], the ion yield of
Al-27[+] falls off rapidly with time. This is shown in Fig. 12.
The sputtered yield of Al-27[+] is enhanced by strongly-bonded
chemical compounds which result from chemisorption of reactive
gases, oxygen in this case. As the primary ion beam sputters the
oxide away, the sputtered ion yield of Al-27[+] decreases. In the
case of aluminum, it is the thin surface layer of aluminum oxide
of perhaps 10-30 Å thickness that provides the initial high yield
of Al-27[+] ions. After the transient peak has passed, the steady-
state output of Al-27[+] depends on the arrival rate of primary ions
relative to the arrival rate of reactive gas molecules in the vacuum
system of the microprobe.

In the second part of Fig. 12, 0-32[+] is the bombarding beam.
The dip in Al-27[+] ion output is believed to be related to the
initial depth distribution of implanted oxygen ions, since relatively fewer ions are deposited near the surface than at the
mean range. Thus, the first few atom layers, as they are sputtered

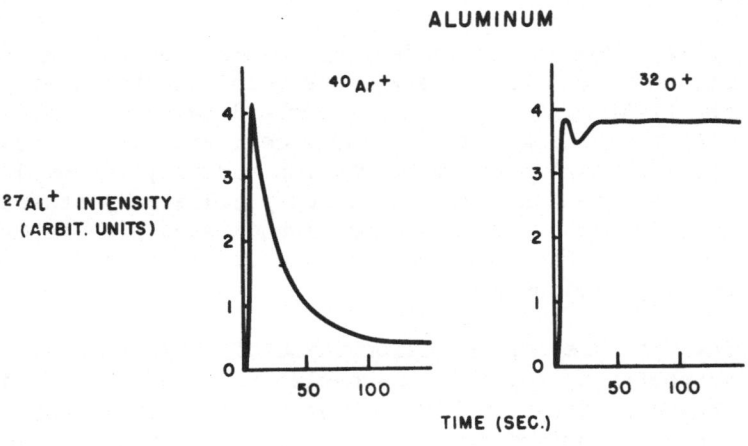

ALUMINUM

Fig. 12.   Anderson's curves of sputtered ion yield of
         Al-27[+] for Ar-40[+] and 0-32[+] bombarding ions.

ALUMINUM

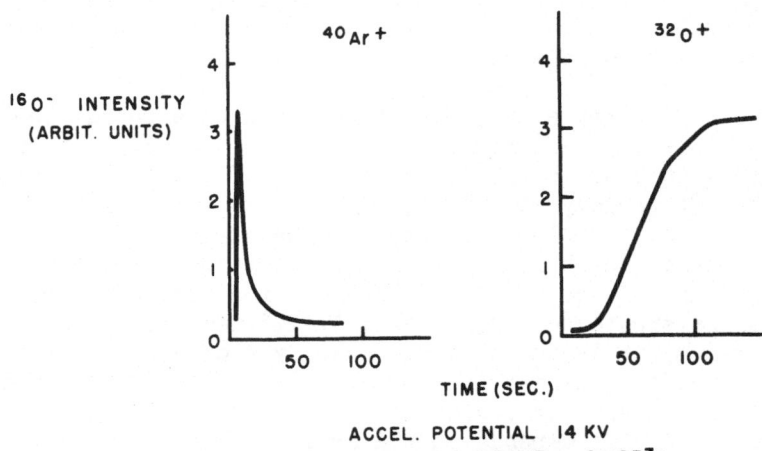

ACCEL. POTENTIAL 14 KV
SAMPLE CURRENT ∼ 5×10⁻⁷A

Fig. 13. Anderson's curves of sputtered ion yields of O-16⁻
for bombarding ion beams of Ar-40⁺ and O-32⁺.

away, are not as highly oxidized as deeper layers and therefore
have lower yield. The intensity of sputtered Al-27⁺ remains constant after the initial mean depth of implanted oxygen ions has
been exceeded.

To demonstrate further the effect of chemisorbed gas on the
target surface, Anderson (22) observed the O-16⁻ sputtered ion
intensity using Ar-40⁺ and O-32⁺ as the bombarding species. As we
have seen in the preceeding figure, bombardment with Ar-40⁺ results
in a characteristic peaking of sputtered ion current, followed by
an approximately exponential decay to a low-yield steady state.
Next, Anderson (22) bombarded this same cleaned spot with an O-32⁺
beam and observed that the sputtered ion output rose steadily to
a high, stable output as before.

In keeping with the ion detection theme of this paper, we present the mass spectra of several ion-to-electron conversion dynodes.
The first of these is shown in Fig. 14 and is for a highly-polished
2S-aluminum dynode, anodized in an electrochemical cell to a depth
of 200 Å of aluminum oxide. We use these dynodes in our electron
multiplier. These results are for a primary beam of O-16⁻ at 16.5
keV impact energy. The beam diameter was several microns and a
raster scan of approximately 300 microns x 300 microns was used.
The aluminum is only about 98-99% pure and impurities of a tenth
% or less of H, C, Mg, Si, Ca, Mn and Fe result in significant peaks.

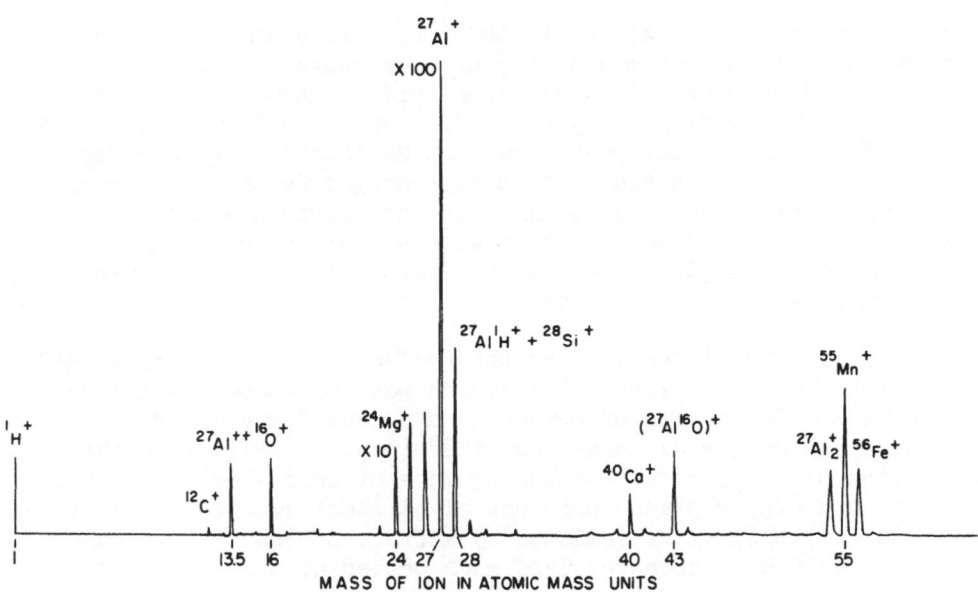

Fig. 14.   Ion microprobe scan of polished 2S-aluminum conversion dynode.

The Al-27+ counting rate is about 1 MHz.   All unmarked peaks have a scale factor of X1.

In Fig. 15 we have an aluminum conversion dynode that was made

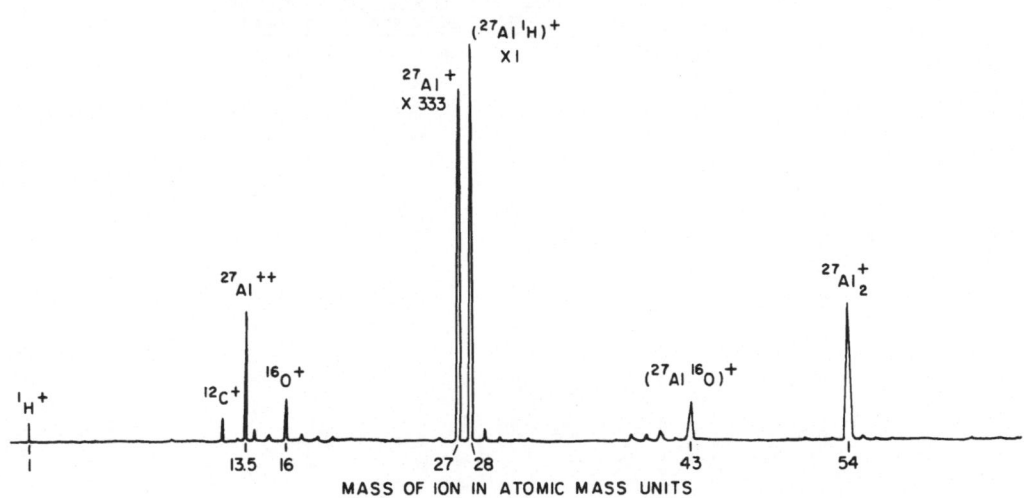

Fig. 15.   Ion microprobe scan of oxidized Al conversion dynode on glass.

by electron beam evaporation of 1000 Å of pure aluminum onto a
section of glass microscope slide, at a pressure less than 10^{-8}
Torr. The freshly-deposited aluminum surface then was exposed to
a partial pressure of pure oxygen to form an oxide layer about 10 Å
thick. Here most of the peaks are combinations of aluminum and
oxygen. Hydride peaks and some doubly-charged peaks are commonly
observed. There also is a small amount of hydrogen and carbon in
the surface of this dynode. The beam diameter was a few microns
and a 300 micron X 300 micron rastered area was used to insure a
low sputtering rate per unit area.

Next, in Fig. 16 we look at the surface of a CuBe dynode used
in our electron multiplier. The dynode was activated (oxidized)
at 630° C in 100 microns of oxygen in a vacuum furnace. This
activation technique has been described (14). Beryllium is the
major peak and represents a counting rate of approximately 1 MHz.
The iron and copper peaks represent significant amounts of these
elements at the surface, followed by a trace of lead. The lead
spectrum could have been resolved much better at a sacrifice in detected ion current by narrowing the solid angle of acceptance of
the mass analyzer. Again, this is a continuous raster scan several
times a second over an area of about 300 microns X 300 microns.

In Fig. 17 we have penetrated several thousand angstroms into
the base material. The mass spectrum has shifted dramatically to
copper as the major peak and beryllium now is the secondary peak.
Also, the mass spectrum is rich in many combinations of copper

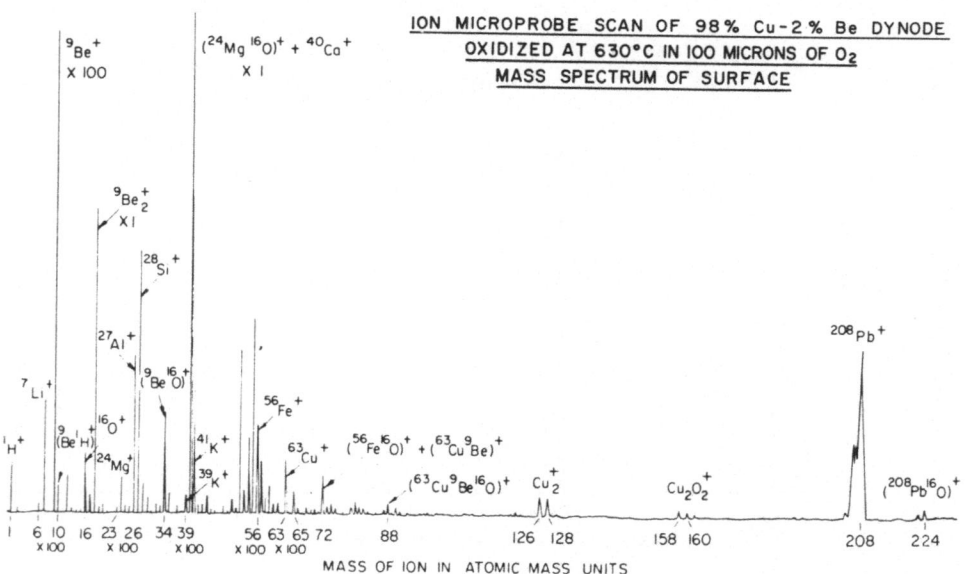

Fig. 16. Ion microprobe scan of surface of CuBe dynode.

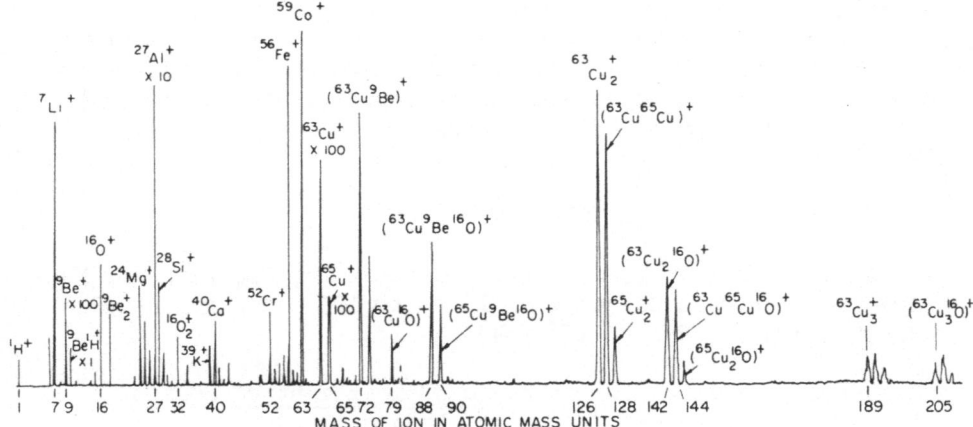

Fig. 17.   Ion microprobe scan below surface layer of CuBe
dynode.

isotopes with oxygen.   Iron is much smaller and lead is absent.
Point-to-point imaging of ion concentrations, displayed on the
oscilloscope, show that most of the chemical elements other than
Cu and Be are quite non-uniformly distributed over the surface.
We can estimate the beryllium concentration at the surface of the
dynode by using a spectrographic analysis of the CuBe alloy to calibrate the Cu and Be ion current intensities relative to each other.
The actual 1.9 weight % Be converts into 13 atom %, so that in the
bulk material there are 5 atoms of Cu-63 for every Be-9 atom.   From
the observed ratios in the bulk material and at the surface it
appears that heat treatment has enhanced the beryllium concentration
at the surface to perhaps as much as 60 atom %.   Our future plans
include the analysis of these types of dynode surfaces in much
greater detail.

In conclusion, the ion microprobe shows great promise for the
elemental analysis of surfaces and certainly is destined to become a prime analytical tool.   It already is superb for qualitative
analysis of surfaces and trace detection down to the ppb level, and
in cases where accurately known standards are used it can be
accurate to a few percent or better.

ACKNOWLEDGMENTS

I am indebted to J. C. Sheffield and J. H. Thomas for construction of the secondary electron emission apparatus and to J. C. Sheffield for making the measurements of secondary electron yields. I also wish to thank J. F. Stevens for the ion microprobe scans of the dynode surfaces and others in the mass spectrometry group who helped prepare these surfaces.

REFERENCES

1. F. A. White and T. L. Collins, "Positive Ion Counting for Mass Spectrometer Beam Currents of 10^{-14} to 10^{-19} Ampere," Appl. Spectroscopy $\underline{8}$, No. 1, 17(1954).

2. G. Carter and J. S. Colligon, Ion Bombardment of Solids, (Heinemann Educational Books Ltd, London,1968).

3. L. A. Dietz, "Electron Multipliers and Statistics of Secondary Electron Emission," 18th Annual Conf. on Mass Spectrometry and Allied Topics, San Francisco, California, June (1970).

4. E. S. Parilis and L. M. Kishinevskii, "The Theory of Ion-Electron Emission," Transl. Soviet Phys.-Solid State $\underline{3}$, 885 (1960).

5. B. L. Schram, A. J. H. Boerboom, W. Kleine and J. Kistemaker, "Amplification Factors of a Particle Multiplier for Multiply Charged Noble Gas Ions," Physica $\underline{32}$, 749 (1966).

6. M. Kaminsky, Atomic and Ionic Impact Phenomena on Metal Surfaces, (Academic Press, 1965), p. 329.

7. J. A. Davies, "The Penetration of keV Projectiles in Solids," Atomic Energy of Canada Ltd., Rept. No. AECL-2757, August (1967).

8. B. Domeij, F. Brown, J. A. Davies and M. McCargo, "Ranges of Heavy Ions in Amorphous Oxides," Can. J. Phys. $\underline{42}$, 1624 (1964).

9. R. E. Collins and L. W. Davies, "The Transport of Hot Electrons in Al-Al_2O_3-Al Tunnel Cathodes," Solid State Electronics, (Pergamon Press, 1964), Vol. 7, p. 445.

10. L. A. Dietz, "General Method for Computing the Statistics of Charge Amplification in Particle and Photon Detectors Used for Pulse Counting," Int. J. Mass Spectrometry Ion Phys. $\underline{5}$, 11 (1970).

11. N. R. Daly, "Scintillation Type Mass Spectrometer Ion De-
     tector," Rev. Sci. Instr. 31, 264 (1960).

12. W. A. P. Young, R. G. Ridley and N. R. Daly, "Fast Counting
     in Mass Spectrometry with the Scintillation Detector," Nucl.
     Instr. Meth. 51, 257 (1967).

13. J. R. Prescott, "A Statistical Model for Photomultiplier Single
     Electron Statistics," Nucl. Instr. Methods 39, 173 (1966).

14. L. A. Dietz, A. B. Hance and L. R. Hanrahan, "Fabrication of
     18-Stage Electron Multiplier," Knolls Atomic Power Laboratory
     Rept. KAPL-3352, June (1967).

15. L. A. Dietz, "Basic Properties of Electron Multiplier Ion De-
     tection and Pulse-Counting Methods in Mass Spectrometry, Rev.
     Sci. Instr. 36, 1763 (1965).

16. F. H. Sawada, "Design Technique for High Speed Pulse Circuitry
     For Surface Mass Spectrometer, "IIIE Trans. Nucl. Sci., Vol.
     NS-12, No. 1, 374, Feb. (1965).

17. A. B. Hance and L. R. Hanrahan, "Further Development of a
     High-Speed Electron-Multiplier Discriminator for Mass Spec-
     trometer Ion Detection," Knolls Atomic Power Laboratory Rept.
     No. KAPL-3122, Feb. (1966).

18. L. J. Rainwater and C. S. Wu, "Applications of Probability
     Theory to Nuclear Particle Detection," Nucleonics, Oct. 1947,
     p. 60.

19. L. A. Dietz, C. F. Pachucki and G. A. Land, "Half Lives of
     Cessium-137 and Cesium-134 as Measured by Mass Spectrometry,"
     Anal. Chem. 35, 797 (1963).

20. H. Liebl, "Ion Microprobe Mass Analyzer," J. Appl. Phys. 38,
     5277 (1967).

21. C. A. Anderson, "Analytic Methods for the Ion Microprobe Mass
     Analyzer. Part II," Int. J. Mass Spectrometry Ion Phys. 3,
     413 (1970).

22. C. A. Anderson, "Progress in Analytic Methods for the Ion
     Microprobe Mass Analyzer," Int. J. Mass Spectrometry Ion
     Phys. 2, 61 (1969).

APPLICATION OF COMPUTERS IN ELECTRON PROBE AND X-RAY FLUORESCENCE
ANALYSIS

Stanley D. Rasberry

Institute for Materials Research

National Bureau of Standards

Washington, D. C. 20234

ABSTRACT

This paper is a review of automation of electron microprobe
and x-ray fluorescence instrumentation. Such a review seems timely
because of the great increase in the application of computer systems
in this field over the past decade. Some of these applications have
been conceived to meet true technological needs while in other cases
they have been undertaken to "keep up with the Joneses." I would
like to show not only what automated systems are now feasible but
also when and how they should be employed. The "when" and "how" of
automation are largely dependent upon the application being considered; in this study, x-ray applications have been divided into
the following classes: (1) on-stream process-control, (2) off-line
quality assurance, (3) routine service laboratory, (4) general-
purpose analytical laboratory. Several phases are present in these
classes, including: specimen preparation and loading, measurement,
data acquisition and transfer, data processing and display, and
finally, archival data storage. Various workers have undertaken the
automation of all these operations in one or the other of the classes
of applications; from a review of their work and by examining details
of each operation within the framework of a given application, we
can now draw conclusions on the extent of desirable automation.

INTRODUCTION

In the beginning, chemical analysis by automated x-ray fluorescence equipment was directed toward increasing the speed of

analysis in order to better support mining and metal processing operations (1-5). These applications stimulated the construction of several automated x-ray fluorescence analyzers and a considerable degree of automation in electron probe microanalyzers. Recent developments have been mainly to improve efficiency or to make possible experiments and analyses which would be difficult by manual operation. On the other hand, some laboratories have been automated in the past decade to demonstrate that this automation is feasible, without any specific analytical application in mind.

When emphasis is not placed on the actual analytical need, what stands to be lost is frequently overlooked in automation. Flexibility of instrument use and automation are often mutually exclusive, and where a large measure of each is required in the same instrument, consummate attention must be given to the system design. Automation also tends to isolate the analyst from the sample, unless care is given to the design of reports and information about the sample which are transmitted to the analyst. Electron microprobers seem particularly susceptible to losses of this kind, perhaps because the wonderfully complex nature of many of the specimens they are requested to investigate demands extraordinarily incisive examination, not only by the machine but also by the man. Automation can be helpful in such work, but again the goals for it must be carefully specified.

To see where the potential gains and losses lie in automating electron probe and x-ray fluorescence equipment, it is important to define the types of application for this equipment and the individual steps in its routine use. The definitions and classifications given in the next two sections are not meant to be universially applicable, but rather are drawn to facilitate the discussion of automated systems.

Several recent reviews have contained sections of bibliographies on automated equipment and computer use in electron probe and x-ray fluorescence analysis (6-8).

CLASSES OF APPLICATION

On-stream Process-control

The aim of on-stream process-control is to govern a manufacturing operation so as to yield a product which meets specifications. When the desired results can be obtained by changing some parameters of the operation (temperature, pressure, or ratio of constituents, for example) there is a need for monitoring the characteristics of the product, selecting adjustments to be made and causing the adjustments to be effected. Where the measurement signals are obtained from automatic sensors (x-ray devices in some cases) and the

adjustment signals are transmitted without manual intervention,
the operation is said to be controlled in a "closed loop" mode.

No use of electron probes in this class of application has been
brought to my attention; however, x-ray fluorescence and absorption
have been applied extensively both as sensors for elemental composition (1-2,4,9), and as gauges for stock thickness or plating thickness (10).

On-stream process-control applications usually require very
limited, if any, flexibility for change in the operational routine
(program). For this reason, the program can be hard-wired into the
controller or computer. Before the recent substantial decreases in
digital computer prices, hard-wired controllers usually had a cost
advantage over the general-purpose computer. Even now, hard-wired
programs, in either controllers or computers, can have advantages
of ruggedness when the environmental conditions are demanding.

Off-line Quality Assurance

Some manufacturers, especially those with production methods
based on batch-operations, have little need for on-stream process-
control. Instead they usually wish to rapidly check their product,
relative to prescribed specifications, at various stages of manufacture. Economical - and frequently non-destructive - final composition analysis also may be important to both the producer and the
buyer where actual performance tests are difficult, expensive or
time consuming (i.e., cement durability).

The needs mentioned in this category have been met with a
variety of automated x-ray fluorescence (3,5,11) and electron probe
(12-14) equipment.

Routine Service Laboratory

In quality assurance work, the objective is not so much
analysis as it is comparison with previous production - in some cases,
the compositions observed remain nearly constant year after year.
Routine service laboratories - examples would include agricultural,
clinical and geological - may also work with unchanging sample types;
however, the range of compositions and accuracy requirements vary
widely. These labs may be required to identify, catagorize, or
analyze specimens in support of either individual analysis of larger
surveys. For this reason, they need greater flexibility in altering
analytical programs and in data treatment. They share many of the
attributes of the final class to be mentioned.

General-purpose Analytical Laboratories

The general-purpose lab is called upon to deal with the non-
routine as well as the routine. Successive problems may be as
diverse as detecting counterfeit money, failure analysis of a rotor
from a crashed helicopter, measuring atmospheric lead content, and
determining the homogeneity, on a micron scale, of castings produced
by a new process. This diversity of problems, plus the research
orientation that is often present in this kind of laboratory, place
great demands of flexibility in automated systems. The automated
instruments and methods (15-20) which have been developed in the
past five years to serve in the general-purpose laboratory are
different in many ways than those developed for the other classes
of applications. Some of these differences will be discussed in
more detail in the next section.

TASKS WHERE AUTOMATION MAY BE USEFUL

Specimen Preparation and Loading

Specimen preparation is not generally employed in process-
control applications. Rather, the x-ray detector directly observes
the stream of material, which may be a liquid or slurry moving
through a pipe with a window for the x-ray detector (2), or a solid
moving in a strip mill (1).

Specimens may be prepared either manually or automatically in
devices normally not connected to the x-ray instrument. Exceptions
to this are automatic briquetting presses arranged to load the
briquettes directly into the x-ray instrument sample chamber.

When automated x-ray instruments are used, specimen preparation
and loading may constitute a large share of the total manual labor.
Specimen identification is an especially important consideration in
automation whenever correlation of analytical results with discreet
specimens is important.

The identification number of a sample in the analytical chamber
must be displayed to the operator and also conveyed to the measurement controller; it is an indispensable item in any record obtained
for the specimen. Also, it is important that the specimen be labeled
and in a known storage position (carousel, tray, sectioned box,
etc.) at all times when not in use. This prevents mix-ups and
facilitates specimen selection for further testing.

Bayard has described the use of duplicate specimen stages, a
technique which Adler used as early as 1959, in non-automated equipment, in conjunction with an automated electron probe (12). Indexing
the cartesian coordinates of interesting test locations is possible

on the alternate stage using an optical microscope and thus can
leave the electron probe available for analysis. When the specimen
is transferred to the stage in the electron probe, the indices of
locations to be tested are entered in the control computer as part
of the specifications of the measurement program.

Measurement - Data Transfer - Data Processing

Programming systems - whether hard-wired or in software - range
between two extremes in operational concept. On one hand, completely
fixed "canned" routines are developed, in advance, for each different
sample type. At the other end of the scale, especially in the
investigative environment, an input scheme is provided which is
oriented toward the operator. This allows him to change his
approach during an analysis.

The fixed routine approach has been used in the automated
electron probe developed and reported by Bayard (12). His goal was
unattended operation for periods in excess of 15 hours (overnight),
for the purpose of determining as many as 42 elements in each of
several sample points on one or more specimens. This requires computer control in positioning the stage to the prescribed locations
at the beginning of analysis for each sample point, positioning the
beam, positioning the spectrometers and operating the measurement
electronics. Measured quantities, including sample current, beam
current, stage position, beam position, spectrometer position, and
ratemeter, timer, and scaler readings are transferred to the computer
via a custom-built interface. Oscillograph-imaged data, such as area
scans and line scans, are recorded, when required, by the use of a
computer-controlled 16-mm camera.

Not all analytical procedures of the fixed-routine type require
as complete automation as that described above. In fact, in an
automated electron probe system described by Heinrich, Myklebust,
Rasberry, and Michaelis (21), no provisions were made for moving
the stage or spectrometers. The goal of this work was unattended
operation while determining the homogeneity, on a microscale, over
very small regions of materials proposed as Standard Reference
Materials for microanalysis. The only functions which required automation were deflection of the electron beam, operation of the measurement electronics, and transfer of analytical data to punched paper
tape. The device is shown schematically in Fig. 1 and its operating
sequence is summarized in Fig. 2; details are given in the reference.
This system has been used by Yakowitz, Fiori and Michaelis in microscale homogeneity testing (22).

When the x-ray measurements are collected from a two-dimensional
grid of points, the digital quantitative information obtained in
single-point analysis is conserved, yet topographic coverage analogous
to continuous area scanning at low resolution is also obtained.

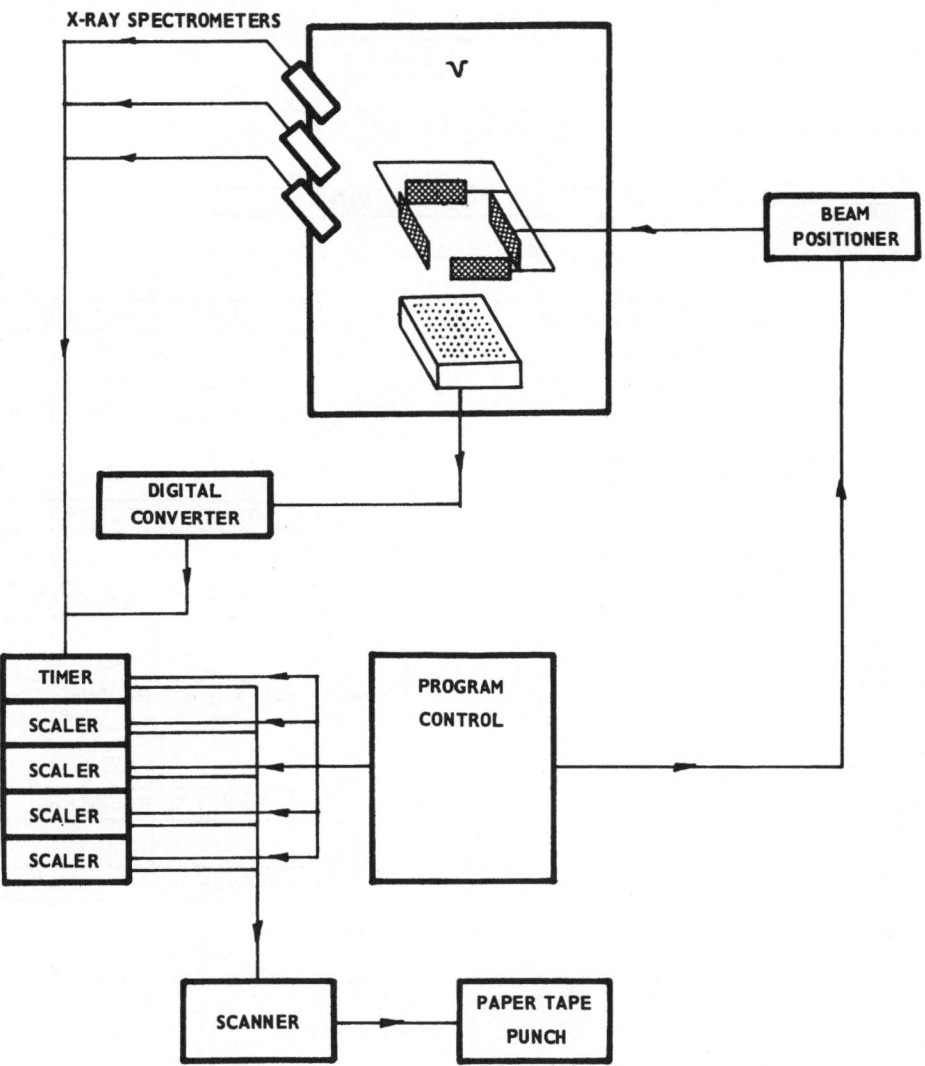

Fig. 1 An automated electron probe data acquisition system
        used in determining composition homogeneity over small
        regions.

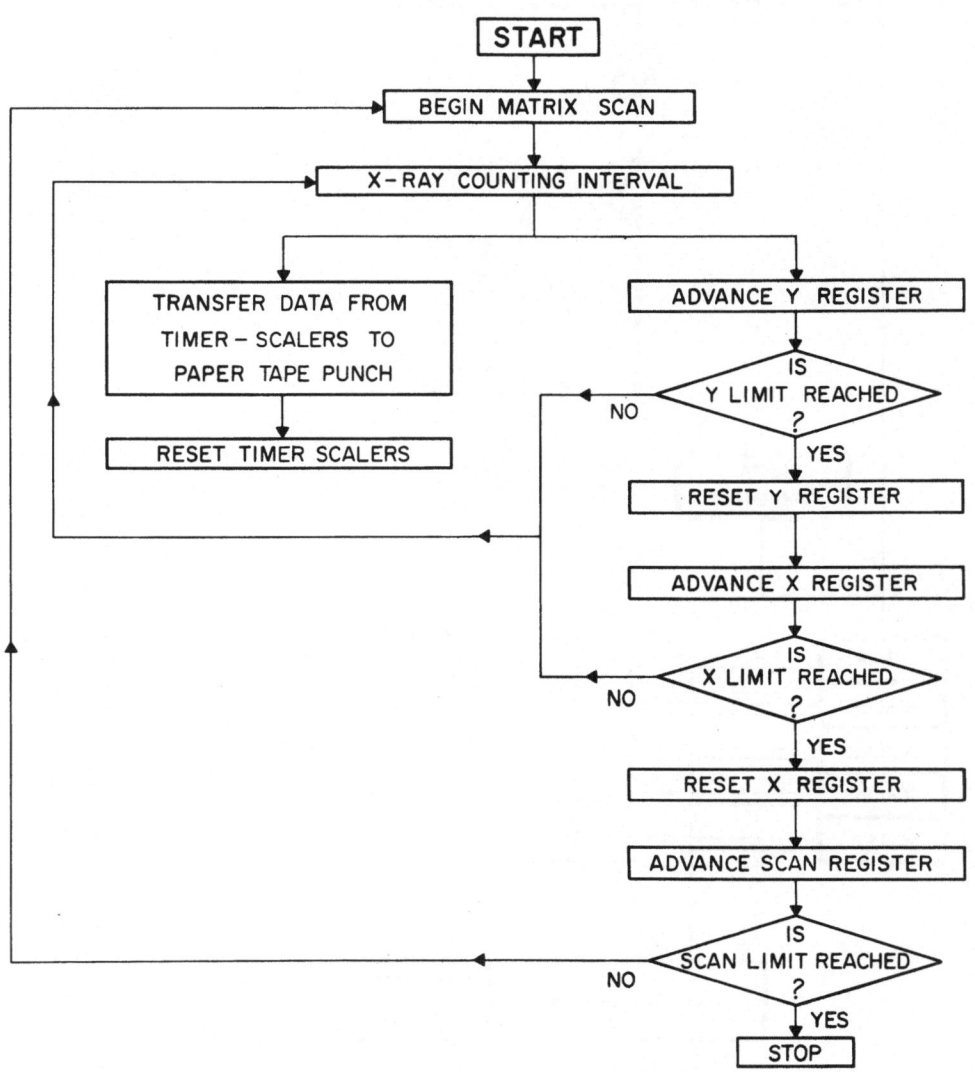

Fig. 2   Operating sequence of the device shown in Fig. 1.

Heinrich, et al. (21) describe some simple approaches to processing and display of topographic data. The topographic concentration images that are available with an electron probe contribute greatly to its usefulness; it is not surprising that increasing attention is being given to computer processing of these images. In several papers co-authored variously by McMillian, Johnson, White, Matson, McKinstry and Görz (23-25), automated data acquisition and computer image processing is described. Yakowitz and Heinrich have discussed the use of digital grid images in identifying inclusions with an electron probe (26).

A great deal of the measurement and control equipment for automating both the fixed and interactive applications has been of modular type, especially using Nuclear Instrumentation Modules (NIM's). This approach has been reviewed by Pedersen (17), who used as a case study the computerized automation of an electron probe. An individual module, which can also be manually controlled by the operator, is dedicated to each required function - stage positioning, spectrometer positioning, x-ray counting, etc. Having a system which can be manually operated is necessary to prevent the total loss of instrument availability during times of computer failure.

Kunz, Eichen, Matthews and Piner (13) have further developed the concepts of mini-computer and NIM usage in an interactive system of electron probe automation. They have developed a very simple computer language called CLASS, for operating and controlling laboratory instrumentation. With the mini-computer they can, by loading programs one at a time, locate x-ray peaks, collect data from standards and unknowns, calculate intensity ratios and statistics and, using the computation program by Colby (27), calculate concentrations.

One of the difficulties with mini-computers or hardware controllers arises from the limitations in size of the memory core available for processing quantitative calculations. Rasberry, Margoshes and Scribner (28) have described transfer of data from these devices, via paper tape, to remote time sharing computers. An alternative solution has been offered by Kirkendall and Varadi (20). They have automated an electron probe by dedicating a mini-computer and NIM's exclusively to control and data acquisition in a highly interactive mode. When the quantity of data obtained approaches the core limit, they are transferred by telephone to a large-scale computer where storage, calculation, and presentation capabilities are many-fold greater. According to Kirkendall and Varadi (29), it is especially useful that the large computer can be used for storage and maintenance of the programs used in the mini-computer, thus facilitating loading it with different programs.

If there is a single rule which must be adhered to in automating the collection and calculation of data, it is that the measurement and computations must be designed to be consistent with physical

principles. No matter how fancy or rapid is the data acquisition
system, it is useless as an analytical tool unless this rule is
observed. In the early days of process-control automation, pre-
occupation with the automatic parts of the equipment reduced the
attention given to the principles of measurement (for example,
limitations of the counting electronics) and the equations used in
quantitation. Now, more consideration is being given to these
"details" which, of course, are basic.

Beaman and Isasi (30-31) have given an elementary introduction
to the theory and practice of quantitative electron probe microanalysis; a more advanced presentation is found in a book edited by
Heinrich (32). Beaman and Isasi (33) have also produced a list of
40 computer programs written since 1963 for use in quantitative
electron probe microanalysis.

In the area of x-ray fluorescence analysis, the reviews of
Campbell and Brown (7), and Campbell and Gilfrich (8) are useful
to finding references which deal with quantitation. For calibration
in those cases where interelement interferences of fluorescence and
absorption are pronounced, Criss and Birks (34) have described and
contrasted fundamental-parameters and empirical-coefficients quan-·
titation methods. Rhodes, Hunter, Kellogg, Sieberg and Furuta (35)
have extended a fundamental parameters calibration method to a
computer-coupled non-diffractive spectrometer used in the analysis
of steels. The subject of automatic processing of solid-state
detector data has also been dealt with by Short and Stephany (36),
Lifshin (37) and Yakowitz, et al. (22).

Reports and Data Storage

Besides the calculations necessary to quantitation, the computer can do useful work in sorting, collating and comparing data,
making statistical tests and error detection checks, monitoring
instrument performance, preparing reports, and arranging data for
archival storage.

Except in the most rudimentary forms, these kinds of jobs have
not been undertaken. They are not as basic as the comparison and
control operations that were developed early, or the more sophisticated quantitation and image-handling capabilities that have been
created more recently. These secondary tasks can require quite
complex computer software, development of which could be prohibi-
tively expensive for a single instrument. Now, however, there is
increasing interchangeability in computer programs and generalized
programs in these areas are becoming available.

THE NBS COMPUTER NETWORK FOR ANALYTICAL INSTRUMENTS

A computer system which will serve the needs of several laboratories for data acquisition, calculation, and instrument control
is being installed in the Analytical Chemistry Division of the
National Bureau of Standards (38). Electron probe and x-ray fluorescence equipment will be interfaced to the system in early 1972.
In both cases the equipment is used for general-purpose investigations so that a high level of interaction with the operator is
planned.

Communications between the analytical instruments and the computer will be via an external digital data bus. This "party line"
of 50 parallel wires is common to all stations. The assignment of
wires permits 8 bits for address, 10 bits for control and 32 bits for
two 16-bit words. Local stations can enter data to the bus or receive information from it only when their address is enabled by the
computer. In order to prevent the loss of data when the network or
computer is busy, limited storage is provided as a buffer at the
computer end and as "hold" circuits at the individual experiments.
Information is transmitted via the bus in a multiplexed format.
Alternate 10-µs time-slices are given to "low-speed data" (up to
1-kHz) with each experiment being serviced in sequence if it has
set a transmit request flag. The remaining alternate time slices
are dedicated, for short periods, to acquiring data from one instrument at data rates up to 50-kHz. A request for this high-speed service can be sent from any one of the instruments to the computer via
the low-speed data path.

The computer has a scheduler provided by the manufacturer which
interchangeably operates a program area for real-time job servicing
(foreground), an area for experimenting with new programs (middle-
ground), and an area in which lower priority calculation and compilation jobs can be undertaken on a time-available basis (background). When individual experimenters tinker with their own data
acquisition programs on a shared computer, interferences frequently
arise in storage and peripheral allocation. A unified data acquisition program, called DACQ (39), has been implemented in foreground
to serve the users of the system by fulfilling all their data acquisition requirements. This should reduce interferences and simplify
use of the system by the experimenters. Other utility programs which
can be shared by all users are being written - this should help
prevent duplication of similar programs on the disk.

This network approach to computer usage has several advantages
for x-ray instruments. It provides more computing power and randomly accessible storage space than is normally present with a mini-
computer without disk storage. Further, it makes available many

utility programs and peripheral devices which are only required
occasionally and would be very expensive for a single instrument.
Several other groups are working on the network approach to laboratory automation; among these, the work of Kennicott, Scavullo,
Sicko and Lifshin (40) is especially interesting because it includes
service to an electron probe as part of the system.

CONCLUSION

This paper is a review of automation of electron microprobe
and x-ray fluorescence equipment. I have attempted to identify
the several classes of application for this kind of equipment in
order to give a framework for understanding the diverse needs of
automation in the field.

Rather than attempt predictions of what is in store for the
future of computer applications in electron probe and x-ray fluorescence analysis, it seems more useful to draw together three
observations made earlier. An automated system can be made truly
successful only when its objectives are clearly defined. The extent
of desirable automation is limited to be the minimum system which
will attain all the objectives set forth. Finally, the design and
operation of any automated system must be in harmony with the underlying physical principles of the measurement process.

REFERENCES

1. R. A. McCune, W. M. Mueller and P. J. Dunton, "Feasibility of
    the X-Ray Spectrograph as a Continuous Analytical Instrument
    for Process Control," in W. M. Mueller (ed.), Advances in X-Ray
    Analysis, Vol. 1, Plenum Press, New York, 1960, p. 399.

2. W. J. Campbell, "Apparatus for Continuous Fluorescent X-Ray
    Spectrographic Analysis of Solutions," Applied Spectroscopy 14,
    26 (1960).

3. C. M. Davis and M. M. Yanak, "Performance of an Unattended
    Automated X-Ray Spectrograph," in: W. M. Mueller, G. Mallett
    and M. Fay (eds.), Advances in X-Ray Analysis, Vol. 7, Plenum
    Press, New York, 1964, pp. 644-652.

4. W. G. Moffat and R. Carson, "Some Instrumental Considerations
    in the Automatic On-Stream Analysis of Pulps for Elemental
    Content," in: W. M. Mueller, G. Mallett and M. Fay (eds.),
    Advances in X-Ray Analysis, Vol. 8, Plenum Press, New York,
    1965, pp. 204-214.

5. E. Davidson, A. W. Gilkerson and W. G. Shequen, "X-Ray Fluorescence Analysis with a New High Speed Multichannel Instrument," Pittsburgh Conference on Analytical Chemistry and Applied Spectroscopy, March 1965.

6. K. F. J. Heinrich, "Electron Probe Microanalysis: A Review," Applied Spectroscopy 22, 395 (1968).

7. W. J. Campbell and J. D. Brown, "X-Ray Absorption and Emission," Anal. Chem. 40, 346R (1968).

8. W. J. Campbell and J. V. Gilfrich, "X-Ray Absorption and Emission," Anal. Chem. 42, 248R (1970).

9. R. W. Deichert and J. P. O'Connor, "A Review of Process Control Instrumentation Development," Norelco Reporter 10, 43 (1968).

10. H. A. Liebhafsky, H. G. Pfeiffer, E. H. Winslow and P. D. Zemany, X-Ray Absorption and Emission in Analytical Chemistry, (John Wiley and Sons, Inc., 1960), pp. 69-71.

11. W. E. Fowler, P. J. Breckheimer and A. J. Hartwick, "The Impact of Digital Computers on Spectrochemical Analysis," Pittsburgh Conference on Analytical Chemistry and Applied Spectroscopy, March 1965.

12. M. Bayard, "Computer Interfacing the Electron Microprobe," The Microscope 17, 169 (1969).

13. F. Kunz, E. Eichen, G. Matthews and S. Piner, "Electron Microprobe Automation," Proceedings Sixth National Conference on Electron Probe Analysis (1971). Copies available from L. Vassamillet, Carnegie-Mellon University, Pittsburgh, Pa. 15213.

14. A. A. Chodos and A. L. Albee, "Quantitative Microprobe Analysis and Data Reduction Using an On-Line Mini Computer," Proceedings Sixth National Conference on Electron Probe Analysis (1971). See 13.

15. J. W. Frazer, "Digital Control Computers In Analytical Chemistry," Anal. Chem. 40, 26A (1968).

16. S. D. Rasberry, "X-Ray Spectrometry," in B. F. Scribner (ed.), Activities of the NBS Spectrochemical Analysis Section, July 1967 to June 1968, U.S. National Bureau of Standards Technical Note 452, U.S. Government Printing Office (Sept. 1968), pp. 38-44.

17. E. Pedersen, "A Modular Approach to Laboratory Instrument Automation," American Laboratory, February 1970 Issue.

18. W. L. Baun and E. W. White, A Vacuum Spectrometer for Studying the chemical Effect of Soft X-ray Spectra," in: B. L. Henke, J. B. Newkirk and G. R. Mallett (eds.), Advances in X-Ray Analysis, Vol. 13, Plenum Press, New York, 1970, pp. 237-247.

19. W. J. Steele, "LRL Computer Program: Microanalysis," Proceedings Sixth National Conference on Electron Probe Analysis (1971). See 13.

20. T. D. Kirkendall and P. F. Varadi, "An Automated Microprobe Under PDP-8 Control Using an IBM 360/65 for Program and Data Storage," Proceedings Sixth National Conference on Electron Probe Analysis (1971). See 13.

21. K. F. J. Heinrich, R. L. Myklebust, S. D. Rasberry and R. E. Michaelis, "Appendix 4 - Automated Techniques for Homogeneity Analysis," in: Preparation and Evaluation of SRM's 481 and 482 Gold-Silver and Gold-Copper Alloys for Microanalysis, U.S. National Bureau of Standards Special Publication 260-28, U.S. Government Printing Office (August 1971), pp. 70-89.

22. H. Yakowitz, C. E. Fiori and R. E. Michaelis, "Homogeneity Characterization of Fe-3Si Alloy," U.S. National Bureau of Standards Special Publication 260-22, U.S Government Printing Office (February 1971).

23. R. E. McMillan, G. G. Johnson, Jr. and E. W. White, "Computer Processing of Binary Maps of SEM Images," The Proceedings of the Second Annual Scanning Electron Microscope Symposium, IIT Research Institute, Chicago, Ill. 60616 (1969).

24. W. L. Matson, H. A. McKinstry, G. G. Johnson, Jr., E. W. White and R. E. McMillan, "Computer Processing of SEM Images by Contour Analyses," Pattern Recognition 2, 303 (1970).

25. E. W. White, H. Görz, G. G. Johnson, Jr. and R. E. McMillan, "Particle Size Distributions of Particulate Aluminas from Computer-Processed SEM Images," The Proceedings of the Third Annual Scanning Electron Microscope Symposium, IIT Research Institute, Chicago, Ill. 60616 (1970).

26. H. Yakowitz and K. F. J. Heinrich, "Inclusion Identification by Means of Electron Probe Microanalysis," Metallography 1, 55 (1968).

27. J. W. Colby, "Magic IV - A New Improved Version of Magic," Proceedings Sixth National Conference on Electron Probe Analysis (1971). See 13.

28. S. D. Rasberry, M. Margoshes and B. F. Scribner, "Applications of a Time Sharing Computer in a Spectrochemistry Laboratory: Optical Emission and X-ray Fluorescence," U.S. Natl. Bur. Stds. Tech. Note 407, U.S. Govt. Printing Office (Feb. 1968).

29. T. D. Kirkendall and P. F. Varadi, Communications Satellite Corporation, private communication.

30. D. R. Beaman and J. A. Isasi, "Electron Beam Microanalysis - Part I," Materials Research and Standards 11, No. 11, 8 (1971).

31. D. R. Beaman and J. A. Isasi, "Electron Beam Microanalysis - Part II," Materials Research and Standards 11, No. 12, 12 (1971).

32. K. F. J. Heinrich (ed.), Quantitative Electron Probe Micro-analysis, U.S. National Bureau of Standards Special Publication 298, U.S. Government Printing Office (October 1968).

33. D. R. Beaman and J. A. Isasi, "A Critical Examination of Computer Programs Used in Quantitative Electron Microprobe Analysis," Anal. Chem. 42, 1540 (1970).

34. J. W. Criss and L. S. Birks, "Calculation Methods for Fluorescent X-Ray Spectrometry - Empirical Coefficients vs. Fundamental Parameters," Anal. Chem. 40, 1080 (1968).

35. J. R. Rhodes, C. B. Hunter, D. L. Kellogg, R. D. Sieberg and T. Furuta, "Applications of Computer-coupled Radioisotope X-Ray Spectrometer to Analysis of Steels," in: C. S. Barrett, J. B. Newkirk and C. O. Ruud (eds.), Advances in X-Ray Analysis, Vol. 14, Plenum Press, New York, 1971, pp. 127-138.

36. J. M. Short and J. F. Stephany, "Storage and Treatment of Energy-Dispersive X-ray Data by Time-Shared Computer," Proceedings Sixth Natl. Conf. on Electron Probe Analysis (1971). See 13.

37. E. Lifshin, "Computer Processing of Solid State X-ray Detector Data," Proceedings Fifth National Conference on Electron Probe Analysis (1970). Copies available from D. Beaman, Dow Chemical Co., Midland, Mich. 48640.

38. J. R. DeVoe, S. D. Rasberry, F. C. Ruegg and R. W. Shideler, "Computer Assisted Measurement in Spectroscopy," Tenth National SAS Meeting, October 1971.

39. J. Arranson of the Computer Services Division of the National Bureau of Standards is the author of the program DACQ.

40. P. R. Kennicott, V. P. Scavullo, J. S. Sicko and E. Lifshin, "Laboratory Automation at General Electric Corporate Research and Development," AFIPS Conf. Proc. 39, 423 (1971).

COMPUTER-CONTROLLED X-RAY AND NEUTRON DIFFRACTION EXPERIMENTS[*]

Melvin H. Mueller

Argonne National Laboratory

Argonne, Illinois 60439

ABSTRACT

The use of on-line computers for control and acquisition of data from x-ray and neutron diffractometers has continuously improved and expanded. Systems vary from a small 4K core computer to a time-sharing system with a medium or large computer. The choice of a single time-shared computer or an individual stand-alone system must be based on one's own particular environment. As large high-speed electronic computers became available, increasingly complex chemical and magnetic structures have been analyzed and solved; this has created a demand for rapid, reliable, and versatile means of obtaining diffraction data. Since small computers have been developed at reduced cost and with increased storage capacity, they must be considered for use in diffraction experimentation. Therefore, in x-ray and neutron scattering, small computers are needed for data acquisition and large computers are needed for data analysis.

The ARgonne Computer Aided Diffraction Equipment (ARCADE) is an example of an effort to develop more automated methods. The system utilizes an IBM-1130 with 4K core memory and a 500K disc storage. The computer is used to (1) control the experiment, (2) collect the data, (3) make decisions during the experiment, and (4) process the data, which will be used in a larger computer to solve the problem.

The computer may be coupled to various parts of the experiment. Both the sample crystal and the detector are positioned by

---

[*]Work performed under the auspices of the U. S. Atomic Energy Commission.

dc stepping motors that are position-controlled by optical encoders to an accuracy of ±0.005°. Pulses from both the detector and clock/monitor systems are recorded by the computer. The computer may also be used to control a rate meter, strip chart, console printer, temperature controller, or rf generator for the flipping of a polarized neutron beam. The data may be recorded on disc, paper tapes, or transmitted to a larger computer.

A generalized software control program is basic to the system. This control program must always remain in core. The experimenter can use the program, in a straightforward conversational mode, by means of a keyboard console or by automatic coupling of a series of programs. The ARCADE system now has a repertoire of approximately 150 programs that involve several hundred subroutines. The programs can be categorized into the following groups: control, orientation, data collection, and data reduction. Although the computer is small, the large disc storage permits programs to be linked together to extend the range of experimental control. Changes and additions are constantly being made to adjust to the experimental needs.

INTRODUCTION

When x-ray or neutron diffraction techniques are used to determine the atomic structure of material, two types of information are needed: (1) the location or position at which diffracted intensity may occur, and (2) the intensity of the diffracted beam. The first type of information is controlled by the crystal system to which the material belongs. For purposes of our present discussion we will assume, in the main, that we have determined the crystal system,(e.g., cubic, hexagonal, or orthorhombic,) and also that we know the size of the unit cell. Therefore, we can easily calculate where possible reflections may occur. To determine the arrangements within the unit cell, we must collect diffracted intensity data and compare with calculated intensities from a model of atom locations. When the calculations from the model and the observed intensities show good agreement we say that the structure is solved. As more complicated structures are studied, a large volume of accurate, single-crystal diffraction data is needed, which is both lengthy and difficult to obtain. Hence, small computers are being used to assist in collecting the data, and larger computers are used to solve the structures by evaluating the observed and calculated intensities.

The ARgonne Computer Aided Diffraction Equipment (ARCADE) is an example of an effort to develop automated methods of performing experiments to improve the data collection process, to eliminate human error, and to redirect the efforts of the scientist.

BASIC MECHANICAL MOTIONS REQUIRED

To obtain single-crystal diffracted intensities, the crystal
may be mounted on an x-ray goniometer and then aligned on a
$\theta-2\theta$ axis of a conventional diffractometer. This instrument
together with a suitable detector can be used to obtain neutron or
x-ray diffraction data. Each desired reciprocal lattice point
(see Fig. 1) of the crystal may be brought into the diffracting

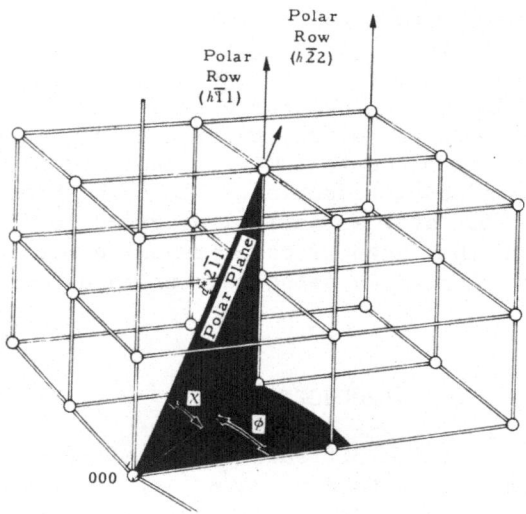

Figure 1.   Model of a reciprocal lattice. The black segments
correspond to the ϕ and χ motions, and the distance
from the origin to the $2\bar{1}1$ point represents the
reciprocal distance d*. (Reprinted by permission
of Furnas[1]).

position by moving the crystal to appropriate phi and chi angles
and the detector to the proper 2-theta position. The mounting
of the crystal and angle motions are shown schematically in
Fig. 2. The sample crystal and the detector are positioned by
dc Slo-Syn stepping motors, as described previously(3,4) with
200 steps per revolution, which is equivalent to 1° rotation of
an axis.

COMPUTER AND INTERFACE

An IBM-1130 computer with a 4K core memory and a 500K disc
storage together with suitable interface is used in the ARCADE
system(5). The sample crystal motions (ϕ and ψ) and the detector
(2θ) are positioned by dc stepping motors computer-controlled
through an appropriate translator interface. Optical digitizers

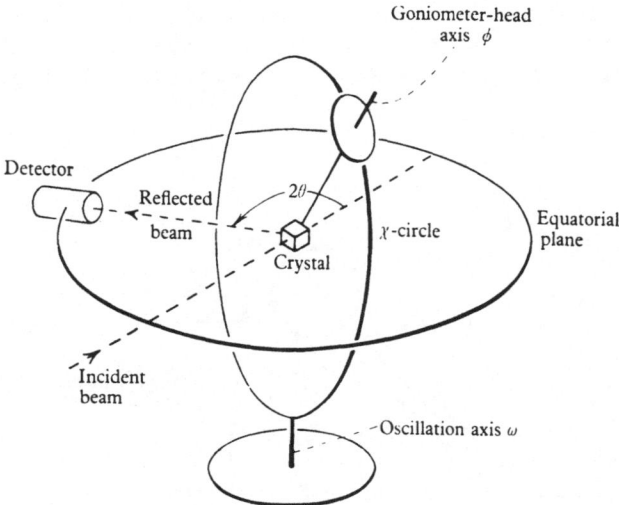

Figure 2.   Normal—beam equatorial geometry. The crystal moves in
            terms of φ and χ, and the detector moves in the
            horizontal plane. (From SINGLE CRYSTAL DIFFRACTOMETER
            (1966), U. W. Arndt and B.T.M. Willis. Used by permis-
            sion of Cambridge University Press.)

are coupled to the motor drives and connected through the computer
interface to enable the computer to know and control each angle to
an accuracy of ±0.005°. The present system uses a self-accelerating,
positive feedback, optical slit encoder system. Basically, a drive
pulse is initiated from the computer, and a feedback pulse initiates
an interrupt when the motor has stepped one increment. The feedback
pulse is derived from an incremental optical slit encorder using
one hundred slits per revolution. A pulse from either of two
space-phased photocells yields positive indication of movement. A
single degree slit provides an interrupt so the present position
and calculated position can be compared. The slew mode uses only
the single degree interrupt because slewing can only be initiated
when the calculated position and the present position differ by 1°
or more. As many as six angles can be moved simultaneously at
individual speeds and gear backlash is eliminated by overtravel.
It is possible to operate the dc Slo-Syn motor stepping system
without the optical encoders by counting the dc pulses; however,
an error may be introduced if the torque is too great. After
trying several expensive self-aligning couplings between the motor
shaft and optical encoder, it was found that two short pieces of
gum rubber tubing (one inside the other) were the most desirable
and reliable.

Pulses from the detector and/or fission monitor (timing device for the neutron system) are counted by a register in the computer or by a separate scaler, as shown schematically in Fig. 3. As

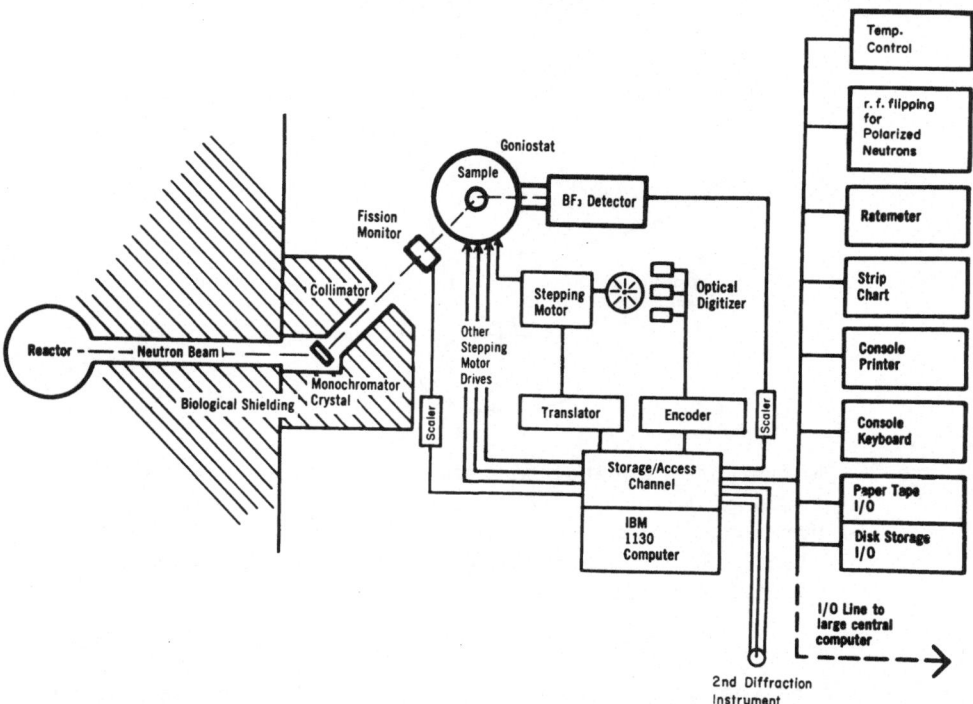

Figure 3. Schematic of the neutron diffraction equipment, located at the Argonne CP-5 Reactor, linked to a small computer for instrument control and data collection. (Fig. 2 of Ref. 5 modified and used by permission of Mueller et al., original Copyright 1968 by F. D. Thompson Publication, Inc.)

noted, the computer also controls auxiliary devices such as rate meter, strip chart, and console printer. Data may be recorded on disc, or paper tape, or transmitted to a larger computer.

The computer system for the x-ray case is similar to that shown in Fig. 3, except that the primary beam is from a x-ray tube and a clock instead of a fission monitor is used as the timing device. During the past year, the same computer has been exchanged quickly and conveniently from the neutron application to an x-ray unit. The x-ray unit consists of a General Electric (GE) Spectrogoniometer with the same motors and optical encoders that were used for the neutron instrument but which are now attached to the quarter circle GE goniometer.

For routine data collection a small, relatively inexpensive computer is sufficient; however, it is much superior to black-box

type controls because of reliability and flexibility. Although
the IBM-1130's are small computers, the large disc storage
permits programs to be linked, which extends the range of experimental control. Large control and analysis programs can now be
used with this type system that were not feasible with a small
computer containing core storage only. Five IBM-1130 computers
are now in use at Argonne; three for neutron diffraction instruments and two for x-rays.

Figure 4 shows two neutron diffraction instruments that use

Figure 4.   IBM-1130 computer installed at the CP-5 reactor that is
            used to control two neutron diffraction instruments.

the same neutron beam but are otherwise independent. The upper
instrument is a modified Picker x-ray unit with a detector arm
that travels in a vertical plane. The unit is ideal for obtaining
thousands of pieces of data at room temperature. The crystal is
mounted on a goniometer in the center of the small circle.

The lower instrument is a heavy duty unit for the support of
equipment for environmental control of the sample such as magnetic
field, and high or low temperature. The large ring on the unit
may support a cryostat as discussed later in this paper.

SOFTWARE FOR COMPUTER CONTROL

A generalized control program is always resident in core and is basic to the system. This program maintains communication with the typewriter for selection of main programs from the disc. In addition, it will initiate subprograms that are available to the experimenters for manual operation of the motor drives, presetting of experimental and crystal parameters in core tables, and inquiry into the present operation of the system. The ARCADE system now consists of nearly 150 programs that involve some two hundred subroutines (a small portion is shown in Fig. 5). The main

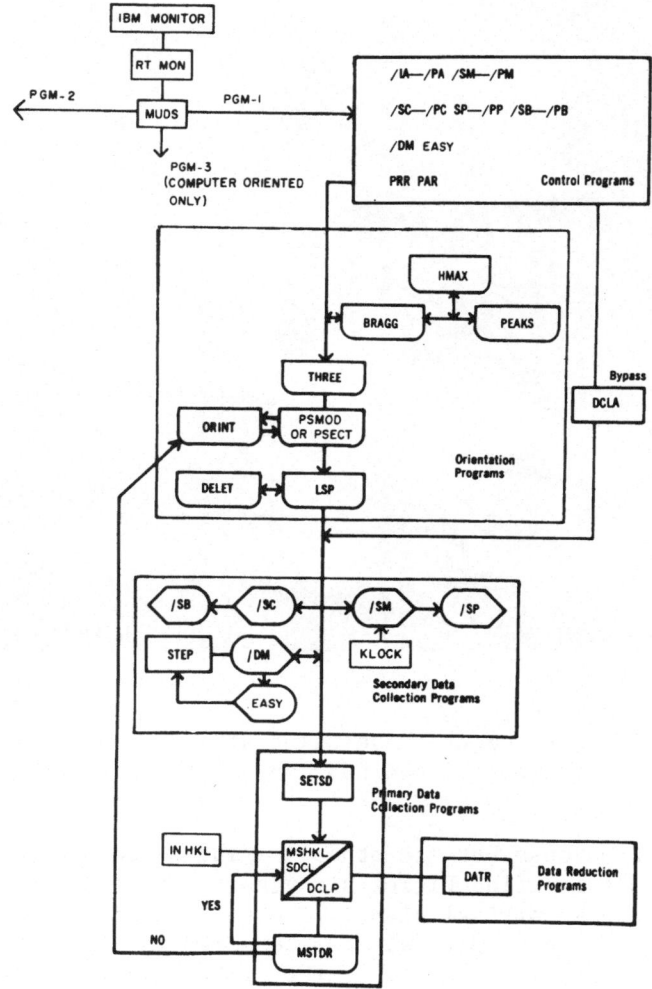

Figure 5. Abbreviated flow diagram of a few of the computer programs used in the ARCADE system. (Fig. 4 of Ref. 5 modified and used by permission of Mueller et al., original Copyright 1968 by F. D. Thompson Publication, Inc.)

portions of the programs can be grouped, as shown in Fig. 5, into several categories, such as control, orientation, data collection, and data reduction.

The system is being modified to link the IBM Monitor with the Argonne-written Real Time MONitor (RTMON) and a Multiple-Unit Diffraction System MUDS so that more than one diffractometer system can be operated from one IBM-1130. Provision is then made to store the data on a disc and thus manipulate the data before it is produced as output on paper tape or typewriter.

The control group (IA, SM, PAR, and PRR), as shown in Fig. 5, permits the operator to enter the starting angle positions, to set the monitor or counting time, and to set the crystal system, symmetry requirements, wavelength of the radiation, and angular range of scans. In addition, the control programs supply keyboard control to provide direct communication between the operator and the experiment, which is so necessary in the early stages.

The orientation programs (ORINT, PEAKS, and LSP) determine the unit-cell constants and describe the orientation of the crystal in terms of unit vectors that lie along the ϕ axis and in the χ plane of the instrument when ϕ is zero (see Fig. 2). In the LSP program, the variations between observed and calculated $\sin^2\theta$ and those of ϕ and χ are shown. From these variations, it is possible to judge whether the orientation is sufficiently well established so that data collection can proceed (see Fig. 6). A program DELET may be used to delete poor observations from the list before re-running the least-squares program. For example, in Fig. 6 numbers 1, 2, 9, and 13 have been deleted. The secondary data collection programs (/SC, /SB, STEP) permit the user to set other conditions such as chart full scale, conditions for obtaining backgrounds, and step size.

The primary data collection programs generate the instrument angles 2θ, χ, ϕ for each hkl reflection permitted under various extinction, octant, and the angle range option selected by the user. During the course of data collection, standard reflections may be remeasured and tested to determine, for example, whether instrument failure or crystal slippage has occurred. A restart from the point of the last standard reflection may be initiated. The following specific programs may be used during data collection: DCLP collects data in the usual sequential way; SDCL, special data collect, allows one to start at any point in the normal sequence; and SETSD permits the introduction of specific reflections to be used as standards. In addition, there are some other collection programs (see Fig. 7) that for example, permit scanning along a zone (ISTEP) within a range of 2θ. The ISTEP program may also be used to investigate the shape of a peak or obtain integrated intensities of a reflection. The combination (INHKL)(MSHKL)

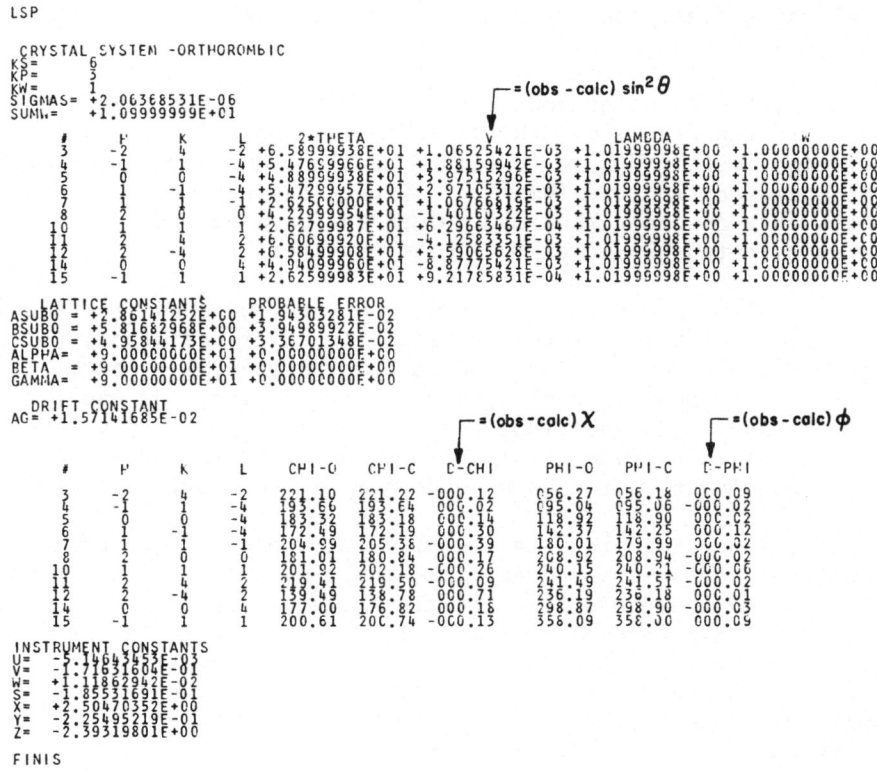

Figure 6.   Least-squares output from alpha-uranium single-crystal alignment. Relatively small values for V, D-CHI, and D-PHI indicate the crystal is well aligned.

D-1: SETSD (Input up to 5 standard reflections).
D-2: MSTDR (Measure the standard reflections).
D-3: DCLP (Data collect program--initial start).
D-4: DCLR (Data collection--reorient).
D-5: SDCL (Special data collection).

E-2: STEP (Simple step scan -one angle).
E-3: ISTEP (Integrated intensities obtained for step scan).
E-4: STEPP (STEP with monitor counts per minute printed).

F-1: MSHKL (Measure selected H, K, L's only).
F-2: INHKL (Input selected H, K, L's for MSHKL).

Figure 7.   A list of some of the Data Collection Programs used in the ARCADE system.

allows one to insert several hundred specific reflections (INHKL) and then (MSHKL) is used to obtain the measured intensities.

The data-reduction programs integrate the peak (unless stationary crystal-stationary counter techniques are used) and subtract the background, calculate the standard deviations of the recorded intensities, make geometric and absorption corrections, and reduce the intensities to F's, which are the amplitudes scattered by the unit cell.

Although the programs are divided into classes according to function, they are interrelated and dependent. Optional routes are often available and may be selected by the computer in its decision making or may be selected by the operator through the sense switches located above the console keyboard (see Fig. 4). A partial list of variations obtained by the sense switches is shown in Fig. 8.

SENSE SWITCH #1	POSITION	PURPOSE
0	DN	PRINT ONLY #, H,K,L
	UP	PRINT #,H,K,L,2θ,χ,φ,ETC.
1	DN	DO NOT MEASURE STANDARDS.
	UP	MEASURE STANDARDS PER PERIOD.
2	DN	PUNCH DATA AT END OF REFLECTION.
	UP	DO NOT PUNCH.
3	DN	DO NOT CHECK χ.
	UP	CHECK χ FOR 230° > χ > 130°
4	DN	
	UP	TREATS PROGRAM AS "D" SPACE FOR CALCULATION ONLY OF ANGLES.
7	DN	USES MOTOR #4 AS χ.
	UP	USES MOTOR #5 AS Σ ROTATION ABOUT THE DIFFRACTION VECTOR.
8	DN	
	UP	DOES A ROUGH ORIENTATION BEFORE ORINT.

Figure 8. A partial list of some of the sense switch options available in the ARCADE system.

EXAMPLES OF INVESTIGATED STRUCTURES

Intermetallics

In an x-ray diffraction investigation of the intermetallic phase URhGe, a tiny, pin-point single crystal was selected from

crushed powder of the material. Since this material was related
to the Fe_2P compound it was presumed to be orthorhombic(6).
Previous x-ray powder patterns had indicated the structure to
be orthorhombic with a possible doubling along two axes. Balanced
filter intensity data were obtained by scanning approximately 266
independent reflections in one quadrant with both the Zr and the
Y balanced filters. The computer scans of several reflections
using the Zr filter are shown in Fig. 9. The two columns of zeros
indicate that no background counts were obtained either before or
after the peak. Background counts are not necessary because
balanced filters were used. The recorder chart shown in Fig. 10

#	H	K	L	2THETA	CHI	PHI	BG1	BG2	INTEN	(F)
107	3	5	2	43.31	43.63	244.79	0	0	33120	271.75
108	2	5	2	37.42	51.18	256.66	0	0	121328	570.58
109	1	5	2	33.48	58.33	278.05	0	0	35080	328.37
110	0	5	2	32.07	60.59	310.72	0	0	123944	633.26
111	0	6	2	37.99	57.27	309.99	0	0	579672	235.69
112	1	6	2	39.21	56.19	284.04	0	0	36144	302.50
113	2	6	2	42.70	51.17	264.38	0	0	70288	399.55
114	3	6	2	48.05	45.09	251.81	0	0	29264	238.59
115	2	7	2	48.32	50.82	270.38	0	0	43384	289.44
116	1	7	2	45.18	54.41	286.08	0	0	40248	291.44
117	0	7	2	44.09	54.97	309.53	0	0	54888	345.78
118	0	7	3	45.32	61.69	311.07	0	0	48312	315.89
119	1	7	3	46.98	60.97	286.09	0	0	74520	386.51
120	3	6	3	49.77	50.43	246.78	0	0	165568	554.37

Figure 9.   A portion of the x-ray diffraction computer output from
a single crystal of URhGe intermetallic.

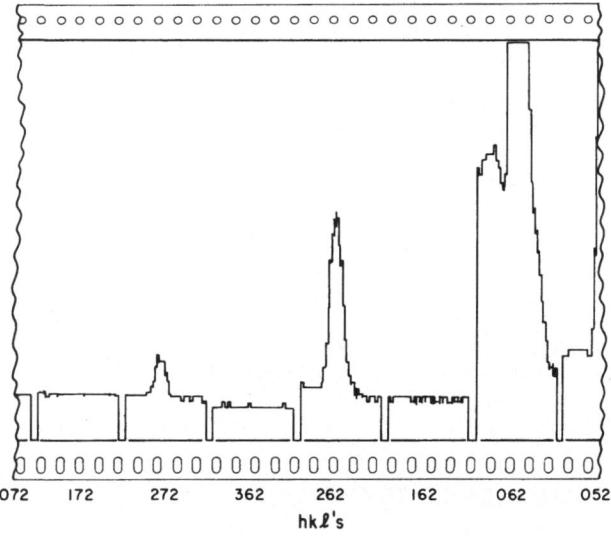

072   172   272   362   262   162   062   052

hkℓ's

Figure 10.   Strip chart output with computer control for the
various reflections shown in Figure 9.

illustrates another typical type of computer output. It has been
found highly desirable to use these chart scans because it is a
good display of what occurs during the accumulation of the
intensity. In other words, we are using the chart scan as an
inexpensive Cathode Ray Tube (CRT) and are thus able to rapidly
judge from the tracing that the integrated intensity (shown by a
single typed number) did not include electronic noise or unresolved
reflections.

To obtain the integrated intensities for the various reflections from URhGe, two scans were necessary, one with the Zr filter
and one with the Y filter. These may follow each other automatically with the computer inserting the proper filter. The
intensities are then subtracted to give the net intensity, which
may be temporarily stored, printed out, or punched on paper tape
suitable for use in the larger computer.

In the case of intermetallics, it is often desirable and
many times necessary to make absorption corrections to equate the
various reflections. This procedure is especially true for heavy
metal material that have a high x-ray absorption. Data for this
type correction can be obtained conveniently by a program called
ISTEP, which permits a small incremental step scan of ϕ as the
crystal plane is rotated about the diffraction vector. Data from
a typical scan is plotted in Fig. 11. It can be noted that the

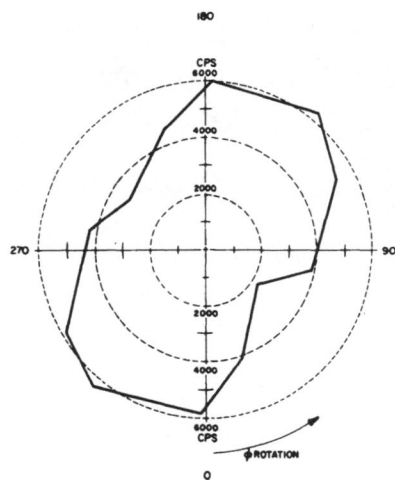

Figure 11. Plot of intensity vs ϕ rotation for a single crystal
            of HoAlNi. Crystal is rotated about an axis perpen-
            dicular to the (210) plane, i.e., about the
            diffraction vector.

intensity varies by a factor of nearly two. Absorption correction
can be applied to the intensity data as a function of the ϕ angle
and can usually be considered independent of ψ.

Alpha Uranium

It might be of interest to briefly trace the steps necessary
to mount a crystal, establish its orientation, calculate the
position of the reflections, collect the intensity data, refine
the structure, and then use the computer to collect additional
low-temperature data with special control applications. For
neutron diffraction, a cube-shape single crystal of alpha uranium,
approximately 3 mm on an edge was attached by low-temperature
expoxy to a vanadium pin for mounting on a conventional x-ray
goniometer. Although the cube faces of this single crystal were
very near the orthogonal faces of the orthorhomic unit cell, it
was not necessary to mount the crystal with principal axes along
the instrument axes, as will be shown later.

After the crystal was mounted within the full circle of the
top unit shown in Fig. 4, the program EASY was used initially to
move angles and detector so that two starting reflections (200
and 111) were located (see Fig. 12). PEAKS was used to determine

```
       THREE

      *INPUT*IANG  *ICHI *WAVE *THMAX
      1     1     2    1.02  90.0

      **A ZERO**B ZERO**C ZERO**ALPHA **BETA   **GAMMA
      2.858   5.877   4.955    90.0     90.0     90.0

      *H(1)*K(1)*L(1)**2THETA**OMEGA **CHI   **PHI
      2     0     0   42.26    0.0    181.05  208.9

      *H(2)*K(2)*L(2)**2THETA**OMEGA **CHI   **PHI
      1     1     1   26.30    0.0    201.9   240.18


          SET ENTRY SWITCH NO. 15 FOR DESIRED OUTPUT
              1. UP  -SINGLE REFLECTION OUTPUT
              2. DOWN-LISTED OUTPUT
      PUSH PROGRAM START TO START PROGRAM

          ENTER DESIRED H,K,L
      *H *K *L
      0  0  4

        2THETA=   48.624   OMEGA=   0.000   PHI=  298.836   CHI=  176.512
      *H *K *L
      -2  4  -2

        2THETA=   65.195   OMEGA=   0.000   PHI=   55.798   CHI=  220.852
      *H *K *L
      0  0  0

      ****** THE END ******
```

Figure 12. Output from THREE used to find and calculate reflec-
            tions during initial alignment. Either two reflections
            with the lattice constants or three reflection posi-
            tions are necessary.

the best angle positions from given starting angles. This
information was then used with THREE to calculate other reflection
positions, such as (004)($\overline{2}4\overline{2}$), and entered into PSMOD. ORINT, with
the optional program ROUGH, was then used to find the best angle
positions for 15 reflections. ROUGH was made operational by
calling Sense Switch 8. This is a single angle step scan using
the input angles from PSMOD file that steps each angle over a
2° range first on φ, then 2θ, and then ψ. These angles at which
the highest detector count are obtained are then stored and used
in ORINT for final refinement. The mode used by ORINT to
establish the best angles is shown in Fig. 13. As shown in Fig. 13,

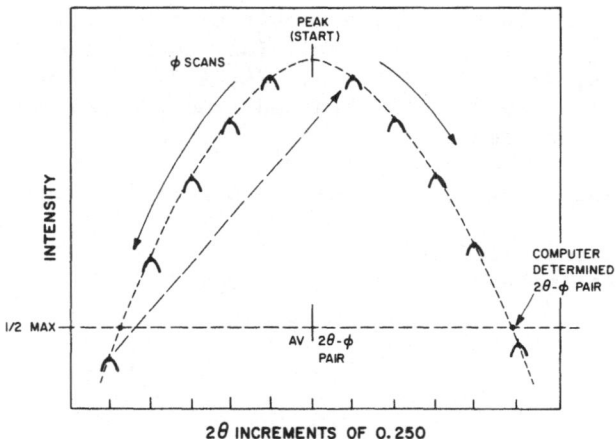

Figure 13. Schematic of 2θ-φ step scans carried out by the ORINT
program to find the best angle positions for a
particular hkl reflection.

the angles 2θ and φ operate as a team and are determined
simultaneously. After starting at the peak position defined by
2θ, φ, and ψ, 2θ reverses 0.25° and stops, φ then reverses 0.15°
and step scans in 0.03° increments over the peak and continues
until it finds six intensities lower than the highest intensity
recorded in the φ scan. The φ angle is driven back to the peak
position, and a count is obtained. This process of 2θ step-φ
scan is continued until the peak count is 12.5% below the 1/2
maximum value of the initial count. The 2θ and corresponding φ
angles that yield the 1/2 maximum intensity are determined. This
entire process is then repeated on the other side of the peak 2θ
position. The two sets of 2θ-φ data are averaged for best 2θ-φ
angles for the particular reflection. The crystal is moved to
this location, an intensity count obtained, and the best χ angle
is determined by an ordinary 1/2 maximum stepping procedure for a

single angle. The three new angles are then printed out with
intensity and corresponding old angles and intensities.

    If for some reason the program cannot find the reflection,
there are fail-safe counters in the program that will stop the
process and print UNABLE TO DO PEAK and proceed to the next
reflection. This type of intensity and angle information is then
stored and used in LSP (see Fig 6) to determine the best lattice
constants and orientation of the crystal. This will then be used
as information to calculate and drive the angles for obtaining
intensity data according to a predesignated order in DCLP. These
intensities converted to structure factor (F) values are then used
in a larger computer program to determine the best least-squares
values for scale factor, atomic position parameter y, and aniso-
tropic temperature factors. The resulting structure of alpha
uranium has four atoms per unit cell (all equivalent) located at
0, y, 1/4; 0, $\bar{y}$, 3/4; 1/2, 1/2 + y, 1/4; 1/2, 1/2 - y, 3/4
(see Fig. 14).   The results obtained from 120 nonequivalent
reflections at room temperature are as follows:

    (displacement along the b axis) y = 0.1025 $\pm$ 0.0002

    radii of thermal ellipsoid
    along the following axes   a = 0.083 (2) A

                                      b = 0.063 (5) A

                                      c = 0.069 (3) A

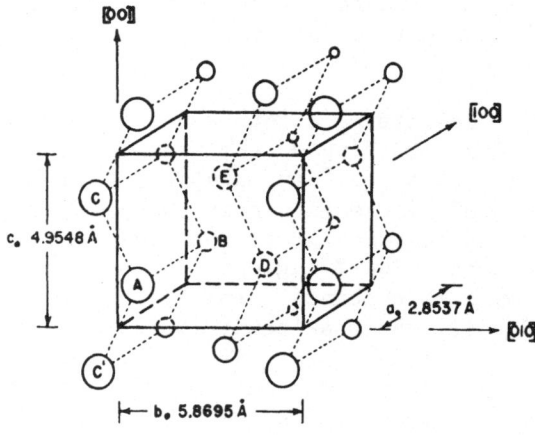

Figure 14.   Alpha uranium crystal structure.   (Reprinted by
             permission of Barrett et al.[7]).

CRYO-ORIENTER FOR LOW TEMPERATURES

The computer may also be used to control the low-temperature experiments with the cryo-orienter(8). A schematic of the instrument is shown in Fig. 15. It consists of a movable inner ring 17 inches in diameter and 6 inches wide that supports a cryostat.

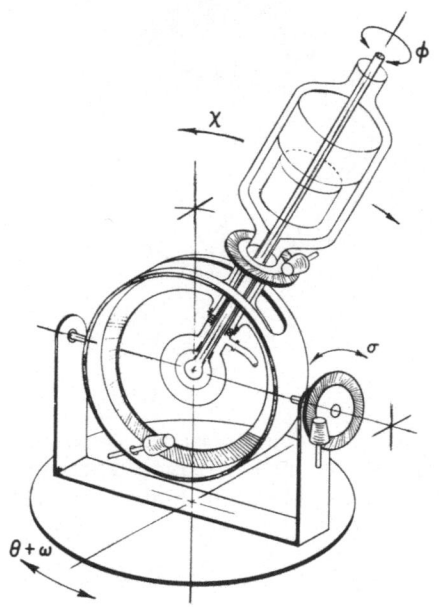

Figure 15.   Schematic view of the cryo-orienter.   (Used by permission of Heaton et al.[8]).

The tail section is inserted through the wide ring. An exchange gas tube extends to the center of the ring. The crystat can be displaced ±50° from vertical (χ) and can be rotated 360° (ϕ). This provides crystal orientation about two axes (χ and ϕ) with ±0.01° precision.

The crystal is attached to a copper heat sink that contains control sensors and trimmer heaters. The temperature of the sample may then be set with computer input and varied from below helium temperatures to room temperature. Temperature can be controlled to ±0.05°K.

The reciprocal lattice sphere shown in Fig. 16 details the octants available with this instrument. For crystals with low symmetry, such as orthorhombic or lower, it is desirable to mount the crystal along a nonmajor zone to provide sufficient data with one crystal mounting. As an example, a low-temperature investigation of alpha uranium was performed with the ARCADE computer control. Note in the computer output shown in Fig. 17, data

M. H. Mueller

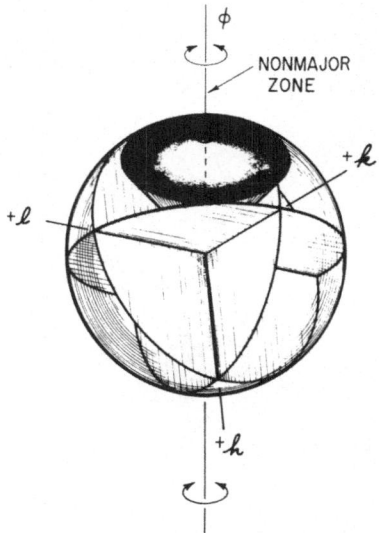

Figure 16.   Reciprocal lattice of alpha uranium showing included
             and excluded (cone) volumes with the cryo-orienter.
             (Used by permission of Heaton et al.[8]).

#	H	K	L	2THETA	CHI	PHI	BG1	BG2	INTEN	(F)
83	2	0	6	92.36	180.66	96.54	25	25	23825	154.29
84	1	1	6	82.10	165.00	94.02	23	22	18202	134.27
85	0	2	6	81.05	147.58	90.70	29	22	2970	54.17
86	2	2	6	95.93	172.17	107.39	15	27	3101	55.54
87	1	3	6	89.25	156.55	106.44	32	30	3035	55.09
88	0	4	6	91.76	140.38	105.17	32	33	18476	135.89
S	4	0	0	92.31	223.29	143.98	28	38	19665	140.18
89	0	2	7	97.19	148.86	88.65	31	32	23216	151.77
90	1	1	7	98.24	163.90	91.79	21	29	10261	100.77
0	-2	0	0	42.46	137.12	323.98	29	44	42214	168.81
1	-4	0	0	92.31	137.12	323.98	26	38	20648	143.64
25	-4	0	2	97.27	130.47	345.37	33	41	20637	143.08
45	-2	0	4	66.98	132.64	36.16	27	44	30604	167.83

SAMPLE OUTPUT FROM CRYO-ORIENTER

Figure 17.   Partial output from alpha uranium in the cryo-orienter.
             Skips in the serial numbers represent positions
             through which the crystal cannot move to due to
             angle limitations.

skips occur in one of the octants from 1 to 25 and 25 to 45. The
control program thus restricts the cryostat motion, which prevents
spilling of the liquid coolants. It was interesting to observe
that many of the low-temperature uranium reflections showed a
large increase of intensity (as much as 50%). A detailed analysis
of this data(9) indicated that this could be explained by an
extinction effect, namely, that lowering the temperature produced
a change in the perfection of the single crystal. This change is
elastic because the intensity is lost again at room temperature.
Several recent experiments at Argonne and Northwestern University
may indicate that several different changes may take place near
40°K. Therefore, further neutron studies at selected low
temperatures may be desirable.

EXAMPLES OF A SOMEWHAT MORE COMPLICATED STRUCTURE

The structure determination of the di- and hexahydrate of
uranyl nitrate was conducted to find the uranium coordination
and locate the water hydrogens(10,11). A neutron diffraction
structure determination was performed recently on a urea uranyl
nitrate hydrate complex that is more complicated than the previous
two structures. Initial x-ray determination had defined the unit
cell to be monoclinic with a unit cell approximately 10 by 15 by
13 A. Neutron intensity data were then obtained from a single
crystal for near 2500 nonequivalent reflections. In refining the
structure, the following facts became evident:

Formula: $[UO_2(H_2O)\{CO(NH_2)_2\}_4](NO_3)_2$

Number of nonequivalent atoms/unit cell: 46

Number of parameters refined in least-squares
treatment of data: 414 (9 per atom - 3 position
and 6 temperature parameter)

Agreement fact R obtained: 4.7%

A partial picture of the structure is shown in Fig. 18. The
uranyl ion is nearly linear and is surrounded by a near plane
of five oxygen atoms, four from urea ligands, and one from a
water ligand. It is evident that the greatest thermal vibration
is perpendicular to the bond direction.

CONCLUSION

Although a moderately complex structure was solved as
illustrated above, it is evident that computer control becomes
even more important when solving complex bio-organic structures
from thousands of pieces of intensity information. The important

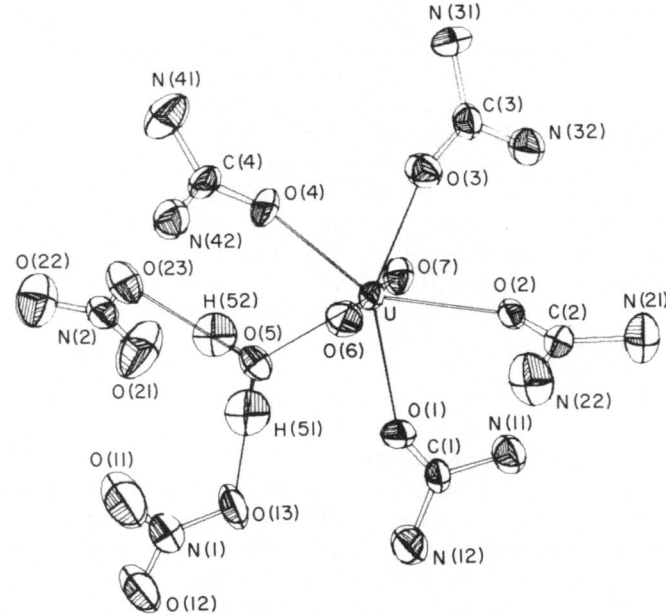

Figure 18.   Arrangement of atoms about the uranyl ion in a uranyl
             nitrate urea complex. Ellipsoids represent anisotropic
             thermal motion of each atom.

aspect of computer control for diffraction experiments is its
versatility and ability to be constantly changed. Space will not
permit us to discuss all the applications that we and others have
used in diffraction experiments. For example, computers are
excellent for setting and measuring temperatures, and have become
invaluable in controlling various aspects of a neutron polarized
beam unit. Several different computers have been coupled with
various diffraction instruments and sold as a package together
with limited software.

ACKNOWLEDGMENTS

     The development of the automated system would not have been
possible without the support of the Applied Mathematics Division,
especially R. Ashenbrenner, L. Amiot, and J. Becker, and the
continued electronic modifications by E. Johanson of the
Electronic Division. All the members of our Scattering Studies
group have contributed, especially R. Hitterman and L. Heaton.

REFERENCES

1. Thomas C. Furnas, Single Crystal Orienter Manual, X-Ray
    Department, General Electric Company, p. 16 (1956).

2. V. W. Arndt and B. T. M. Willis, Single Crystal Diffracto-
    metry, Cambridge Press, 1966, p. 8, Fig. 3.

3. M. H. Mueller, L. Heaton, and E. W. Johanson, "Stepping
    Mechanism for X-ray and Neutron Diffractometers and
    Spectrometers," Rev. Sci. Inst. 32, 456 (1961).

4. L. Heaton, M. H. Mueller, and E. W. Johanson, "θ-2θ Stepping
    Motion without Gears," Nucl. Instr. Meth. 24, 411 (1963).

5. M. H. Mueller, L. Heaton, and L. Amiot, "A Computer Controlled
    Experiment," Research/Development, p. 34-37, Aug. 1968.

6. A. E. Dwight, M. H. Mueller, R. A. Conner, J. W. Downey, and
    H. Knott, "Ternary Compounds with the Fe_2P-Type Structure,"
    Trans. Met. Soc. AIME 242, 2075 (1968).

7. C. S. Barrett, M. H. Mueller, and R. L. Hitterman, "Crystal
    Structure Variations in Alpha Uranium at Low Temperatures,"
    Phys. Rev. 129, 625 (1963).

8. L. Heaton, M. H. Mueller, M. F. Adam, and R. L. Hitterman,
    "Neutron Diffraction Cryo-Orienter," J. Appl. Cryst. 3, 289
    (1970).

9. G. H. Lander and M. H. Mueller, "Neutron Diffraction Study
    of α-Uranium at Low Temperatures," Acta Cryst. B26, 129 (1970).

10. J. C. Taylor and M. H. Mueller, "A Neutron Diffraction Study
    of Uranyl Nitrate Hexahydrate," Acta Cryst. 19, 536 (1965).

11. N. K. Dalley, M. H. Mueller, and S. H. Simonsen, "A Neutron
    Diffraction Study of Uranyl Nitrate Dihydrate," Inorg. Chem.
    10, 323 (1971).

AN AUTOMATED TWO-CRYSTAL SPECTROMETER EMPLOYING DIRECT ANGULAR POSITIONING AND READOUT

T. K. Gregory and P. E. Best

Physics Department and Institute of Materials Science

University of Connecticut, Storrs, Connecticut 06268

ABSTRACT

A unique approach has been used in the construction of a vacuum-two-crystal spectrometer. The instrument uses digital shaft encoders directly coupled to the analyzing crystal axes to give accurate absolute angular measurement over a 360° revolution of each crystal. Each axis is driven directly by a torque motor tachometer combination which allows precise speed and position control. The complete two-crystal carriage assembly rotates about the first crystal axis, thus permitting the x-ray tube to be fixed to the vacuum chamber. The carriage and detector rotations are driven through antibacklash gearing by smaller shaft encoder, torque motor, tachometer assemblies. Because the complete unit is housed within the vacuum chamber, there are no alignment changes on system pump-down. The spectrometer is part of a completely automated computer controlled system. The computer is used for data acquisition and analysis as well as for controlling the angular position of the analyzing crystals, detector, and carriage.

Although the techniques described are applicable to other areas of x-ray physics, the spectrometer is being used for x-ray studies of metals and alloys. Initially x-ray absorption spectra of Fe-Ni-Al alloys will be measured in an investigation of their electronic band structures.

INTRODUCTION

With recent advances in our understanding of the subject,

x-ray spectroscopy is becoming more powerful as a tool in the study of electronic structures of solids. At the University of Connecticut, the extension of present studies of electronic structures of alloys required an instrument which could measure spectra in the 2-20Å region. (1,2) The vacuum two-crystal instrument described in this paper was built for this task.

The basic two-crystal instrument is, of course, unchanged, (3-7) and in this paper we emphasize those technological features which are novel to this application. These features evolved from a number of design objectives, perhaps the main one of these arising from the well known fact that most x-ray spectroscopy involves long counting times. (8) To maximize statistical accuracy by obtaining the longest time usage, a completely automated instrument was called for, and this was the first objective. Secondly, it was desired that the instrument retain maximum flexibility, with the ability to measure emission and absorption spectra, as well as isochromat curves. Thirdly, if compatible with the previous two requirements it was considered desirable to eliminate mechanical feeds through the vacuum chamber wall. In particular this arrangement removes the possibility of misalignments due to warpage caused by evacuation of the chamber. With these three considerations in mind the following instrument was constructed.

INSTRUMENT DESIGN

The heart of the instrument is the direct drive arrangement used to position each crystal. (9) The axis drive consists of a torque motor and tachometer generator joined by a bellows coupling to an incremental shaft encoder and crystal holder (see Fig. 1). The drive uses no gearing but relies on the ability of the motor to move in fractional arcsecond increments. The bellows coupling allows alignment flexibility, but maintains angular coherence between the two shafts, independent of temperature.

The detector arm and effective x-ray tube motions are also controlled by the use of torque motor, tachometer generator, encoder combinations. However, since neither of these axes require the high accuracy of the crystal axes, the components used are of lower precision and a worm gear reducing system is employed in the final drive.

Crystal Axis Design

The torque motor used in the crystal axis is a multi-winding

T. K. Gregory and P. E. Best

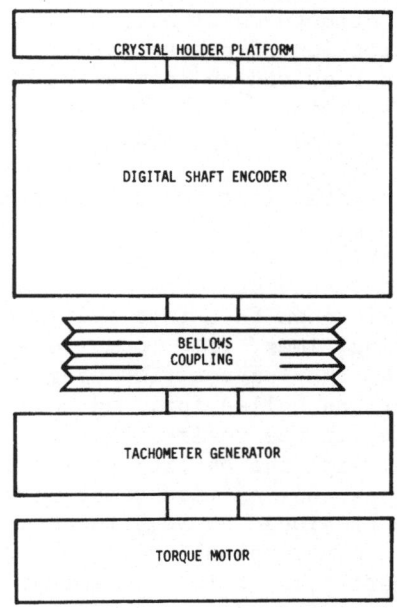

Figure 1. Crystal axes drive and readout design.

direct current motor. The particular motor (10) used was chosen for the high torque linearity and low torque ripple required for accurate control.

The tachometer generator (11) produces a voltage which is proportional to the angular velocity of the shaft. The unit used in this instrument has a sensitivity of 10 volts/radian/ second. It also was chosen for good linearity and low ripple content.

The torque motor and tachometer are mounted on a single shaft and are located in an aluminum housing. The combination when used in conjunction with the computer controlled servo-system is capable of moving at speeds as low as one arcsecond per second.

The optical shaft encoder is a 524,288 bit incremental type. (12) That is, it produces a pulse for each angular motion of 2.47 arcseconds. The pulses are coded so that there is a separate output line for each rotational direction. In addition a pulse output is provided to supply a fixed zero angle reference. Finally, two analog signals (sine and cosine) are provided from the encoder.

Each analog signal completes 16,384 cycles in one revolution of the axis, the sensitivity being 79.2 arcseconds per cycle.

Detector and X-Ray Axes Design

The torque motor-tachometer combinations for these axes are similar to those used for the crystal axes, but both units are physically smaller. (13) The tachometer generator in this case has a sensitivity of 0.235 volts/radian/second.

The shaft encoder is again of the incremental type having a resolution of 4096 bits/revolution. (14) Because of the lower accuracy required for these axes no sine or cosine outputs are provided. The encoder is coupled directly to the torque motor-tachometer combination. The axes, however, are driven through a 100:1 worm gear reducer. Ignoring, for the moment, the errors introduced by the worm gear drive, the sensitivity of the tachometer and encoder with respect to the axis motion is 23.5 volts/radian/second and 0.053 arcminutes/pulse, respectively. In practice the positioning accuracy is limited by errors introduced by the 100:1 reducer.

System Description

The system operates with vertical crystal axes. The crystal axis assemblies are mounted by a differential spring arrangement to a 1" thick cast iron plate which forms the top of the main carriage. The carriage carries the two crystal axes along with the detector axis. It also houses the drive mechanism for the "x-ray tube" motion. Because this motion is simulated by the main carriage rotating about the first crystal axis, the x-ray tube position is fixed with respect to the vacuum chamber. Two crystal holders similar to those described by Simson and Deslattes (15) are set on platforms attached to each axis. The spectrometer is mounted on a 26" vacuum baseplate which is part of a 6" liquid nitrogen trapped diffusion pump system. A stainless steel chamber encloses the spectrometer unit.

The x-ray tube is a 4kW demountable Henke tube. (16) It is attached directly to the vacuum chamber. The tube operates from a .05% regulated high voltage supply. The filament supply is controlled by electron beam current and has a regulation of .01%. A twelve sample turret type holder is mounted in the chamber and is operated by a stepping motor. The detector is a gas flow proportional counter. Alignment of the instrument is performed in the manner suggested by Schnopper. (17)

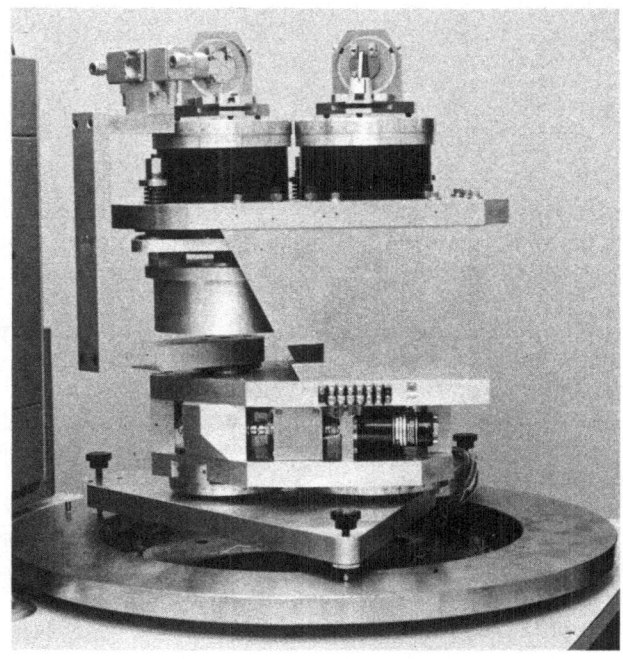

Figure 2.   Two-Crystal Spectrometer

COMPUTER AUTOMATION

The spectrometer is integrated into a computer controlled
system which directs the positioning of each axis, accepts the
detector pulse data, controls sample positioning, monitors
possible error conditions, and processes the acquired data.

System Configuration

The computer system is based on an 8-K PDP-12 computer. It
has as standard equipment two magnetic tape transports, 16
channels of analog to digital (A/D) conversion, a digital to
analog (D/A) converter, six relays, and an oscilloscope display.
In addition, a programmable real time clock has been added.

The spectrometer is interfaced to the computer by six
4-channel multiplexers, which choose the axis to be examined.
Three of the multiplexers are digital and choose the appropriate
clockwise, counterclockwise, and zero pulse channel. Two are
analog multiplexers which select the appropriate sine or cosine
channel. The remaining multiplexer is also analog, but acts as
a distributor of the output from the D/A converter to the

appropriate torque motor. The four tachometer signals are fed
directly into 4 A/D channels of the computer.

The detector signal, after passing through an amplifier and
single channel analyzer, enters a 100 count interface buffer.
The buffer keeps the count rate entering the computer to a
reasonable level. The computer may interrogate the buffer
count status at any time. Although it is anticipated that the
detection system will operate in the fixed count mode, provision
has also been made for operation in the fixed time mode.

Finally the interface provides conditioning for the digital
error status signals. These signals, formed by contact closures,
check for such failures as low water flow, main vacuum failure,
x-ray tube vacuum failure, high voltage failure, spectrometer
collision, etc.

Axis Motion Control

The crystal axes are controlled in two modes: speed control
and position control. In the speed control mode the output of
the tachometer is compared against a computer generated set
point speed. A digital error signal is calculated and outputed
through the D/A converter to the driving motor. The error signal
is digitally generated on the basis of the equation:

$$V_{out} = K_1(S_1-S_2) + K_2\int(S_1-S_2) \; dt \tag{1}$$

where S_1 is the digital set point speed, S_2 is the tachometer
signal and K_1 and K_2 are constants chosen to maximize system
control.

The position control mode for the two crystal axes is operated
by comparing either the sine or cosine encoder outputs with a
computer generated set point. As in speed control, the error
signal is generated on the basis of equation 1. In this case
S_1 is the digital position set point, and S_2 is the sine or
cosine encoder output.

When a command is given to move one of the crystal axes,
the following sequence takes place. The angular motion requested
is translated into the integral number of pulses to be moved and
the final position on the sine and cosine wave is calculated.
The axis is then moved under speed control. The speed is set at
one degree/second until the axis is within 200 pulses (8.22
minutes of arc) of the final goal. At this point a proportional
speed reduction sets in until the speed reaches a minimum of
one arc minute/second. When the correct number of pulses have

been received the speed is held constant until the proper final
sine or cosine value is reached whereupon the mode changes to
position control and the axis is held at that point. Whereas
normal digital encoder systems can position an axis to within
±one pulse of a desired angle, the hybrid type of control which
is used in this system has the advantage of allowing angular
placement of the axis between encoder pulse locations. The
accuracy of such placement is limited by the accuracy of the
encoder's sine or cosine output.

Control is based on that signal, sine or cosine, which will
have the smaller absolute value at the final position. This
procedure allows the maximum positioning accuracy. The crystal
axes are constantly monitored and controlled, assuring positional
stability even if mechanical shocks or vibrations occur.

The detector and x-ray axes operate only in the speed
control mode since they have no sine or cosine output. Their
operation is similar to that described above, except that the
final position is determined by the nearest pulse. As previously
noted this type of digital control has an inherent possible error
of ±one pulse. In this system this error, combined with gear
inaccuracy, translates to ± 1 minute of arc which is well below
the accuracy required of either axis.

The Software System

The software operating system chosen for this instrument
is one based on a modified version of Focal-12. (18) Focal is a
conversational language similar to Basic. Focal-12 is an expanded
version which includes data acquisition commands as well as
commands to record programs and data on magnetic tape.

The Focal-12 system has been modified to operate in the
background of the data acquisition and control routines. Focal-12
commands have been added to position each axis, read each axis
angle, and control data acqustion. All alignment, operating,
and analysis routines are written in the Focal language. This
allows great flexibility for program modification without
requiring the user to be familiar with machine language coding.

In operation the system is initialized using a Focal program
which requests sample identification, angular region to be
investigated, angular step size, detector counts per step, etc.
Focal operation is then suspended while the spectrometer angles
are set up, and the first sample correctly positioned. When
data taking begins Focal is reinstated and analysis routines
or other programs may be executed. At any time while Focal is

operative the user may request the current status of any or all
of the experimental parameters. When all samples have been
tested at a particular angle Focal will again be suspended until
a new angular position is reached, and the previous data is
stored on magnetic tape. Data taking proceeds in this manner
until the region of interest has been investigated. At the
conclusion of the experiment Focal will be reinstated, and
data analysis programs may be automatically initiated.

If at any time an error signal occurs Focal is suspended
and an error detection program is called to save all pertinent
parameters for later analysis. The data acquisition is then
halted and Focal is reinstated.

Since the operating system is based on program transfers
to and from magnetic tape, delays of several seconds may occur
during program transfers. However, in normal operation these
transfers occur only infrequently and pose no real problem.
It should be noted that even during these transfers the angular
position control routines still operate to insure angular
position integrity.

INSTRUMENT PERFORMANCE

The performance of the instrument for x-ray spectroscopy
depends critically on the ability to position each crystal
accurately.

Inaccuracies in position can be separated into three types:
systematic errors for large angular changes with respect to a
fixed reference, relative errors in moving a small amount about
a known position, and control errors introduced by the digital
control system. The investigation of positioning accuracy was
performed with a 12 sided mirror and an autocollimator. The
mirror was specified accurate to ±0.15 arcseconds, and the
autocollimator had a working resolution of ±0.3 arcseconds

Systematic errors occur due to masking imperfections in the
optical encoding disc. In these encoders the disc is optically
sampled from four symmetric positions. Thus the maximum error
must occur within ninety degrees, and will be repeated with a
four-fold symmetry.

Systematic errors were tested for by making angular changes
in multiples of thirty degrees. At no time were deviations
greater than the resolution of the autocollimator observed. The
motions were performed in both the clockwise and counterclockwise
direction to check for possible hysteresis effects. Here, too,

any error was below the resolution of the detection equipment.

Since small angular motions depend on the accuracy of the sine and cosine outputs of the encoder a study of these waves was made. The waves are formed from a summation of hundreds of slits on the encoder disc. (17) Thus small inaccuracies in disc manufacture tend to be averaged out. The precision with which a particular position may be reached depends on the amplitude, offset, purity and phase of the sine and cosine waves. It was found that the amplitude varied by as much as 10% over large angular motions. From cycle to cycle, however the amplitude was essentially constant. An offset of about 7% did exist, but made only small changes over large angular variations. An attempt was made to measure the harmonic distortion of the waves by moving the encoder at constant angular velocity while examining the waves with a computer based Fourier analyzer. The results were dependent on the uniformity of the velocity, which could not be independently confirmed. Even with that reservation, however, the purity of the waves was good. Only a percent of fourth harmonic distortion could be measured. Finally the phase of the cosine wave with respect to the sine wave was checked and found to be ninety ±3 degrees.

The errors discussed in this section can be compounded by errors caused by time and temperature variations. In particular, temperature variations may cause distortion of the glass encoder disc and drift in the electronics of the encoder. Room temperature variations of 1 1/2°C during the measurements had no observable effect. Time variations may cause drift in the light output of the encoder bulbs which will also cause variations in sine and cosine amplitudes.

The control programs were designed to remove errors caused by the offset and amplitude variations. This was accomplished by observing the maximum and minimum of an adjacent wave and normalizing the results. Positions were then calculated on the basis of a pure sine or cosine wave.

Short term positioning accuracy was again limited by the resolution of the autocollimator. That is, no deviations greater than ±0.3 arcseconds could be seen. In long term measurements, periods of 16 hours, errors as large as 0.6 arcseconds could be seen. Upon renormalization, however, these errors were removed.

Finally, no drift will occur in the control system since it is of a digital nature. There is, however, the ±1 bit uncertainty inherent in the A.D converter which processes the sine and cosine wave. The A/D converter has a resolution

of one part in 1000. This uncertainty will allow a maximum
position drift of ±0.03 arcseconds.

CONCLUSIONS

While final testing of the instrument will be made using
x-ray rocking curves, the novel features of the design have
been adequately tested by the procedures outlined here.

The direct drive principle is more than adequate to the
task at hand. In its present form it can give angular changes
reproducible to within 0.06 arcseconds. This reproducibility
is adequate for other tasks in x-ray physics; namely low
angle scattering, rocking curve measurements, etc.

With the implementation of a more accurate A/D converter
and the inclusion of a Fourier approximation of the sine wave,
it is believed that the system would have an accuracy of 0.004
arcseconds over scans of a few arcminutes.

Complete automation of this instrument has proved feasible
and has resulted in a convenient and flexible system. It is
only by using the computer that the encoder can be utilized to
its greatest extent. In general, the spectrometer that has been
built has lived up to the expectations that were formulated in
the beginning.

ACKNOWLEDGEMENTS

This project was initiated by Professor L. V. Azaroff
whose continued interest and support is gratefully acknowledged.
Thanks are also due to Mr. Roy Thomas of the Charles Supper
Company who helped in the mechanical design, and whose careful
supervision of the construction contributed to the success of
the instrument. The work of Mr. Li Tseng in much of the interface
design and construction is greatly appreciated.

REFERENCES

1. L. V. Azaroff and R. J. Donahue, "X-Ray K Absorption Edges in Binary Solid Solutions of Cobalt, Iron, and Nickel," Proc. Colluq. International du C.N.R.S., No. 196, Sept. 1970, In Press.

2. L. V. Azaroff and H. N. Murty, "X-Ray K Absorption Spectra of Iron-Aluminum Alloys," Acta Met. 15, 1655 (1967), and references therein.

3. M. A. Blokhin, The Physics of X-Rays, 2nd ed. (State Publishing House of Technical-Theoretical Literature, Moscow, 1957) translated by U. S. Atomic Energy Commission, AEC-TR-4502.

4. R. D. Deslattes, "Two-Crystal, Vacuum Monochromator," Rev. Sci. Instr. 38, 616 (1967).

5. L. V. Azaroff, "Two-Crystal X-Ray Spectrometer Attachment," in G. R. Mallett, M. Fay, and W. M. Mueller, Editors, Advances in X-Ray Analysis, Vol. 9, Plenum Press, New York, 1966, pp. 242-250.

6. A. E. Sandstrom, Handbuch der Physik, S. Flugge, Editor (Springer Verlag, Berlin, 1957), Vol. 30, pp. 78-245.

7. E. Suoninen, M. Karras, and J. Levoska, "A Double-Crystal Soft X-Ray Spectrometer," Acta Polytech. Scand., Ph 71 (1970).

8. L. G. Parratt, "Electronic Band Structure of Solids by X-Ray Spectroscopy," Rev. Mod. Phys. 31, 616 (1959).

9. R. D. Deslattes, private communication to L. V. Azaroff.

10. Inland Motors Corporation, T-4436.

11. Inland Motors Corporation, TG-4014.

12. Wayne-George Company, RI-19/55C.

13. Inland Motors Corporation, T-1352, TG-1312.

14. Wayne-George Company, RI-12/15CPZ.

15. B. G. Simson and R. D. Deslattes, "Kinematic Locator for Crystal Alignment," Rev. Sci. Instr. 37, 300 (1966).

16. Amperex Electronic Corporation, 408546.

17. H. W. Schnopper, "Spectral Measurements with Aligned and Misaligned Two-Crystal Spectrometers," I and II, J. Appl. Phys. <u>36</u>, 1415 (1965).

18. Digital Equipment Corporation, DEC-12-AJAA-D.

19. Wayne-George Company, Application Note 6401-1A.

A PAPER TAPE CONTROLLED X-RAY DIFFRACTOMETER FOR THE MEASUREMENT OF RETAINED AUSTENITE

Carol J. Kelly and M. A. Short

Scientific Research Staff, Ford Motor Company

Dearborn, Michigan 48121

ABSTRACT

The instrumentation and software for performing X-ray intensity measurements with a paper tape controlled diffractometer are described. The hardware includes two addressable axis positioners which control Slo-Syn stepping motors on the 2θ and ω axes of the diffractometer, an addressable scaler-timer, a multi-axis programmer and a Teletype, in addition to the normal counting electronics. This system may be manually controlled with front panel switches or with instructions entered on the Teletype. In the automatic off-line mode instructions for motor speed, motor direction, starting angle, final angle, angular increment and scaler preset (time or counts) punched on paper tape are read and executed in sequence. A Teletype output of 2θ and ω angles, time and counts is obtained at each step. This off-line system was used for the measurement of austenite in H12 hot work die steel austenitized at various conditions and which contained a maximum of 13% austenite. A helium chamber was used to extend the limit of detection to 0.4% austenite. The X-ray analysis involved measuring the areas of the (200) austenite diffraction line and the (200) martensite line. For each of these lines, the system was programmed to integrate the counts over angular intervals corresponding to a low-angle background, the peak and a high-angle background using the ability of the axis positioner to stop the scaler-timer at the end of each angular interval. The additional capability of slewing rapidly between the various diffraction lines reduced the time required for automatic data collection. The present off-line system can be used to simplify other types of X-ray diffraction analysis such as residual stress and microstrain/particle size determinations, since the manual data handling can

be eliminated with the computer compatible punched paper tape
output. Future development of this instrumentation includes
direct computer control of the diffractometer and computerized
data reduction, with the advantage of a paper tape back-up system.

INTRODUCTION

Some areas of X-ray powder diffractometry, such as those
involving repetitive intensity measurements of specified
diffraction lines for a number of similar samples, are particularly
suitable for automation. One such area is the measurement of
retained austenite in steels, a problem frequently encountered in
metallurgical laboratories. In view of the potential saving in
operator time, an X-ray powder diffractometer has been provided
with paper tape controlled automation hardware and, with
appropriate software, has been used to measure retained austenite
in a series of steels. In addition to the automation, the
sensitivity of the technique has been improved by the addition of
a helium chamber to the sample stage.

INSTRUMENTATION

Most X-ray diffractometers may be readily automated to a
certain extent by the addition of an electronic control to the 2θ
drive of the goniometer. For the present work a Picker
diffractometer was selected for two reasons. Firstly, accessible
countershafts are provided for both the 2θ and ω axes to which
suitable motors can be attached. Secondly, the $2\theta/\theta$ drives are
connected through a differential rather than a mechanically
operated clutch which might be difficult to automate. To drive
the 2θ and ω axes, Slo-Syn stepping motors are connected to the
diffractometer countershafts using universal connectors. These
connectors, which are selected so as to introduce no significant
backlash, eliminate additional torque requirements due to shaft
misalignment between the motor and the countershaft. With
remotely controlled motors attached to both the 2θ and ω shafts,
the diffractometer can be used for residual stress measurements
in addition to the retained austenite determinations discussed
here.

The automation electronics[*](hardware), which include two axis
positioners, a dual counter-timer, and a multi-axis programmer are
shown in Figure 1. A schematic diagram showing the interconnection
of the hardware with the diffractometer and the X-ray counting
chain is shown in Figure 2.

---

[*]   Canberra Industries, Meriden, Connecticut

Figure 1.   Automation hardware.

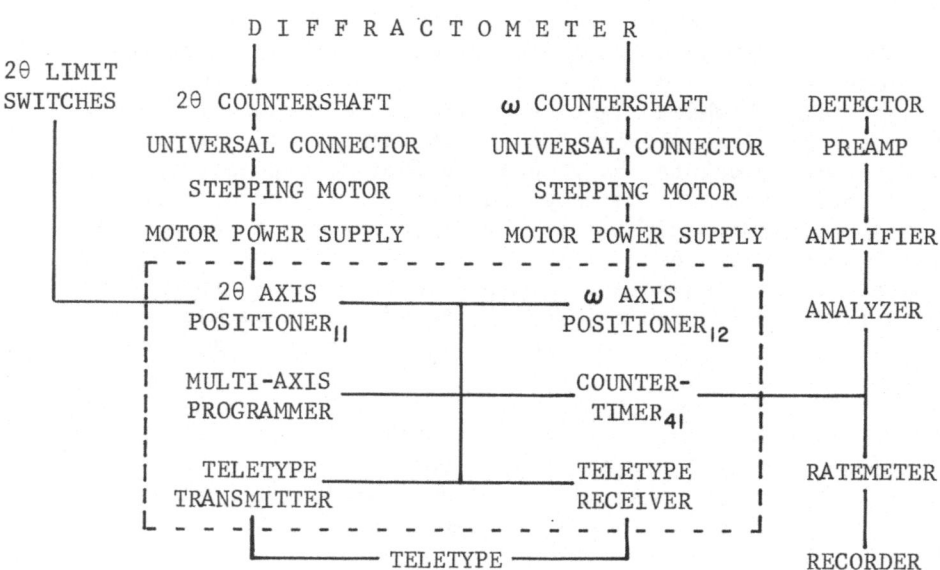

Axis Positioners. These units, which may be controlled either manually or remotely from the Teletype, position the 2θ and ω axes of the diffractometer in steps of $0.01°$. Pulses from the axis positioners control the stepping motors on the diffractometer through motor power supplies. Pulses returning from the motor power supplies are counted by the axis positioners which display the angles on Nixie tubes. The angles may also be printed and, if desired, punched on the Teletype. To permit remote control, an "address" is set on each module such that instructions for a particular motor are received only by its associated axis positioner. For example, the 2θ axis positioner has the address 11, to distinguish it from the ω axis positioner which has the address 12.

Dual Counter-Timer. This module contains two counters one of which can be preset manually or remotely. Both counters are designed to accept either pulses from the usual X-ray counting chain or from an internal clock. X-ray or time pulses are fed as desired into the counter which is to be preset. For remote operation the counter has the address 41, which permits presetting with Teletype instructions. The reset, start and data output at the end of the preset are also controlled with Teletype instructions. A digital display of time and counts on Nixie tubes is provided.

Multi-Axis Programmer. The programmer receives instructions from the Teletype, decodes and stores this information, controls the operation of the axis positioners and the dual counter-timer and returns data to the Teletype. The data consist of the 2θ and ω angles from the axis positioners and the accumulated time and counts from the dual counter-timer. A switch is provided on the multi-axis programmer which allows manual operation of the axis positioners and the counter-timer when desired. For computer controlled operation, the multi-axis programmer may be replaced with a Telecomputer Interface which is designed to permit either Teletype paper tape control similar to that provided by the multi-axis programmer or computer control.

Teletype. The Teletype and the associated Teletype Transmitter and Teletype Receiver are used in the automatic mode to instruct the automation electronics and provide a punched paper tape and/or printed data output.

With the paper tape controlled mode of operation, instructions on punched paper tape are read by the programmer, carried out and the collected data are printed and/or punched on paper tape at the Teletype. At the completion of each instruction, the next instruction is read from the paper tape automatically.

SOFTWARE

The automation hardware is designed to accept instructions through the multi-axis programmer from the Teletype tape reader or the Teletype keyboard. There are a number of acceptable instructions.

Execute. "Carry out the preceding instruction" command, is denoted by a colon :

Read. This command instructs the system to read out the angular positions from the axis positioners and the time and counts from the counter-timer. The code for this command is R and the complete instruction would be R:, the colon being included to start execution of the read command.

Write. An instruction which is used to preset the counter-timer and is denoted by the symbol W. For example, to preset a time of 200.00 seconds (or 20,000 counts) the complete instruction to be entered from the Teletype is

Fixed Input. An information instruction is provided so information or comments can be included on the punched paper tape or printed output without affecting the operation of the automation hardware. The code for this command is F, and it also is terminated with a colon. For example

F MARTENSITE (200) LINE :

End. This instruction, which is used to signify the end or completion of a program or series of instructions, is denoted by the symbol E. It tells the multi-axis programmer that the program has been completed. The programmer then disables the system so no further instructions can be accepted until the programmer is reset.

Position. This instruction, denoted by the letter P, is used to control the 2θ and ω stepping motors on the diffractometer through the axis positioners. A typical instruction would be:

---

*   The counter has a seven decade capacity and all decades are printed or punched, however the first decade is not displayed on the Nixie tubes. The preset decade relates only to those decades displayed.

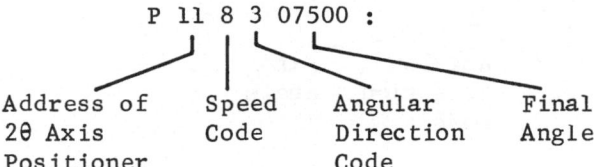

Address of   Speed   Angular   Final
2θ Axis   Code   Direction   Angle
Positioner   Code

This instruction calls for the 2θ axis positioner to move the 2θ
goniometer at 60°/minute in an increasing 2θ direction and to stop
when 75.00° has been reached. There are eight available motor
speeds ranging from 1/8 to 60 degrees/minute, each speed having a
different speed code. The goniometer can be moved in either an
increasing or a decreasing angular direction by specifying the
appropriate code. The code also includes information as to
whether the angle specified is the final position or an angular
interval.

    Step Scan. This is an extension of the position instruction.
Denoted by the letter S, it specifies additionally the step size.
For example, to step scan from 75° to 80° 2θ, counting for 200
seconds at 0.10° intervals, the diffractometer would be positioned
to 75° using the Position instruction and the counter-timer would
be preset to 200 seconds using the Write instruction. The step
scan instruction would then be entered as follows:

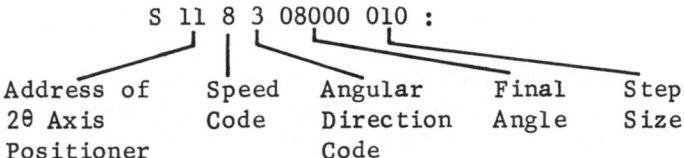

Address of   Speed   Angular   Final   Step
2θ Axis   Code   Direction   Angle   Size
Positioner   Code

After each step, the scaler is started automatically. At the end
of the preset time, the accumulated counts, time and angles are
read out on the Teletype. It should be emphasized that the first
data point is not accumulated at the starting angle but at the
angle after the first step is taken. This ensures that any
diffractometer backlash is taken up before the accumulation of
data is commenced.

    Integrate. This mode of data accumulation, in which counts
are accumulated while the goniometer is in motion, is frequently
used in X-ray diffractometry. The instruction is denoted by the
letter I. For example, to make an integral scan from 75° to
80° 2θ the diffractometer would first be set to 75° using the
Position instruction. Any existing counter-timer preset is then
removed by entering a zero preset using the Write instruction
since the axis positioner is the controlling module. The
Integrate instruction would then be read in as follows:

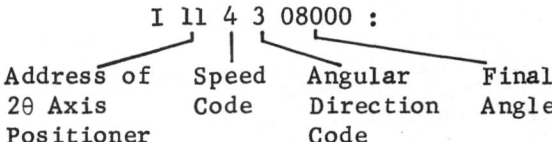

In the integrate mode the scaler is stopped at the end of the
angular range rather than after some preset time. This provides
great versatility in selecting angular ranges for integration of
diffraction lines.

MEASUREMENT OF X-RAY DIFFRACTION INTENSITIES

One technique for determining the relative amounts of
austenite and martensite is based on the measurement of the
relative intensities of the (200) austenite and the (200)
martensite diffraction lines. Instructions on punched paper tape
of the type described above are used to control the scanning of
the diffractometer and the accumulation of counts in the counter-
timer for the integrated peaks and the backgrounds for these two
lines.

The paper tape program, shown in Figure 3, is placed in the
Teletype reader and the reader started. The first instruction
tells the operator to enter the sample number on the Teletype
keyboard and to restart the paper tape. Then the 2θ axis of the
diffractometer is moved to the angle 74° 2θ and the scaler is
preset to zero, which implies that the timer is not preset to stop
after a specific time. The next three instructions control
integral scans for the background, the (200) austenite peak, and
another background respectively. At the completion of each data
accumulation, the final angles, the accumulated counts and
accumulated time are printed at the Teletype and the next
instruction is read in by the multi-axis programmer.

At the completion of the integration of the (200) austenite
peak and backgrounds, an instruction is read in to position the
diffractometer to 98° 2θ. Integral scans are then made for the
backgrounds and peak intensity of the (200) martensite line. The
angles, accumulated counts and time are again printed on the
Teletype. Typical output data are shown in Figure 4. The data
presented for the integral scan from 74° to 76° 2θ are interpreted
as follows:

```
     407600,  800000,  0011999   0055349
```

Motor	Final	Motor	Final	Accumulated	Accumulated
Speed	2θ	Speed	ω	Time in units	Counts
Code	Angle	Code	Angle	of 0.01 sec.	

At the completion of the data collection for the sample, the paper tape instructions ring the Teletype bell several times to call the operator's attention to the fact that the program has finished. A final instruction returns the diffractometer to 73° 2θ, so that it is ready to start a measurement on another sample. For repeated measurements on a single sample the paper tape can be looped to repeat the program.

Because the diffractometer can be slewed rapidly between peaks, one sample can be examined in 26 minutes using a scanning speed of 1°/minute over the peaks and background measurements without operator intervention. Using conventional diffraction equipment this would require 40 minutes, although this time could be reduced by manual slewing of the diffractometer. An even greater saving in time is made when two austenite lines are integrated, as is the usual procedure.

The tape controlled system will also readily permit the changing of scanning speeds so that a slow scan could be used for the weak austenite lines and a faster scan for the more intense martensite line. No change of gears is required as with a conventional diffractometer.

```
F PROGRAM FOR MEASUREMENT OF RETAINED AUSTENITE

ENTER THE SAMPLE NUMBER ON TELETYPE KEYBOARD
PUSH RESET & START ON PROGRAMMER TO RESTART PAPER TAPE:

E:

F PROGRAM SETS TWO THETA POSITION & SCALER PRESET:
P 11 8 3 07400:  W 41 00:

F PROGRAM INTEGRATES (200) AUSTENITE LINE:
I 11 4 3 07600:
I 11 4 3 08100:
I 11 4 3 08400:

F PROGRAM SETS TWO THETA POSITION & SCALER PRESET:
P 11 8 3 09800:  W 41 00:

F PROGRAM INTEGRATES (200) MARTENSITE LINE:
I 11 4 3 10200:
I 11 4 3 11000:
I 11 4 3 11400:

F PROGRAM RINGS THE TELETYPE BELL 6 TIMES:

F PROGRAM RETURNS TWO THETA TO STARTING POSITION:
P 11 8 2 07300:

E:
```

Figure 3. Paper tape program.

```
F PROGRAM FOR MEASUREMENT OF RETAINED AUSTENITE

ENTER THE SAMPLE NUMBER ON TELETYPE KEYBOARD
PUSH RESET & START ON PROGRAMMER TO RESTART PAPER TAPE:

E::         1650 F WATER QUENCH

F PROGRAM SETS TWO THETA POSITION & SCALER PRESET:
P 11 8 3 07400:  W 41 00:

F PROGRAM INTEGRATES (200) AUSTENITE LINE:
I 11 4 3 07600:
407600  800000  0011999  0055349
I 11 4 3 08100:
408100  800000  0029999  0145662
I 11 4 3 08400:
408400  800000  0017999  0079377

F PROGRAM SETS TWO THETA POSITION & SCALER PRESET:
P 11 8 3 09800:  W 41 00:

F PROGRAM INTEGRATES (200) MARTENSITE LINE:
I 11 4 3 10200:
410200  800000  0023999  0106845
I 11 4 3 11000:
411000  800000  0047999  0306873
I 11 4 3 11400:
411400  800000  0023999  0107116

F PROGRAM RINGS THE TELETYPE BELL 6 TIMES:

F PROGRAM RETURNS TWO THETA TO STARTING POSITION:
P 11 8 2 07300:

E:
```

Figure 4. Data output.

THE HELIUM CHAMBER

The standard technique for the X-ray diffraction measurement
of retained austenite (1-3) uses filtered chromium radiation.
Because of the relatively high absorption of CrKα radiation in
air and of the low level of austenite in some steels, it was
decided that at least a part of the air path should be replaced
by a helium path (4). To provide this helium path a Picker X-ray
fluorescence spectroscopy chamber was modified so the roll-up
window was on the normal diffraction side of the diffractometer.
This chamber is used in conjunction with the spectroscopy crystal
holder, on which the samples are mounted. To increase the length
of the helium path, a beam tunnel is added to the diffracted beam.
This helium chamber and beam tunnel are shown in Figure 5. The
helium path was found to increase the observed diffracted X-ray
intensity by a factor of 2. This increase in diffracted intensity
is illustrated by the chart records shown in Figure 6 which display
the (200) austenite and (200) martensite lines for a sample
containing 6.3% austenite obtained with and without the helium
path.

Figure 5. Helium chamber and beam tunnel.

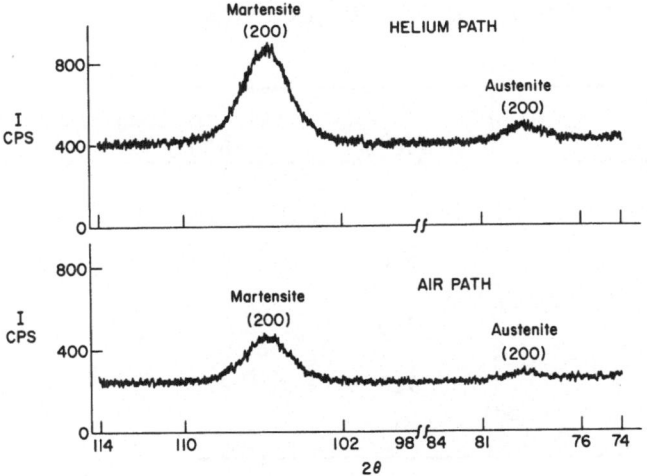

Figure 6.   Diffractometer scans of 6.3% austenite in
            martensite with air path and helium path.

EXPERIMENTAL RESULTS

Using the tape control program shown in Figure 3 for the
measurement of the integrated peak and background intensities, a
series of H12 hot work die steels which had been austenitized at
1650, 1800, and 1950°F using different quench rates were examined
on the helium path diffractometer. The volume percentage of
austenite was calculated from the integrated X-ray intensities
using the following equation (5)

$$\% \text{ Austenite} = \cfrac{100}{1 + \cfrac{(R_m(P_a - B_a))}{(R_a(P_m - B_m))}}$$

where P_a, B_a, P_m, B_m are the integrated peak and background
intensities for austenite and martensite, and R_a and R_m are the
appropriate R factors (5).

The austenite contents of these steels, calculated from the
X-ray measurements, are shown in Table 1.

Table 1.   VOLUME PERCENT AUSTENITE IN H12 STEEL

Quench Medium	Austenitizing Temperature		
	1650°F	1800°F	1950°F
Air	12.7	8.0	6.7
Water	2.9	2.3	2.0
Oil	7.0	5.7	12.8*
Air (Sample Encapsulated)	9.6	9.8	5.6

REPRODUCIBILITY AND LIMIT OF DETECTION

Some idea of reproducibility may be obtained from the results of three consecutive measurements made on a low level austenite sample. The results obtained were: 1.8, 2.1, and 2.0% austenite.

The limit of detection for austenite was calculated for the integration procedure described above based on the usual assumption that the limit of detection is equivalent to that amount of austenite which would give a diffracted signal of twice the standard deviation of the background. The calculated limit was found to be 0.4% using a helium path and 0.7% using the air path.

DISCUSSION

It has been shown that an off-line paper tape or Teletype keyboard control may be readily added to an X-ray diffractometer system. The automatic control has resulted in a considerable saving of operator time in the routine analysis of retained austenite in steels. The technique could easily be extended to include other diffraction analyses such as the measurement of lattice parameters, microstrain, particle size, and so on. The addition of a computer, with a Telecomputer Interface, would enable a calculation of the retained austenite (or other parameter) to be made directly from the data collected. The result would be printed directly on the Teletype. Since the Telecomputer

---

* The high retained austenite content is apparently due to a higher carbon content caused by the solution of carbides.

Interface also incorporates the off-line features of the Multi-Axis Programmer, paper tape control can be used if the computer is not available.

ACKNOWLEDGEMENT

We wish to thank Dr. C. A. Stickels for preparing the samples and for helpful discussions during the course of this work.

REFERENCES

1. R. J. Brincks, "Evaluation of the X-Ray Diffraction Method for Determining Volume Percent of Retained Austenite in Carbon Steel," Technical Report #68-1105, U. S. Army Weapons Command, Rock Island Arsenal Research Engineering Division (1968).

2. R. E. Ogilvie, "Retained Austenite by X-Rays," Norelco Reporter, 5, p. 60-61 (1959).

3. H. R. Erard, "Technique of Measuring Low Percentages of Retained Austenite Using Filtered X-Ray Radiation and an X-Ray Diffractometer," in W. M. Mueller, G. Mallett, and M. Fay, Editors, Advances in X-Ray Analysis, Vol. 7, p. 256-264, Plenum Press (1963).

4. R. R. McCune, "Helium Path Diffractometry and Its Application to Determination of Retained Austenite and Macrostress in Steel," in W. M. Mueller and M. Fay, Editors, Advances in X-Ray Analysis, Vol. 6, p. 85-89, Plenum Press (1962).

5. B. L. Averbach and M. Cohen, "X-Ray Determination of Retained Austenite by Integrated Intensities," Trans. AIME, 176, p. 401-415 (1948).

AUTOMATED X-RAY DIFFRACTION LABORATORY SYSTEM

Armin Segmüller

IBM Thomas J. Watson Research Center

Yorktown Heights, New York 10598

ABSTRACT

An IBM 1800 time-sharing system is used in our X-ray laboratory to control a four-circle diffractometer for structure research, several powder diffractometers, a pole-figure goniometer and a microdensitometer along with other instruments outside the diffraction area. A survey of the computer system is given and the hardware necessary to automate the diffractometers is discussed. The computer supervision ranges from simple data-logging with a minimum of control to complete control of all actions depending on the diffractometer and the requirements of the experiment. Also described is the use of the computer to process the data and to perform background jobs.

INTRODUCTION

Crystallographic computations were among the earliest scientific applications of electronic computers. Soon X-ray diffractometers were operated under various degrees of automation. In our laboratory, automation started with a four-axis diffractometer for structure research controlled by an IBM 1620 computer (1). After replacement of the 1620 by an IBM 1800 process control computer, the opportunity presented by the added capability was used to control other X-ray diffractometers. Shared computer control is well suited for X-ray diffraction data acquisition. It makes available the resources of a relatively large computer to the individual diffractometer. Data rates normally are low and the data to be acquired does not disappear if the computer cannot respond immediately while serving another experiment. Besides data acquisition, the computer can also be used in time-sharing,

for background jobs such as data processing, general computations, compilations of new programs, and operating displays.

LABORATORY AUTOMATION SYSTEM

Our present system comprises an IBM 1800 computer with 32K of core, three disk drives, card reader-punch and printer keyboard. The digital front end has 16 words (16 bits each) of digital output for contact closures (Electronic Contact Operate or ECO), 6 words of register output for driving scopes, 2 words of pulse output for the communication adapter, 32 words of digital input to read scalers, encoders, keyboards, etc. and 6 words of interrupts for service requests by the individual experiments. Our system does not have analog input or output. This is not a serious restriction because most X-ray diffraction devices are digital. Where analog instruments, like a ratemeter or photomultiplier, are used the analog/digital conversion is performed locally at the instrument by a digital voltmeter (DVM). For the storage scopes, a fast, central digital/analog converter is used.

Connected at present (July 1, 1971) are a four-axis diffracto-meter for crystallographic data taking (1), a pole-figure spectrometer (2), three powder (one-axis) diffractometers, one of them described at the Denver Conference 1969 (3), a microdensito-meter (4) and ten, physical and chemical experiments.

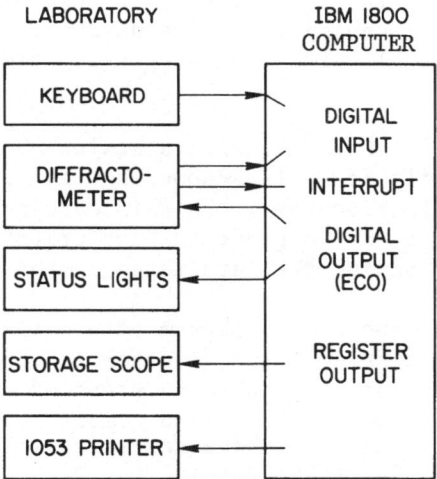

Figure 1.   Interaction between diffraction laboratory and computer.

Communication between the individual experiments and the computer is provided by two alphanumeric and ten numerical keyboards, eight storage scopes, three 1053 printers and numerous status lights at the laboratory. Figure 1 shows a block diagram of a typical X-ray diffractometer/computer installation. Generally a printer and/or storage scope is shared by several experiments at the same location.

A communication adapter with two data phones allows the attachment of experiments via a 1050 terminal and the telephone line. A data channel repeater connected to the IBM/360-67 computer provides data transmission to and from our computer center.

The system at present, and as described here, is run under the TSX III monitor and LAB DIRECTOR (5,6). The expansion of the system to 64 K of core and the change to the MPX III monitor is under way in order to provide better service to more users.

AUTOMATED DIFFRACTOMETERS

A commercial diffractometer may be automated in several ways, depending on electronics and driving motors used. If the diffractometer is equipped with a synchronous motor for the 2Θ-drive a cam mounted on the driving shaft and operating a microswitch may be used as a timing device to request the reading of a scaler or ratemeter by the computer. Figure 2a shows this arrangement used in our automatic pole figure spectrometer (2). The motor is running continuously. However, a time-out is initiated by the cam-microswitch which stops the motor if it is not reset by the computer. This feature prevents loss of data if the computer is momentarily unable to read the data. A ratemeter/DVM is used for the pole figure data acquisition. The data rate of about 1 data/sec is comparable with the integration time of the ratemeter.

If higher precision is required a scaler should be used which may be read on the fly if 2Θ is scanned continuously as it is done on the four-axis diffractometer. In this case an external timer is used rather than a cam to generate high priority interrupts to read the scaler containing the integrated number of counts (7).

Figure 2b shows the block diagram of a diffractometer used for automatic step scanning. The detector is manually set to the initial position and data are taken in steps of defined length. X-ray quanta are counted up to a fixed number of counts or during a fixed time interval. Each time the cam-microswitch is closed, the motor is stopped, an interrupt occurs, the step count is incremented and scaler and timer are reset and started. The angle is defined by the starting position, the step length and the step

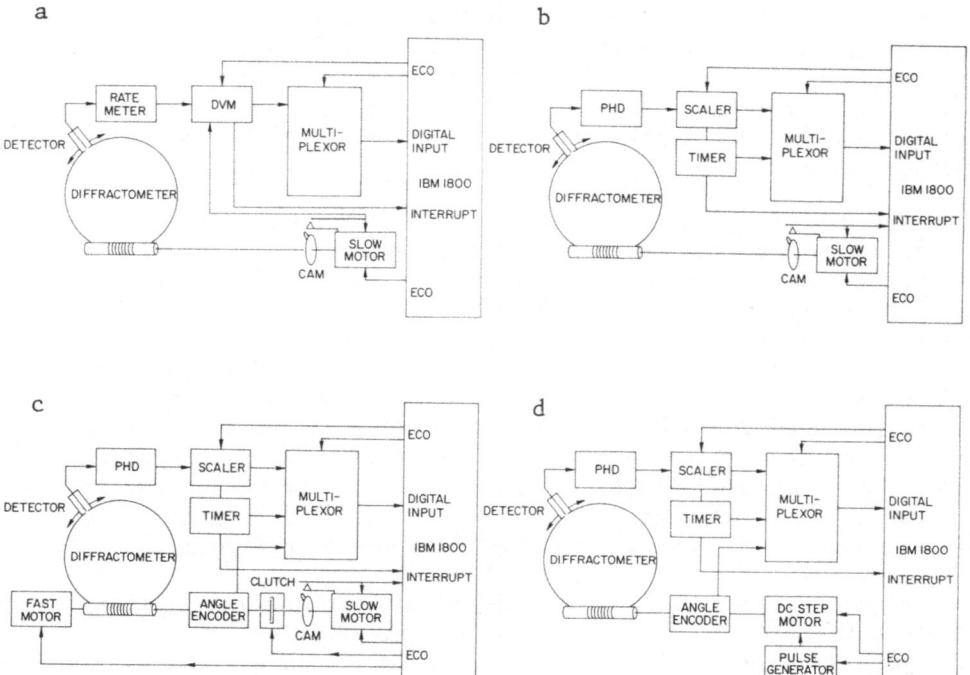

Figure 2.   Automated Diffractometers.
        a.   Ratemeter/DVM and cam/a.c. motor arrangement.
        b.   Scaler/timer replaces ratemeter/DVM in a.
        c.   Fast motor and shaft angle encoder added to b.
        d.   Fast/slow motor in c. replaced by d.c. step motor.

        Configurations a, b and c require cam to initiate
        a motor-time-out or to stop the motor independently
        of the computer.
        PHD: pulse height discriminator, DVM: digital
        voltmeter.

count, a parameter in the control program. If one uses the
logic provided by the diffractometer manufacturer all the computer
has to do is to update the step count upon a step cam interrupt;
to read the scaler and/or timer upon a preset count/time interrupt
and to enable the 2θ motor to step to the next position. The use
of two interrupt lines, one from the step cam and the other from
the scaler/timer facilitates recognition of errors.

     If we add a fast driving motor and a shaft angle encoder we
arrive at the arrangement shown in Figure 2c and described in a
paper at this conference two years ago (3). The fast motor and

the angle encoder allow slewing the detector to the starting
position. The angle encoder measures the angle absolutely, it
may be read at each step of the scan providing the best supervision of the diffractometer scan. This driving motor arrangement is used for the ω and 2Θ-drives of the four-axis diffractometer (1); however, instead of a cam the logic of the angle
encoder is used to generate interrupts.

The use of a d.c. step motor driven either by a computer-
controlled pulse generator or by pulses generated by the digital
output of the computer and a programmed timer simplifies the
diffractometer hardware considerably without restricting the
possibilities of the system discussed previously. The pulse
generator is used to drive the motor with a speed comparable to
that of the fast motor in Figure 2c, whereas computer-generated
pulses are used for step scanning. Figure 2d shows this arrangement implemented in the ϕ and χ circle of the four-axis diffractometer (1). Our microdensitometer (4) uses the driving motor
arrangement of Figure 2d, with the photomultiplier signal being
read through a DVM, similar to the arrangement in Figure 2a.

Figure 2 shows various degrees of automation ranging from
simple data-logging with a minimum of control (reset of time-out)
in Figure 2a to a complete control of motors, scalers and timers
by the computer in Figure 2c and d. It also shows the advantage
of using d.c. step motors to drive shafts if reliable control is
assured by shaft angle encoders. Comparing Figures 2c and d it
is evident that a d.c. step motor drive requires less hardware
and software than a fast/slow a.c. motor combination.

Multiplexors are also shown in Figure 2. They may be used to
send one decimal digit at a time over to the computer thus
requiring only four bits of digital input for a diffractometer.
If several instruments are to be read this serial reading might
take too much time. In this case all five or six digits of a
scaler are read at the same time, but, devices are read serially.
Multiplexing is very suitable for step scanning since the data
to be read is stationary. For the four-axis diffractometer
continous scanning is used and the scaler and shaft angle encoder
are read on the fly. Here, each device is wired directly into a
unique digital input word. Multiplexing does save wiring, but it
is slower and requires more programming.

DATA ACQUISITION

Interrupts from instruments are served by core resident
Interrupt Service Subroutines (ISS) with immediate response.
These programs perform simple basic tasks, for instance read
a scaler and encoder, convert the reading from BCD code to a binary

number, store it in a data buffer in the core memory, start the
step motor etc. When more service is required, a larger program
in executable form is loaded into a different core area from disk
storage under the supervision of LAB DIRECTOR (5,6), and executed.
This so-called Mainline Program stores a full data buffer on disk,
initiates parameters, interprets keyboard entries or processes the
data. The ISS's or parts thereof are often shared by several
experiments with similar requirements.

The data rates range from 1 data/sec to 1 data/min for the
diffractometers. The data rate of the microdensitometer is about
25 data/sec. Since the ISS needs only a few milliseconds to
service an experiment, there is still considerable CPU time
available.

The data acquisition is supervised by a "watch-dog" routine
which is called periodically. If it detects that the data
acquisition is not progressing properly it causes the mainline
program to be called to diagnose and, if possible, to correct
the error condition or to abandon the run. The data acquisition
can be suspended and continued from the system keyboard at any
time to allow maintenance of the computer.

Disk files to store the data are allocated dynamically in a
big disk file pool when the run is started. The system allocates
only the exact number of disk sectors (320 words) needed for a
specific run and it frees the disk space when the data is not
needed anymore. Some experiments require large disk files for
data storage, but in a research environment they may be run only
infrequently. Therefore, the file pool allows an economical use
of the available disk space by all experiments (8).

DATA PROCESSING

As soon as the data acquisition is terminated, the data is
easily accessible for further processing. Programs for
smoothing (2), background approximation (4) and peak search
(3,4) are available. A storage scope is used to display the data
between processing steps (4,9). A cursor can be moved over the
display by operating a set of switches and its present position
can be recorded (7). This feature, similar to a "light pen",
allows the operator to mark significant portions of the data,
like peaks, or obviously wrong data points. A literal language
is used by the operator to communicate with the data processing
program at the system keyboard (10,4).

Processing of the four-axis diffractometer data includes the
following steps.

Determination of the centroid to calculate the d-spacing
Integration of the rocking curve, Lorentz and polarization
 correction to determine the structure factor $|F|$
Normalization of $|F|$ to obtain $|E|$ factors
Determination of the overall temperature and scale
 factor using Wilson's method.

The data is then shipped to the IBM/360 computer via the communication adapter to perform least-squares analysis or Fourier synthesis. The structure can be displayed on the storage scope in stereo view, as shown in Figure 3, and it can be rotated to obtain an optimal view (7).

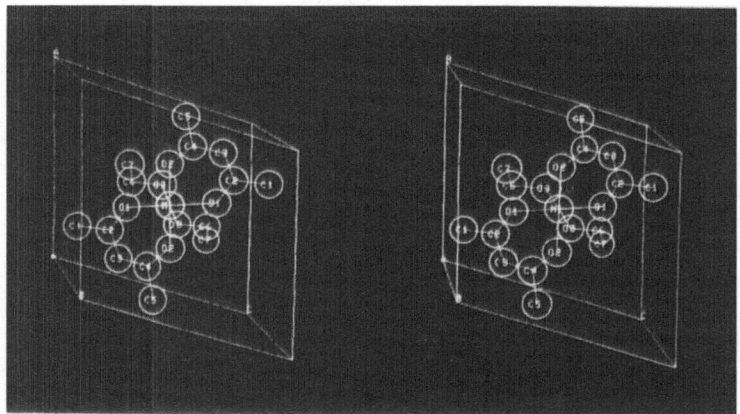

Figure 3.   Typical crystal structure, displayed on storage scope
            in stereo view (7).

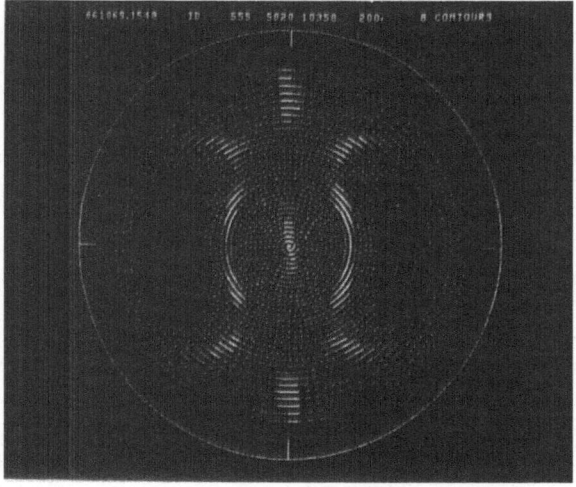

Figure 4.   {200} pole figure of hard rolled copper, displayed on
            storage scope.

Figure 4 shows the display of a pole figure on the storage scope. As the electron beam writes the spiral trace, its density is modulated with the diffracted intensity (9).

The storage scope is useful to other crystallographic displays. Figure 5 shows the 1$\bar{1}$0 plane of the reciprocal lattice of GaAs. The large circle comprises all reflections in this plane that can be obtained with Cu-K$_\alpha$ radiation. The reflection within the two small circles can be obtained only in transmission (Laue case) on a (001) wafer, whereas the reflections outside the two small circles are obtained in reflection (Bragg case).

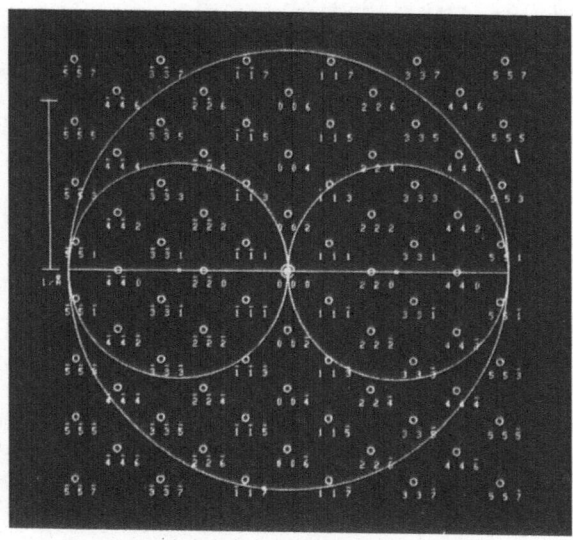

Figure 5. Display of [1$\bar{1}$0] reciprocal lattice plane of GaAs.

ACKNOWLEDGEMENTS

The author is indebted to A. L. Bednowitz and H. Cole for many suggestions and discussions and to J. Angilello, R. V. Dobransky and C. G. Wood for their valuable assistance in implementing the automation of the powder diffractometers.

REFERENCES

1. H. Cole, Y. Okaya and F. W. Chambers, "Computer-Controlled Diffractometer," Rev. Sci. Instr. **34**, 872-876 (1963).

2. A. Segmüller and J. Angilello, "Automated Pole Figure Evaluation," J. Appl. Cryst. **2**, 76-80 (1969).

3. A. Segmüller, "Automated Lattice Parameter Determination on
    Single Crystals," in B. L. Henke, J. B. Newkirk and G. R.
    Mallet, Editors, <u>Advances in X-Ray Analysis</u>, Vol. 13, p. 455-
    467, Plenum Press (1970).

4. A. Segmüller and H. Cole, "Procedures to Run an Automated
    Micro-Densitometer on a Shared Computer System," in C. S.
    Barrett, J. B. Newkirk and C. O. Ruud, Editors, <u>Advances
    in X-Ray Analysis</u>, Vol. 14, p. 338-351, Plenum Press, (1971).

5. H. Cole, "Computer-Operated X-Ray Laboratory Equipment,"
    IBM Journal <u>13</u>, 5-14 (1969).

6. A. L. Bednowitz, "Programming for Automatic Diffraction Data
    Acquisition in a Time-Shared Environment," in F. R. Ahmed,
    Editor, <u>Crystallographic Computing</u>, p. 336-342, Munksgaard
    (1970).

7. A. L. Bednowitz, private communication.

8. H. M. Gladney, private communication.

9. A. Segmüller, "Cathode Ray Tube Display of Automatically
    Recorded Pole Figure Data," J. Appl. Cryst. <u>2</u>, 259-261 (1969).

10. F. W. Chambers and A. L. Bednowitz, "A Literal Language for
    Laboratory Automation," paper presented at the Pittsburgh
    Conference on Analytical Chemistry and Applied Spectroscopy,
    Cleveland, Ohio, March 1-6, 1970.

X-RAY DIFFRACTION TOPOGRAPHY-DIFFERENTIAL OMEGA SCANNING TECHNIQUE

Ray L. Silver and Jack C. Turner

Delco Electronics Division, General Motors Corp.

Kokomo, Indiana 46901

ABSTRACT

The Differential Omega Scanning Technique (DOST) was developed to improve topographs obtained from silicon wafers which generally possess process induced flexure during the manufacture of semiconductors products. A commercially available Lang camera has been electronically instrumented to provide an automatic adjustment of omega motion of the specimen on the goniometer axis. The position of the specimen relative to the primary beam (theta) is continually altered by DOST to assure maximum, or near maximum, diffraction intensity with changes of flexure of the wafer as the specimen is scanned. The resulting topographs have near complete images and appear to have improved quality. Using techniques developed within the x-ray laboratory, routine acceptable topographs are obtained in only the time required to scan the diameter of the specimen one time.

INTRODUCTION

The Lang x-ray diffraction technique is very sensitive to strain within single crystals and readily records localized changes of crystal curvature resulting from this strain. One effect of this curvature is to misorient the diffraction planes of the crystal relative to the primary beam to an extent that diffraction will not occur.

123

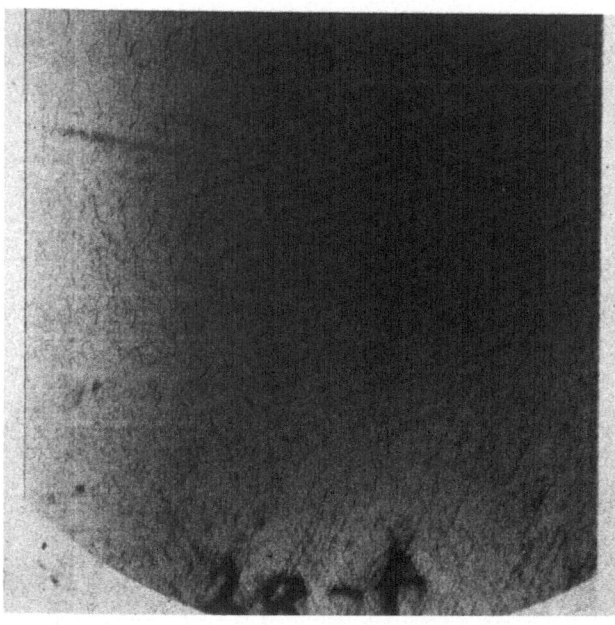

Figure 1.   Single scan topograph (Lang technique) of silicon wafer with minimum of flexure.

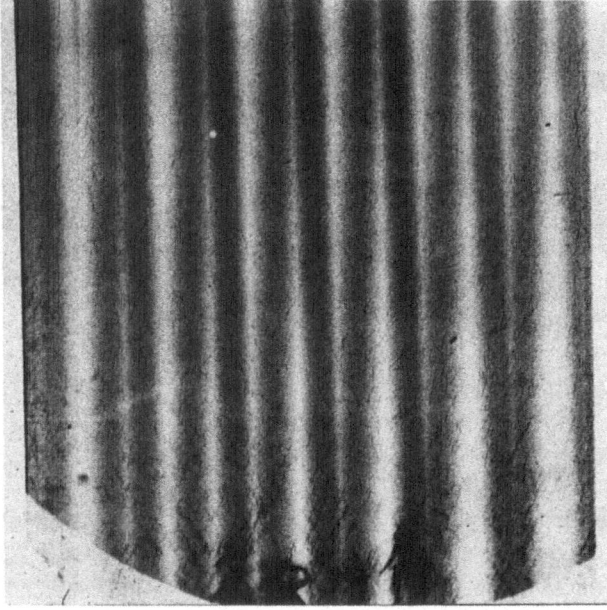

Figure 2.   Single scan SOT topograph of same specimen used to obtain Figure 1.

Figure 3.   Single scan DOST topograph of specimen used to
obtain Figures 1 and 2.

Unfortunately, large diameter crystal slices, processed for
semiconductor fabrication, generally have induced strain and
localized variations in curvature which correspondingly produce
void areas in the topographic images. Anticipating this difficulty
at the time the topographic facility was established, this
laboratory purchased a commercial Jarrell-Ash Model 80-050
Lang camera instrumented to perform the scanning oscillator
technique (SOT). This additional capability helped in efforts to
obtain topographic images more nearly complete than originally
afforded by the Lang camera.

SOT and Lang topographs are generally obtained by
multiscanning the specimen in order to achieve optimum image
density. Single scan topographs (Figure 1) are made possible by
varying the processing of the plate; however, SOT topographs
(Figure 2) are not routinely used because they exhibit striated
images, which the authors have descriptively named "venetian
blind effect" images.

The venetian blind effect is caused by the oscillations of the
crystal specimen through the angle omega (or theta - the
rotation angle omega is coincident with theta about the vertical

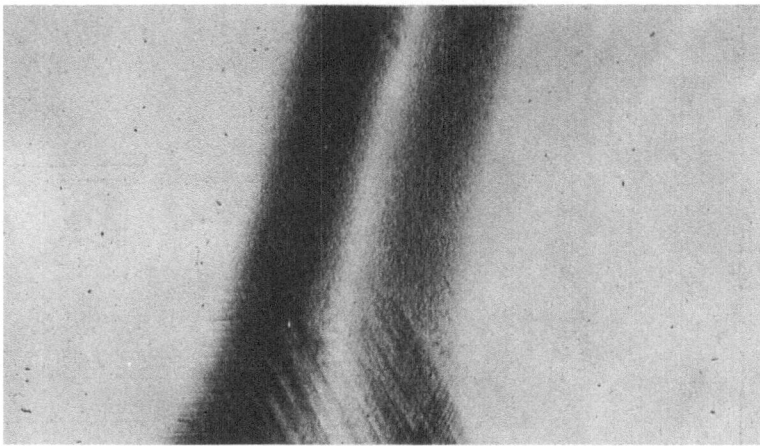

Figure 4.   Single scan topograph (Lang technique) of silicon wafer with considerable flexure.

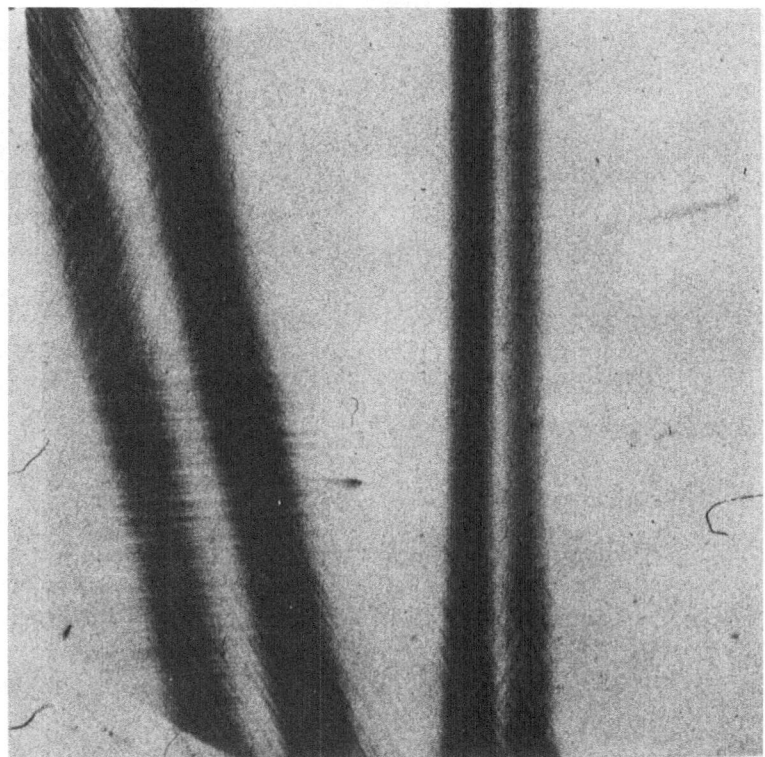

Figure 5.   Single scan SOT topograph of specimen used in Figure 4.

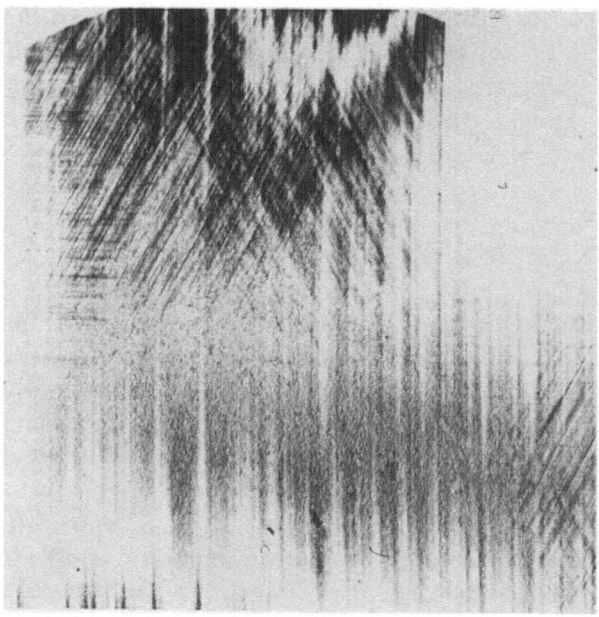

Figure 6.   Single scan DOST topograph of specimen used in
Figures 4 and 5.

axis of the goniometer). The only time an image is formed on the
plate is when the angle omega is correct for diffraction to occur
in accordance with Bragg's Law. Unfortunately, for SOT,
crystals with large curvature (flexure) require large oscillatory
excursions of omega which cause areas of image in the
topographs to be small relative to the overall exposed area of the
crystal. The occurrence of the proper omega angle position is
momentary.

The major contributing causes of the venetian blind effect
are the flexure of the specimen and the specific design configuration of the SOT instrumentation on the Lang camera. The
instrumental problem arises from the electromechanical
oscillator programmer which is limited to a minimum oscillatory
omega angle of twenty seconds of arc. Also, the oscillatory
angle can be increased by properly setting the limit switches on
the programmer to include all possible omega angles required to
permit diffraction to occur over the span of specimen. However,
for crystals with large flexure, the limit switches must be set to
provide large excursions of omega angle which unfortunately
accentuates the venetian blind effect (Figure 5).

Vertical void striations appear in the images of SOT topographs because there is no responsive control within the instrumentation system; that is, the range of omega oscillator angle is preset and is time dependent rather than dependent on the rate of change of flexure within the crystal.

DIFFERENTIAL OMEGA SCANNING TECHNIQUE

The differential omega scanning technique (DOST) senses the change of flexure of the specimen and correspondingly changes the omega angle to maintain a maximum, or near maximum, diffraction intensity of the K alpha radiation which exposes the plate. The venetian blind effect is minimized (Figure 3) and in some cases does not exist.

DOST is a concept based on the continuous and automatic adjustment of the position of the crystal (angle omega) for each increment of change of flexure in the sample crystal. Adjustment is achieved by special electronic circuitry used in conjunction with the commercial Lang camera and scan programmers. An additional standard translation scan programmer is altered to switch its normal operation to DOST operation.

Input into the DOST circuitry is obtained from the two theta readout detector employed normally on the Lang camera. A Victoreen Model 489-35 GM probe and a Victoreen Model 490 Thyac III survey meter are used to monitor the diffracted radiation level originating from the crystal specimen.

The GM detector probe is normally used to determine the goniometer's zero, theta and two theta diffraction positions for Lang and SOT topography. In the DOST mode of operation, the detector probe also serves as a monitor of x-ray diffraction beam intensity from the crystal.

As the crystal is traversed through the primary x-ray beam, flexure of the crystal will cause the specific diffraction planes to alter their position with respect to the primary x-ray beam. This change of theta angle position will cause a corresponding decrease or increase of diffracted x-ray beam intensity for a fixed two theta position of the detector in accordance with Bragg's Law. Therefore, flexure affects diffraction intensity which affects the output from the two theta detector. The change

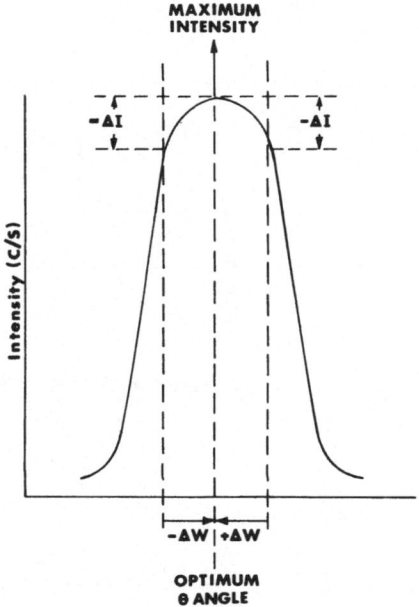

Figure 7.   Intensity vs. omega profile.

of output of the detector provides the means of establishing the change of theta position electronically.

The DOST electronic circuitry, which obtains its input from the GM probe and survey meter, senses the change of theta position and correspondingly activates triac devices in an output circuit to control the omega scan programmer. This programmer activates the standard electromechanical drive system on the Lang camera and alters the position of the goniometer to the correct omega position.

DOST is a continuously seeking system which purposefully controls the omega scan programmer thereby changing the omega goniometer adjustment so that a near maximum diffracted x-ray beam intensity is maintained from the crystal. As the goniometer omega motion is constantly being changed, the crystal is rotated on the goniometer vertical axis in the primary x-ray beam. The diffracted intensity monitored by the detector probe, set at a fixed two theta position, will increase and decrease as a function of change of theta position. This is graphically illustrated in Figure 7. The maximum possible diffracted intensity occurs when the optimum theta angle is realized through action of the scanning

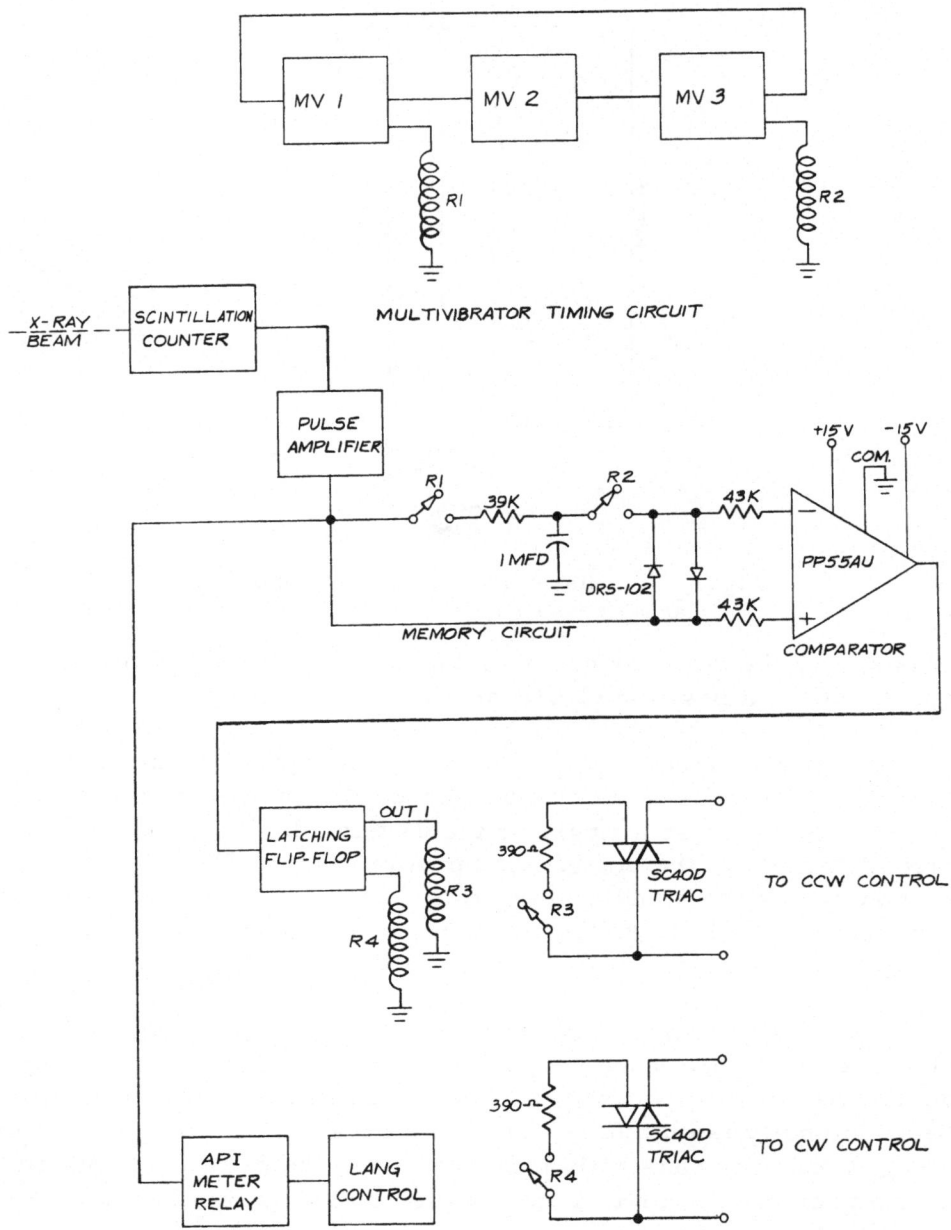

Figure 8. DOST circuit block diagram.

omega drive. This intensity level is registered in the DOST memory circuit and a change in intensity ($-\Delta I$) is achieved by the constantly driven omega programmer. The change in intensity is compared with the previously recorded maximum intensity level and the DOST comparator circuit correspondingly changes the omega angle drive direction to maintain a maximum diffraction intensity. Thus DOST provides a topograph that has an image formed at or very near the maximum diffraction intensity across the entire area of the crystal scanned.

DOST, through timing, memory and comparator circuits, senses the change of intensity with respect to time, $\Delta I/\Delta T$. The time duration determines the minimum frequency with which the crystal will pass through the optimum omega (theta) angle position. The time span can be electronically altered from 0.5 to 2 seconds duration by appropriate switching of circuitry.

DIFFERENTIAL OMEGA SCANNING CIRCUITRY

DOST utilizes a fully transistorized, solid state, printed circuit board configuration. It is housed in a steel cabinet of the approximate size of the standard scan programmer.

The circuit is composed of three monostable multivibrator sections, a latching flip flop section and a pulse amplifier section. Delco Electronics type DS-67 and DS-83 transistors are used. Included in the circuit are a Philbrick 55 AU comparator and three General Electric SC-40D triac A.C. switching devices.

Operation

The specific circuits and their interconnections are shown in a block diagram configuration in Figure 8.

X-ray photons entering the counter tube ionize the gases contained therein and are in effect changed into electric impulses. These impulses are fed into the Victoreen Thyac III detector survey instrument; there the impulses are channeled into two different circuits simultaneously.

One circuit is the internal circuitry of the Victoreen survey instrument, which amplifies the impulses and provides a visual (microammeter) and audio (headphones) readout.

The other circuit is the DOST pulse amplifier which contains multivibrator, amplifier and filter sections. The incoming impulses from the survey meter trigger the multivibrator, which effectively increases the width of the pulses. The widened pulses are amplified and then filtered to obtain a DC output.

The output of the DOST pulse amplifier circuit is fed into the memory circuit. The incoming DC voltage from the pulse amplifier is applied to the memory capacitor through one channel to relay #1, which is controlled by a 50 microsecond pulse from the monostable multivibrator circuit.

The monostable multivibrator circuit is used for a timing function to the memory circuit. When the main DC power supply is turned on, the first multivibrator is almost immediately triggered because of the design parameter providing a short time constant to this section of the circuit. Each successive multivibrator section is triggered from the previous stage; thus the complete multivibrator circuit yields phase cycles of 2.5 milliseconds, 50 milliseconds, 0.5 second, 1.0 second, 1.5 seconds, 2.0 seconds. The circuit actuates relay R1 every 50 milliseconds and relay R2 every 0.5 second, 1.0 second, 1.5 seconds, or 2.0 seconds, depending on the setting of the selector switch used for adjusting the frequency at which the memory capacitor is discharged. The action of relays R1 and R2 controls the DOST memory circuit.

The memory circuit, containing two channels from the pulse amplifier to the PP55 AU comparator, provides both a negative and positive voltage to the comparator circuit. The action of relay R2 applies the negative input from the memory circuit into the comparator circuit every 0.5 second, 1.0 second, 1.5 seconds or 2.0 seconds, depending on selection of desired frequency of action of the comparator circuit. The output of the pulse amplifier is also continuously applied to the comparator as a positive voltage.

The comparator circuit compares the incoming voltages, both positive and negative. The continuous change of the omega position of the diffracting crystal, brought about by the omega scan programmer, will always eventually cause the comparator circuit to yield a negative output voltage with time. The time required is dependent on the change of omega position and flexure of the crystal. A negative voltage output from the comparator circuit will trigger the latching flip flop circuit.

The activated latching flip flop circuit will activate the triac switching circuit, which, in turn, switches the omega scan programmer to drive in a reverse direction.

The continuous reversal of direction of the omega scan programmer will assure optimization of the crystal omega position for maximum diffraction intensity.

A safety feature has been incorporated into the DOST circuitry to prevent damage to the omega scan programmer in case of loss of intensity to the detector and input to the DOST circuitry. An API meter relay controller (50 microammeter) is incorporated into the circuitry to sound an alarm or shut off the omega scan programmer if the intensity falls below a preset level. The preset intensity level is adjustable by knob and indicator on the microammeter in the front panel. Input to the DOST circuit is monitored at all times by this meter. Choice of audio alarm or shut-off of omega scan programmer is made by appropriate setting of a toggle switch.

Operational Procedure

DOST circuitry is activated by one pushbutton. Lock-in on maximum diffraction intensity from the crystal is automatic. Setup time is less than one minute in addition to that required for standard Lang topography. This includes the time required to engage the electromechanical drive to the omega drive assembly.

DOST is most effective for crystals with flexure along the same rotational axis as the angle omega, as is the case for the standard SOT method. Crystals with flexure not along this axis can be brought into position by special crystal holders designed for this purpose. The crystal holder now in use is a dual ring design, such that one ring fits into the other and can be rotated and indexed every 30°.

SUMMARY

Lang and SOT topographs are normally obtained by multi-scanning the specimen. In most instances, flexure in the crystal will cause void areas in the topographic image. By changing the plate developing process, single scan topographs are possible. Single scan SOT topographs are undesirable because of the venetian blind effect.

DOST minimizes the effects of flexure within the specimen, minimizes the venetian blind effect and achieves the single scan requirement. The technique has provided improved topographs in less time than other techniques. The rapidity and ease of operation makes it suited for applications within quality control and production areas of semiconductor manufacture.

A MODULAR AUTOMATIC X-RAY ANALYSIS SYSTEM

M. Slaughter, Dept. of Chemistry
Colorado School of Mines
Golden, Colorado 80401

Davis Carpenter
Technical Equipment Corporation
Denver, Colorado 80204

ABSTRACT

A modular x-ray and electron beam system has been constructed to accomplish most standard types of x-ray analysis automatically. The system includes a vertical powder diffractometer, a vacuum spectrograph, a four-axis single-crystal diffractometer and an electron microprobe with four spectrometers. The system is controlled by a single computer with auxilliary drum storage. The powder diffractometer has automatic goniometer drive, automatic sample changing with computer cataloging of samples and automatic variable divergence slit. The x-ray spectrograph is a standard unit with added 100 KV capability, automated goniometer drive, eight or 32 position sample changer, six-analyzing-crystal changer, collimator changer, and detector changer. All functions including sample number cataloguing are computer controlled. The four-axis single crystal diffractometer has all functions computer controlled and includes peak search option. The electron microprobe/scanning electron microscope has all spectrometers, x, y, z stage motions and sample changer controlled by computer. X-Ray channels, beam current and sample current are automatically read-out. Control of probe functions and correction of mechanical errors in stage operation are aided by a digital beam scanner with a light pen. Counting channels and computer interfacing are modular, and channel components are interchangable by the operator. Computer hardware may be expanded and the software is modular. Any combination of instruments up to eight may be operated.

The computer operates all instruments simultaneously in a time-sharing mode. Programs consist of operating system programs,

individual instrument programs and data processing programs. The
operating system drives all x-ray hardware, fetches programs and
data from the drum, allocates data storage, queues final results
for printing and performs other necessary executive functions. For
qualitative analysis, spectra for each instrument are collected automatically in several modes and are stored on the drum. A variety of
methods are used to remove background, strip unwanted peaks and find
peak position and intensity. A.S.T.M. diffraction, spectrographic
and special files are maintained on the drum. Quantitative analysis
is executed automatically using a version of "MAGIC" on the electron
microprobe and a multiple linear regression method for the spectrograph and diffractometer.

INTRODUCTION

    Computer operated analytical x-ray systems can speed data
collection and relieve tedium. Equally important, a computer operated x-ray system can collect data of quality heretofore impractical
to achieve by manual means. It can execute complicated control procedures, extensive mathematical corrections, and look-up and extraction functions that give a new dimension to analytical results.

    To gain the advantages of computer operation of a single crystal
x-ray device without major sacrifices, the computer usually must have
a configuration larger than minimal and perhaps much larger than many
x-ray analysts realize. It is of questionable value for some analysts to automate x-ray or electron beam equipment with a minimal
computer configuration. The speeding of data collection or relief
from tedium afforded by a minimal computer system attached to an
x-ray device has trade-offs for the analyst. An obvious sacrifice
is the loss of much of the information contained in the x-ray strip-
chart. An adequate computer attached to an x-ray device is moderately expensive and because of costs not associated with manufacturing, computers may not get much cheaper for several years.
Combining an adequate computer configuration with a single x-ray
device such as a powder diffractometer could risk an unwise expenditure of resources.

    It is the purpose of this paper to describe a modular automatic
x-ray analysis system having an adequate computer configuration to
drive multiple x-ray and electron beam devices. The system, developed at the Colorado School of Mines and Technical Equipment Corporation, Denver, with industrial and government support, provides
comprehensive automation of multiple and optional combinations of
x-ray (or other) devices, all operable simultaneously. The higher
cost of an adequate computer is thus spread among several instruments.

    We shall first describe the hardware of this system, then the
programs and finally some of the operation modes of two instruments.

AUTOMATIC X-RAY HARDWARE

X-Ray Powder Diffractometer (vertical)

The simplest of all units to automate data collection is the powder diffractometer. This unit has the standard drive motor replaced by a stepping motor and driver kit with controls allowing manual or computer operation. Speed changes for manual operation are made by turning a dial rather than the usual method of changing gears. The counting channel components are the same as in a manual unit except that one of the two outputs from the P.H.S. is input to a 24 bit binary computer-controlled scaler.

The diffractometer in its practical configuration has a 32-position magazine-type sample changer, rigidly fastened to the diffractometer ω-shaft. As each 32-sample cassette is added the code can be read by the computer. The computer can also read each sample position and load specimens either in sequence or randomly. The computer or operator may also return to and reload any specified specimen. Sample and cassette numbers are displayed on the sample changer control panel. A variable slit coordinated with the goniometer drive maintains equal radiation area on the sample with changing 2θ angles.

Vacuum X-Ray Spectrograph

The goniometer on this unit is driven by the same type of kit as that of the powder diffractometer. The two-position collimator and the eight-position sample changer are rotated into position by stepping motors. The proportional detector is moved from advanced to normal position by a small stepping motor mounted in the scintillation detector arm. The standard manual two-crystal holder is replaced by a six-crystal holder driven by a stepping motor inside the changer. All motors are driven automatically or manually.

The x-ray chamber is a new casting sufficiently shielded for operation up to 100 KV. This allows operation on both ends of the periodic table with KV discrimination for improved signal to noise ratio on the low end of the periodic table.

Although the counting channels have components standard in a manual instrument, the computer may, through D/A converter modules, control any of the following four parameters: P.H.S. baseline, P.H.S. window, counter voltage, amplifier gain.

A 32-position automatic sample changer which we do not yet have installed in our system will allow interchange of samples between diffractometer and spectrograph.

Four-Axis Single-Crystal Diffractometer

This unit, designed for molecular structure and pole-figure analysis, is similar to other four-axis units. The motions 2θ, ω, X, ϕ are driven by separate motors, in this instance, stepping motors, controllable either manually or automatically. The unit has a diffracted beam monochromator. A sector-wheel brings to the x-ray port attenuators, Ross filters or a beam stop for protection of the specimen during slewing. The sector-wheel, mounted on the tube housing, is driven by a stepping motor. A second collimated counting channel is mounted on the tube housing opposite the goniometer. This channel views the same area of tube target as the diffraction collimator and monitors the x-ray flux during peak scan, acting as a scaling standard. The standard channel uses a zirconium attenuator to insure that diffracted and standard beams are from the same target depth.

Electron Microprobe

The initial design of the electron microprobe was suited to automation with few kit additions. Stepping motor drives were standard on all four spectrometers and the x,y stage drives. A BCD readout of sample current and beam flux was available at the digital voltmeter on the console.

Mechanical additions introduced a stepping-motor driven z-motion to the stage and a synchronous-motor driven slide sample changer. The major device added to the electron microprobe was a digital scanner. The scanner, coupled with the computer, has the following functions:

1. Computer controlled x,y beam positioning through two ten-bit D/A converters.
2. Computer readable beam and light pen coordinates using 256 16-bit words of memory in the scanner.
3. Display of secondary and primary electron back-scatter and sample current through alphacon-tube buffer storage.
4. Operator control of mechanical x,y stage movement using a light pen.
5. Operator control of spectrometer and beam scan with a light pen.
6. Display of results of elemental analysis area scan, either contoured or color contrasted.
7. Monitoring and correction of position of the beam with respect to sample.
8. Presentation of computer sharpened images.

Counting and Control Electronics (NIMS, CAMAC)

The counting channel electronics are standard NIMS modules and need not be commented on further. The CAMAC system electronics are computer controllable modules which execute all data collection and control functions upon computer command. Each module fits into a "station" and may be positioned in any of 24 "stations" of a "crate". Each module is of standard design like NIMS and each has its own computer controlled functions. All modules within multiple bins are operated by the only dedicated CAMAC device, the computer-mated controller. The modules in our system are of the following types: 1) Scalers, 24 bit binary, 2) Motor drivers, delayed pulse generators, 3) Drivers, 16 bit, 4) D/A converters, 8 bit, 5) Parallel input registers, 16 and 24 bit, 6) Clock, 7) Pre-set scaler, 24 bit, 8) Controller. We have more than one manufacturers' CAMAC modules, and have found them to be of high quality.

Computer System

The computer has 8K of 16-bit words and hardware arithmetic functions. Subsidiary storage is supplied by a 262K drum, accessible by the computer without interrupting computation. Input is via a card reader and teletype. Output is on an ASR 35 teletype. The drum was chosen as the subsidiary storage device because of its speed and reliability, though more expensive than an equivalent disc. Figure 1 diagrams automated components of the modular system.

PROGRAMMING

A modular system, where instruments could be added, changed, or deleted at will with all able to run simultaneously, necessitated writing a special operating system. Figure 2 diagrams the main parts of this program and Figure 3 diagrams the minimum 8K memory allocation used to drive the instruments.

Executive. The Executive program controls all instruments. The individual instrument program gives the Executive information. The Executive contains all drivers for scalers, motors, etc. A complete diffractometer scan can be accomplished, for example, by sending to the Executive as few as 10 data words. The Executive and Executive Service then operate the instrument and complete the scan including assigning of the data pages and filing of the completed pages on the drum.

Executive Service and Interrupt Processor. Executive Service routes to and from the Executive to start or shut-down operation of an instrument and to assign scaler data storage. It re-routes delayed requests from the Executive and instrument or data processing

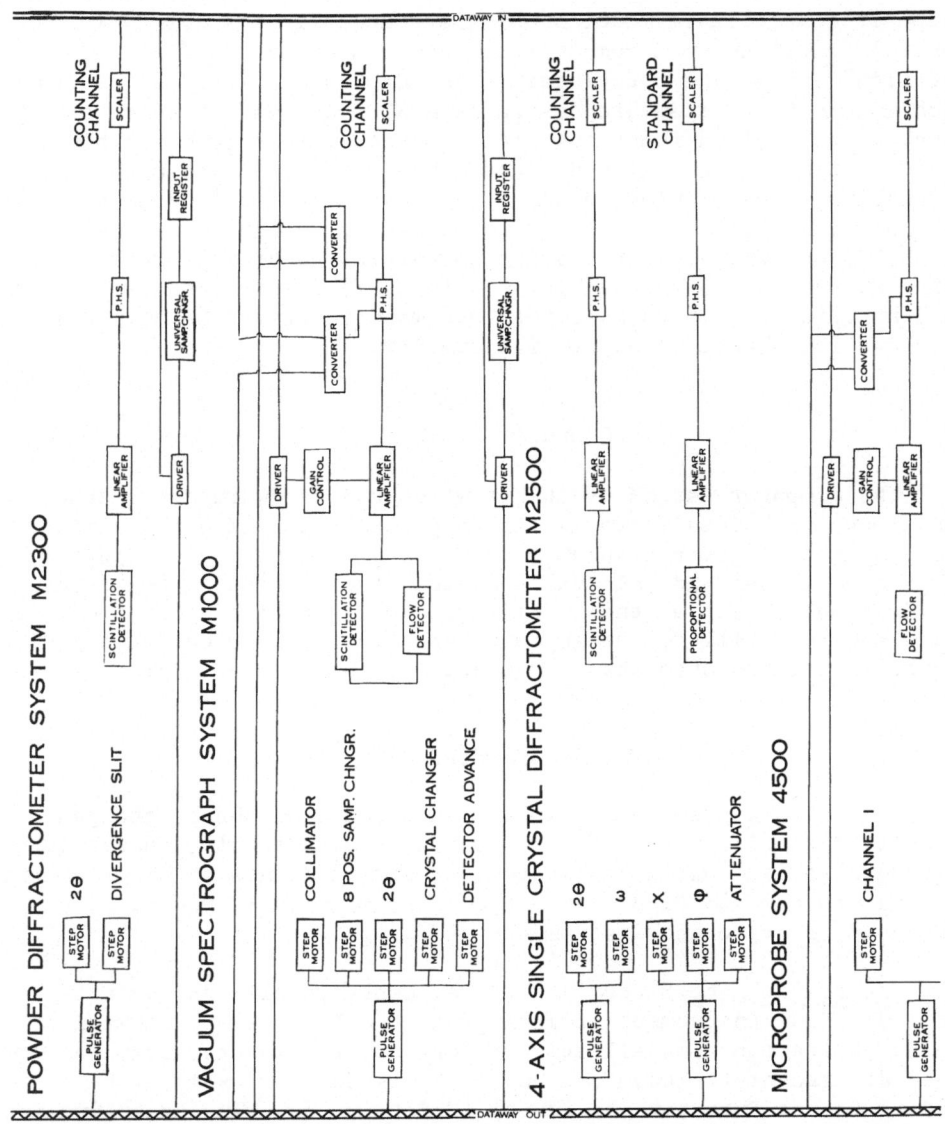

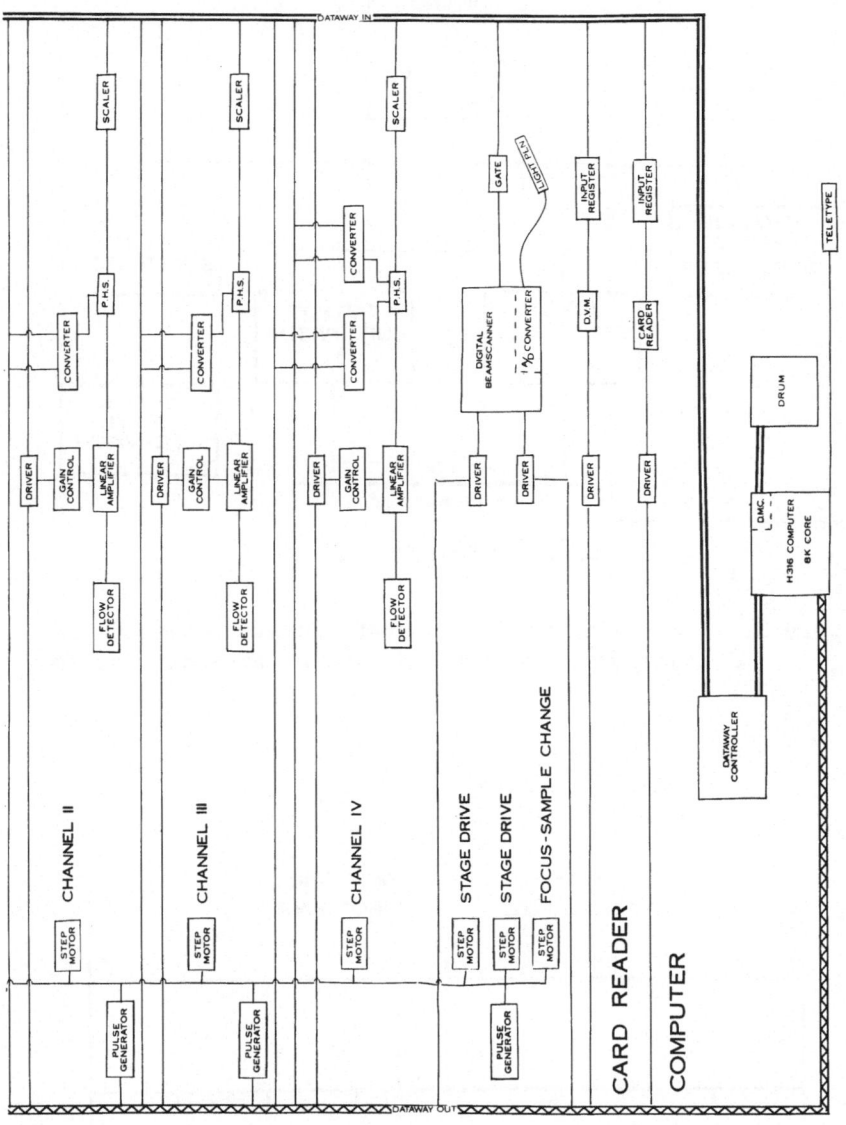

Figure 1.

Block diagram of automatically controlled modules of the x-ray analysis system.

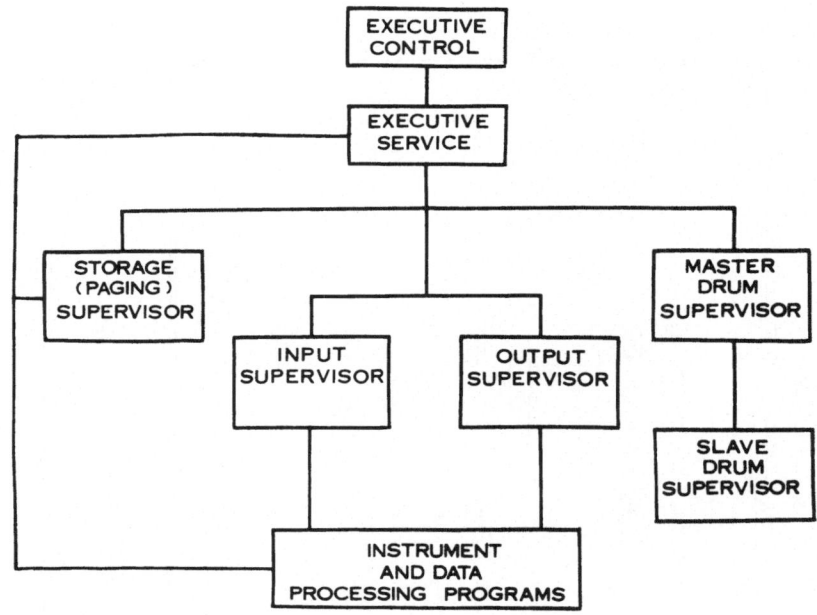

Figure 2.   Block diagram of resident operating system and
instrument control and data processing programs.

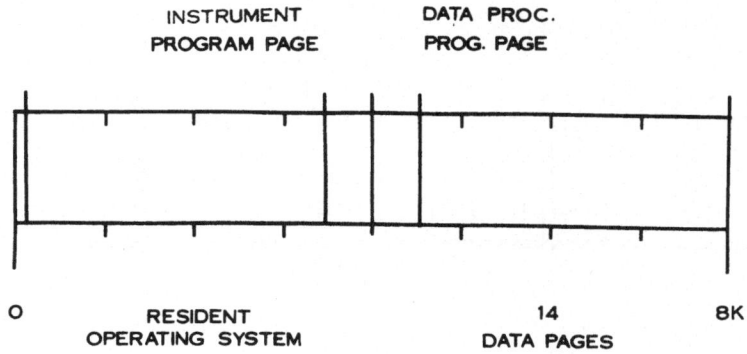

Figure 3.   Memory map of computer in instrument control
configuration.

programs. This program also enters the drum, input, and output supervisors on a regular schedule.

The Interrupt Processor services interrupts from a clock and the card reader. The card read operation is handled so that card reading may take place simultaneously with system operation. The clock interrupt normally causes routing to execute the necessary instrument function every 15 msec.

Storage and Master Drum Supervisors. The Storage and Master Drum Supervisors service all requests for data storage, data files, permanent files such as A.S.T.M. powder diffraction file and programs. The Storage Supervisor maintains directories of allocated storage for both data and programs. The Master Drum Supervisor executes all requests for data and programs to and from the drum. It also calls the Slave Drum Supervisor to bring in large data processing programs such as "MAGIC" (1) for the microprobe.

Input and Output Supervisors. The Input Supervisor handles all card input while the system is running, does decimal to binary conversions and initiates new instrument calls or instrument shutdowns if requested by the operator. The OUTPUT Supervisor queues all output from instruments and prints the data on a first completed - first served basis. It also does the necessary binary to decimal conversions of numerical data and prints trouble messages.

Slave Drum Supervisor. This program is called by the Master Drum Supervisor to execute programs independently of the operating system. It maintains all files on the drum and under operator control reads and writes programs and data in various modes.

Configurator. This routine allows the analyst to reconfigure the Executive, to add, delete or change an instrument function. The analyst for example may desire to add an instrument. He inserts the proper CAMAC module into a station of his choice and reconfigures the Executive to include that device. Reconfiguration normally requires less than one minute.

Instrument Programs. Instrument programs are in 512 word segments and may cause execution of an instrument or process data from the instrument. These programs reside on the drum and are called when needed. After execution they are either discarded or paged to a swapping area on the drum, from which they may be recalled for further use. The control of an instrument usually requires several segments for operation. For example, the basic four-axis diffractometer control program requires seven segments while the spectrograph requires three to five segments for each of three qualitative analysis modes. Table 1 lists major programs in the system.

Table 1. System Programs

Instrument Programs
  Powder Diffractometer
    Qualitative Analysis
      Normal Scan
      Specific Compound Scan
      Peak-Seek Scan
      Identification (2)
      Quantity Estimation
    Quantitative
      Data Collection
      Standard Preparation
      Multiple Regression
  Spectrograph
    Qualitative Analysis
      Normal Scan
      Specific Element Scan
      Peak-Seek Scan
      Identification
      Quantity Estimation
    Quantitative
      Data Collection
      Standard Preparation
      Lucas-Tooth & Pyne
      Method (3)
      Multiple Regression
  Four-Axis Diffractometer
    Data Preparation (Fortran)
    Data Reduction (Fortran)
      Slew-Scan Mode
      Slew-Search-Scan Mode
    Cell Refinement (Fortran)
    Listing
  Electron Microprobe
    Qualitative
      (See spectrograph)
    Quantitative
      Data Collection
      MAGIC (1) (Fortran)
    Scanner Control
      (incomplete)

System Programs
  Operating System
    Executive
    Executive Service
    Master Drum Supervisor
    Slave Drum Supervisor
    Input Supervisor
    Output Supervisor
    Configurator
  Languages
    Assembler
    Fortran IV(ASA)
    Basic
  Loaders
    System
    Relocatable Object
    Spectrographic Files
    Diffraction Files
  Maintenance
    Drum Test
    CPU Test
    Teletype Test
    Card Reader Test
    CAMAC Test

Files
  A.S.T.M. Diffraction
  Custom Diffraction
  A.S.T.M. Spectrographic

Miscellaneous
  A.S.T.M. Diffraction File
  Preparation (Fortran)

SYSTEM OPERATION

System time is divided into 15 msec intervals. The time
interval coupled with the mechanical gearing of the instruments
allows scan rates in degrees per minute. At the start of the time
interval, all active instrument functions are executed with corrections for small time errors between cycles. Following execution
of control function storage and programs, requests generated during
the executive cycle are serviced. Routine card reader, printer
and drum requests are then serviced along with waiting requests
for storage, etc. Finally, interrupted programs from the instrument
program area or data processing area are executed until the next
interrupt. Subsequent clock interruption causes return to the
Executive. Drum and card reader functions are more or less autono-
mous and may occur any time during the 15 msec cycle. Allocation
of 256 word data pages in core or on the drum is dynamic and
independent of instruments.

It is the combination of operating system functions and the
type of interfacing that allows any combination of eight instruments except electron microprobe to be attached to the system.
Only two fully equipped electron microprobes may run simultaneously.

INSTRUMENT OPERATION

Powder Diffractometer and Vacuum Spectrograph

For brevity we shall describe examples of the operation of the
powder diffractometer and spectrograph. Table 1 summarizes the
programmed modes of operation for the system.

To begin execution of the diffractometer or spectrograph the
instrument is called by the analyst with a single standard-format
card. This card is followed by cards punched with the control
parameters, i.e., scan range, scan speed, operation mode, etc.
Sample numbers may be assigned on separate cards or the computer
may assign them. File type, i.e., A.S.T.M., mineral and positive
or negative chemical type may be specified on cards.

Normal Scan Mode. The scan is automatically initiated and
data collected and stored for the whole 2θ interval. At the end
of scan, data processing programs retrieve the data and determine
and subtract the background. The programs extract peaks with
stripping of overlapped peaks where possible and determine
d-spacings or wave-lengths and intensities.

Spacings or wave-lengths and intensities are then used to
identify compounds or elements. The diffraction or spectrographic

file on the drum is accessed. A primary list of possible compounds
or elements is accumulated using the methods of Johnson and Vand (2),
allowing fairly wide margins of error in d or λ and I values. For
spectrographic identification there are usually three or four
possibilities for each element in the unknown. The correct elements
have never failed to appear in this list. The primary list of
diffraction compounds extracted from the A.S.T.M. file may be much
larger, ten or more per compound in the unknown. There is no
certainty that all compounds in the unknown will appear in the
diffraction list although they usually do.

The correct elements or compounds must be in the primary list.
Although elemental identification appears to be no problem, compound
identification is a problem. The difficulty in using the A.S.T.M.
diffraction file to identify compounds is the same when done automatically as when done manually. Therefore we have a special diffraction file, contiguous with the A.S.T.M. file, on the drum but
containing our own data. This file, separately accessible can have,
for example, data for end and intermediate members of a solid solution series, data for compounds which show orientation effects such
as exhibited by clay minerals, etc. Reliable identification then
becomes feasible. For final identification we use some of the tests
of Johnson and Vand with additional ones of our own. We are still
experimenting with final methods.

After identification, rough estimates are made of amounts of
each component. Impirical intensity ratios of pure compounds to a
reference standard, and linear absorption coefficients for fixed KV,
ma, etc. are stored as part of the custom diffraction file. These
intensities and absorption coefficients are used to estimate the
amounts. The estimation procedure is more complicated for spectrographic analysis.

Specific Elements or Compounds Scan. This qualitative mode
allows the operator to ask if specific elements or compounds are
present. Elements or compounds are specified on input cards. Before
the scan, the appropriate data are accessed from the drum. The
proper crystals, collimator, etc., are set automatically on the
spectrograph and the goniometer slews to measure specific peaks and
background. The peak is searched, analyzed and recorded and the
goniometer slews to the next peak.

Slew and Search Mode. This mode has the most limited application. It is begun with a fast scan. During the fast scan the
background is analyzed. When the count-rate above background is
statistically significant and when peak width and slope criteria
are met, the scan slows to a pre-selected speed until the peak is
passed. Fast scanning then begins again and continues until the
next peak is sensed. Identification proceeds as in the normal
scan mode.

The disadvantage of this mode is that only part of the information is available at any moment to make decisions about background, peak, etc. Sensitivity is lost and small quantities may go undetected. It is thus a mode for quick scanning of the major elements or compounds.

CONCLUSION

We have presented some of the features of a modular automatic x-ray system. The system was designed and built so that instruments and individual devices could be added or deleted easily. No single instrument on the system could be controlled comprehensively with a minimum computer configuration. By integrating the system we have lowered the computer cost per instrument and increased the capability over a single-computer-per-instrument system.

ACKNOWLEDGMENTS

The development program described above was made possible with the cooperation of Philips Electronic Instruments, 750 South Fulton Avenue, Mount Vernon, New York and the National Science Foundation (NSF 102-517).

REFERENCES

1. J. W. Colby, "MAGIC - A Computer Program for Quantitative Electron Microprobe Analysis," Advances in X-Ray Analysis, Vol. 11, p. 287-305, (1968).

2. G. G. Johnson and V. Vand, "Computerized Multiphase X-Ray Powder Diffraction Identification System," Advances in X-Ray Analysis, Vol. 11, p. 376-384, (1968).

3. H. J. Lucas-Tooth and E. C. Pyne, "Accurate Determination of Major Constituents by X-Ray Fluorecent Analysis in the Presence of Large Interelement Effects," Advances in X-Ray Analysis, Vol. 7, p. 523-541, (1964).

AN AUTOMATED ELECTRON MICROPROBE SYSTEM

F. Kunz and E. Eichen

Ford Motor Company

Dearborn, Michigan

H. Matthews and J. Francis

Canberra Industries

Meriden, Connecticut

ABSTRACT

An automated electron microprobe system using an on-line 8,000 word minicomputer has been designed. The general principles and need for such a system is discussed. The microprobe-computer interface hardware consists of computer addressable axis positioners, scalers, timer and a digital to analog converter. The axis positioners are used to position the spectrometers and specimen stage, whereas the digital to analog converter is used to position the electron beam in a 1000 X 1000 point matrix. The digital to analog converter is also used to drive a X-Y recorder for computer plotting of calibration curves, pulse amplitude distributions, X-ray line profiles and wavelength scans. System software is written in the interactive CLASS language which was designed specifically for laboratory instrument control. The language permits large computer programs to be chained through a minicomputer while passing variables from one function to the next. The operating software consists of automatic peak location, sequencing of data collection, statistical analysis of data, and correction of data for fluorescence, absorption and atomic number effects. An automated quantitative analysis of several copper-gold alloys and a computerized test for specimen homogeneity is described to demonstrate system operation. Future automation considerations are also discussed.

148

INTRODUCTION

Analytical instruments have in recent years become more and more complex as well as very expensive. This has required the need for operators who are scientists in their own right rather than the technicians who used to operate analytical instruments in the past. In order to obtain maximum utilization of this expensive equipment as well as minimizing the tedious manual operations, modern laboratory instruments are being automated. This automation pioneered with the X-ray fluorescence spectrometer (1) has been shown to effectively reduce errors, optimize data accumulation, reduce operator tedium and increase specimen output. Traditionally, computer automation as opposed to hard-wired automation, has been characterized by sophisticated design efforts in both hardware and software so as to make a system perform exactly as desired. To accomplish this task the user has had to become familiar with logic circuit and system design, computer organization, timing and interfacing, and machine language level software generation. Very few analytical scientists could devote the necessary time to the development of such a complex system, therefore, few have attempted to apply on-line computer automation to their application. Furthermore, many users have felt that automation, as typified by a few systems, prevented them from interacting with the instrument. With a unique hardware/software philosophy which will be described in this paper, most of the interfacing problems have been overcome, bringing the power of

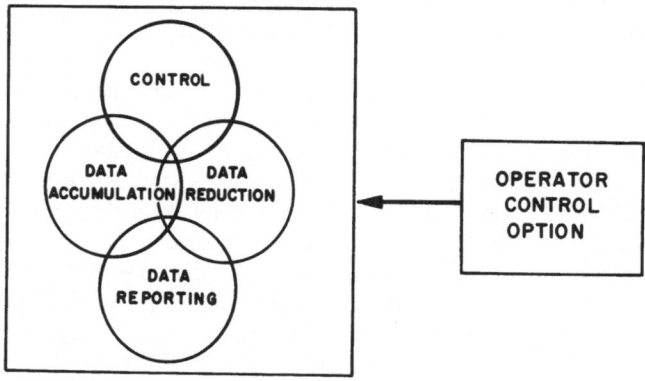

Figure 1. Functions to be Automated

the automated system to where it will truly aid the user, not just
in running the system, but in making the system do what the user
wishes. This automation philosophy can best be described by referring to Figure 1 which shows how the control, data accumulation,
data reduction and reporting functions are used to construct the
universal automation set while maintaining the external operator
control option. The automation of the electron microprobe is an
excellent example of a multiparameter instrument on which this
generalized philosophy may be applied. As will be described in
this paper, the system is designed such that the operator can
interact with the computer at any point in the four functions.

Since the development of the electron microprobe by
Castaing (2) the need for and use of computers to correct electron
microprobe data for interelement effects has been well established (3-4), however, this usage has been in most cases limited
to off-line computer processing of data for quantitative analysis.
It has been only recently that on-line minicomputers (8,000 words
core memory or less) have been used to provide microprobe control
in addition to data accumulation and reduction (5-7). This paper
describes our efforts to automate an electron microprobe using a
minicomputer to provide the necessary control, data accumulation,
data reduction, and outputting of corrected concentrations.

GENERAL PRINCIPLES

The electron microprobe employs from one to four semi-focusing
or focusing spectrometers to detect the energy and intensity of
X-rays which have been produced by electron beam bombardment of a
suitable specimen. The emitted X-rays provide information as to
the specimen's composition at the point of electron beam impact.
To accumulate the X-ray data from the irradiated area, the spectrometers must be accurately moved and positioned, the X-rays must
be detected, amplified, analyzed for pulse height and counted. In
applications which require semiquantitative or quantitative analysis,
the X-ray data is usually obtained from many points on the specimen
surface by deflecting the electron beam or translating the specimen
under the beam. The intensities from the specimen are then compared
to the intensities recorded from a selected set of known standards
and corrected for interelement effects such as fluorescence,
absorption, and atomic number effects. From these basic concepts
the requirements for an automated electron microprobe system can be
defined and are given in Table I. Not all controllable parameters
are included in this table, however, others will be considered in a
later section.

TABLE I

ELECTRON MICROPROBE PARAMETERS AND FUNCTIONS TO BE AUTOMATED

A. CONTROL
  Spectrometers
  Electron Beam
  Specimen Stage

B. DATA ACCUMULATION
  X-ray Peak Location
  Standard and Specimen Count Rates
  Tests for Statistical Significance
  Point Counting
  Line Profiling
  Matrix

C. DATA REDUCTION
  Correction for Deadtime
  Interelement Effects
    Fluorescence
    Absorption
    Atomic Number Effect
  Multiple Correction Methods

D. DATA REPORTING
  Concentration Reports
  Print Out or Suppression of Constants
  X-Y Plot
  Calibration Curve

SYSTEM HARDWARE

The system hardware necessary to achieve the desired degree of automation can best be described by referring to Figure 2 which shows a block diagram of the complete system. Of principal concern, however, is the module which interfaces the microprobe to the man/computer source of system control. This module was designed to meet the following requirements:

a. It must be sufficiently flexible to interface other instruments such as the X-ray fluorescence spectrometer or X-ray diffractometer to a computer.
b. It must provide manual control options as well as complete computer control.
c. It must minimize software requirements for routine programming.
d. It must be reliable in operation and data transmission.
e. It must allow man/instrument interaction at all times.

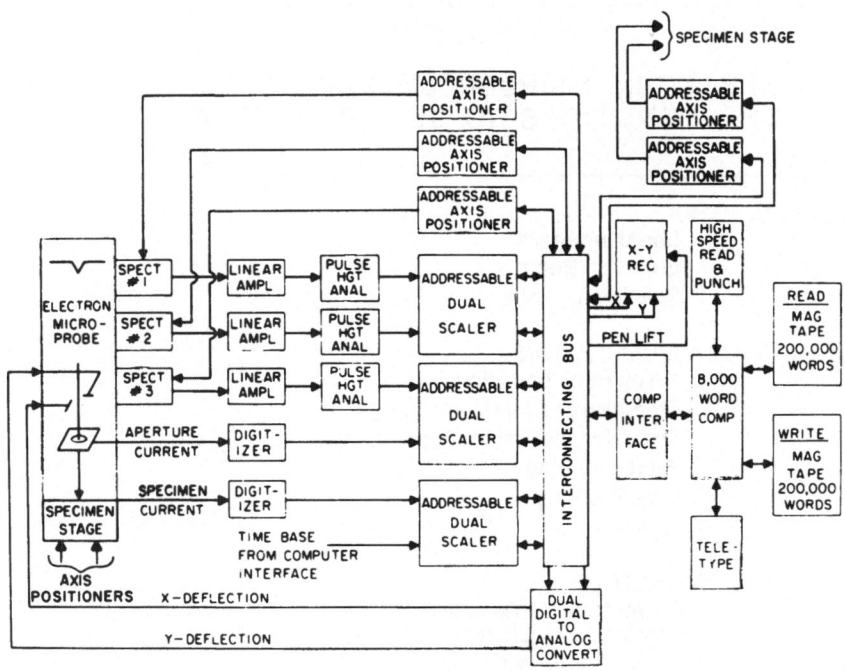

Figure 2. Block Diagram of an Automated Electron Microprobe

While data displays and manual controls provide interaction with the modules by the operator, each module is designed as a self interfacing, computer organized system component. Each module, for example, is given a unique two digit ADDRESS which allows it to be called upon anytime regardless of its locations in the system. When called upon, it will respond to a one digit OPERATION CODE which accompanies the ADDRESS to specify the operation desired for that module. For example, if the scaler which counts the X-ray intensity for spectrometer #1 is assigned ADDRESS 42, that scaler may be started automatically by generating within the computer or from the teletype keyboard the three digit number 42-2 where the -2 corresponds to the START operation code. Similarly, the axis positioner which moves and positions spectrometer #1 may be started by ADDRESS/ OPERATION CODE 11-2. In addition to its unique ADDRESS, each module will respond to a GENERIC address which it holds in common with other modules of the same class, i.e. all scalers for example. To start all scalers in the system simultaneously, the ADDRESS/OPERATION CODE 40-2 is generated where 40 is the GENERIC address for all scalers.

Spectrometer and specimen stage movement is accomplished with Slo-Syn Model HS25E stepping motors, an example of which is shown in Figure 3. The axis positioners which record the spectrometer position may be preset to one of eight different speeds with the

Figure 3.   Slo-Syn Stepping Motors Used to Drive Electron
Microprobe Spectrometers

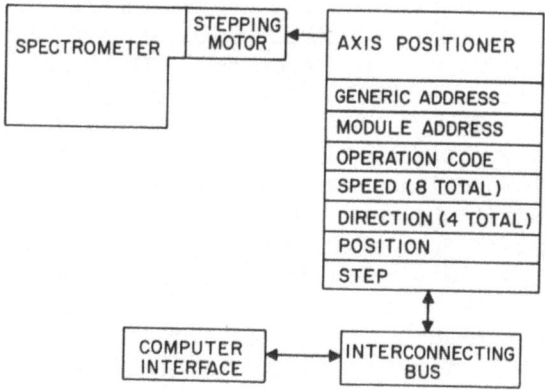

Figure 4.   Block Diagram of Electron Microprobe Spectrometer
Control

direction also preset to either negative or positive in an absolute or incremental mode. A block diagram of spectrometer control is given in Figure 4.

SYSTEM SOFTWARE

The software necessary to operate the electron microprobe and other time shared instruments has been written in CLASS* language which was designed specifically for operating and controlling laboratory instruments. The language was structured to meet the following requirements:

    a. The language must be capable of controlling instruments such as multichannel analyzers, X-ray diffractometers, and electron microprobes.

    b. It must offer full mathematical computation and data manipulation in a high-level user oriented language.

    c. The language must function in a minicomputer environment (4,000 words core memory minimum) and yet be expandable to function in a 16,000 word environment.

    d. The language must be designed for multi-user time sharing so that several instruments can share the same computer.

    e. It must provide for teletype keyboard, punched paper tape, DECtape, disc, casette or stored program input and output of programs and data.

    f. The language must provide for desk calculator type of usage while providing instrument control, data accumulation or data reduction by stored program.

The CLASS language can be viewed as a conversational and control language which is a cross between FORTRAN, FOCAL and BASIC. The most important feature is the ability to chain large programs together. Without this chaining feature, quantitative electron microprobe analysis combined with instrument control would be virtually impossible to accomplish in a 8,000 word minicomputer. The three commands which make chaining possible are DELETE (function name), CHAIN, and ERASE (variable name). The DELETE (function name) command permits the deletion of a named function (stored program), whereas the CHAIN command deletes all functions currently in core memory while retaining all variables until they are individually erased. The ERASE (variable name) command is used to erase variables after they are no longer needed. It is these commands which permit sections of programs to be brought into core memory, processed, erased and other sections read in. An example of how the three commands are used is given in Figure 5. The first entry defines PROG1 as a function which sets X, Y and A to various values and then types the value of A on the teletype. The PROG1 entry executes the function named PROG1.

*CLASS is an acronym for Canberra Laboratory Automation Software System.

```
DEFINE    Prog I <
          X = I.64
          Y = I.32
          A = X + Y
          Type A, ! >


PROG I
DELETE    Prog I

DEFINE    Prog 2 <

          C = X + Y
          Type C, ! >


PROG 2
DELETE    Prog 2
ERASE     C
TYPE      A, !


ERASE     All
CHAIN
```

Figure 5. Example of Program Chaining in CLASS Language

After execution the PROG1 function is deleted from core memory by
the DELETE PROG1 command. A second function PROG2 is defined and
then executed by the PROG2 command. PROG2 is then deleted by the
DELETE PROG2 command. The variable C is erased by the ERASE C
command and then the variable A is typed out on the teletype. At
this point the variables X, Y and A are still in core memory and can
be retrieved by a simple TYPE command. The command ERASE ALL will
then erase all variables in core memory and the CHAIN command then
deletes all programs and instructions in core memory. The CLASS
commands, except the ERASE, DELETE and CHAIN commands, are summarized in Figure 6.

OPERATING SYSTEM

The operating system consists of the CLASS language interpreter
and CIMSS* which when combined permits the storage of programs, data
and constants on DECtape and essentially extends the 8,000 words of
core memory by 400,000 words. The system is divided into two categories which are program and file type commands. Once a program is
written and defined as a function (DEFINE PROG1) a SAVE (function
name) command is given which will automatically write the program
on a selected DECtape. If the program is to be used within another
defined function, a DECLARE (function name) command is given which

*CIMSS is an acronym for Canberra Industries Mass Storage System.

COMMAND	EXACT FORMAT
TYPE	TYPE (any desired information)
TYPE (for text)	TYPE "desired text"
SET	SET NAME = (any desired expression)
DO	DO (count) any desired expression
DEFINE	DEFINE any desired expression
IF	IF (relationship (variable I, variable2)) then do
GOTO	GOTO name
LOOP	NAME:
STOP	STOP
DIMENS (for array)	DIMENS name (number, number, number)
ARG	ARG (number of argument)
READ	READ (device) variable I, variable 2, ...
DUMP	DUMP user defined function name
EDIT	EDIT user function name
PRESET	PRESET (device number) data, data, data
READOUT	READOUT (device number) data
ZEROSTART	ZEROSTART (device number)
HALT	HALT (device number)
RESET	RESET (device number)

Figure 6. Summary of CLASS System Commands

then "tags" the function as being a program on DECtape. When invoked, the program is automatically retrieved from DECtape, loaded into core memory and executed. The second group of commands are used to build files of data, constants or programs. A file is created by typing the CREATE (file name, record length) command, which will set-up on DECtape a named file with a finite record length. When the file is to be updated the APPEND (array name): file name (record length) command is given which adds the array name contents to the end of the named file. To move data from core memory to DECtape and from DECtape to core memory the COPY commands are used. A summary of the operating system commands is given in Figure 7.

APPLICATIONS

As an example of a typical automated analysis, several National Bureau of Standards SRM 482, copper-gold electron microprobe standards were chosen. The results of these analyses are shown in Figure 8. The analyses consisted of an initial dialogue entry from the teletype which includes the accelerating voltage, elements of interest, X-ray line to be used and crystal-spectrometer selection. A typical entry would be typed in as CU;KA;LIF;1 where the first entry is the chemical symbol, KA is the X-ray line selected and LIF;1 is the crystal-spectrometer selection. After the dialogue entry the computer calculates the module addresses based upon the

PROGRAMS

$U	(DECtape Number)
SAVE	Program I, Program 2, ...
LOAD	Program I, Program 2, ...
DECLARE	Program I, Program 2, ...
RUN	Program I

FILES

CREATE	File I (Record Size)
APPEND	Array : File I (Core ➔ Tape)
OPEN	File I, File 2, ...
COPY	Array : File I (R) (Same as Write)
COPY	File I (R) : Array (Same as Read)
REMOVE	File I, File 2, ...

Figure 7. Summary of CLASS and CIMSS Operating System Commands

	SRM 482 TRUE CONCENTRATION	CALCULATED (ZAF) PROBE RATIO, k	AUTOMATED* PROBE RATIO, k	CORRECTED** WEIGHT FRACTION
GOLD	0.801	0.749	0.751	0.801
COPPER	0.198	0.229	0.219	0.194
GOLD	0.603	0.531	0.525	0.594
COPPER	0.396	0.430	0.420	0.386
GOLD	0.401	0.333	0.331	0.399
COPPER	0.599	0.630	0.633	0.602
GOLD	0.201	0.158	0.147	0.187
COPPER	0.798	0.816	0.819	0.802

* Corrected for Deadtime and Background
** Corrected for Fluorescence, Absorption and Atomic Number Effect

Figure 8. Results of an Automated Analysis of Several Copper-Gold NBS Alloys

spectrometer selection and obtains from a table of wavelengths the correct wavelength for the element and X-ray line combination. The spectrometer is then moved by the computer to a position 0.0050 angstroms higher than the theoretical value. The computer is programmed so that all wavelengths are approached from the same direction, however, if the next higher or lower atomic number standard is to be used to accumulate background information, an option can be selected.

X-Ray Peak Location

Two methods have been used to locate the X-ray peak of interest. The first method consists of automatically stepping across the peak in 0.0002 angstrom steps and recording the intensity and wavelength at each point. The maximum intensity is determined by the computer and its corresponding wavelength is then taken to be the peak wavelength. The second method consists of recording the X-ray intensity at three equally spaced wavelengths and calculating the peak wavelength using a Lorentzian, Gaussian or Parabola model. The equations are:

Lorentzian

$$\lambda_{peak} = \lambda_1 + \frac{\Delta\lambda}{2} + \frac{\Delta\lambda}{1 + \dfrac{I_1(I_2 - I_3)}{I_3(I_2 - I_1)}} \tag{1}$$

Gaussian

$$\lambda_{peak} = \lambda_1 + \frac{\Delta\lambda}{2} + \frac{\Delta\lambda}{\left[1 + \dfrac{\ln\left(\dfrac{I_2}{I_3}\right)}{\ln\left(\dfrac{I_2}{I_1}\right)}\right]} \tag{2}$$

Parabola

$$\lambda_{peak} = \lambda_1 + \frac{\Delta\lambda}{2}\left(\frac{3a + b}{a + b}\right) \tag{3}$$

where

$$I_n = \text{intensity for wavelength, n}$$

$$\lambda_n = \text{wavelength, n}$$

$$a = I_2 - I_1$$

$$b = I_2 - I_3$$

Experimental data indicates that by using the above equations to calculate the peak wavelength, the calculated peak is usually displaced from the absolute peak which was located by stepping across the peak. Such a displacement has been previously reported by Merrill and DuMond (8) in their work on precision measurement of X-ray lines using a two crystal spectrometer, however, the application of their techniques to electron microprobe X-ray peaks requires further investigation.

Data Accumulation and Reduction

Once the X-ray peak is located for each standard, a total of five counts are accumulated for the high background, then low background and peak. The counting times for the individual counts are determined by presetting the scaler used to monitor the digitized aperture current and the computer program is written so that each time the X-ray count is read into memory, a deadtime correction is automatically applied and the counts subjected to statistical evaluation (9). The system is so designed that probe ratios are immediately printed out on the teletype. The probe ratios are then entered into a chained version of MAGIC (10) which corrects the ratios for fluorescence, absorption and atomic number effects. The output from this program for one NBS alloy is shown in Figure 9.

```
AU-CU NBS                    MARCH 29, 1971

     SUBMITTED BY: F. KUNZ

          MEAN CHEMICAL COMPOSITION ON TWO
          SIGMA LIMITS BASED ON  4 ANALYSES


                          WEIGHT              ATOMIC
          ELEMENT         PERCENT             PERCENT

            AU      80.101 +- .245        57.169 +- .062
            CU      19.359 +- .042        42.830 +- .062


     ------------------------------------------------------------

          MEAN INTENSITY RATIOS AND TWO SIGMA LIMITS
          ELEMENT                  K
            AU                   .7589 +- .00216
            CU                   .2188 +- .00042


     ACCELERATING VOLTAGE     30.0 KEV

     X-RAY EMERGENCE ANGLE    36.5 DEGREES



  JOB FINISHED
```

Figure 9. Teletype Output from MAGIC for the 80% Copper and 20% Gold NBS Electron Microprobe Standard

True Concentration to Probe Ratio

Also shown in Figure 9 are the results of entering the true concentrations into a ZAF (11) computer program which is used to calculate probe ratios before starting an analysis. It is obvious that the exact probe ratios cannot be calculated due to the iterative nature of the FAZ (12) type of correction, however, the results do indicate the direction in which the experimental probe ratios are to be corrected. By using this program, the microprobe operator can to a certain degree estimate what interelement effects can be expected before attempting the actual analysis.

Determination of Specimen Homogeneity

Homogeneity characterization of material by electron microprobe technique has been the subject of much work (13-17) and most recently by Varshneya and Cooper (17). Each of the reported methods requires large data sets (normally X-ray intensities) to be accumulated and subjected to statistical tests for the final homogeneity characterization. The National Bureau of Standards method (15) necessitates the accumulation of X-ray data from each point in a 100 x 100 matrix from several areas on the specimen. The Midwest Probe Users Group method (16) is to accumulate 20 X-ray intensities from one area on the specimen and then from 20 random areas, refocusing the specimen after each translation. The method that we report is basically that of Varshneya and Cooper (17) which has been altered to include beam deflection rather than specimen translation.

The method consists of automatically stepping the electron beam over the specimen surface and accumulating X-ray data at each point of dwell. As shown in Figure 10, the data is collected from each point on a line in one direction and then from the same points in the opposite direction. Normally, a total of 10 intensities are recorded from each point and used to construct a 10 x 10 matrix. On application of the two variables of classification, analysis of variance (18) to this data, the F-ratio for the column means reflects the variation due to inhomogeneity and Rowland circle effects, whereas, the F-ratio for the row means reflects the variation due to time dependent drift (17), i.e. filament drift. The variation due to Rowland circle effects is shown in Figure 11 which is a plot of X-ray image magnification versus the F-ratio for the column means. Experimental data indicates that if the magnification is maintained at 500X or greater, the F (column) ratio will reflect the inhomogeneity contribution to the ratio, that is, the contribution from Rowland circle effects are negligible. The method is then usable to characterize material for its homogeneity. It is estimated that the computerized system reduces the homogeneity

NO. OF COLUMNS = No. OF POSITIONS = K

NO. OF ROWS   = No. OF REPETITIONS AT
                 THE SAME POSITION = n

X_{11}	X_{12}	X_{13}	------	X_{1K}	$\bar{X}_{1j}$
X_{21}	X_{22}	X_{23}	------	X_{2K}	$\bar{X}_{2j}$
X_{31}	X_{32}	X_{33}	------	X_{3K}	$\bar{X}_{3j}$
—	—	—	------	—	—
—	—	—	------	—	—
X_{n1}	X_{n2}	X_{n3}	------	X_{nK}	$\bar{X}_{nj}$

MEAN $\bar{X}_{i1}$ $\bar{X}_{i2}$ $\bar{X}_{i3}$ ------ $\bar{X}_{iK}$ | $\bar{X}_{ij}$

THE SEQUENCE OF DATA COLLECTION IS:

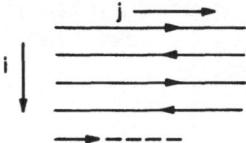

Figure 10. Sequence of Data Accumulation for Automated
Electron Microprobe Homogeneity Analysis

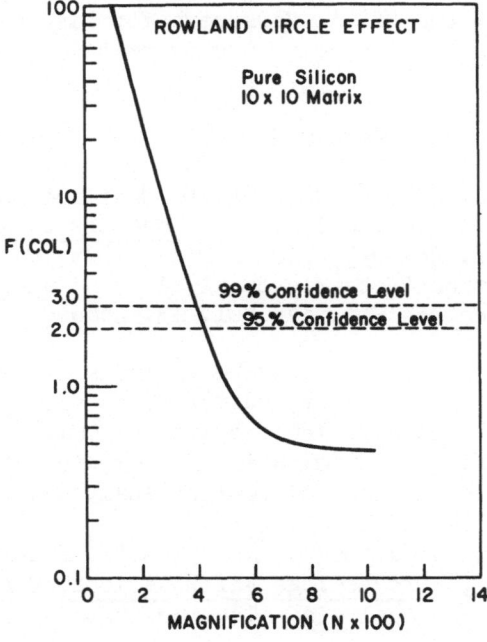

Figure 11. Plot of X-Ray Image Magnification Versus
F (Column) Ratios

analytical time from 9 hours to just under 20 minutes for a 10 second per point dwell time.

SUMMARY

An automated electron microprobe system has been described which uses an on-line minicomputer to control the instrument, accumulate and correct the data, and print out the corrected concentrations. Future automation considerations include an interactive nondispersive/dispersive analytical capability, filament adjustment, specimen stage Z-axis translation, condensor lens adjust and electron beam focusing. Monitoring of housekeeping functions such as water flow, vacuum, and high voltage power supplies are being considered. At the present it is not known whether any of the above parameters or systems can be automated, however, with an on-line computer system such as described within this paper, many methods may be investigated in the quest for a completely automated system.

ACKNOWLEDGEMENTS

The authors would like to thank S. Piner for his assistance in software development, C. Kelly for her valuable discussions concerning X-ray peak location and Dr. A. Varshneya for his contributions to the statistical F-test for specimen homogeneity.

REFERENCES

1. D. C. Miller, "Results Obtained with the Modified Norelco Autrometer," in W. M. Mueller, Editor, Advances in X-ray Analysis, Vol. 1, p. 283-295, Plenum Press (1957).
2. R. Castaing, "Electron Probe Microanalysis," Advances in Electronics and Electron Physics, Vol. 13, Academic Press, New York (1960).
3. J. W. Colby, "Quantitative Microprobe Analysis of Thin Dielectric Films," in F. Vratny, Editor, Thin Film Dielectrics, The Electrochemical Society, Incorporated (1968).
4. D. R. Beaman, "A Critical Examination of Computer Programs Used in Quantitative Electron Microprobe Analysis," Anal. Chem. Vol. 42, p. 1540-1568, 1970.
5. E. Eichen, F. Kunz, H. Matthews, "An Automated Electron Microprobe Analyzer," Proceedings Fifth National Conference on Electron Microprobe Analysis, Electron Probe Analysis Society of America, (1970).
6. R. Wolf and A. Saffir, "A Computerized Electron Microprobe," ibid., Paper No. 6.

7. W. F. Chambers, "A Computer Assisted Microprobe Laboratory,"
    ibid., Paper No. 7.

8. J. Merrill and J. DuMond, "Precision Measurement of L X-Ray
    Wavelengths and Line Widths for $74 \leq Z \leq 95$ and Their
    Interpretation in Terms of Nuclear Perturbations," Annals
    of Phy. Vol. 14, p. 166-228, 1961.

9. R. Evans, The Atomic Nucleus, McGraw-Hill Book Company,
    New York, N.Y. (1955).

10. J. Colby, MAGIC: Version III, personal communication.

11. J. Philibert and R. Tixier, "Some Problems with Quantitative
    Electron Probe Microanalysis," in K. F. J. Heinrich, Editor,
    Quantitative Electron Probe Microanalysis, p. 14, NBS Special
    Publ. 298 (1968).

12. _____, ibid., p. 15.

13. H. Yakowitz, D. Veith, and R. Michaelis, "Homogeneity
    Characterization of NBS Spectrometric Standards III: White
    Cast Iron and Stainless Steel Powder Compact," NBS Misc.
    Publ. 260-12, (1966).

14. H. Yakowitz, D. Veith, K. Heinrich and R. Michaelis,
    Homogeneity Characterization of NBS Spectrometric Standards II:
    Cartridge Brass and Low-Alloy Steel," NBS Misc. Publ. 260-10,
    (1968).

15. H. Yakowitz, R. Michaelis and D. Veith, "Homogeneity
    Characterization of NBS Spectrometric Standards IV: Prepa-
    ration and Microprobe Characterization of W-20% Mo Alloy
    Fabricated by Powder Metallurgical Methods," Advances in X-Ray
    Analysis, Vol. 12, p. 418, Plenum Press (1969).

16. Minutes of Midwest Probe Users Group, dated March 14, 1968.

17. A. Varshneya and A. Cooper, "Inhomogeneities and Iron
    Diffusion in a Thailand Tekite," J. Geophys. Res., Vol. 27,
    p. 6845, 1969.

18. W. Dixon and F. Massey, Introduction to Statistical Analysis,
    (second edition), McGraw-Hill Book Company, New York, (1957).

RAPID QUANTITATIVE ANALYSIS BY X-RAY SPECTROMETRY

Robert D. Giauque and Joseph M. Jaklevic

Lawrence Berkeley Laboratory, Univ. of Calif.

Berkeley, California 94720

ABSTRACT

An x-ray fluorescence analysis method applicable to the case of fluorescent spectra excited with monoenergetic x-rays has been developed. The technique employs a minimum number of calibration steps using single element thin film standards and depends upon theoretical cross sections and fluorescent yield data to interpolate from element to element. The samples are treated as thin films and corrections for absorption effects are easily determined. Enhancement effects, if not negligible, are minimized by sample dilution techniques or by selective excitation.

INTRODUCTION

The ability to obtain rapid multielement x-ray fluorescence spectra with energy dispersive x-ray spectrometers makes it necessary to have available convenient and accurate methods to convert the spectral intensity data to meaningful elemental concentrations. Although most present methods employ extensive standard calibration samples (1,2), there have been attempts to formulate more general methods based on the calculation of absolute fluorescent intensities (3,4). The present method is similar to that of reference (3), but is derived as an extension of the technique for the determination of film thickness using monochromatic excitation such as described by Liebhafsky et al. (5).

In order for the technique to be applicable, it must be possible to prepare the sample in the form of a thin homogeneous specimen of uniform thickness. If monochromatic primary radiation

164

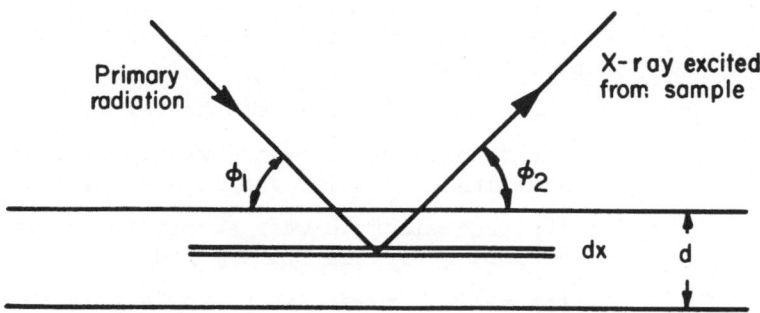

Figure 1. Schematic diagram of x-ray method.

is used to excite the characteristic x-rays from the sample, and
if the critical thickness for the characteristic x-ray of energy
E_j has not been reached, the problem may be described as shown in
figure 1. If enhancement effects can be neglected, the probability
of exciting the K x-ray of energy E_j from element j in the thickness dx and detecting the x-ray equals the product of three probabilities which may be written as follows:

$$P_1 = G_1 \exp(-\mu_1 \csc \phi_1 \rho x) \tag{1}$$

which is the probability that the primary radiation will reach a
depth x

$$P_2 = \tau_j \rho_j \csc \phi_1 [1 - (1/J_K)] \omega_K f \, dx \tag{2}$$

being the probability that element j will absorb the primary radiation in the thickness dx and emit a K x-ray of energy E_j

$$P_3 = G_2 \exp(-\mu_2 \csc \phi_2 \rho x) A_b \in \tag{3}$$

which in turn is the probability that the K x-ray of energy E_j will
reach the detector and be detected where

G_1 and G_2 are geometry factors

μ_1 and μ_2 are the total mass absorption coefficients
                     (cm^2/gm) of the sample for the primary and the
                     characteristic, E_j, radiations

ϕ_1 and ϕ_2 are the angles formed by the primary and the emergent characteristic radiations with the sample surface

ρ is the density (gm/cm^3) of the sample

ρ_j is the density of element j within the sample considering the entire sample distribution

τ_j is the photoelectric mass absorption coefficient of element j for the primary radiation

J_K is the ratio between the photoelectric mass absorption coefficients at the top and the bottom of the K absorption edge; $[1 - (1/J_K)]$ is the fraction of photoelectric events which occur in the K shell

ω_K is the fluorescent yield for the K x-rays from element j

f   is the fraction of the K x-rays of energy E_j with respect to the total K x-rays emitted

A_b is the absorption for the air path, if present, plus the absorption of the detector window, both of which can be calculated

$\in$ is the detector efficiency for x-rays of energy E_j.

The values of μ_1, μ_2, τ_i, J_K, ω_K, and f can be obtained from literature (references (6,7,8,9)). The values of $\in$ can be measured or calculated from the detector thickness. In the case of L x-rays the appropriate substitutions for J_L, ω_L, etc. can be made.

If the intensity of the primary radiation is I_o, then the x-ray intensity due to element j at a depth x is:

$$dI = I_o \ G \ \csc \phi_1 \ \tau_j \ \rho_j [1 - (1/J_K)] \ \omega_K \ f \ A_b \in [\exp(-a\rho d)] \ dx \qquad (4)$$

where

$$G = G_1 \ G_2 \ \text{and} \qquad\qquad (4a)$$

$$a = \mu_1 \ \csc \phi_1 + \mu_2 \ \csc \phi_2 \qquad\qquad (4b)$$

Integrating over the sample thickness d and multiplying the numerator and the denominator by d one obtains:

$$I = I_o \ G \ \csc \phi_1 \ \tau_j \ \rho_j \ d[1 - (1/J_K)] \ \omega_K \ f \ A_b \in [1 - \exp(-a\rho d)]/a\rho d$$
$$(5)$$

The value ρ_j d is the concentration (gm/cm^2) of element j which is to be determined. The quantity $exp(-a\rho d)$ is the product of the total absorption of the primary radiation ($\mu_1 csc \phi_1$) and the total absorption of the characteristic radiation ($\mu_2 csc \phi_2$) in the total sample thickness. This quantity can be determined by measuring the relative intensity with and without the specimen of the K x-ray of a pure target of element j, located at a position adjacent to the back of the thin specimen.

Since the values of τ, J_K, ω_K, and f are known for each element, the value of $I_o G csc \phi_1$ may be determined from a single element thin film standard for which the absorption effects are negligible. (The value $[1 - exp(-a\rho d)]$ essentially equals the value $a\rho d$ when the value $a\rho d$ is small.) Thus, from a thin film standard, at constant primary radiation intensity and constant geometry, equation (5) will reduce to

$$I = CK_j \rho_j d[1 - exp(-a\rho d)]/a\rho d \qquad (6)$$

where

$$C = I_o csc \phi_1 \text{ and}$$

$$K_j = \text{constant for the element j}$$

In effect, theoretical values of relative excitation and detection efficiencies for various x-ray lines are calculated. The values are calibrated in units of gm/cm^2 using a convenient element for which a thin film standard is available.

Thus, using the above procedure, one has an excellent method for determining the concentration of many elements after simply calibrating from a single element thin film standard. Other than determining the intensity of the individual element x-ray lines, only absorption measurement corrections need be made.

DISCUSSION

Either characteristic x-ray tubes (10) or radioisotope source-target assemblies (11) are used to provide primary characteristic K x-rays to be employed in the analysis. The primary radiation is treated as two monochromatic x-ray beams corresponding to the Kα and Kβ x-rays. The Kα/Kβ intensity ratio can be easily measured from the spectrum obtained by scattering the primary radiation from light element material.

The experimental x-ray fluorescent yield data commonly are
several percent in disagreement with the theoretical fluorescent
yield data (7,8) used in the calculations. To obtain higher
accuracy in analysis, thin film standards for each of the elements
of interest may be prepared. The standards should be thin enough
such that absorption effects are negligible. Using this procedure,
the values of K_j of equation (6) can be determined empirically for
the individual elements of interest.

Simultaneous absorption measurements of many elements can
often be made by using a multielement uniform reference material.
However, in order to use this procedure, enhancement effects
between the sample and the reference material must be negligible.
For thin samples with similar element concentrations, repetitive
absorption measurements often need not be made if the thin samples
prepared are of approximately equal total mass concentrations
(gm/cm^2). For some thin samples, particularly when using higher
energy x-rays for analyses, absorption effects are essentially
negligible.

Since matrix enhancement effects are not included in the
analysis, it may be necessary to select the primary radiation of
energy below the absorption edges of major constituents to eliminate their enhancement effects. Since monoenergetic excitation
is used, it is often desirable to use several different incident
energies to obtain high excitation efficiencies.

EXPERIMENTAL

The data were obtained using a low-background guard-ring
detector (12) with pulsed light feedback electronics (13). The
total resolution of the system, FWHM, was approximately 200 eV at
5.9 keV. Excitation was provided by a transmission x-ray tube (10)
with ϕ_1 and ϕ_2 equal to approximately 45°. Due to wide variations
in count rates, corrections were applied for both pile-up rejection
in the amplifier and the multichannel analyzer dead time. Variations in the tube intensity were taken into account by monitoring
the total integrated charge over the duration of the run.

Standardization

The thin film standards were prepared by evaporation of the
elements onto thin aluminum films. Typically, the evaporated
layers were in the mass range of 50 to 150 $\mu g/cm^2$. The aluminum
films weighed approximately 800 $\mu g/cm^2$. Thin film standards may
also be prepared by precipitation of the elements on thin
filters (14). The total area used in standardization and analysis
was approximately 2.5 cm^2.

Sample Preparation

Biological Specimen. Weighed samples are either lyophilized
or oven dried, reweighed, and pulverized. Thin pellets are pressed
at 15,000 p.s.i. and weighed. Typically, the pellets are 1" in
diameter, weigh 150 mg, and are approximately 0.03 cm thick. If
the prepared specimen is not self-binding, cellulose powder is
used as a binding material.

Rock, Glass, and Pottery Specimen. Samples must be pulverized
such that particle size effects are negligible for the analyses to
be made. For many analyses, grinding the sample to pass through a
325 mesh screen (less than 44 microns) is sufficient. Weighed
samples are mixed with weighed amounts of cellulose powder, thin
pellets are pressed and weighed. The ratio of cellulose powder to
sample specimen should be large enough to minimize possible enhancement effects.

Alloy Specimen and Solutions. A weighed amount of the sample
is put into solution, a fraction of the solution is absorbed on a
known weight of cellulose powder and is dried at 80°C (15). The
mixture is weighed, pulverized, and a portion of the mixture is
pressed into a thin pellet and weighed. To obtain higher sensitivity in analysis one may precipitate the element of interest
along with a carrier and collect the element on a filter (14).

Air Filter Specimen. Since absorption corrections are not
necessary for many elements collected on air filters, the contents
of these elements may simply be determined by measuring the intensities of the characteristic x-rays. This is also often true when
preconcentration procedures have resulted in the material being
collected on a filter paper.

RESULTS

The following are some results obtained using a molybdenum
transmission x-ray tube to provide the primary radiation. The
tube was operated at 42 KV, and 250 μamp. Only one standard,
101 μg Cu/cm^2, was used for calibration. The fluorescent yield
values used in the analyses were the theoretical values (7,8).

Table I shows a comparison of the results obtained by x-ray
fluorescence and neutron activation on a pottery specimen. 50 mg
of pulverized pottery were mixed with 150 mg of cellulose powder
and a 1" diameter pellet was pressed. Total analysis time,
including absorption measurements, was 1 hour. The precisions
listed are for one standard deviation.

Figure 2 shows the spectrum of a dried plant specimen obtained
in 30 minutes. 75 mg of the plant specimen were mixed with 75 mg
of cellulose powder and a 1" diameter pellet was pressed. Table
II shows a comparison of the results.

Table I
Analysis of Pottery

	X-ray Fluorescence	Neutron Activation
Ti	.76% ± .02	.78% ± .03
Cr	114 ppm ± 6	115 ppm ± 4
Mn	47 ppm ± 5	40.9 ppm ± 0.5
Fe	1.05% ± .01	1.017% ± .012
Ni	301 ppm ± 6	279 ppm ± 20
Cu	55 ppm ± 2	60 ppm ± 8
Zn	60 ppm ± 2	59 ppm ± 8
Ga	40 ppm ± 2	44 ppm ± 5
As	29 ppm ± 2	30.8 ppm ± 2.2
Rb	58 ppm ± 2	70.0 ppm ± 6.3
Sr	123 ppm ± 3	145 ppm ± 22
Pb	31 ppm ± 2	——

Table II
Analysis of Plant Specimen

	X-ray Fluorescence	Neutron Activation
Ti	121 ppm ± 5	< .01%
Cr	26 ppm ± 1	23.8 ppm ± 0.9
Mn	60 ppm ± 2	49.3 ppm ± 1.4
Fe	.186% ± .002	.201% ± .006
Ni	8 ppm ± 1	13.8 ppm ± 3.0
Cu	21 ppm ± 1	——
Zn	80 ppm ± 1	84 ppm ± 8
Br	48 ppm ± 1	42 ppm ± 1
Rb	7 ppm ± 1	7.0 ppm ± 1.4
Sr	97 ppm ± 2	236 ppm ± 66
Pb	206 ppm ± 3	——

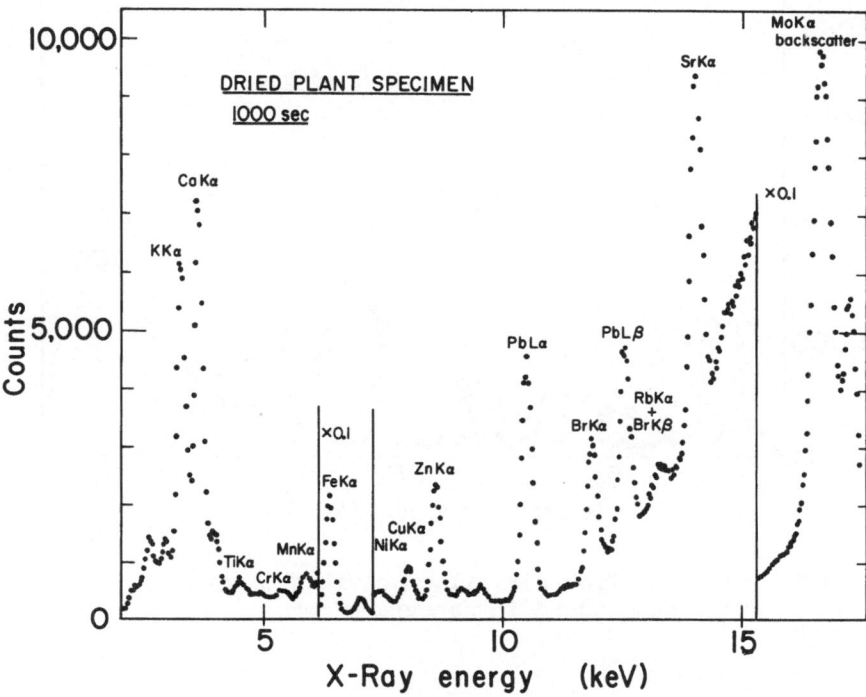

Figure 2. Spectrum from dried plant specimen.

Table III shows the results obtained on NBS Steel 121A. After dissolving 125 mg of the sample, 5 mg of the alloy in solution was absorbed on 750 mg of cellulose powder and dried at 80°C. The mixture was weighed, pulverized, and 150 mg was pressed into a 1" diameter pellet. The pellet contained only 1 mg of the original sample. Total analysis time was 45 minutes. The results are slightly high, principally due to enhancement effects caused by scattering of the primary radiation within the pellet.

Table III

Analysis of NBS Steel 121A

	X-ray Fluorescence	NBS
Ti	0.37% ± .05	0.36%
Cr	19.50% ± .25	18.69%
Mn	1.43% ± .04	1.28%
Ni	10.92% ± .13	10.58%

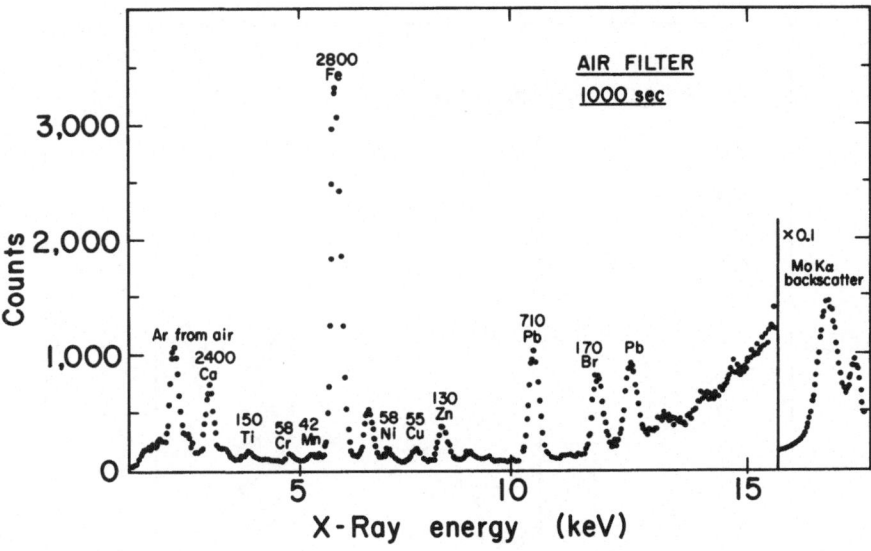

Figure 3.   Spectrum from air filter of mass 3 mg/cm^2.   Concentrations listed in nanograms/cm^2.

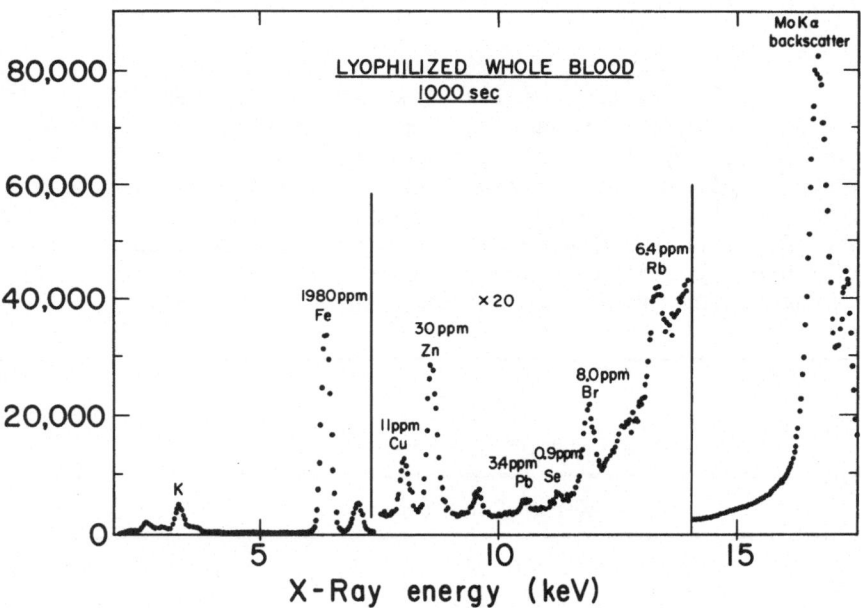

Figure 4.   Spectrum from lyophilized whole blood specimen.   Concentrations listed are for the lyophilized specimen.   This preparation gave a concentration factor of five.

Figure 3 shows the spectrum and results obtained on an air pollution filter of mass 3 mg/cm^2. The spectrum was taken in 20 minutes and the results are reported in nanograms/cm^2.

Figure 4 shows the spectrum and results from a human whole blood specimen which had been lyophilized, pulverized, and of which 150 mg were pressed into a 1" diameter pellet. Since the concentration factor obtained by lyophilizing the specimen was five, the concentrations of the elements in the original specimen are one-fifth the values shown. The spectrum was taken in 30 minutes. The content of lead in this specimen is several times the normal value.

Table IV shows a comparison of the calculated and measured values of relative excitation and detection efficiencies, K_j, of various x-ray lines. The calculated values were determined using theoretical fluorescent yield data. The measured values were determined from the average of three thin films of approximate mass 50, 100, and 150 μg/cm^2.

The calibration method used has been successfully applied to trace concentrations in light element matrices. When employed in conjunction with low background detector systems and x-ray tube excitation, it can provide rapid and accurate multielement analyses down to less than 1.0 ppm. For a more detailed discussion of detection limits and sensitivities, see reference (12).

CONCLUSION

The technique of using a single element thin film standard to calibrate for the analysis of many elements is an excellent procedure if the following conditions are met. A thin uniform sample must be prepared. Enhancement effects must be negligible. The critical thicknesses for the x-ray lines to be used in analyses

Table IV
Relative Excitation and Detection Efficiencies

Line	Calculated	Determined
CrKα	.380	.371 ± .007
NiKα	.887	.906 ± .009
CuKα	1.000	1.000 ± .014
SeKα	1.89	1.78 ± .04
PbLα	.815	.789 ± .016

must not have been reached. Absorption correction measurements,
if not negligible, must be made.

The analyses performed provide excellent agreement with
results obtained by other means.

ACKNOWLEDGMENTS

The authors wish to express gratitude to Mr. Fred Goulding
for contributions to this work and to members of the Nuclear
Chemistry Instrumentation Group for developing and supplying much
of the equipment used. The authors are grateful to Mr. Gordon
Steers and Mr. Karl Scheu for preparing the thin film standards.
The authors also wish to thank Dr. Frank Asaro for performing the
neutron activation analyses, and Mr. William Searles for assistance
in assembling and maintaining the equipment.

REFERENCES

1. L. S. Birks, X-ray Spectrochemical Analysis, p. 71-79, John
    Wiley and Sons, Inc. (1969).

2. H. A. Liebhafsky, H. G. Pfeiffer, E. H. Winslow, and P. D.
    Zemany, X-ray Absorption and Emission in Analytical Chemistry,
    p. 179-191, John Wiley and Sons, Inc. (1960).

3. T. Shiraiwa and N. Fujino, "Theoretical Formulas for Film
    Thickness Measurement by Means of Fluorescence X-rays," in
    C. S. Barrett, J. B. Newkirk, and G. R. Mallett, Editors,
    Advances in X-ray Analysis, Vol. 12, p. 446-455, Plenum Press
    (1969).

4. R. Barbier, "Analytical Chemistry - On a Method of Absolute
    Quantitative Elemental Analysis by X-ray Fluorescence
    Spectrometry," Proceedings Academy Science, Paris, Vol. 270,
    p. 1581-1584 (May 11, 1970).

5. Reference (2), p. 153-158.

6. W. H. McMaster, N. K. Del Grande, J. H. Mallett, and J. H.
    Hubbell, "Compilation of X-ray Cross Sections," UCRL-50174,
    Section II, Revision I, Lawrence Livermore Laboratory, Univer-
    sity of California, Livermore, California (1969).

7. R. W. Fink, R. C. Jopson, N. Mark, and C. D. Swift, "Atomic
    Fluorescence Yields," Rev. Mod. Phys. $\underline{38}$, 513-540 (1966).

8. E. J. McGuire, "Atomic L-Shell Coster-Kronig, Auger, and
    Radiative Rates and Fluorescence Yields for Na-Th," Phys. Rev.
    $\underline{A3}$, 587-594 (1971).

9. C. M. Lederer, J. M. Hollander, and I. Perlman, <u>Table of Isotopes</u>, 6th Edition, p. 570-571, John Wiley and Sons, Inc. (1967).

10. J. M. Jaklevic, R. D. Giauque, D. F. Malone, and W. L. Searles, "Small X-ray Tubes for Energy Dispersive Spectrometers," <u>Advances in X-ray Analysis</u>, Vol. 15, Plenum Press (to be published).

11. R. D. Giauque, "A Radioisotope Source-Target Assembly for X-ray Spectrometry," Anal. Chem. <u>40</u>, 2075-2077 (1968).

12. F. S. Goulding, J. M. Jaklevic, B. V. Jarrett, and D. A. Landis, "Detector Background and Sensitivity of X-ray Fluorescence Spectrometers," <u>Advances in X-ray Analysis</u>, Vol. 15, Plenum Press (to be published).

13. D. A. Landis, F. S. Goulding, R. H. Pehl, and J. T. Walton, "Pulsed Feedback Techniques for Semiconductor Detector Radiation Spectrometers," <u>IEEE Transactions on Nuclear Science</u>, Vol. NS-18, No. 1, p. 115-124 (1971).

14. C. L. Luke, "Determination of Trace Elements in Inorganic and Organic Materials by X-ray Fluorescence Spectrometry," Anal. Chim. Acta <u>41</u>, 239-250 (1968).

15. M. E. Salmon, "An Improved X-ray Fluorescence Method for the Analysis of Museum Objects," in B. L. Henke, J. B. Newkirk, and G. R. Mallett, Editors, <u>Advances in X-ray Analysis</u>, Vol. 13, p. 94-104, Plenum Press (1970).

ON THE METHOD OF VARIABLE TAKE-OFF ANGLE FOR QUANTITATIVE X-RAY FLUORESCENCE ANALYSIS (XRFA)

H. Ebel and M. F. Ebel

Institut für Angewandte Physik

Technische Hochschule Wien, Vienna, Austria

ABSTRACT

Two years ago a paper (1) on a new method of X-ray fluorescence analysis was presented. This method uses a variable take-off geometry and permits a quantitative analysis without calibration curves or the knowledge of the primary X-ray spectrum(2). As references either chemically analysed samples or pure elements are used. In case that only primary excitation exists, under the aspect mentioned above, the method is to be regarded as "absolute". For secondary excitation a simple parabolic approximation has been introduced, which requires a calibration. This paper gives an outline on the progress obtained during the last two years.

SECONDARY EXCITATION

Shiraiwa and Fujino(3) presented an exact treatment of secondary excitation. Using this equation we calculated this effect for combinations of different elements and variable geometry. We succeeded in finding an efficient approximation(4) for a take-off angle $\beta = 90^\circ$. The characteristic radiation of the element j excites the radiation of the element i which is to be measured. The relation r_{is} between secondary excited i-radiation of the sample and i-radiation of the pure element i is expressed by equ.1.

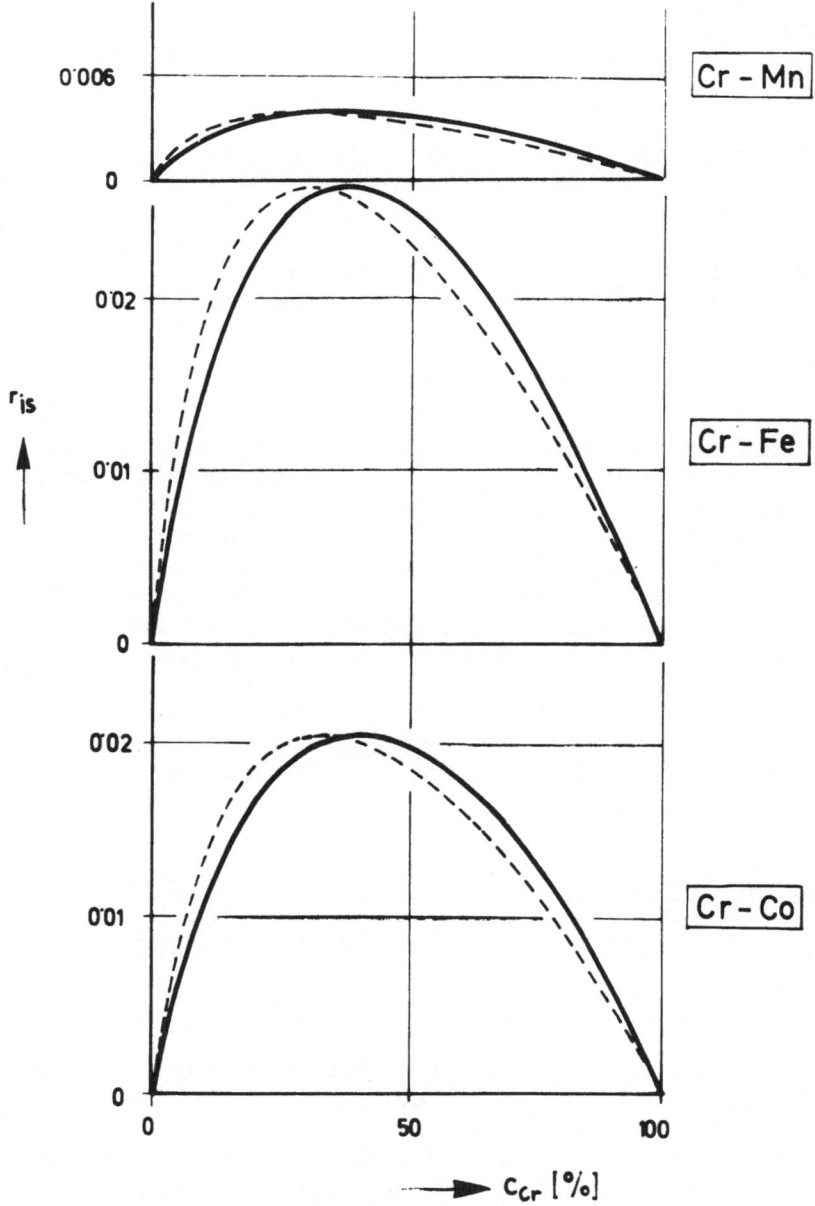

Figure 1. Calculated ratio r_{is} of secondary excited
Cr-Kα-radiation from different binary chromium alloys
and Cr-Kα-radiation from pure chromium (α =0°, β=90°,
20 kV). The full lines mean exact treatment and the
dotted lines mean approximation.

$$r_{is(\beta=90°)} = \frac{c_i \cdot \frac{\mu_{ii}}{\rho_i}}{\sum_k c_k \cdot \frac{\mu_{ik}}{\rho_k}} \cdot \frac{1}{5} \cdot \sum_j \left\{ \frac{c_j \cdot \frac{\mu_{ji}}{\rho_j}}{\sum_k c_k \cdot \frac{\mu_{ik}}{\rho_k}} \cdot \frac{S_{Kj}-1}{S_{Kj}} \cdot W_j \cdot p_j \cdot \frac{\frac{\mu_{jj}}{\rho_j}}{\frac{\mu_{ii}}{\rho_i}} \right\} \tag{1}$$

The utility of this approximation is shown by
figure 1. The diagrams point out the dependence of
secondary excited Cr-Kα -radiation on concentration for
the binary systems Cr-Mn, Cr-Fe and Cr-Co and β =90°.The
results gained by exact treatment deviate from the
results of the approximation for a maximum amount of 0,2%.

TERTIARY EXCITATION

Based on the equation for tertiary excitation(5)
we calculated the contributions for the system Cr-Fe-Ni
and varied geometries. Tertiary excitation here is produced in succession Ni-K-, Fe-K-, Cr-K-radiation. In
figure 2 the result of the most unfavorable case of
tertiary excited contributions in the ternary system
Fe-Cr-Ni in dependence of β is presented. It is clear
that for our method this contribution is not to be taken
into account. The induced error is generally below 0,2%,
since for quantitative analysis the ratio of tertiary
radiation to pure element radiation is used, which for
our example is an order of magnitude smaller than
n_{it}/n_{ip}.

EXCITATION CAUSED BY SCATTERED WHITE RADIATION -

COHERENT SCATTERING OF CHARACTERISTIC RADIATION IN THE

TAKE-OFF DIRECTION

By modifying the equation for secondary excitation
(3) a theoretical mode of treatment is obtained, to describe the increase of characteristic radiation by
scattering-processes(6). The results of calculations for
different pure elements, are depicted in figure 3. An
X-ray tube voltage of 30 kV was used. It is interesting
that the contribution to the primary excited intensity
ranges from 1% to 3% at β =90°.

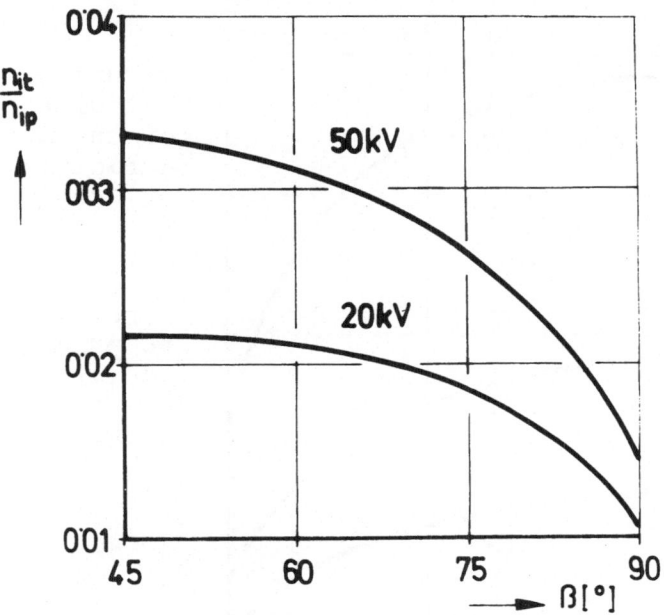

Figure 2. Calculated ratio n_{it}/n_{ip} of tertiary (n_{it}) and primary (n_{ip}) excited Cr-Kα -radiation of a ternary chromium alloy ($c_{Fe} = c_{Ni} = 45$ wt%, $c_{Cr} = 10$ wt%, $\alpha + \beta = 90°$).

EXCITATION OF CHARACTERISTIC X-RAYS BY PHOTOELECTRONS

According to their kinetic energy photoelectrons from the K-shell ejected by X-rays with an energy $E > 2E_K$ are able to excite K-radiation themselves (E_K means the K-ionisation energy). The Cu-Kα -radiation excited by this mechanism has been calculated in dependence of X-ray tube voltage(7).The result is shown in figure 4. The contribution does not exceed 0,3% of primary excited intensity.

QUALITY OF SAMPLE SURFACE

Experiments with flat samples of equal composition but different surface-quality were done. It was performed that a roughness of 5 μm shows no measurable deviations from the X-ray intensities of surfaces with a roughness of 0,2 μm. A flatness to 5 μm can be achieved in most cases.

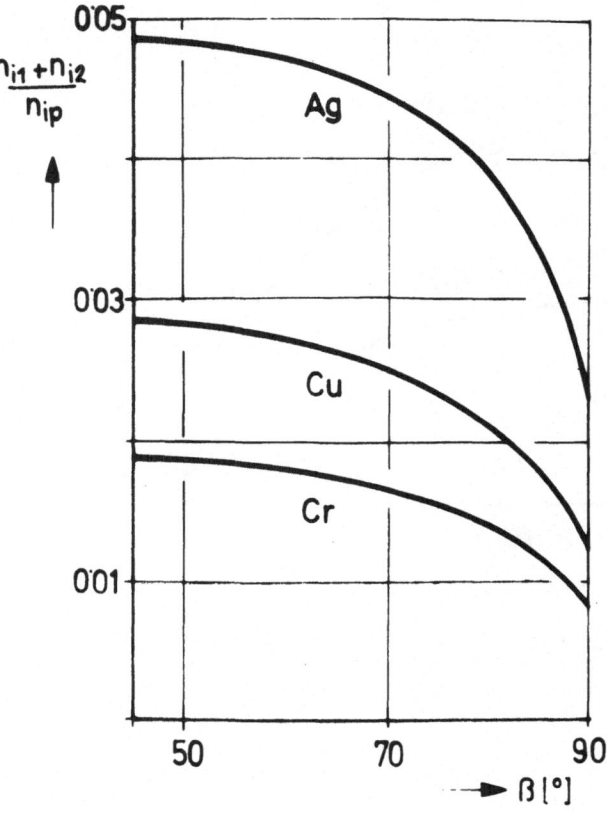

Figure 3.
Calculated ratio
of characteristic
K-radiation excited
by scattered white
radiation (n_{i1})
and that scattered
in the take-off
direction (n_{i2}),
to primary excited
radiation for
different pure
elements ($\alpha + \beta = 90^\circ$,
30 kV).

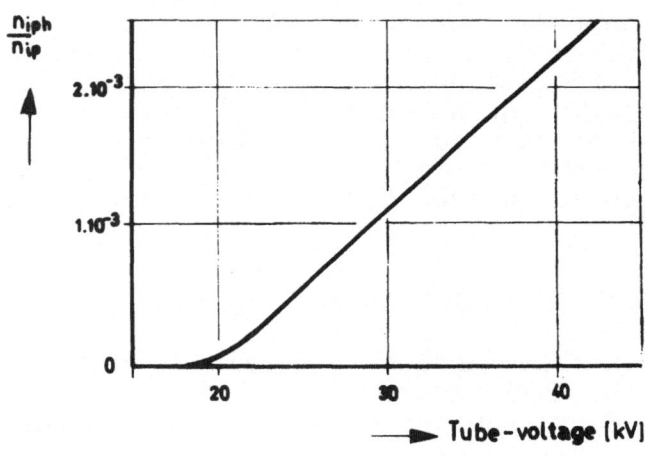

Figure 4.
Calculated ratio
of Cu-Kα -radiation
excited by photo-
electrons (n_{iph})
and by white radi-
ation (n_{ip}) for
pure Cu ($\alpha = \beta = 45^\circ$).

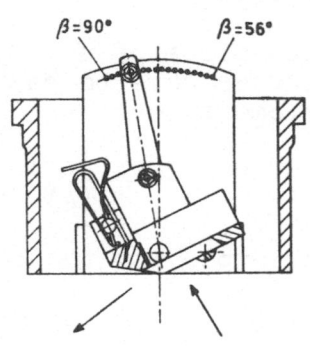

Figure **5.** Sample holder for the variable take-off angle technique ($\alpha + \beta$ = const). The arrows indicate the directions of incidence and take-off.

SAMPLE-HOLDER

Figure 5 provides a schematic drawing of a sample-holder, developed for a Philips X-ray spectrometer. A variation of take-off angle β , in gradual steps of 2^0, can be realized. The accuracy in β is better than $\pm 0,1^0$. As material for the sample-holder plastic was used.

EVALUATION

The intensity ratio r_i (sample radiation - reference radiation) for pure element standards can be written

$$r_{i\,(\beta=90^\circ)} = \frac{c_i \cdot \frac{\mu_{ji}}{\rho_i}}{\sum_k c_k \cdot \frac{\mu_{ik}}{\rho_k}} \cdot \left[1 + \frac{1}{5} \cdot \sum_j \left\{ \frac{c_j \cdot \frac{\mu_{ji}}{\rho_i}}{\sum_k c_k \cdot \frac{\mu_{jk}}{\rho_k}} \cdot \frac{S_{Kj}-1}{S_{Kj}} \cdot W_j \cdot P_j \cdot \frac{\frac{\mu_{ji}}{\rho_j}}{\frac{\mu_{ii}}{\rho_i}} \right\} \right] \qquad (2)$$

Values of r_i are measured for different take-off angles, and $r_i(\beta \equiv 90^0)$ is found by extrapolation of this curve towards $\beta = 90^0$. $r_i(\beta)$ is represented by the following equation:

$$r_i(\beta) = K_1 + K_2 \cdot ctg\,\beta + K_3 \cdot ctg^2\beta + K_4 \cdot ctg^3\beta \qquad (3)$$

In equ.3 $r_i(\beta=90^o)$ is equivalent to K_1. The coefficients K_1, K_2, K_3 and K_4 are calculated from measured $r_i(\beta)$-values applying a least square method(8).

The sample which is to be analysed consists of n elements. Therefore, n different $r_i(\beta)$-curves have to be measured. Using the K_1-values for each element together with equ.2 and the following equ.4

$$\sum_k c_k = 1 \tag{4}$$

the unknown concentrations are determined, employing the iteration method according to Criss and Birks(2).

EXAMPLE

Analysis of a Cr-Ni-Steel

	chem. analysis	XRFA
Cr	18,6 wt%	18,8 wt%
Fe	71,3	70,8
Ni	10,1	10,4

C-content smaller than 0,06 wt%.

ACKNOWLEDGMENTS

This research work has been sponsored by "Österreichischer Forschungsrat, Forschungsprojekt Nr.775".

REFERENCES

1. H. Ebel, "Quantitative X-ray fluorescence analysis with variable take-off angle," Advances in X-ray Analysis, Vol. 13, 68-79 (1970).

2. J. W. Criss and L. S. Birks, "Calculation methods for fluorescent X-ray spectrometry," Anal.Chem. 40, 1080-1086 (1968).

3. T. Shiraiwa and N. Fujino, "Theoretical Calculation of fluorescent X-ray intensities in fluorescent X-ray spectrochemical analysis," Jap.J.Appl.Phys. 5, 886-899 (1966).

4. H. Ebel, J. Derdau and G. Pollai,"Die Sekundäranre-
    gung bei der quantitativen Röntgenfluoreszenzanalyse
    mit variablem Beobachtungswinkel," Spectrochim. Acta
    26 B, 237 - 259 (1971).

5. G. Pollai and H. Ebel, "Die Tertiäranregung in der
    quantitativen Röntgenfluoreszenzanalyse," Spectro-
    chim. Acta B (in press).

6. G. Pollai, M. Mantler and H. Ebel, "Der Einfluß der
    Streuung auf die Röntgenfluoreszenzintensität,"
    Spectrochim. Acta B (in press).

7. H. Ebel, F. Landler and H. Dirschmid, " Zur Anregung
    charakteristischer Röntgenstrahlung durch Photoelek-
    tronen," Zeitschr. f. Naturforschg. 26a, 927-928
    (1971).

8. H. Ebel and M. Mayr, "Praktische Anwendung der Rönt-
    genfluoreszenzanalyse mit variabler Strahlengeome-
    trie," Spectrochim. Acta 26 B, 291-299 (1971).

SYMBOLS

r_i ratio of i-countrates (e.g. Cu-Kα) of the
           unknown and the reference sample.

c_i concentration of the element i (e.g. Cu) in the
           unknown sample, wt%.

S_{Ki} K-photoabsorption jump of the element i
           (e.g. S_{KCu} =10).

μ_{ii}/ρ_i mass-absorption coefficient of the element i for
           characteristic i-radiation (e.g. Cu-Kα in Cu,
           $\mu_{CuK\alpha,Cu}/\rho_{Cu}$=51,2 cm^{2}g^{-1}). In literature sometimes
           only μ_{ii} is denoted.

ρ_i density of the element i (e.g. ρ_{Cu} =8,93 gcm^{-3}).

μ_{ij}/ρ_j mass-absorption coefficient of the element j for
           characteristic i-radiation (e.g. Cu-Kα in Cr,
           $\mu_{CuK\alpha,Cr}/\rho_{Cr}$ = 253,1 cm^{2}g^{-1}). In literature sometimes
           only μ_{ij} is denoted.

W_i efficiency of the element i (e.g. W for Cu-Kα-
           radiation is 0,48).

p_i transition probability (e.g. for electrons from
           higher levels into the K-level of Cu, $p_{CuK\alpha}$=0,6,
           $p_{CuK\alpha2}$=0,3, $p_{CuK\beta}$ =0,1).

DISCUSSION

R. JENKINS (Philips Electronic Instruments): Have you considered
the effect of Compton shift on the effective exciting spectrum?
Is it not so that this would give a shift of almost 10 keV at 60
keV over the range of your take-off angles? I assume that this
would introduce a significant error in the excitation of the
shorter characteristic wavelengths.

H. EBEL: Fluorescence radiation is mainly caused by primary excitation. There exists, of course, a contribution from Compton-
scattered white radiation, but for low- and middle-Z elements, the
effective wavelength is in a range where Compton shift can be
neglected. High-Z elements in a light matrix need a consideration
of this effect, and we shall calculate the error which arises from
a neglection of the wavelength shift due to Compton scattering.

AN AUTOMATIC X-RAY ANALYTICAL INSTRUMENT FOR THE CHEMICAL LABORATORY

A.L. Gray

Applied Research Laboratories Limited

Wingate Road, LUTON, BEDFORDSHIRE, ENGLAND

ABSTRACT

Although isotope excited non-dispersive X-ray analysis has shown considerable promise for use in portable instruments and in on-line control applications, its use as a general analytical tool in the chemical laboratory has not developed as well as expected. Much of this is thought to be due to lack of flexibility in the instruments available and to difficulties with their use by unskilled operators. Wide element coverage is essential and many analytical problems arise at the light element end of the periodic table. While developing on-line techniques for light elements it was found that a measuring head using a very compact source-sample-detector geometry was well suited to a number of routine analytical laboratory problems, and enabled high precision to be obtained with minimum sample preparation for rapid single element analysis. Sensitivities in a light matrix from 100 p.p.m. for silicon to 10 p.p.m. for heavy elements were attainable. An instrument specifically for chemical laboratory use was therefore designed to accept solid, liquid and powder samples for single element analysis of silicon and all heavier elements. The instrument can give direct numerical readout of concentration and achieves high stability and precision. In its basic form it is equipped with a range of isotope excitation sources but will also accept a small X-ray tube which has been specially developed for it to provide higher sensitivity and precision. The measuring head may be removed for remote use, and multiple sample changing facilities with printout of results can be added, permitting unattended automatic routine analysis.

INTRODUCTION

Instrumental methods have now become firmly established in
the routine chemical analysis laboratory and there seems to be no
inherent reason why an isotope excited non-dispersive X.R.F.
instrument should not be acceptable alongside other single element
analytical techniques such as atomic absorbtion. Attempts have
been made to apply existing, primarily portable, isotope
instruments in the laboratory but these have often not proved
flexible enough or simple enough in use for operators unfamiliar
with nucleonic instruments. In designing a new instrument the
needs and practices of analytical laboratories were therefore
considered paramount. The resulting instrument appears perfectly
at home in the analytical laboratory and has shown that this
simple technique can have considerable analytical advantages.

INSTRUMENT DESCRIPTION

The original impetus to produce an instrument for this
market arose during the course of the development of an on-line
measuring head for light elements down to phosphorus. To achieve
an adequate performance at these energies an annular source was
adopted, fitting closely round the window of a small, thin window,
proportional counter, which gave a high geometrical efficiency
together with a very short air path. Experimental work in the
laboratory showed that this geometry was very suitable for use with
thin window sample cups for the analysis of liquids and loose
powders. Considerable interest in this head, for the routine
analysis of solutions and light matrices, began to develop and a
target specification was drawn up for an instrument aimed at the
chemical laboratory. The main features of this specification,
which it was felt had not been offered simultaneously before,
were as follows:-

(1) Capable of precisions of the order of 0.1%, with a
stability of the same order, and short analysis times.

(2) Able to analyse samples in liquid, solid and powder
form with simple sample insertion without vacuum system
complications.

(3) Numerical readout with facilities for printout.
Capable of being set to give direct readout in percent
or p.p.m. where the calibration curve is linear.

(4) Minimum number of controls.

(5) Safe for use by radiation unclassified workers.

(6) Tolerant of an unstable environment.

(7) Able to accept an automatic multiple sample presenter.

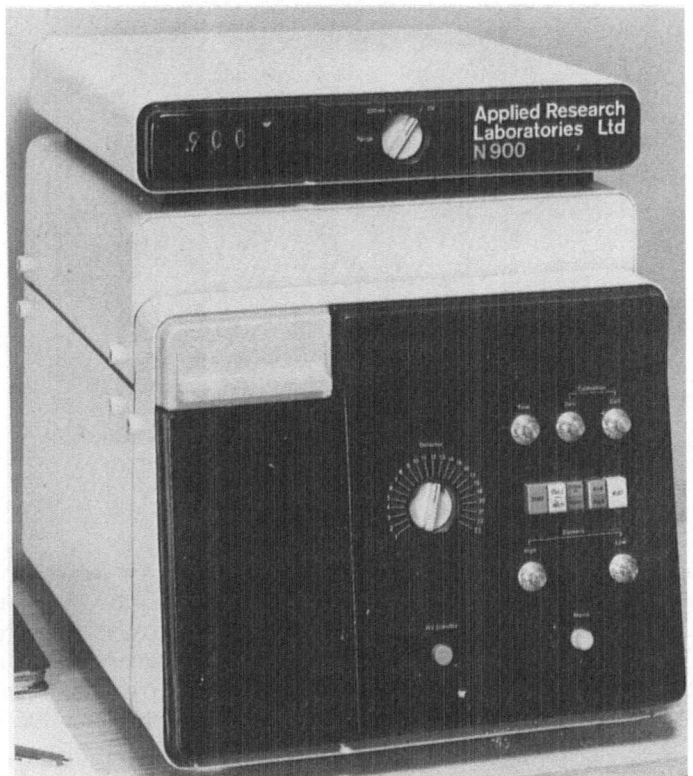

Figure 1 - The N900 Non-Dispersive X-Ray
Fluorescence Analyser.

The instrument that resulted from the design process that
followed is shown in figure 1. Here the general layout of the
instrument can be seen. The central switch sets the detector high
tension to one of a number of fixed values specified for each
detector supplied and covers the whole range of proportional and
scintillation counters. The row of buttons to the right of this
select the mode of operation and above these are the duodials for
setting the measurement time and for the two calibration controls,
zero bias and gain. Below the buttons are the upper and lower
thresholds for the single channel analyser. Readout is displayed
on the digital voltmeter at the top left and immediately below
this is the sample tray. This sample tray may be seen open in
figure 2 with a sample cup in position.

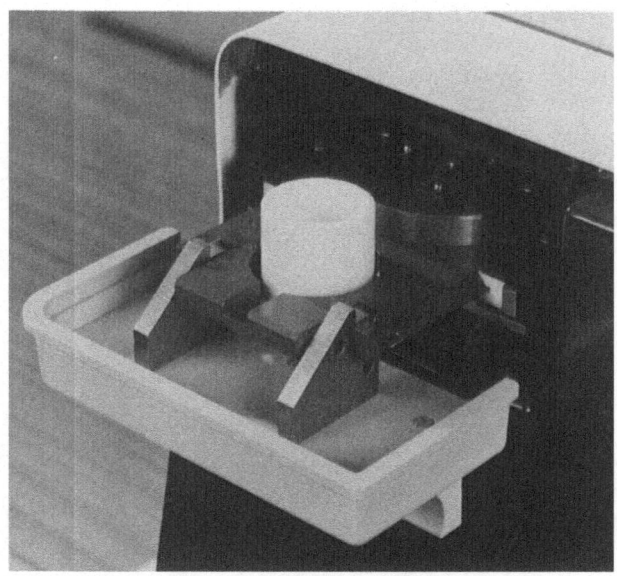

Figure 2 - Sample Slide Showing Cup and
Source Cassette.

The most important part of the instrument in determining performance is of course the sample-source-detector geometry and this is illustrated in figure 3. Here the sample can be seen above the detector with the annular source positioned closely beneath it round the detector window. The sample slide carries both the main part of the source cassette and the sample cup and figure 2 shows how these are arranged.

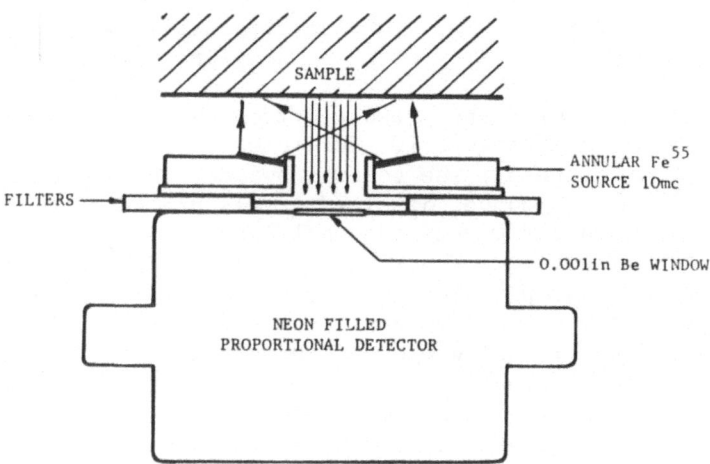

Figure 3 - Source-Sample-Detector Geometry.

When the slide is open the source is shielded by the top of the
cassette which covers it. The source may be drawn up into the
cassette which may then be removed for source changing. The
interchangeable detectors are mounted in a removable detector
unit, shown in figure 4, from which the detector can be extracted
as necessary. A small carriage is pivoted above the detector for
filters which may be placed in one or both of the apertures,
which may then be located at choice above the window. Movement
of this carriage by the small solenoid is controlled automatically
when the correct mode of operation is chosen.

Figure 4 - Removable Detector Unit showing
Proportional Counter & Filters.

In order to provide excitation over the whole periodic table
a number of alternative sources with increasing X-ray energies are
supplied, iron 55, plutonium 238 and americium 241, being the most
used. Others such as promethium-aluminium and cadmium 109 can
also be supplied. Three proportional counters with alternative
fillings of neon, argon and xenon and a caesium iodide
scintillation counter are provided. These permit the best
detector to be chosen over the whole periodic table for either
the K or L emissions. There is of course considerable overlap
between the useful energy ranges of the detectors and
discrimination against unwanted lines may be aided by careful
choice of detector. The operator is guided in the choice of both
detector and source by two periodic table charts.

READOUT

To offer the type of readout which is both convenient and
familiar to the analyst signal handling is rather different from
that employed in the usual isotope instrument. The detector output pulses are fed from the pre-amplifier and the main amplifier to
the pulse height analyser in the usual way and then the normalised
pulses from the analyser fed to an integrating D.C. amplifier.
This amplifier provides an output D.C. level proportional to the
total count obtained. It is provided with a zero bias and gain
control so that both the starting level and slope of calibration
lines may be adjusted. The resulting D.C. level is then displayed
on a digital voltmeter. When filters are used it is sometimes
necessary to subtract a reading from the preceding one. This may
be done automatically, when the appropriate function button is
pressed, by inverting the pulses from the P.H.A. before
integrating them for the second measurement.

The use of a D.C. measurement and display in this way
provides the analyst with the facilities with which he is familiar
on other instruments, and makes the technique more acceptable. It
does however impose more stringent requirements on the stability
of the system in order to achieve the desired precision. Similar
requirements are imposed on the timer which also uses an integrating
D.C. amplifier to provide a range of 15 seconds to 10 minutes.
Adequate overall stability is however obtained to achieve a short
term precision of $\pm 0.1\%$ at constant temperature. Within the range
$10\text{-}30^{\circ}C$ the temperature coefficient is $0.04\%/^{\circ}C$. Long term drift
per 8 hours is less than 0.1% of reading at constant temperature.

A binary coded decimal signal is available from the digital
voltmeter which may be used to feed a variety of standard print out
devices.

PERFORMANCE

It was found possible to obtain an adequate performance for
silicon in a light matrix with air present between the sample and
the detector although at the low energy involved the operator must
check calibration after any ambient temperature change which will
affect the air path absorption. Using the maximum integrating
time available the limiting sensitivity for silicon, defined as
twice the standard deviation at zero concentration, is better than
100 p.p.m. in a light matrix. At higher atomic numbers this
sensitivity improves to better than 10 p.p.m. for elements like
zinc. These sensitivities represent the most favourable figures
obtainable. Heavier or more complex matrices may reduce them
considerably and this of course makes it very difficult to quote
general figures for any X-ray fluorescence method. It is however
considered that this method of quoting sensitivity in a light
matrix such as polypropylene or polystyrene at least gives

reproducible figures which are a measure of the effective physical
parameters of the system. The source strengths chosen to give
these figures represent, it is felt, a reasonable compromise
between sensitivity, cost and shielding needed.

As with most instrumental analytical methods the instrument
requires calibration with the unknown element in the same matrix
and this also enables the operator to set up his instrument for
the correct channel. The zero bias and gain controls may then
also be set so that the numerical reading of the digital voltmeter
corresponds to the concentrations in p.p.m. or percent.

The instrument performance is of course complicated by
interfering elements which cannot always be simply rejected by
the P.H.A. Interferences from elements of higher atomic number
may be reduced by, where possible, using a low energy exciting
source and/or a detector with a light gas filling. Where this
does not help Ross balanced filters may be used. By selecting
the mode "FILTERS A-B" the difference signal may be displayed or
alternatively by using the "A & B" mode, the two signals may be
successively displayed. This can be useful where the contribution
of an adjacent element is better dealt with by applying a
correction to the main reading.

Where repeat measurements are required for determining
standard deviations the 'MULTI' mode provides continually
repeated measurements, each reading being displayed for 20 seconds.

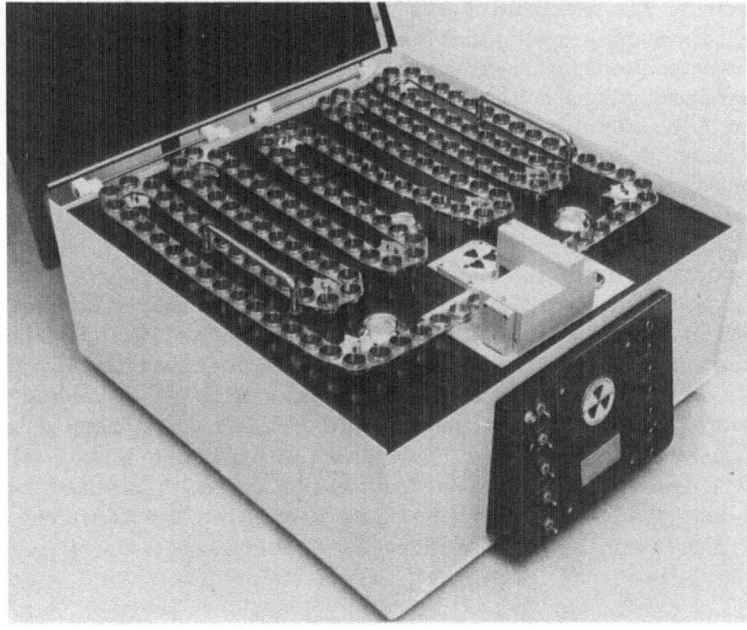

Figure 5 - Automatic Multi-Sample Presenter.

Once the usefulness of the basic instrument had been established a multiple sample presenter was designed which in conjunction with printout equipment, would enable the instrument to be loaded for unattended operation. Existing designs for liquid scintillation counting were unsuitable because of the type of sample cup needed with its thin bottom window. An effective window diameter of 22 mm was found to be the minimum which could be simply positioned and not affect accuracy and this required a cup diameter of at least 28 mm. To obtain the highest sensitivity the window is made the lowest part of the cup and could therefore not be allowed to touch the baseplate. A design based on a self supporting chain was adopted which holds the cups suspended between the sprockets (figure 5). Where the cups are brought up to the measuring position the length of chain between supports permits the cup to be pressed down until it touches the table. It is held there above the detector window until the next sample is called up when it is released and the chain moves on. Rather than arrange the sample presenter around the instrument it is built as a separate entity and the detector unit and source are removed from the main instrument and installed in the sample presenter. The presenter is then controlled by the logic in the main instrument. The presenter shown has a capacity of 150 samples but this is a larger version for a special application. Normally 50 sample capacity is provided.

REMOTE HEADS

The instrument was designed with the removable detector unit for maximum flexibility and soon after its introduction the applications came round full circle to the need for an on-line head. Consequently a separate remote head was designed to take the detector unit, source and head amplifier, so that it may be used in any orientation either as a remote bench detector or for measurements on a process stream. This head may be operated at distance of up to 100 metres from the main instrument.

X-RAY TUBE EXCITATION

In order to provide the greater flexibility of tube excitation a small 30 kV X-ray tube has been developed for this application. This tube fits behind the detector unit and irradiates the sample surface obliquely through a tubular collimator. Because of the oblique illumination sample rotation is necessary and this is provided in a modified sample slide. The tube power supplies are stabilised for voltage and current and in addition the tube output is monitored by a second detector in the tube mount, whose output is used to vary the integrating time of the measuring channel to maintain a constant sensitivity.

The greatly increased signal pulse rates using tube excitation create problems in information handling. In the original design the pulse amplifier and analyser designs were chosen to have high pulse rate capability with resolving times better than 1 μs throughout. By using 0.25 μs shaping time constants in the amplifier and choosing the optimum counter and amplifier gain, peak shift from all causes is negligible when the total count rate is in excess of 120,000 cps in the peak.

The tube, which may be supplied with alternative anode materials, is operated with an earthed cathode so that window bombardment is avoided and initially 0.01 in beryllium windows are used. A small air blower cools the external surface of the anode stem.

TYPICAL ANALYTICAL RESULTS

In general the instrument shows the expected advantage and drawbacks of an X-ray fluorescence technique in comparison with other instrumental methods. In particular its main drawbacks are a limit of detection of 10-100 p.p.m. and the usual interelement effects. There are many applications however where these are not restricted and its advantages are very attractive. The method is of course completely non-destructive and the few millilitres of sample required may be retained for re-measurement at will. Sample preparation is extremely simple, most liquids can be accepted without modifications, many powders can be measured as received, and others simply briquetted without further treatment. Liquids too viscous to handle at room temperature may be measured at elevated temperatures. A wide range of solution concentrations can be dealt with, without non-linearity, and the choice of solvents is very wide. Cross contamination is avoided by the use of cheap sample cups which can be disposed of as needed.

It would be inappropriate here to quote details of the extensive application studies that have been completed for the instrument but some typical examples show the general method and potential.

With the current emphasis on pollution control, measurement of sulphur content of fuel oils is in demand. This can readily be performed directly with either isotope or tube excitation and the measurement is independent of oil density over a useful range. Figure 6 shows a plot of 14 different oils (two of which were N.B.S. standards, with densities ranging from 0.83 to 0.97 g/cc. This plot was obtained using tube excitation, giving in a measuring time of 1 minute, a standard deviation of 0.004% at a concentration of 0.2% (i.e., 2% relative). Tube excitation was used here mainly to reduce measuring time and the figure obtained is still well

above the limit of performance. To obtain the same accuracy with
an isotope source would require 5 minutes or more integration.

   A typical analytical problem in the polymer industry is the
determination of stabilisers or additives in solid or powdered
polymers which are very difficult to dissolve. The determination
of calcium (as stearate) in polypropylene is such a problem and
for non-dispersive analysis is complicated by the presence of a
variable titanium content which causes some interference. By
using a calcium filter the interference can be effectively
eliminated for up to 250 p.p.m. of titanium. On powder poured
directly into the sample cup and over the range 0-500 p.p.m. of
calcium, a straight line plot is obtained, with a standard
deviation at 250 p.p.m. of 6.7 p.p.m. with a 5 minute measurement.

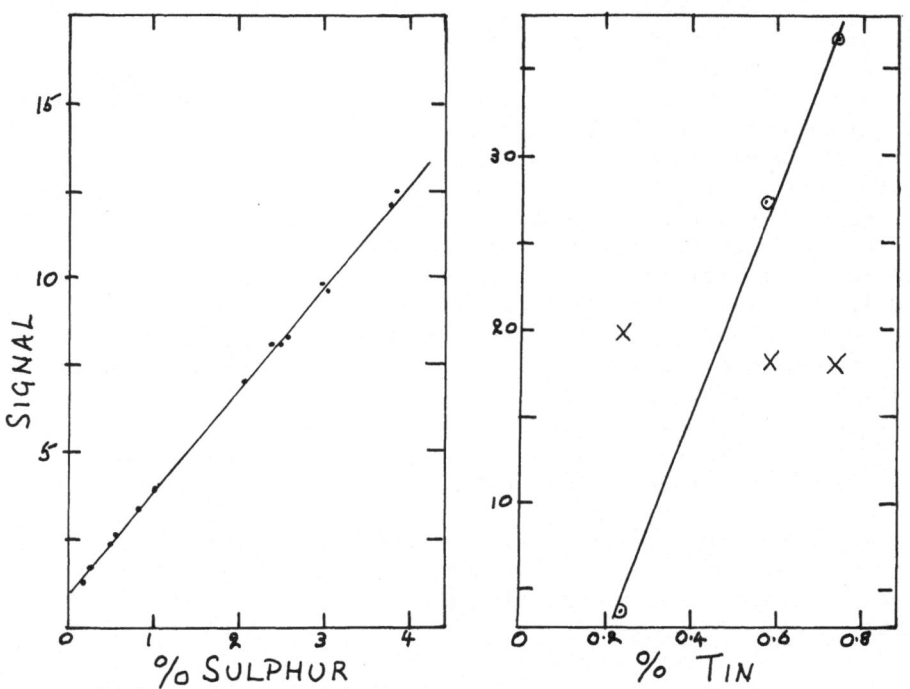

Figure 6 - Sulphur Content   Figure 7 - Tin in Tailings Before
     of Fuel Oil.   and After Correction for Absorbtion

   A more complex problem is the analysis of tin in ores, where
absorbtion of the tin radiation within the sample by varying
concentrations of other elements may prevent a simple linear plot
being obtained. Thus in figure 7 the crosses show uncorrected
results of tin tailings obtained on briquette samples containing
equal weights of material. A correction can be made for tin

radiation absorbtion in these samples however by making a measurement on a piece of tin foil with and without the sample in front of
it. Correcting the original results with an absorbtion factor
derived from the tin foil measurements gives the points shown as
circles in figure 7. Counting statistics alone on this measurement
give a standard deviation of 0.025% tin, and all the points lie
within this deviation from the line shown.

A final illustration of the use of the instrument in an
analysis which is slow by conventional methods is given by the
determination of iron in slags. This is complicated by varying
manganese contents in the samples supplied. The peaks for iron
and manganese are too close to resolve and there is a significant
contribution to the iron signal from manganese. Balanced filters
cannot easily be used to remove an overlapping signal from a lower
energy peak so in this case a separate measurement was made on each
sample of the manganese content, using a chromium filter to
suppress the iron peak and changing the pulse height analyser
window for manganese. The readings obtained for iron, shown plotted
by crosses in figure 8 were corrected using a factor derived from
the manganese reading, and these corrected values plotted as
circles in the same figure. A standard deviation at the 10% level
of 0.09% (1% relative) is thus obtained.

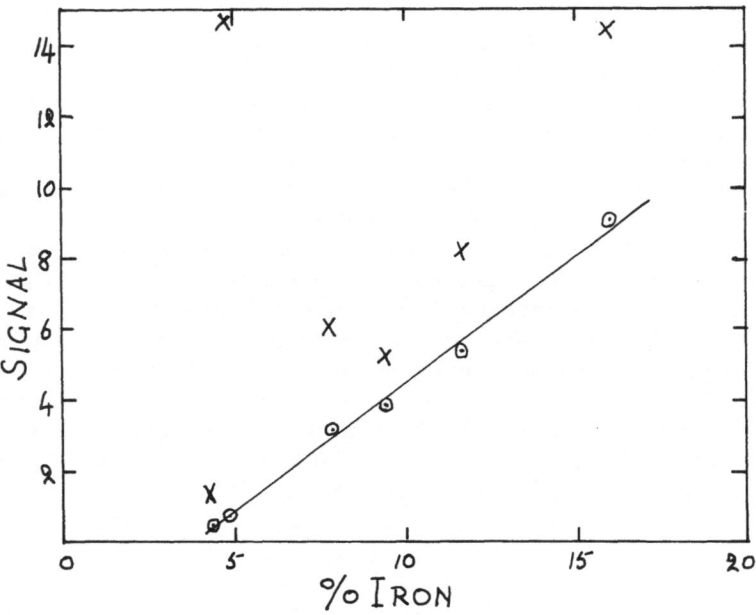

Figure 8 - Iron in Slags before and after
Correction for Manganese.

CONCLUSIONS

By keeping the needs and problems of the analytical chemist
foremost in mind a simple X-ray fluorescence instrument for the
chemical laboratory has been produced, capable of rapid analysis
and high precision. Additional facilities of remote heads, a
multiple sample presenter and data printout make it ideal for
routine analysis. A choice of tube and isotope excitation is
available.

ACKNOWLEDGEMENTS

The author acknowledges with thanks the help of his
colleagues D. Hagger, T.R. Sanger and R.A. Yates and of the design
consultants Allied International Designs Ltd., who jointly created
the instrument. He also freely acknowledges inspiration from the
numerous authors in the field of non-dispersive X-ray fluorescence
whose work is listed in standard bibliographies but is too
extensive to refer to here.

DISCUSSION

R. JENKINS (Philips Electronic Instruments): Have you observed an
influence of changes of atmospheric pressure on the long-term
intensity stability of the longer wavelengths such as S Kα?

A. L. GRAY: We have observed changes in the intensity of long
wavelength X-rays at the detector due to changes in absorption in
the air gap between it and the sample. These can arise from
changes in both atmospheric pressure and temperature, and in normal
circumstances the latter are more likely to be observed. For phosphorus K$_\alpha$ the change is +0.3% per °C, while for titanium it is
only +0.02% per °C, and it is negligible for elements higher in the
periodic table. Changes in atmospheric pressure are usually slower
than local temperature changes and therefore less likely to be
noticed except on long-term calibration drift checks. The effect
of pressure changes has been noticed in a case where the gap was
purged with an air flow when the intensity of P K$_\alpha$ was seen to
change when the purge was removed. These effects are, of course,
predictable, and where analysis is limited to one particular element, compensation can be provided. Routine calibration checks,
however, are normally sufficient to correct for them.

ENERGY DISPERSIVE ANALYSIS FOR ADJACENT ELEMENTS USING TWO SINGLE CHANNEL ANALYZERS

Hubert K. Chow

Instrument Products Department

General Electric Company, Lynn, Massachusetts 01910

ABSTRACT

Energy dispersive x-ray analysis has become an extremely useful analytical tool. The technique provides for the direct observation of x-ray emission spectra, eliminating the need for a dispersive crystal. The purpose of this reported investigation was to study the use of the technique with a simple pulse height analyzing system and to develop a routine method for correcting interferences due to adjacent element spectral overlap and matrix effects.

The analyzing system consists of a radioisotope source, a lithium drifted silicon detector, a preamplifier, an amplifier, two single channel analyzers and two digital ratemeters. In order to obtain results suitable for quantative measurement, a two-step empirical method was employed for the correction of peak overlapping and matrix effects. If two peaks in a spectrum overlap at their tails, one can set up a channel width of the analyzer to a region where there are no overlapping pulses. It is then possible to calibrate the ratio of the intensity obtained from this channel to that obtained from the whole peak in its pure state, i.e. without the appearance of a neighbor peak. The actual intensity of the peak in the overlapping spectrum is, therefore, the observed counts multiplied by the ratio. The next step is the correction of matrix effect by means of conventional empirical methods using standard samples. Two types of the samples, Zn-Cu powder mixtures and Fe-Cu in aqueous solutions, were studied to illustrate this method. The usefulness of applying the analyzing system and technique to industrial measurements, either on-line or batch, will also be discussed.

INTRODUCTION

The conventional method of x-ray emission analysis uses a dispersive crystal in which the x-rays emitted from the sample will be diffracted by passing through this crystal. Different x-ray wavelengths are dispersed to different angles and lose a large amount of their intensity. The x-ray spectrum is then obtained by intensity versus diffraction angle measurement. This method provides excellent resolution, especially when two crystals are used. For instance; Richtmeyer and Barnes (1),(2), determined "full width at half maximum" (FWHM) for the tungsten K-series by means of a two-crystal x-ray spectrometer. The value obtained for FWHM was 43ev. However, all dispersive systems require a strong x-ray source and a precise goniometer. It turns out that this kind of equipment is big and heavy and, moreover, expensive.

In the non-dispersive technique, the x-rays emitted from the sample are detected directly without the use of an analyzing crystal. The energy of the x-rays can be distinguished by means of balanced filters, selective excitation, or pulse height analysis. In pulse height analysis (PHA), the pulses generated at the detector are proportional to the energy of incident x-ray photons. Therefore, different x-ray energies can be separated by an electronic devices. This method is often referred to as "energy dispersive analysis". It provides several advantages:

1) It requires only a small source for sample excitation. In most cases, a few milliCurie radioisotope is enough. This source is small, stable, easy to maintain, and inexpensive relative to an x-ray generator.

2) Since this technique provides direct observation, the intensity of the signal is high and consequently the data acquisition time is short.

3) The total size of the system is small; it is compact, easy to move,and easy to operate.

However, for the energy dispersive analysis, resolution of the detecting system plays an important role. In many cases, the energy peaks in the spectrum overlap due to insufficient resolution. This complicates the characteristic of quantitative calculations. One of the most common methods to unfold the overlapping peak is to use a multichannel analyzer for obtaining a spectrum and to feed the data into a computer. The computer is programmed to resolve the area under each individual peak. This method is convenient and accurate for a laboratory measurement. Some investigators, for example; Dollby (3) and Birk et. al. (4), use a mathematical equation to unfold the overlapping peaks.

In Birk's measurement, using a xenon proportional counter, a multichannel analyzer system showed better results than those obtained by selecting pulses through a single-channel analyzer. However, these methods require a technically trained, high-level operator and expensive equipment. Such a system is not attractive for routine industrial applications.

With the improvement of detector resolution, some of the above complications should be avoided by using a single-channel analyzer system. This system works well when the peaks are well resolved, but the precision becomes poor if the peaks are over-lapping. The work described here was to develop a technique for routine analysis of adjacent elements using this system. For simplicity, we will discuss the method only as it applied to two single-channel analyzers.

PEAK SHAPE ANALYSIS

1.   The Nature of the Distribution of X-ray Lines

Hoyt (5) and Weisskopf et. al. (6), (7) have calculated an intensity function for atomic emission lines based on the Dirac quantum theory of radiation field. They obtain that

$$I(\nu) \ d\nu = \frac{\gamma \ h \ \nu_0 \ d\nu}{(\gamma/2)^2 + 4\pi^2 \ (\nu-\nu_0)^2} \ , \qquad (1)$$

where γ is the Einstein probability coefficient of a spontane-ous transition, ν is the frequency of the emission line and h is Plank's constant. The intensity versus energy relationship can be obtained from the equation (1) by considering that

$$I(E) \ dE = I(\nu) \ d\nu \ ,$$

where E is the x-ray photon energy. Then, equation (1) becomes;

$$I(E) \ dE = \frac{\gamma \ h \ E_0 \ dE}{(\gamma h/2)^2 + 4\pi^2 \ (E - E_0)^2} \ , \qquad (2)$$

which can be written as;

$$I(E) = \frac{a}{1 + (1/b)^2 (E-E_\circ)^2} ,$$ (3)

where a is the maximum amplitude of the peak and b is the half width at half maximum. This function is a type of Lorentz equation.

2.   The Gaussian Distribution

The experimentally observed shape of the x-ray peak is dominated by the instrument used since the natural width of an x-ray line is very small compared to the detector resolution. For instance; the natural width of the copper K-alpha 1 line determined by Spencer (8), is 3.14 eV, but the resolution of a good solid-state detector for the same peak is about 300 eV.

In our energy dispersive system, the shape resembles that of Gaussian distribution function, which is;

$$I_a(E) = A \exp(-\alpha^2 E^2) ,$$ (4)

where α is a constant which can be determined by the resolution of the detecting system, R, i.e.;

$$\alpha = 1.3862 / R ,$$ (5)

and R is defined as the full width at half maximum (FWHM) of the Gaussian peak.

Fig. 1 shows a computer curve fitting for a Gaussian equation with an experimental determined copper K-alpha peak. The small discrepancy at the left tail is probably due to the "edge effect" at the detector diode. Using equation (3), a Lorentz equation is plotted for comparison. The Lorengian curve deviates from the experimental curve largely at the shoulder of the peak.

The instrumental deformation is mainly due to the random fluctuation of the pulses generated at the detector. Some contribution to this deformation is also due to the noise in the preamplifier and amplifier. The FWHM, in this case, can be expressed mathematically (8);

$$R = \sqrt{R_n^2 + R_d^2} ,$$ (6)

where R_d is the FWHM at the detector which is given by;

$$R_d = 2.35 \sqrt{\epsilon EF} ,$$ (7)

ε is the average energy required to produce a single electron-hole pair, F is the Fano factor, and R_n is the electronic noise given by;

$$R_n = R_o + M\ C_d \tag{8}$$

R₀ is the FWHM when there is no detector diode, M is the slope of the energy versus external capacitance curve, and C_d is the capacitance of the detector.

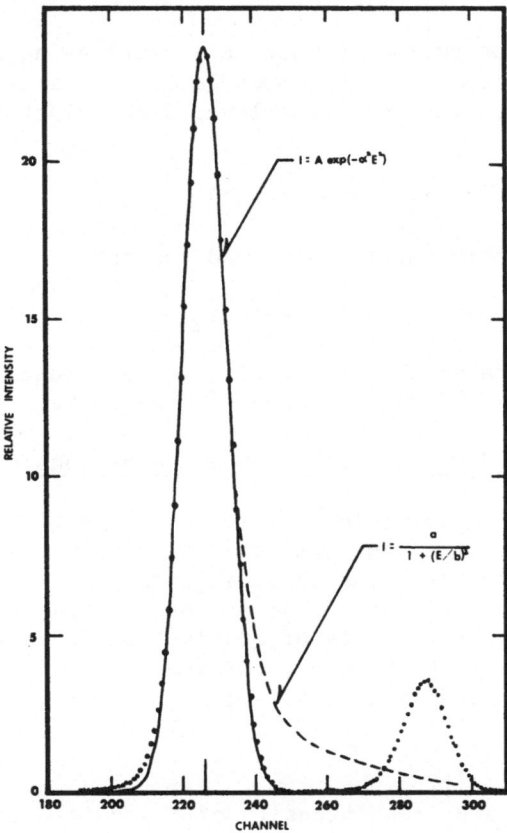

Fig.1   The Computer Curve Fitting for Copper K-alpha X-ray
        Peak.

3.   The Area Under A Peak

    The area under a Gaussian peak can be calculated from equation (4); thus,

$$T_a = \int_{-\infty}^{\infty} I(E)\ dE = (\sqrt{\pi}\ /\ 1.3862)\ R\ A \tag{9}$$

If we collect only the pulses whose corresponding energy falls in the range $E_1 \leq E \leq E_2$ the total number of the pulses will be:

$$T_b = (A \sqrt{\pi} / 2\alpha) [erf(\alpha E_1) + erf(\alpha E_2)] \qquad , \qquad (10)$$

where erf (αE) is an error function defined by the equation:

$$erf(Z) = (1/\sqrt{\pi}) \int_{-Z}^{+Z} exp(-Z^2) dZ \qquad , \qquad (11)$$

Let us consider two peaks which are overlapping at one side shown in Fig. 2. If we set the lower level discriminator (LLD) to negative infinite and the upper level discriminator (ULD) to E_2, we have;

$$T_b = (A_1 \sqrt{\pi}/2\alpha) [1 + erf(\alpha E_2)] \qquad , \qquad (12)$$

The ratio between this partial area and the total area is;

$$r = T_b/T_a = (1/2) [1 + erf(\alpha E_2)] \qquad , \qquad (13)$$

Equation (13) indicates that the ratio, r, is a function of αE_2 which is independent of the amplitude of the peak.

4.   The Optimized Setting for a Single Channel Analyzer

If we choose, as shown in Fig. 2, the value of E_2 as far as possible from peak 2, we can avoid the inaccuracy contributed by this peak when we determine the area under the first peak. Then however, we sacrifice the loss of total intensity and consequently increase the percentage of statistical deviation. A standard statistical deviation σ, is usually considered as the square root of the total counts, i.e.;

$$\alpha = \sqrt{T_b}$$

On the other hand, the intensity of the pulses contributed by the second peak can be calculated from the equation;

$$T_2 = A_2 \int_{-\alpha}^{+E_2} exp[-\alpha_2^2 (E - E_3)^2] dE$$

$$= (A_2 \sqrt{\pi}/2\alpha_2^2) [1 - erf\alpha_2(E_3 - E_2)] \qquad , \qquad (15)$$

Therefore, the percentage contribution to the first peak from the second is;

$$c = T_2 / T_a = (A_2/2A_1) [1 - erf\alpha_2(E_3 - E_2)] \qquad , \qquad (16)$$

or;

$$\text{erf}\alpha_2(E_3 - E_2) = 1 - 2 A_1 c / A_2 \quad , \tag{17}$$

where E_3 is the location of peak 2. Both E_3 and α_2 can be
determined experimentally. A_1/A_2 can be estimated roughly from
the ratio of concentrations of the elements corresponding to
peak 1 and peak 2. The allowable contribution,"c", will be discussed later. Using the calculated value of E_2 from equation (17),
we can estimate the best set of ULD and LLD for a single channel
analyzer.

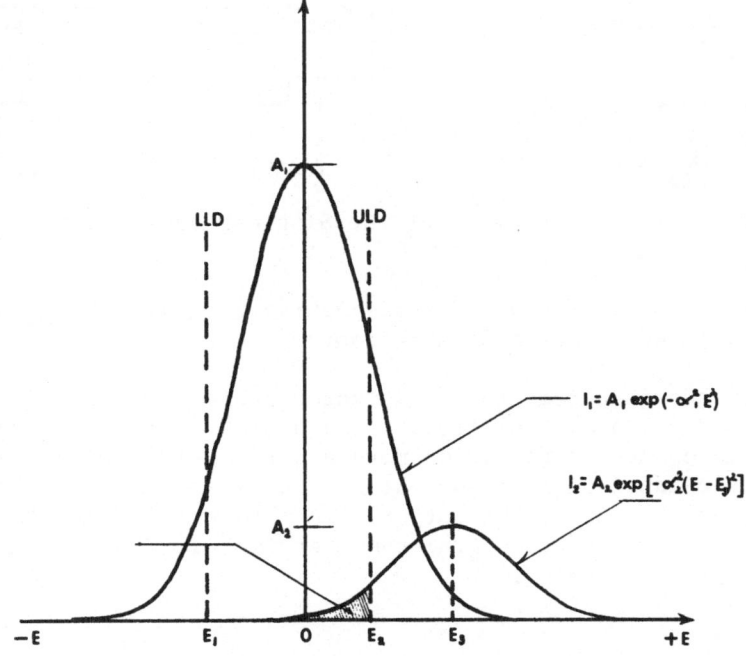

Fig.2   The Overlapping Guassian Curves.

THE MEASUREMENT

1.   Instrumentation

As shown in Fig.3, this system consists of a radiosotope source,
a lithium drifted silicon detector, a preamplifier, an amplifier,
two single-channel analyzers, and two digital ratemeters.   A 90mC
plutonium-238 source was separated and sealed into three discs.
These discs were mounted in an aluminum source holder in a back-
scattering geometry.   Two types of solid state detector were
used.   One was made of a 30 mm^2 diode (active area), with an
optofeedback amplifier and the other one had a 200 mm^2 diode with
an ordinary linear amplifier.   The first one was used to measure
the samples which contained zinc and copper, and the second was
used to measure iron and copper in aqueous solutions.   Their

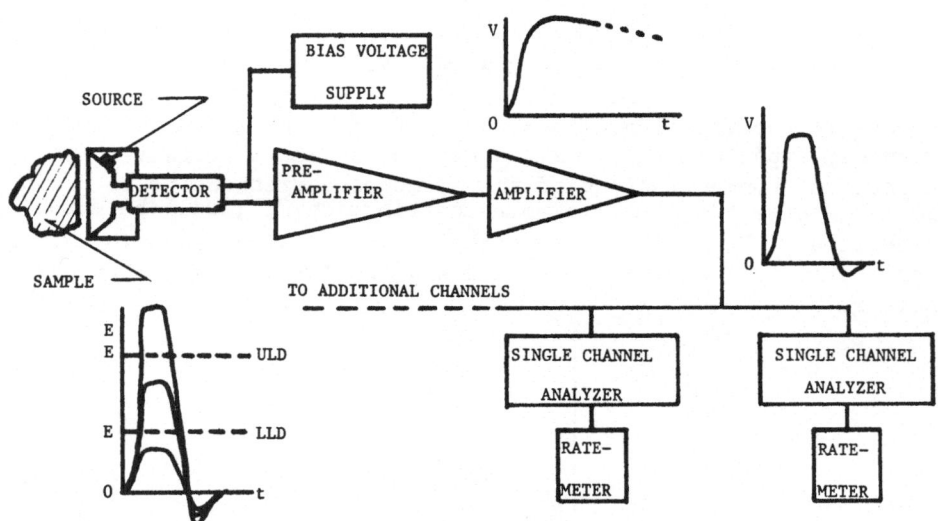

Fig.3   The Block Diagram of the Single-channel Analyzer
         System.

resolutions were 190eV and 320eV respectively.   Both detectors
were manufactured by Kevex Corporation.

The pulses generated in the detector are amplified via a preamplifier and a linear amplifier.   The output from the linear
amplifier is connected to two or more single-channel analyzers.
Each single-channel analyzer output is attached to a ratemeter. The
data from the counter can be output to a printer or the counter can
be directly interfaced to a mini-computer for automatic on-line
industrial process control.

2.   Samples

Two kinds of samples have been used for the demonstration of
this measuring technique.   One, a powder sample, was a mixture of
copper and zinc oxide, and the other was an aqueous solution of
ferrous sulfate ($FeSO_4.7H_2O$) and cupric sulfate ($CuSO_4.5H_2O$).

The sample preparation procedure is a standard chemical method
which will not be repeated here.   The zinc concentration in the
powder samples ranged from 1.95 to 9.55% by weight and that of
copper from 88.95 to 97.68%.   For the aqueous solutions, copper
and iron concentrations ranged from 0.0087 to 0.1942% and 0.16 to
0.29% respectively.

3.   Calibration Procedure

3-1   Discriminator calibration: For the powder sample, we
obtained about 200,000 counts per 100 seconds of copper x-ray and

10,000 counts per 100 seconds of zinc x-ray. The statistical
deviation (square root of the total counts divided by the total
counts) is thus, about 0.22 to 1%.

Let us consider that the "allowable contribution" in equation
(17) must be less than or equal to σ/T_a, ie. c = 0.01. On the
other hand, the ratio of intensity of copper and zinc is approximately 8 to 1 (estimated according to the concentration of the elements
in the sample). The resolution (FWHM) of copper and zinc peaks are
309.5 eV and 315.7 eV respectively.

Accordingly, the calculation for LLD and ULD of the zinc $K\alpha$
channel by using equation (17) yields 8.69 and 8.88 keV respectively.
Similarly, we obtain 7.29 and 8.59 keV for copper $K\alpha$ in the solid
samples. For the liquid samples, the near-neighbor peak of iron
$K\alpha$ was iron $K\beta$. The copper $K\alpha$ had two near-neighbors, iron $K\beta$ and
copper $K\beta$. The calculation results were; 7.42 to 8.27 keV for
copper and 5.80 to 6.81 keV for iron.

3-2 Matrix effect calibration: The x-ray intensity from an
element is not always a linear function of its concentration since
secondary emission and absorption occur due to other elements in the
sample. In the last two decades, many methods have been used to
correct matrix effects. One of the common techniques is the so
called "empirical coefficient method" (9) in which the concentration
of the elements in a sample system can be calculated from a set of

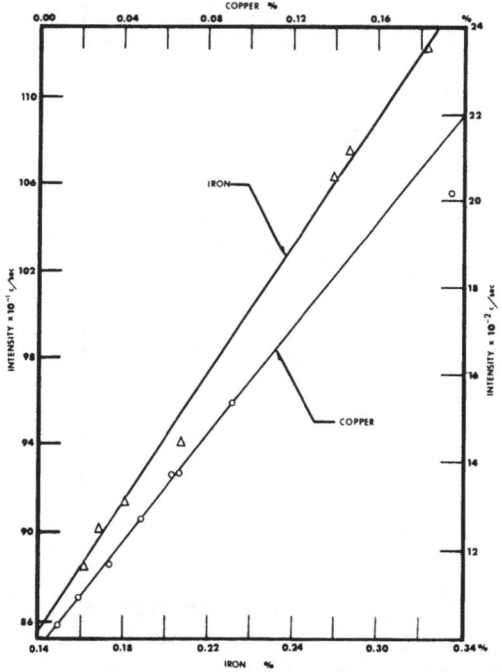

Fig.4 The Calibration Curves for Powder Samples.

linear equations.   The coefficients in the equations are determined
by using a number of standard samples.   For our binary system, we
plotted a calibration curve using the data from the standard samples,
as shown in Fig.4 and Fig. 5.   In the concentration range in question, the curve is approximately linear.

4.   The Measurement and Results

For convenience, the data acquisition time of each sample was
100 sec.   Table-1 and Table-2 are the lists of measurement results
determined directly from the raw data.   For solid samples, the
background and noise level in the copper channel is less than 6
counts per second and in the zinc channel is about 10 counts per
second.   It is as high as 100 counts per second for both the
copper and iron channels in the liquid sample measurement due to
Compton scattering.

CONCLUSION

The results of these measurements are quite satisfactory and
reveal   that an energy dispersive system, using a high resolution
solid state detector and several single-channel analyzers, is very
useful for industrial applications.   Such a system is comparatively inexpensive and easy to operate.   With properly calibrated

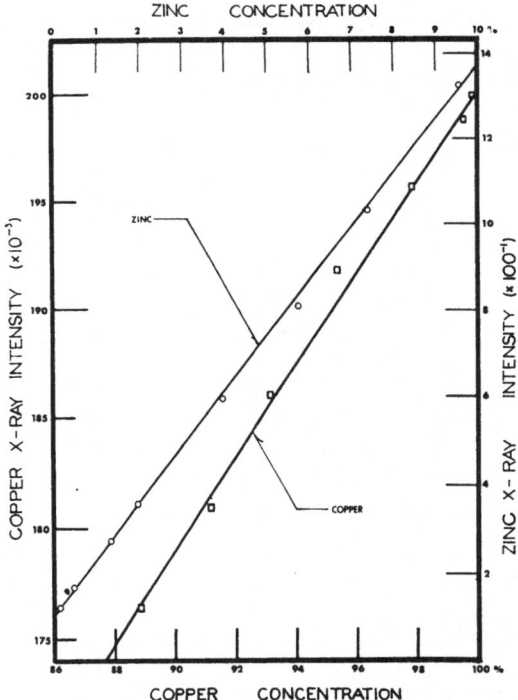

Fig.5   The Calibration Curves for Aqueous Solutions.

Table-1   The Results of the Measurements for Powder Samples

Sample	Copper			Zinc		
	Actual Prepared %	Measured ± S.D. %	Difference %	Actual Prepared %	Measured ± S.D. %	Difference %
A-1	99.81	99.80 ± 0.22	0.01	0.16	0.17 ± 0.04	0.01
A-2	99.45	99.26 ± 0.22	0.19	0.46	0.45 ± 0.04	0.01
A-3	97.68	97.40 ± 0.22	0.28	1.93	1.90 ± 0.05	0.03
A-4	95.32	95.90 ± 0.22	0.58	3.90	3.80 ± 0.06	0.10
A-5	93.12	93.38 ± 0.21	0.26	5.74	5.34 ± 0.07	0.40
A-6	91.17	90.97 ± 0.21	0.20	7.37	7.43 ± 0.08	0.06
A-7	88.95	89.02 ± 0.21	0.07	9.55	9.35 ± 0.09	0.20

Table-2   The Results of the Measurement for Aqueous Solutions

Sample	Copper			Iron		
	Wet. Chem. Analysis %	Measured ± S.D. %	Difference %	Wet. Chem. Analysis %	Measured ± S.D. %	Difference %
B-1	0.0479	0.0455 ± 0.0008	0.0024	0.2074	0.1874 ± 0.0039	0.0200
B-2	0.0905	0.0959 ± 0.0016	0.0054	0.1619	0.1630 ± 0.0034	0.0011
B-3	0.1942	0.1810 ± 0.0025	0.0132	0.1697	0.1797 ± 0.0037	0.0100
B-4	0.0623	0.0658 ± 0.0011	0.0035	0.1619	0.1575 ± 0.0034	0.0044
B-5	0.0653	0.0671 ± 0.0011	0.0018	0.1808	0.1708 ± 0.0036	0.0100
B-6	0.0087	0.0087 ± 0.0002	0.0000	0.3227	0.3227 ± 0.0064	0.0000
B-7	0.0338	0.0327 ± 0.0006	0.0009	0.2784	0.2662 ± 0.0051	0.0122
B-8	0.0192	0.0202 ± 0.0004	0.0010	0.2861	0.2795 ± 0.0053	0.0066

channels, it can be used as a batch analyzer for routine quantitative analysis.   However, for industrial on-line measurements,
the environmental problems such as temperature and vibration may
be encountered.   In general, the industries with applications in
which energy dispersive analysis can be used are, for example;
mining, chemical, paper and petroleum industries.

ACKNOWLEDGEMENT

    The author wishes to thank Dr. B. Bowen and Mr. E. Symes for
their valuable suggestions and discussions. Many thanks are due
to Kevex Co. for letting us use their large area detector.

REFERENCES

(1) Richtmeyer, F. K., and Barnes, S. W.; "The Nature Width of
K-Series of W (74)", Phys. Rev. 46, 352 (1934).

(2) Barnes, S. W., and Richtmeyer, F. K.; "A Direct-Reading,
Two-Crystal X-Ray Spectrometer", Rev. Sci. Inst. 5 (1934)

(3) Dolby, R. M.; "Some Method for Analyzing Unresolved
Proportional Counter Curves of X-Ray Line Spectra", Proc. Phys.
Soc. 73, No. 1, P-81 (1961)

(4) Birk, L. S., Labrie, R. J., and Criss, J. W.; "Energy
Dispersion for Quantitative X-Ray Spectrochemical Analysis",
Analy. Chem. 38, 701 (1965)

(5) Hoyt, A.; "The Shape of an X-Ray Line", Phys. Rev. 40,
477 (1932)

(6) Weisskopf, Von V., and Wigner, E.; "Berechnung der Natulichen
Linienbreite auf Grund der Diracschen Lichttheorie", Zeit f. Physik
63, 54 (1930)

(7) Weisskopf, Von V.; "Die Breite der Spektrallinien in Gasen",
Phys. Zeits. 34, 1 (1933)

(8) Lifshin, E; "Solid-State X-Ray Detectors for Microprobe
Analysis and Scanning Electron Microscopy", General Electric Co.
Report No. 69-C-346, 1969.

(9) Criss, J.W., and Birks, L.S.; "Calculation Methods for
Fluorescent X-Ray Spectrometry" Analy. Chem. 40, No. 7, 1081 (1968)

DETERMINATION OF ZIRCONIUM, HAFNIUM, NIOBIUM, TANTALUM, MOLYBDENUM AND TUNGSTEN IN AQUEOUS SOLUTIONS BY RADIOISOTOPIC EXCITED X-RAY FLUORESCENCE

Frank L. Chan

Aerospace Research Laboratories

Wright-Patterson Air Force Base, Ohio 45433

W. Barclay Jones

Yale University, New Haven, Connecticut 06511

ABSTRACT

Previous investigations on the quantitative determination of sulfur, chlorine, potassium, calcium, scandium and titanium in aqueous solutions by a radioisotopic excited fluorescent spectrometer has been extended to include other elements which are very difficult to separate and determine quantitatively by chemical methods. Six elements taken for the investigation and some of the results to be presented in this paper are: (1) zirconium, (2) hafnium, (3) niobium, (4) tantalum, (5) molybdenum and (6) tungsten. As in previous investigations, aqueous solutions have been used because of the ease in obtaining exact concentrations and homogeneous mixtures of the elements under investigation.

In the earlier investigations which have been reported in this conference, lighter elements (atomic numbers ranging from 16 to 22) were used for the investigation. In the present studies, however, comparatively heavier elements have been used. Therefore a radioisotope such as iron 55 used earlier is not suitable because it cannot excite the K x-ray of these elements. To excite the K and L of these elements, we use the radioisotope iodine 125. The advantage of using this radioisotope is that it is inexpensive and commercially available although its half-life is comparatively short.

The spectrometer used with further improvements has been described and presented earlier. We used a multi-channel analyzer of 1000 in the present investigation. A liquid cell was specially designed for this study. Chemicals used for preparation of solutions were of reagent grades. Some of them had to be specially prepared. For example, hafnium, often contaminated with zirconium, was specially prepared and checked spectroscopically. Some difficulties have been encountered in preparing concentrated solutions such as niobium and tantalum due to the inherit characteristics of these elements to form insoluble compounds. Procedures will be described for the preparation of these solutions. Instruments used and results will be presented in this paper.

INTRODUCTION

The six elements dealt with in this paper have played an important role in the advancement of material science in recent years. For instance, zirconium discovered by Klaproth in 1789, has a low absorption cross section for neutrons and is used for nuclear energy applications. Because of its resistance to corrosion by acids, alkalis and sea water, zirconium has been used in a number of chemical industries. It has been used extensively as an alloying agent in steel, vacuum tubes, ceramics, rayon spinnerats, lamp filaments and in many others.

Commonly associated with zirconium in mineral is hafnium, discovered by D. Coster and G. von Heresey in 1923. Zirconium minerals usually contain 1 to 5% hafnium. The chemical properties of these two elements are so similar, thus making their separation and chemical analysis by conventional methods very tedious if not impossible. Compared to zirconium, hafnium has about 600 times absorption cross-section for thermal neutrons. It is used for reactor control rods. The U.S. Bureau of Mines at Albany, Oregon, had actively engaged in the production of zirconium and hafnium two and one-half decades ago, using the Kroll process. The production was followed up by the Wah Chang Corporation in the same locality.

The Aerospace Research Laboratories has for many years been interested in the chemistry of tantalum and niobium because of their useful and exceptional properties. It is used in carbon and alloy steels and in nonferrous alloys. Alloys containing niobium have improved strength and other desirable properties. Niobium has a low capture cross-section for thermal neutrons and has superconductive properties. Niobium-zirconium wire has been used to fabricate superconductive magnets which retains its superconductivity in strong magnetic field.

Tantalum has been widely used in nuclear reactors, aircraft and missile parts and chemical and surgical equipment. The principal ore is columbite-tantalite. It has recently been found in northern Brazil. Tantalum oxide has been used to fabricate special glass for lenses and many other uses.

It has been claimed that the pure tungsten ore came from southeast China before World War II. Large quantities had come from that part of the world. This element as well as its alloys are widely used in electric lamps, electronic and television tubes, x-ray targets and in high speed steel. Tungsten disulfide has been used as a lubricant up to $500^{\circ}C$.

Molybdenum has also been used in nuclear energy applications and for aircraft and missile parts. Molybdenum sulfide is useful as a lubricant at high temperature.

So useful are these elements and their compounds, it is no wonder that analytical chemists are greatly concerned with regard to their chemical behavior, separations and their qualitative and quantitative determinations. In general, the chemical separation and analysis by the conventional methods are tedious and extremely difficult. For instance, the dissolution, separation and quantitative determination of tantalum and niobium is not an easy task (1). Scientist in Aerospace Research Laboratories has written a treatise in regard to the analytical chemistry of tantalum and niobium (2). Newer methods for the determination of molybdenum (3) tantalum and niobium (1, 2) have been reported in the literature.

Because of the difficulties in separation by conventional chemical methods of these elements, scientists, especially analytical chemists, have investigated other methods such as the conventional x-ray fluorescence method. For tantalum and niobium, difficulties were encountered in the early stage of the x-ray fluorescent method (4). In the conventional x-ray fluorescence method, some time the second-order K lines of the lighter elements occasionally overlap the first order L lines. Thus, the presence of these elements in the sample give rise to the problem of resolution of x-ray lines. One typical example is the presence of niobium and tantalum in a sample taken for analysis. Niobium has $K\alpha_1$ and $K\alpha_2$ lines whose wavelength is slightly less than half that of the $L\alpha_1$, line of tantalum. As a result, the second-order K lines of niobium are difficult to resolve from the $L\alpha$ line of tantalum which is the most convenient line for analytical purposes. In the early days of x-ray fluorescent analysis Birk and

Brooks overcame it by making measurements at several angles
and by comparing the integrated intensity of the unresolved
tantalum-niobium lines with that of single niobium line.
Subsequent introduction of silicon and germanium analyzing
crystals to the conventional x-ray fluorescent analysis in
which the second-order lines do not appear further improved
the x-ray fluorescence method.

Recently a number of reviews as well as original papers
have appeared in the literature describing the energy dispersion system of x-ray fluorescent analysis (5, 6, 7, 8, 9,
10).

With the use of radioisotopic source for x-ray fluorescent excitation coupled with the lithium-drifted silicon
detector in conjunction with energy dispersion system a new
chapter has been written for the determination of such pair
of elements as zirconium and hafnium, niobium and tantalum
or molybdenum and tungsten. Count rate and resolution by
this system at the present is not as good as the conventional
method. However, by the energy dispersion system there are
many advantages as cited in earlier publication (9).

EXPERIMENTAL

Instrumentation

Radioisotopic Excited Fluorescent Spectrometer. The
spectrometer used for the determination of the elements in
solution has been mentioned and described in several of the
publications by these authors (7, 8, 9, 10). It consists of
a vacuum cryostat, a liquid nitrogen reservoir, a radioactive
source, a lithium-drifted silicon detector, a preamplifier
with associated power supplies. Additional electrical units
are a multi-channel pulse height analyzer, x-y recorder, a
digital printer and a punched tape. The dimensions and
specifications of this spectrometer have been previously
described (10).

For this study, the radioactive source is iodine 125
having a half-life of 57.4 days (7, 8, 9). The tellurium
x-ray upon disintegration of the radioactive source has sufficient energy (27.47 keV) to excite the K spectra of
zirconium, niobium and molybdenum and the L spectra of hafnium,
tantalum and tungsten present in solutions prepared for this
study. The characteristic x-ray fluorescent radiation emitted
from the sample is directed to the detector. The electrical
pulses are then amplified and converted to voltage signals
for the pulse height analyzer. After sorting, the pulse

height analyzer displays the resulting spectra and can also be recorded with an x-ray recorder or with a digital printer or a punched tape.

The Liquid Sample Cell. The liquid sample cells are open or closed cups, circular in shape and made of plastic materials. The dimensions of the cup are 30 mm in diameter, 10 mm in depth. The wall thickness is 3 mm and for the closed cups the bottom has a thickness of 2 mm. A collar fits tightly on each cup and has the same depth as the inner cup and the wall thickness is 3 mm. After filling the cup with the solution for analysis, a quarter mil Mylar foil is placed on top and the collar put in place. The cell can then be inverted for its content to be irradiated by the radioactive source.

Preparation of Solutions

The chemicals used for the preparation of solutions were the purest compounds that could possibly be obtained. Many of these chemicals have been used with the development of new analytical procedures based on entirely different methods. Some of these methods are being cited in the references (1, 2, 3). Stock solutions were prepared by weighing calculated amounts of the pure chemicals and dissolving them in appropriate reagents as the case may be. For solution of tungsten, one-normal sodium hydroxide was used. Molybdenum solution was prepared with either one-normal sodium hydroxide or one-normal sulfuric acid. For niobium and tantalum solutions, concentrated sulfuric acid and saturated ammonium are being employed after the pentoxides were fused with potassium bisulfate (1). When potassium heptafluorotantalate was being used for preparation of a clear solution, 48% hydrofluoric acid was used. Zirconium and hafnium were kept in solution with one-normal sulfuric acid.

RESULTS AND DISCUSSION

The decay scheme of the radioisotopic source iodine 125 has been described (7, 8, 9). The radioactive iodine 125 is mounted in a tellurium cylinder about half an inch in diameter and half an inch long. A hole drilled in the tellurium cylinder is large enough to accept a sealed iodine 125 seed. The holder is cemented to a lucite ring which fits over the aluminum housing. This places the source in front and below the detector in such a fashion that samples can be analyzed by being mounted a few millimeters away from the source opening. The sample is then able to view both the source and the

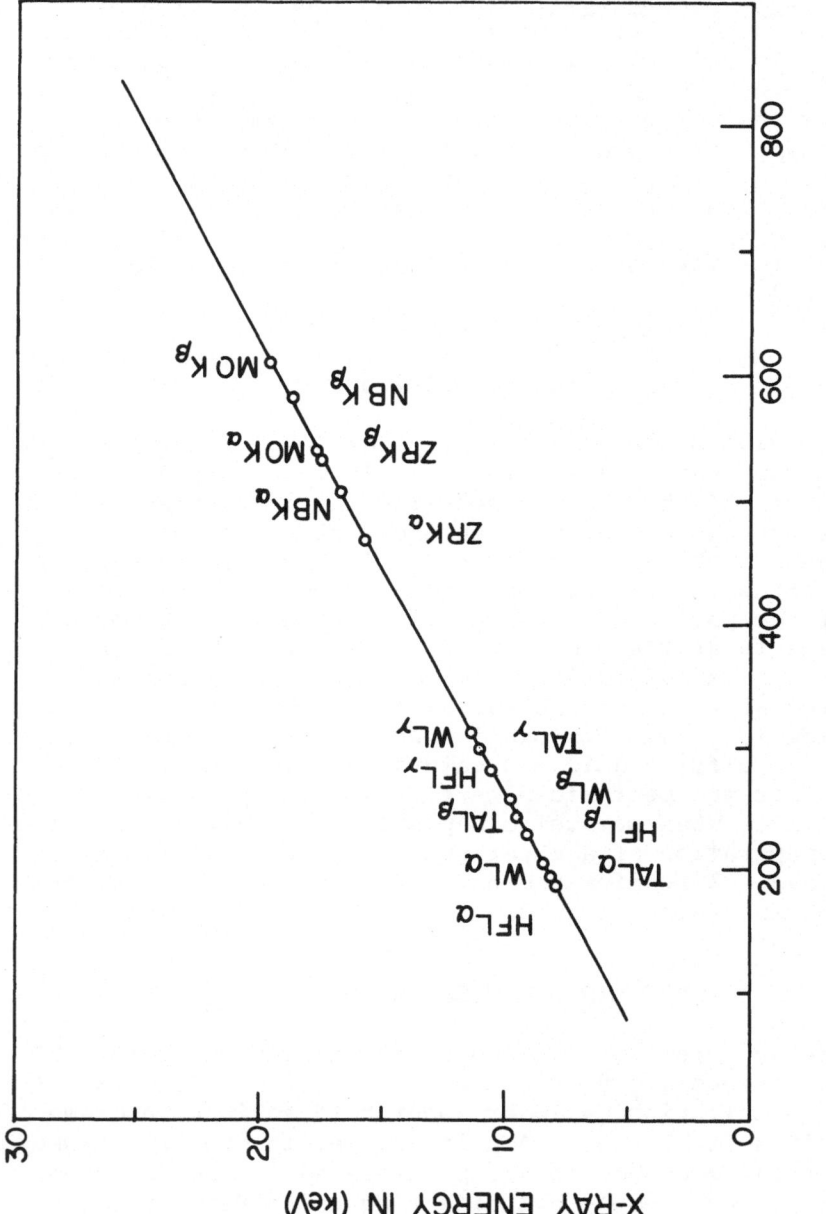

Figure 1. A plot of excitation energy of an element versus its K_α, L_α, L_β or L_γ peak position appearing in the channel of a pulse height analyzer.

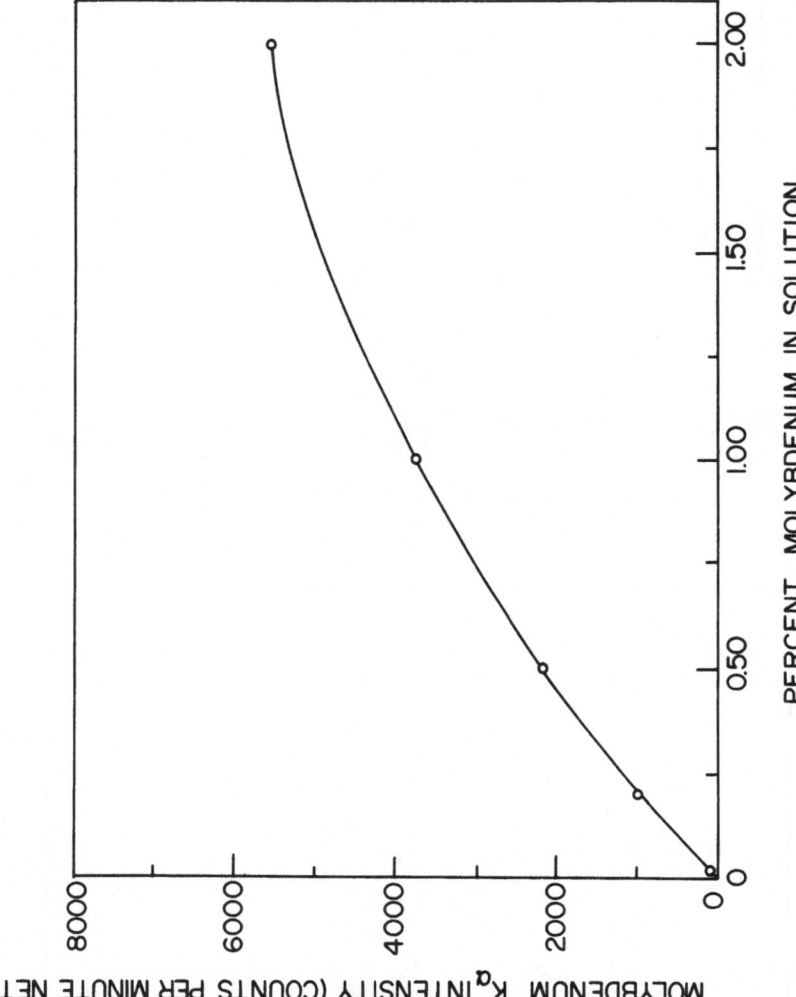

Figure 2. Determination of zirconium (following page, upper) niobium
(following page, lower), and molybdenum (above), in aqueous solutions by
radioisotopic excited x-ray fluorescence using a lithium-drifted silicon
detector; radioisotopic source, I125; beryllium window, 0.005" thick.

F. L. Chan and W. B. Jones

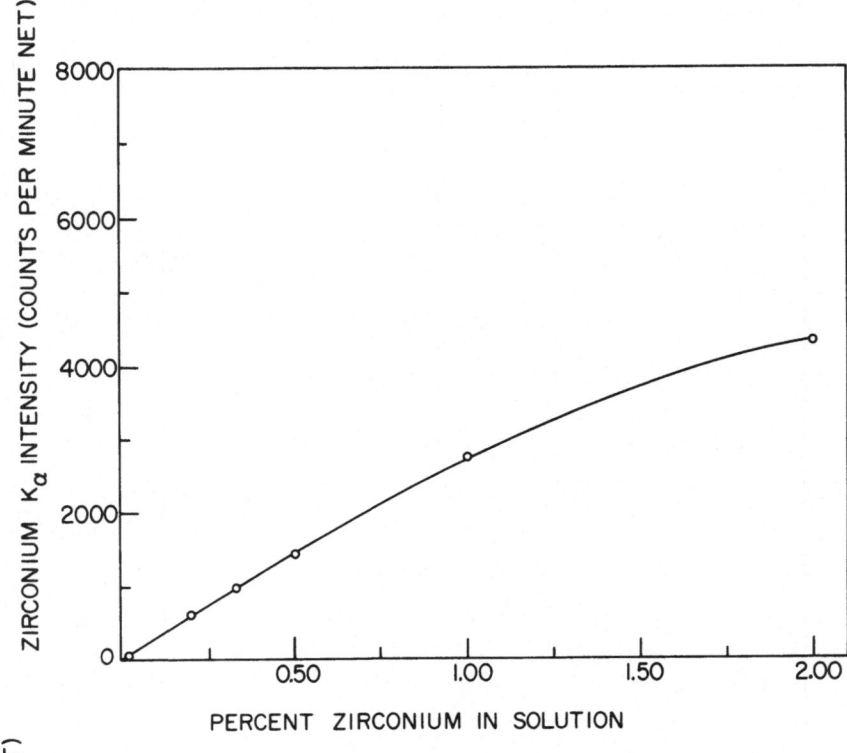

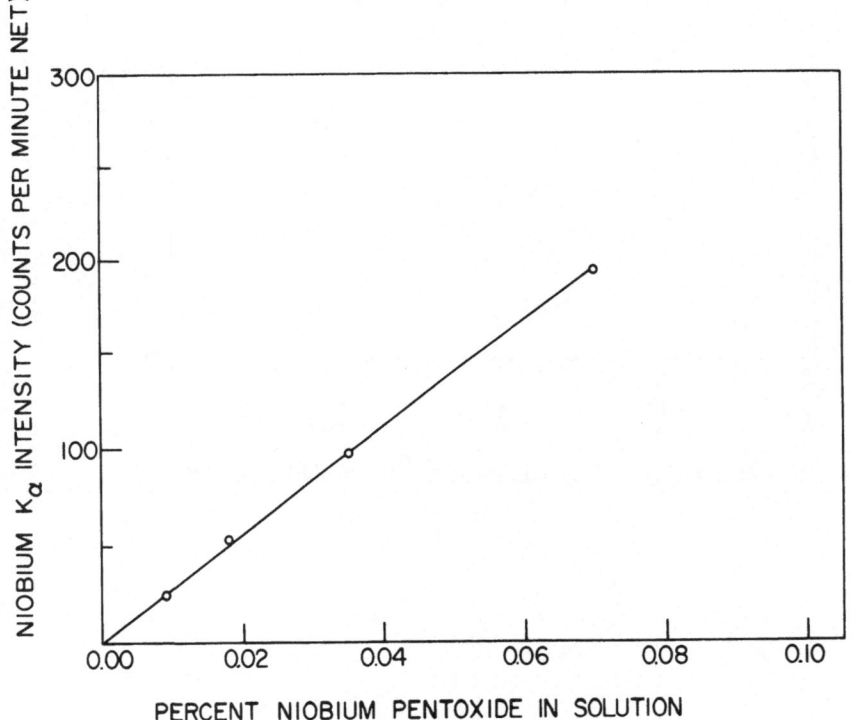

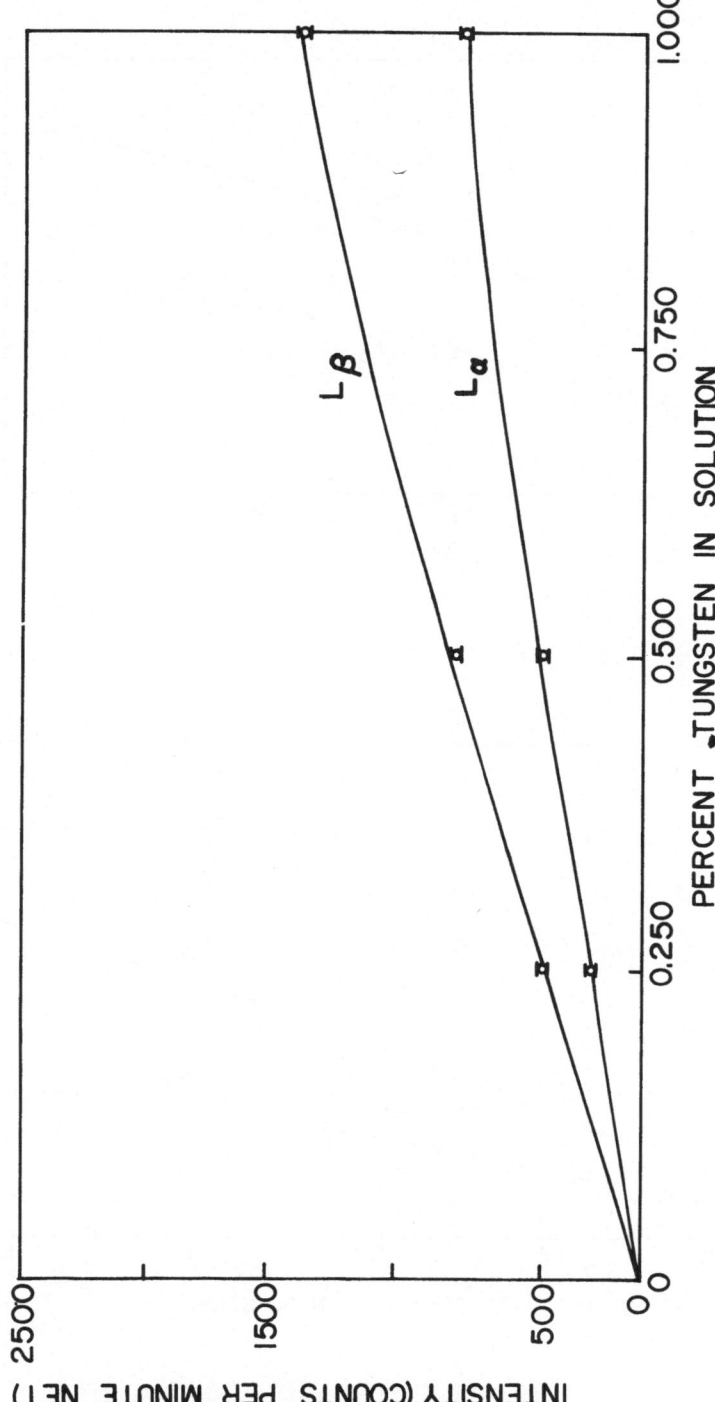

Figure 3. Determination of hafnium (following page, upper), tantalum (following page, lower), and tungsten (above), in aqueous solutions by radioisotopic excited x-ray fluorescence using a lithium-drifted silicon detector; radioisotopic source, I125; beryllium window, 0.005" thick.

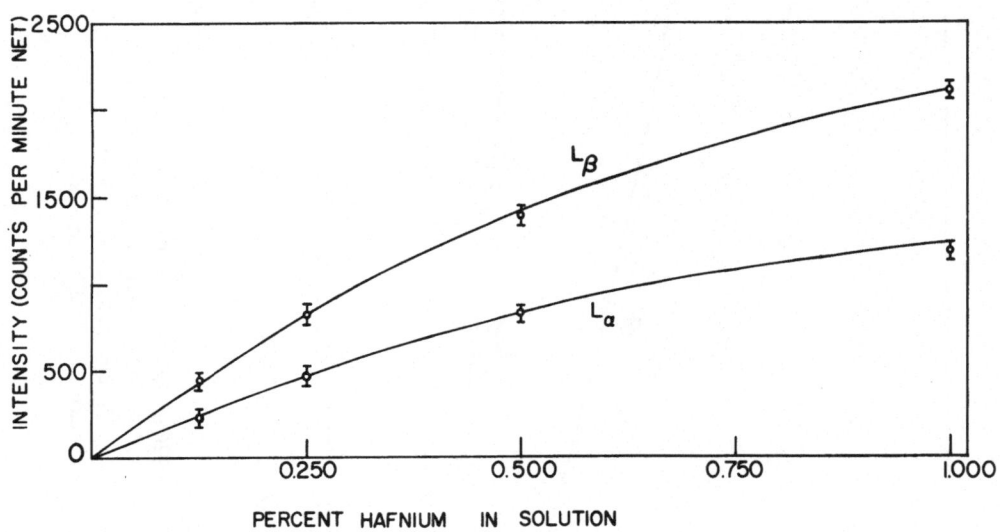

PERCENT HAFNIUM   IN   SOLUTION

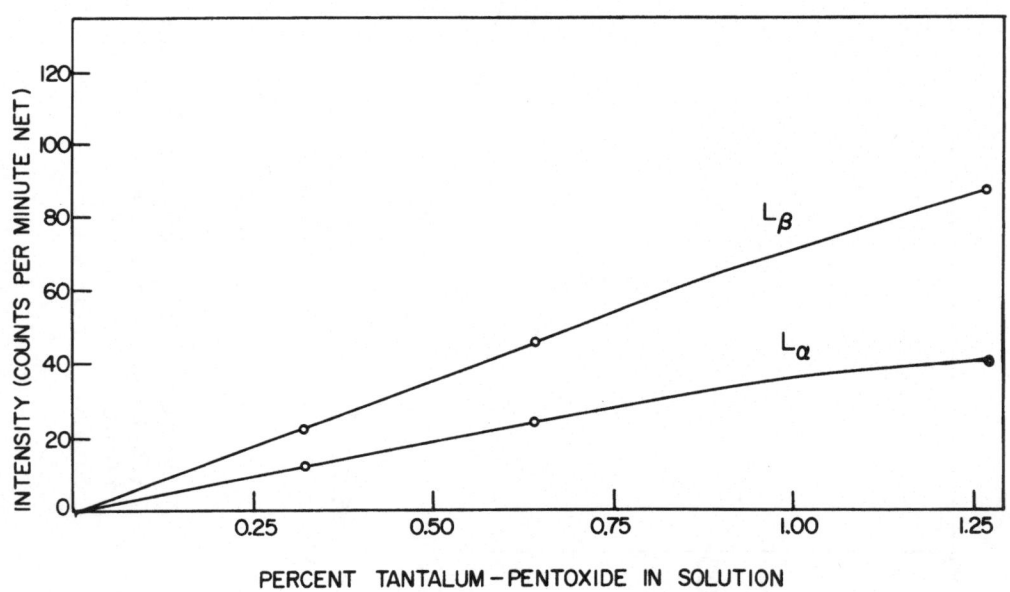

PERCENT TANTALUM—PENTOXIDE IN SOLUTION

detector whereas the detector cannot receive signals from the source. The iodine 125 source have about 2×10^9 disintegrations per minute. It is an electron capture isotope with a half-life of 57.4 days and a transition energy of 35.6 keV which is highly converted. Although there are 27.47, 30.99 and 35.6 keV energies available the predominate radiation is 27.47 keV. This radiation impinging on the sample causes emission of the x-rays in the sample.

A plot of excitation energy versus Kα, Kβ, Lα, Lβ and Lγ peak positions appearing in the channel of a pulse height analyzer is shown in figure 1. The predominant radiation from the disintegration of radioactive iodine 125 has sufficient energy to excite the Kα and Kβ of zirconium, niobium and molybdenum. Results obtained from solutions of these elements are shown in figure 2. In like manner, the intensities Lα and Lβ of hafnium, tantalum and tungsten versus the concentration are shown in figure 3.

Figure 4 shows typical graphs from the x-y recorder. The solution used for the graphs contains 0.009% niobium pentoxide. To obtain a clear solution for the x-ray fluorescent analysis by the energy dispersion system, it was necessary to fuse the pentoxide with potassium bisulphate. The melt was dissolved in concentrated sulfuric acid and saturated ammonium oxalate (1). Digital print-outs obtained simultaneously with the graphs is shown in Table I.

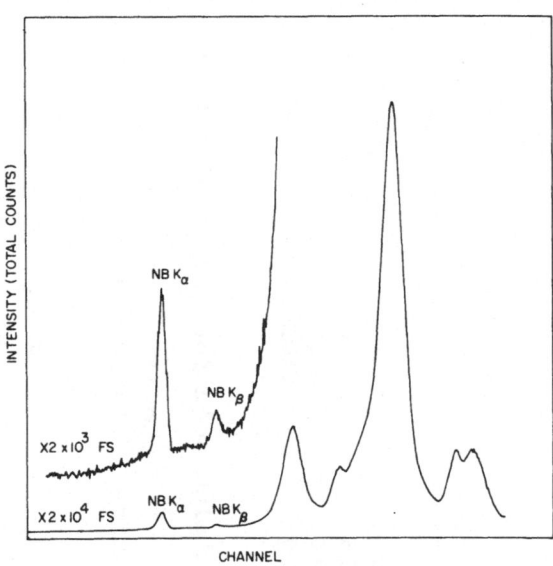

Figure 4. A portion of the graph from analysis of 0.009% niobium pentoxide solution. (See digital print-out data on table I.)

TABLE I

Some typical data from digital print-out on the analysis of niobium pentoxide and tantalum pentoxide solutions.

(A) Niobium solutions (containing 18N sulfuric acid and saturated ammonium oxalate).

% Pentoxide	Count time (minutes)	FW 0.25 Max. $K\alpha$	BG $K\alpha$	Channels in peak $K\alpha$	Net Count* $K\alpha$	Standard error $K\alpha$
0.07	5	1057	5.8	19	192	±0.9
0.035	50	6319	88	16	98.2	±2.4
0.018	50	4260	88	18	53.8	±2.1
0.009	1107.7	57360	1600	19	24.8	±0.4

(B) Tantalum solutions containing potassium heptafluorotantalate in 48% hydrofluoric acid.

% Pentoxide	Count time (minutes)	FW 0.25 Max. $L\alpha$	FW 0.25 Max. $L\beta$	BG $L\alpha$	BG $L\beta$	Channels in peak $L\alpha$	Channels in peak $L\beta$	Net Count* $L\alpha$	Net Count* $L\beta$	Standard error $L\alpha$	Standard error $L\beta$
1.27	20	1358	2532	33	33	17	25	40.5	86.8	±2.1	±2.9
0.64	20	1025	1732	32	33	17	25	24.1	45.9	±2.0	±2.5
0.32	50	2489	4003	110	115	17	25	12.4	22.5	±3.0	±1.6

*Counts per minute net after corrected for decay time.

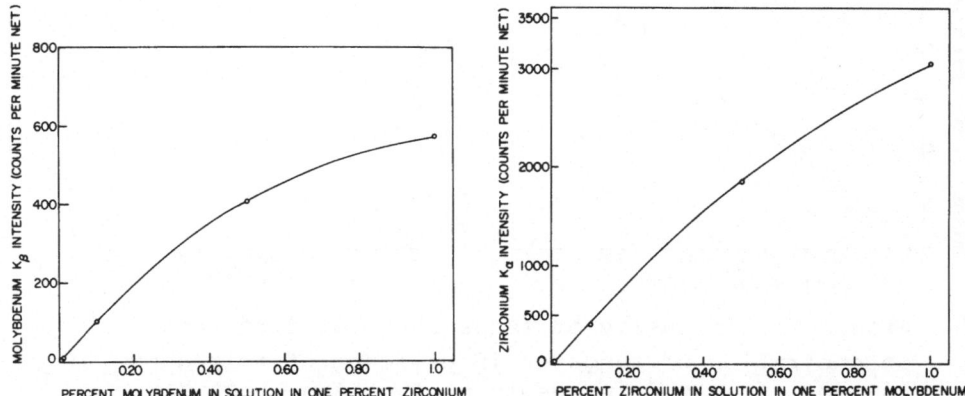

Figure 5.   Intensity of molybdenum K_β at different
concentration of molybdenum in presence of one-
percent zirconium, (upper left), and intensity of
zirconium K_α at different concentration of zirconium
in presence of one-percent molybdenum, (upper right).

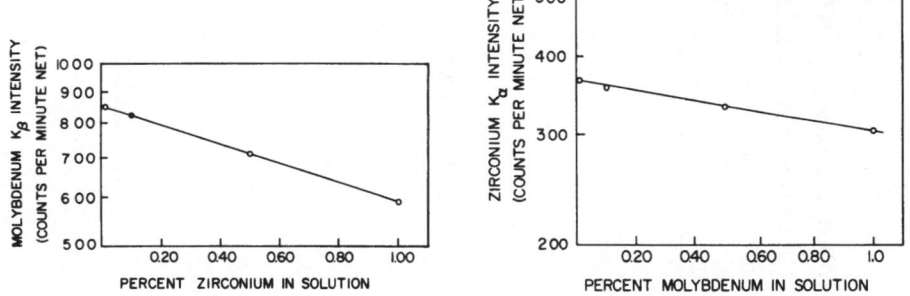

Figure 6.   Effect of intensity of molybdenum K_β
in the presence of zirconium, (lower left) and the
effect of intensity of zirconium in the presence of
molybdenum, (lower right).

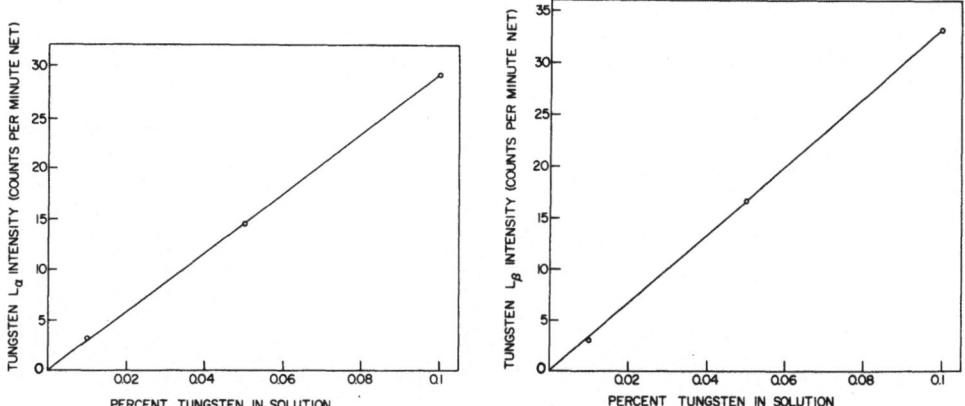

Figure 7. Intensity of tungsten L$_\alpha$ at different concentration of tungsten in presence of molybdenum, (upper left), and intensity of tungsten L$_\beta$ at different concentration of tungsten in presence of molybdenum, (upper right).

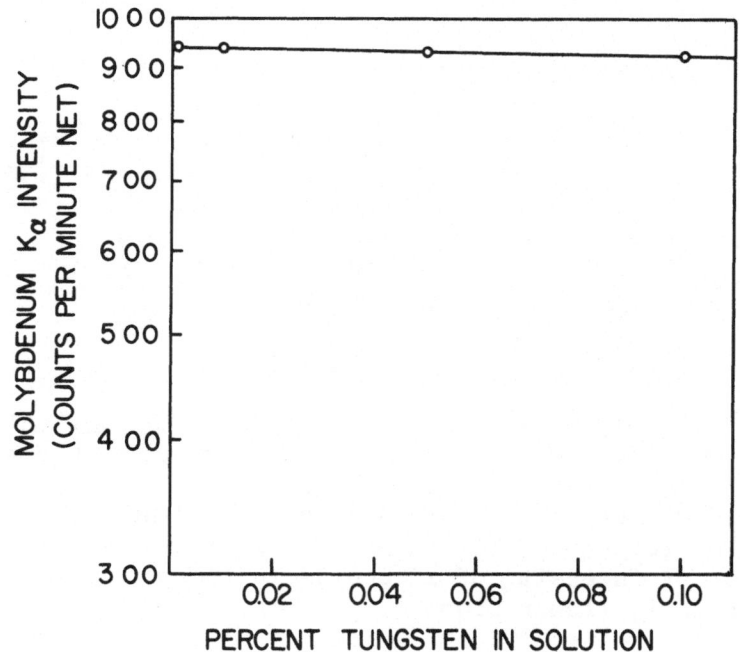

Figure 8. Effect of molybdenum K$_\alpha$ intensity in the presence of tungsten in solution, (lower left).

To study the enhancement and absorption effects, solutions containing zirconium and molybdenum were prepared. In
one series of solutions the zirconium concentration remains
the same with varying concentration of molybdenum. In a
second series the molybdenum concentration remains the same
with varying concentration of zirconium. The results are
shown in figures 5 and 6. In like manner tungsten and
molybdenum mixtures were also prepared and studied. The
results are shown in figures 7 and 8. As the results indicate,
a mixture of tungsten and molybdenum show no significant
enhancement effect. However, some enhancement is evident
in the mixture of zirconium and molybdenum; this is because
the K absorption edge of zirconium is below the Kβ of molybdenum. The Kα line of molybdenum is below the K edge of
zirconium so we do not have the maximum enhancement which
evidenced in a previously reported paper (7).

The L x-rays (in keV) of tantalum, tungsten and hafnium
are shown in the following:

hafnium	Lα	7.898	tantalum	Lα	8.145	tungsten	Lα	8.396
	Lβ	9.021		Lβ	9.341		Lβ	9.670
	Lγ	10.514		Lγ	10.892		Lγ	11.283

Thus, with this system, hafnium and tungsten are easily separated but tantalum and either hafnium or tungsten would be
more difficult. Newer systems are quoted as having resolutions of 130 eV and therefore these detectors would easily
separate all of these lines.

Although the intensity of the Lα line should be greater
than the Lβ line, the absorption by the other elements of
the solution (oxygen, sulfur, sodium etc.) is far greater
for the Lα lines than the Lβ lines so the Lβ line appears to
be more intense. As an example, the mass absorption coefficient of sulfur for the Lβ of tungsten is 48 cm^2/g.,
72 cm^2/g. for Lα, so it can easily be seen that Lα line
will be absorbed much more than the Lβ line.

The energies of the K x-ray (in keV) of zirconium and
molybdenum are as follows:

zirconium	Kα_1	15.774	molybdenum	Kα_1	17.478
	Kα_2	15.690		Kα_2	17.373
	Kβ_1	17.666		Kβ_1	19.607
	Kβ_2	17.969		Kβ_2	19.964

At these energies (15 - 19 keV) the system resolution is
inadequate to separate the Kα_1 from the Kα_2, or the Kβ_1 from

the $K\beta_2$. It is also incapable of separating the $K\beta$ of zirconium from the $K\alpha$ of molybdenum, approximately 0.19 keV apart. In as much as these peaks could not be separated we used the $K\alpha$ of zirconium and the $K\beta$ of molybdenum which are very widely separated (almost 4 keV). The $K\beta$ intensity is about one-sixth or one-seventh that of the $K\alpha$ line so the molybdenum $K\beta$ intensity is much lower than the same concentration of zirconium where the $K\alpha$ line was used.

The limit of sensitivity of this system cannot be directly evaluated; the instrument has the stability to operate continuously for many, many hours without evidence of drift or change in resolution. This means that it is possible to expose a sample for at least several days (3000 minutes) with the total number of counts over the range of channels integrated following the equation:

$$N_T = Ct \pm \sqrt{N_T}$$

where $\quad$ N_T = total counts,

$\qquad$ C = counts per minute (counting rate),

$\qquad$ T = time.

This means that the significant source of error is the statistical spread due to the Poisson distribution.

If the total counts, N_T, has the background subtracted, N_B, then the net counts N_N gives the counts due to the sample; if $\sqrt{N_T}$ is the deviation then as N_N increases linear with time and $\sqrt{N_T}$ increases as $\sqrt{\text{time}}$; our ability to detect is limited only by how long we are willing to count.

The advantages of x-ray fluorescent analysis by the energy dispersion using the system described should be pointed out. The instrument used is compact and could be adopted to on-stream analysis. Samples subjected to analysis are not destroyed and all spectra lines are displayed simultaneously with monotonic background over wide energy range. There is minimal or no sample preparation required and no instrumental spectra line interference. The cost of the instrument is comparatively low. The disadvantages of the system, as compared with the conventional x-ray fluorescent analysis, is its low counting rate, and inferior resolution. These disadvantages will be improved in the future. The half-life of some of the radioisotopic source such as iodine 125 is short but, if need be, corrections can be made for the results taken at different intervals.

SUMMARY

1. A number of solutions containing zirconium, hafnium, niobium, tantalum, molybdenum and tungsten have been prepared. To keep these solutions from precipitation and flocculation hydrofluoric acid, sulfuric acid, ammonium oxalate and sodium hydroxide are used depending on the chemicals used for the preparation of these solutions.

2. For niobium and tantalum solutions, concentrated sulfuric acid and saturated ammonium oxalate are used after the pentoxides have been fused with potassium bisulfate. When potassium heptafluorotantalate was used for preparation of a clear solution, 48% hydrofluoric acid was used. Tungsten forms insoluble tungstic acid in acids such as hydrochloric acid and sulfuric acid and therefore a clear solution cannot be prepared. In earlier experimentations one normal sodium hydroxide was used for both tungsten and molybdenum and for mixtures of the two. The remainder of these elements, namely zirconium and hafnium, were prepared and kept in solutions with one normal sulfuric acid.

3. Conventional chemical analysis dealing with mixtures of tantalum and niobium, molybdenum and tungsten, and zirconium and hafnium are difficult if not impossible. However using the nondispersion technique (energy dispersion) together with radioisotopic source, analysis of these mixtures can be carried out with a spectrometer of simple design. Cost of this instrument is low. Its lightweight and compactness enable one to carry it with ease.

4. For the six elements under study a radioisotopic source (iodine 125) was used. The tellurium x-ray emitted in the disintegration of iodine 125 has sufficient energy to excite the $K\alpha$ and $K\beta$ spectra of zirconium, niobium and molybdenum. It also can excite the $L\alpha$, $L\beta$, and $L\gamma$ spectra of hafnium, tantalum and tungsten spectra. Using a pulse height analyzer having one thousand channels the $K\alpha$, $K\beta$ and the $L\alpha$, $L\beta$ and $L\gamma$ are located in different parts of the channels versus energy dispersion graph. Analysis of a pair of mixtures by this system presents little difficulty. However, iodine 125 has a short half-life compared with other radioisotopes used for this purpose. For consistent results taken at intervals, a correction factor can be applied since the decay scheme for this radioisotope is well known.

5. Enhancement and absorption effects have been studied and typical results are presented.

6. The use of solutions for the present study has the

inherit advantage of obtaining homogeneity of samples used for analysis. By the dilution technique different concentrations of the interested elements can readily be achieved.

REFERENCES

1. Frank L. Chan, "Precipitation and Determination of Tantalum and Niobium from Homogeneous Solution with 3:3': 4':5:7-Pentahydroxyflavanone," Talanta, Vol. 7, pp. 253-263 (1961).

2. Ross W. Moshier, "Analytical Chemistry of Niobium and Tantalum," Pergamon Press, London (1964).

3. Frank L. Chan and Ross W. Moshier, "Spectrophotometric Determination of Molybdenum in Steel with 3:3':4':5:7-Pentahydroxyflavanone," Talanta, Vol. 3, pp. 272-276 (1960).

4. L. S. Birks and E. J. Brooks, "Hafnium-Zirconium and Tantalum and Columbiam System," Anal. Chem., Vol. 22, p. 1071 (1950).

5. William J. Campbell, "Energy Dispersion X-ray Analysis Using Radioactive Sources," X-ray and Electron Methods of Analysis, H. van Olphen and W. Parish eds., Plenum Press, New York, pp. 36-54 (1968).

6. K. G. Carr-Brion and K. W. Payne, "X-ray Fluorescence Analysis," The Analyst, Vol. 95, No. 1137, p. 997 (1971).

7. W. Barclay Jones and Robert A. Carpenter, "Sensitivity of a Nondispersive X-ray Fluorescent Spectrometer for Multielement Trace Analysis." Paper presented at the Second International Symposium on Nucleonics in Aerospace held in Columbus, Ohio, 12-14 July 1967.

8. W. Barclay Jones and Robert A. Carpenter, "Nondispersive X-ray Fluorescent Spectrometer," in John B. Newkirk, Gavin R. Mallett and Heinz G. Pfeiffer, Editors, Advances in X-ray Analysis, Vol. 11, Plenum Press, New York, pp. 214-229 (1968).

9. Frank L. Chan, "Dispersive and Nondispersive X-ray Fluorescence Methods for the Measurement of the Thickness of Films of Cadmium Sulfide and Other II-VI Compounds," in E. L. Grove and Alfred J. Perkins, Editors, Developments in Applied Spectroscopy, Vol. 7A, Plenum Press, New York, pp. 3-30 (1969).

10. Frank L. Chan and W. Barclay Jones, "Quantitative
     Determination of Sulfur, Chlorine, Potassium, Calcium,
     Scandium and Titanium in Aqueous Solutions by Radio-
     isotopic Excited Fluorescent Spectrometer and by Con-
     ventional X-ray Spectrometer," in J. B. Newkirk and
     C. O. Ruud, Editors, Advances in X-ray Analysis, Vol. 14,
     pp. 102-126, Plenum Press, New York (1971).

FLUORESCENCE ANALYSIS USING AN Si(Li) X-RAY ENERGY ANALYSIS

SYSTEM WITH LOW-POWER X-RAY TUBES AND RADIOISOTOPES

G. R. Dyer, D. A. Gedcke, and T. R. Harris

ORTEC, Incorporated

Oak Ridge, Tennessee 37830

X-ray fluorescence spectroscopy has been in use since the early days of the twentieth century, when Moseley confirmed the order of the chemical periodic table (1). However, fluorescence spectroscopy until recently has depended on diffraction methods to obtain sufficient resolution. Intrinsic resolution of ionization chambers, scintillation detectors, and proportional counters is inadequate for discrimination of lines due to adjacent elements of low atomic number. The advent of solid-state detectors, especially those using lithium-compensated silicon and low-noise electronics, has recently brought intrinsic energy resolution to the point where lines from adjacent elements as light as carbon and nitrogen can be resolved in theory; and detection of K radiation from elements as light as sodium is practical. Thus the solution to the long-standing problem of an adequate detector is at hand, and energy-dispersive spectrometers are now feasible.

In fact, the last three or four years have seen several laboratories engaged in work with radioisotope-excited spectrometers; results have been interesting and impressive. Several companies now offer radioactive source-type spectrometers with Si(Li) detectors, and tube-excited spectrometers with these detectors have very recently become available. This paper is intended to compare a tube-excited and a source-excited Si(Li) spectrometer, pointing out advantages and drawbacks of both, and reporting some results to be expected with each type of system.

Both systems have in common the Si(Li) detector system with its analog and data-processing electronics. The detector (block diagram, Figure 1) is basically a small crystal of silicon, especially processed by "drifting" lithium ions through the lattice to compensate for electrical impurities (P-type doped

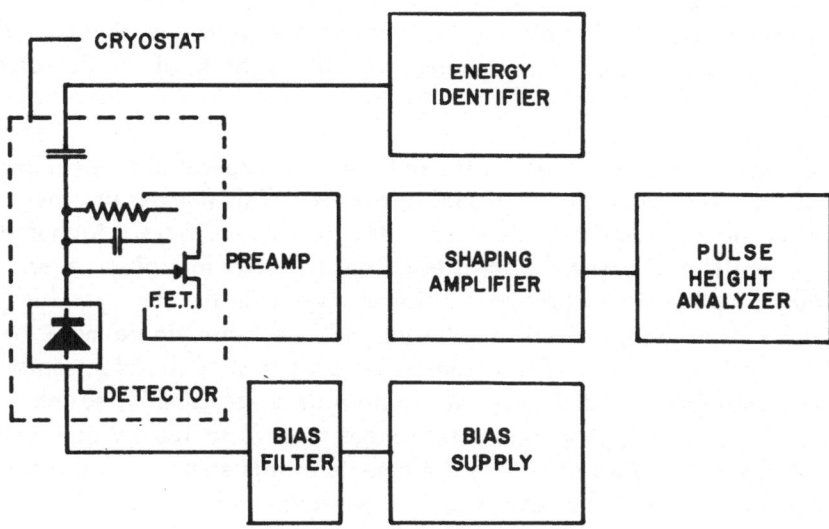

Figure 1.   Diagram of Basic Si (Li) Spectrometer.

silicon is the starting material).   When the lithium-drifted crystal is provided
with evaporated electrodes, connected as a reverse-biased diode, and cooled
to liquid nitrogen temperature, it forms a low-leakage, sensitive volume which
can be ionized by incident radiation.   The average energy of ionization, about
3.8 eV per ion pair, is about one-eighth that required for gaseous ionization
chambers; and the Fano correlation (2) for ionization due to a single photon
gives an effective ionization energy of about one-eighth the above value, so
that the cooled silicon crystal has an intrinsic resolution given by

$$\text{FWHM} = 2.35 \sqrt{\epsilon \text{FE}} \text{ , where } \begin{array}{l} \epsilon = 3.8 \text{ eV/pair} \\ F \approx 0.125 \text{ for present devices} \\ E = \text{photon energy.} \end{array} \quad (1)$$

Thus at 1 keV, the intrinsic resolution of present silicon detectors is approximately 52 eV; at 10 keV, intrinsic resolution is about 162 eV.   At the present
time, detectors of good resolution can be made from both silicon and germanium.   Silicon is used for fluorescence spectrometers for several reasons.
Germanium has a more pronounced efficiency change about its K absorption
edge (at $\approx$ 11 keV) than does silicon (at 1.75 keV), so silicon makes a
better-behaved detector in this important energy region.   Germanium has
a higher atomic number; therefore, problems with entrance windows and
dead layers are more severe with germanium detectors than with silicon.
Finally, silicon detectors at present offer superior resolution at low energies.
However, silicon begins to lose efficiency at higher x-ray energies; a 3 mm

thick detector has only about 15% efficiency at 50 keV, and efficiency falls rapidly above that energy. Thus silicon fluorescence spectrometers usually cover energies up to about 40 keV. The lower energies (down to about 1 keV) are limited by the beryllium entrance window (0.5 mil thick) of the detector cryostat (3).

For a 1 keV photon absorbed in the detector, the 264 electrons released must be collected and the resulting signal amplified. This number of electrons is a small signal, requiring the state of the art in amplifiers. A charge-sensitive preamplifier converts the total electrons released in a given event into a voltage, adding some electronic noise of its own in the process; this noise is usually in the range of 80 to 120 eV, with the latter figure more common for practical systems. The preamplifier adds most of the electronic noise to the signal (noise added by other electronics is very small), so the resolution of a typical Si(Li) spectrometer ranges from 95 to 130 eV at 1 keV, adequate to resolve K_α lines of adjacent elements in this energy range. Separation between sodium and magnesium K_α's is approximately 213 eV.

The preamplifier is followed by a shaping amplifier (whose bandwidth determines system resolution to a great extent; above results are typically attained with a gaussian pulse shape and a time constant of 5 to 15 μsec), a single- or multichannel pulse height analyzer, a scaler or memory unit, and a readout device. Peripheral equipment such as sample changers, energy identifiers, and various system monitors and controllers may also be included. The systems to be discussed here use a multichannel pulse height analyzer; this is most convenient for general analytical use, since an entire range of energies can be monitored at once. This enables a rapid assay of major elements contained in a given sample to be done in seconds or minutes, and gives the operator some idea of what to do to optimize results of a longer quantitative run.

To make a fluorescence analyzer system, a source of exciting radiation must be combined with the Si(Li) spectrometer. Most systems until now have used radioactive isotopes as sources of radiation; and radioactive sources have several advantageous features: the radioisotope fluorescence analyzer can be simpler, less expensive, and more portable than analyzers using other radiation sources. In fact, all the equipment needed for a great deal of useful work is a spectrometer, a suitable radioisotope radiation source, and a sample to be analyzed.

Selecting a suitable radioisotope and geometry for a radiation source involves several compromises. Our experience indicates that best performance is obtained with an annular source configuration using the materials and activities in the following Table I (4).

Table I

Radioisotope	Source Activity	Half-life	Radiation Energy	Range of Elements Stimulated to K Fluorescence
Fe–55	50 mCi	2.6 Y	5.9 keV	Na – V
Cd–109	5 mCi	1.29 Y	(88 keV), 22 keV	Ti – Ru
Am–241	25 mCi	458	≈ 60 keV	Fe – Tm

If a sample-holding arrangement and radiation protection for the user is included in the fluorescence analyzer design, a system such as shown in Figure 2 is finally obtained. This arrangement provides for four samples of maximum diameter two inches and height two inches to be rotated into position over the source and spectrometer. The annular source provides good count rate, high take-off angle, and relative insensitivity to sample surface roughness. The access port for sample changing is diametrically opposite the source and detector, so the user is shielded from source radiation while changing samples.

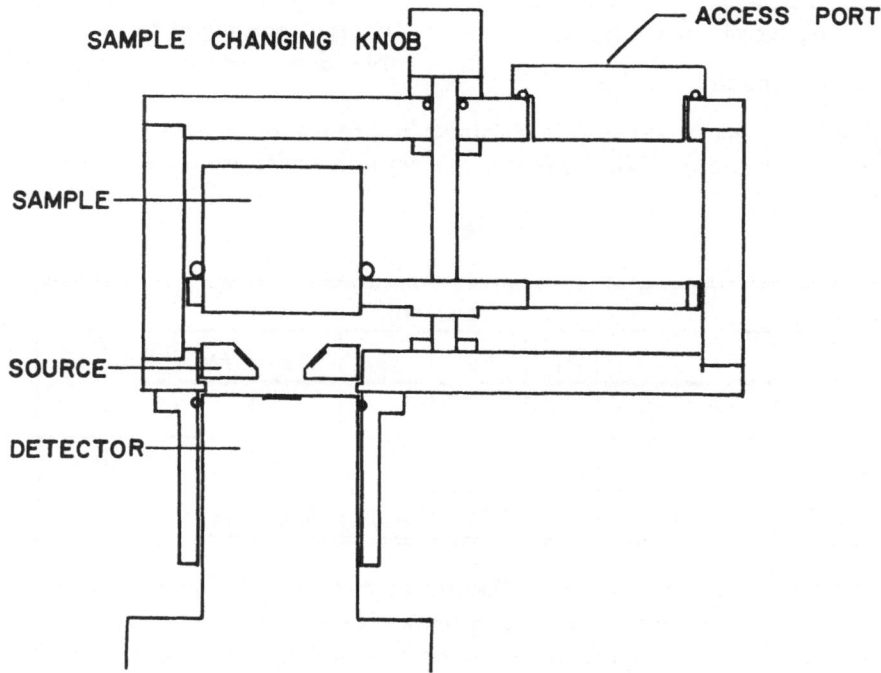

Figure 2.   Source-Excited Fluorescence Analyzer Geometry.

With the preceding system, a series of ten-minute counts has been made on a number of pure elements. For these results we have used the following calculation to predict minimum detectable limits:

$$c_{MDL} = \frac{3}{\sqrt{\frac{N_P}{N_B}\, N_P}} \qquad (2)$$

Figure 3 illustrates the quantities involved in the formula.

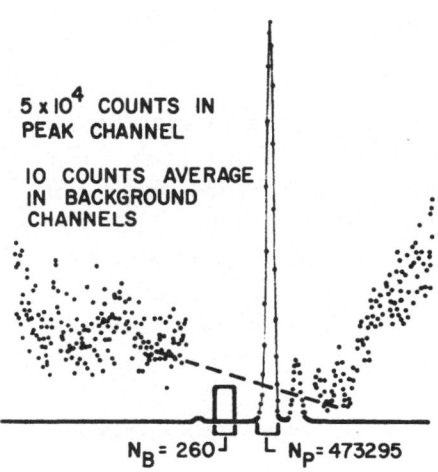

5 x 10^4 COUNTS IN PEAK CHANNEL

10 COUNTS AVERAGE IN BACKGROUND CHANNELS

N_B= 260 N_P= 473295

Figure 3.   Spectrum of Iron.

c_{MDL} = percentage of the pure element which would have to be present to give a peak height at least three times the standard deviation of the background level.

N_P = total counts under the K_α peak down to 1/100 of maximum.

N_B = total counts in an energy span equal to that of N_P, due to average background.

Table II summarizes results obtained in a ten-minute counting period, using a 6 mm diameter Si (Li) detector; detectable limits are given in parts per million.

Table II

Isotope	Element								
	Al	S	Ti	Fe	Cu	Ge	Mo	Cd	Ce
Fe-55	1440	250	72						
Cd-109				152	71	54	64		
Am-241				5340	1433	682	145	90	165

For better detectable limits, the source of radiation needs to be variable in intensity and energy so as to optimize fluorescent excitation for a given element. These requirements can be satisfied to a large extent with an x-ray tube as a source of radiation. The tube-excited fluorescence analyzer has other advantages, also. The tube can be turned off, so that there is no radiation hazard while the operator changes samples or works on the machine. Tubes can be built with different anodes, so that a variety of characteristic lines is available to the operator; and a well-designed tube-type generator can be made very stable, so that need for frequent calibration is eliminated. Of course, the price for these features is a more complex machine and slightly more complicated operation.

Figure 4 illustrates the major components of a fluorescence analyzer used to obtain the results presented below. The machine incorporates an Si (Li) x-ray spectrometer, an electronically stabilized x-ray generator, an automatic sample-changing mechanism holding up to 12 samples, and a controlled sample environment (air, helium, or vacuum).

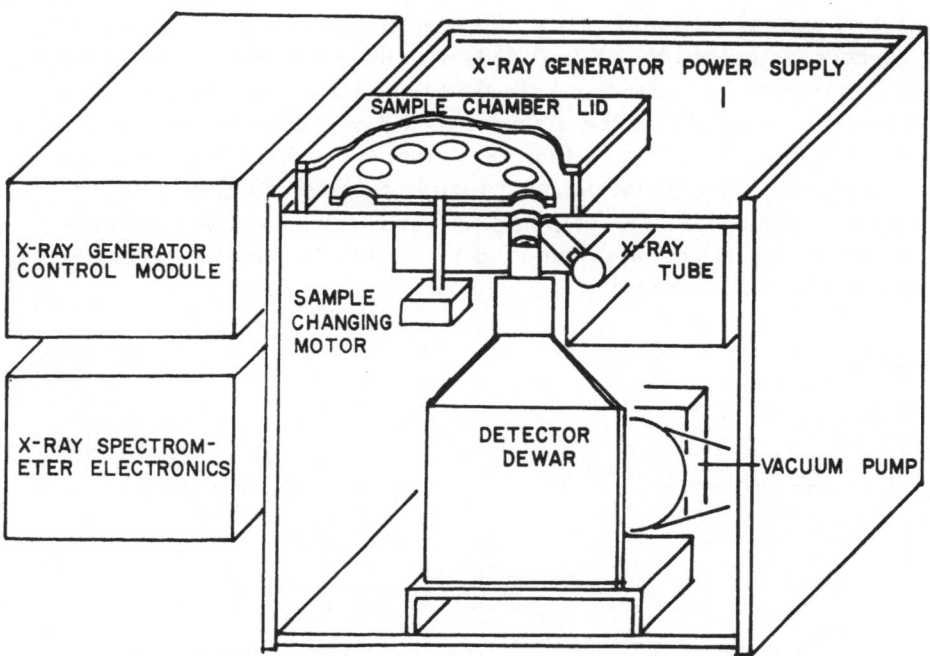

Figure 4.   Schematic Drawing of Tube-Excited Analyzer.

The x-ray generator is built into an L-shaped, oil-filled tank and includes the high-voltage anode supply and low-voltage supplies for the electron gun in the x-ray tube. These supplies are stabilized against line voltage changes over a range of 100 to 130 volts, and are controlled by a module mounted at the operator console. Anode voltages of 5 to 50 kV in 5-kV steps and anode currents of 1 to 200 μA in a 1,2,5 sequence are selected by front-panel switches. Both anode voltage and anode current are regulated against precision zener references.

An example of the stability to be expected from the x-ray generator is given by Figure 5. To obtain this plot, an iron sample was put in the machine and the generator run at 25 kV, 2 μA with a tungsten anode. Counting rate was 1100 counts/second, and a 1000-second counting time gave over 10^6 counts in each point. Data were accumulated over a four-day time span; the points were plotted, suppressing the two most significant figures. Note that all but two of the points fell inside plus-or-minus three times the standard deviation of the average count, indicating that most of the variation in the data was caused by normal statistical fluctuations in count rate. If successive groups of eight data points are summed and plotted, the resulting curve is within ± 0.1% of the average value, with no long-term drifts or variations indicated. To obtain these data, x-ray generator current and voltage had to remain constant to better than 0.1% over the running time of the machine. After leaving the machine off for 24 hours, a second set of data was taken; both the first 1000-second count and the average of this 24-hour run agreed to within 0.15% with the average of the previous data. Experience indicates that both short- and long-term repeatability and count-rate stability can be specified as within ± 0.25% of the mean value. It is interesting to note that for a four-day period, the variation in count rate due to the natural decay of radioisotopes would amount to 0.59% for cadmium-109 and 0.29% for iron-55. The x-ray generator's stability should prove very helpful in eliminating the daily recalibration necessary with machines available to date.

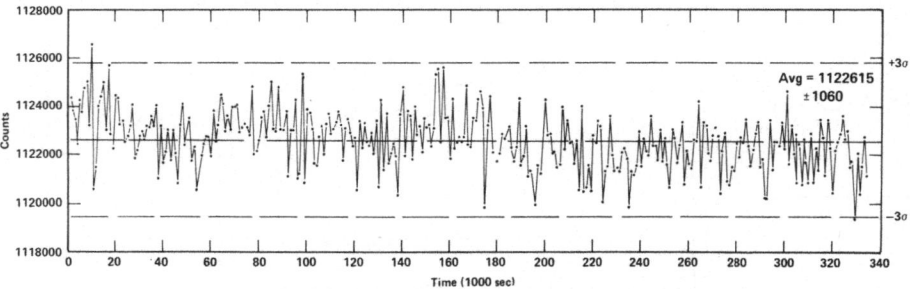

Figure 5. Count Rate Stability of Tube-Excited Fluorescence Analyzer.

Along with constant output, the x-ray generator features selectable anodes. Pushbuttons at the operator's position select steering voltages which deflect the focused electron beam to either of two anodes inside the x-ray tube. Tungsten and molybdenum are standard at present, though other materials (chromium, copper, etc.) may be obtained on special order. The operator is able to enhance by anode selection and operating conditions the K lines of molybdenum (≈ 17 keV), the L lines of tungsten (≈ 8 and 10 keV), or the L lines of molybdenum (≈ 2.3 keV). By using these lines, especially with a proper filter in between tube and sample, the operator is able to excite fluorescence in any of three desired energy ranges. Of course the bremsstrahlung spectrum of the tube extends to the highest energy of the electron beam, so elements up to curium ($K_\alpha \approx 40$ keV) may be fluoresced using the bremsstrahlung spectrum and tungsten target.

The x-ray tube is supplied with a 0.005-inch beryllium window, for good transmission of low-energy radiation. Count rate on a pure aluminum sample is greater than 1000 counts/second in aluminum K radiation for a molybdenum anode, 10 kV anode potential, and 200 µA anode current.

The sample chamber of the tube-excited analyzer is designed to hold 12 samples, each two inches in diameter by one and one-half inches high. The sample-holding wheel is motor-driven, so the operator can select a given sample from the operating console; a Nixie readout shows the sample in position at any given time.

Two modes of operation are provided. In the Manual mode, the machine operator selects which of 12 positions he wishes to analyze, by turning a selector switch; the sample wheel advances to that position and stops. In the Automatic mode, the operator moves the wheel to a starting position as described above, switches to Auto, and sets the selector switch to a final position. He then starts his data collection process (the machine will handle any number of single-channel analyzers set over specific energies, or a multichannel analyzer to cover a whole span of energies); at the end of each data-taking period, the data will be printed out, the sample wheel will be advanced to the next position, and the machine will repeat the data acquisition and print-out cycle until the last sample position selected has been analyzed. This feature should be especially useful for routine analyses and unattended operation.

Only minimal sample preparation is needed, since the analyzer does not depend on a critical focus being maintained. In normal operation, the x-ray tube illuminates an elliptical area of the sample approximately one-half by one inch; the sample-to-source distance is approximately three inches, and sample-to-detector distance approximately two inches. These comparatively long working distances allow considerable variation in sample position, both within the illuminated area and in vertical dimension, without seriously affecting count rate. Rough or curved samples such as geological specimens, pottery, and machine parts, can be analyzed with no difficulty.

The top surface of the sample wheel is the nominal working plane for the analyzer; if a sample larger than two inches in diameter needs to be analyzed, it can be placed on top of the sample wheel so that the desired surface is illuminated by the tube. In this way, tools, coffee cups, large rocks, or what have you can be assayed in a few seconds for gross elemental content; a few minutes' run can show trace elements to a fraction of a percent.

For analyses of radiation much below 6 to 8 keV in energy, air attenuation becomes quite significant (sample-to-detector distance in this machine is two inches) so two options have been provided to allow low-energy analysis. The entire sample chamber is vacuum-tight, and a vacuum pump supplied with the machine provides 100–200 μ vacuum, suitable for elements down to sodium ($\approx$ 1 keV). For volatile liquids, or other problem samples, a helium tank and regulator can be added; the helium path is only slightly worse in attenuation than vacuum for aluminum K radiation, and allows analysis of water, blood, motor oil, etc., for light elements.

A table of detectability limits for the tube-excited analyzer was compiled, using pure elements and the formula already given for C_{MDL}. The detectability limits in parts per million for several elements covering energies from 1.5 to 40 keV are given below. In all cases, analysis time was ten minutes.

Table III

Anode	Element				
	Al	S	Fe	Ge	Cd
Mo	173				
W	504	179	47	40	198

These results were obtained using direct radiation from the x-ray tube on pure elements, and give a rough idea of the capabilities of the machine. In some cases, interference between adjacent lines will make detectability limits considerably worse than given above. For example, traces of iron in a sample predominantly manganese will be masked by the manganese K_β peak, so that 47 ppm will not be attainable. In other cases, these limits may be considerably exceeded. For example, a light matrix containing a heavy element can be analyzed quantitatively to better than 10 ppm in ten minutes' running time, especially if a filter technique is used to decrease the bremsstrahlung spectrum of the x-ray tube and enhance the characteristic lines (5). Figure 6 shows the spectrum of molybdenum radiation at 30 kV anode potential scattered from a lucite block, without and with a molybdenum filter to suppress bremsstrahlung. Figure 7 shows the spectrum obtained in ten minutes for a water sample containing 10 ppm of mercury by weight (as Merthiolate). Figure 8, A and B, shows a ten-minute count for a demineralized water sample, without and with a 10 ppm addition of chromium (as $K_2Cr_2O_7$)

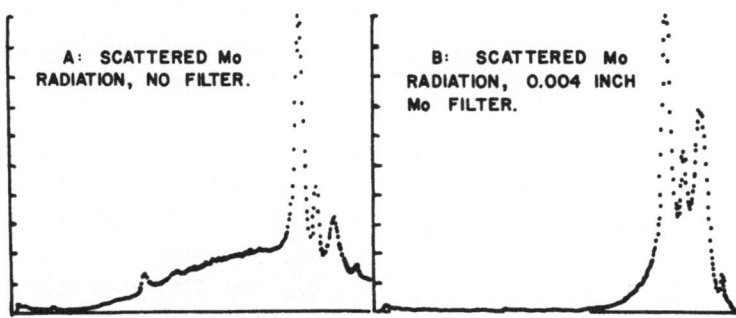

Figure 6.   Bremsstrahlung Suppression by Filtration.

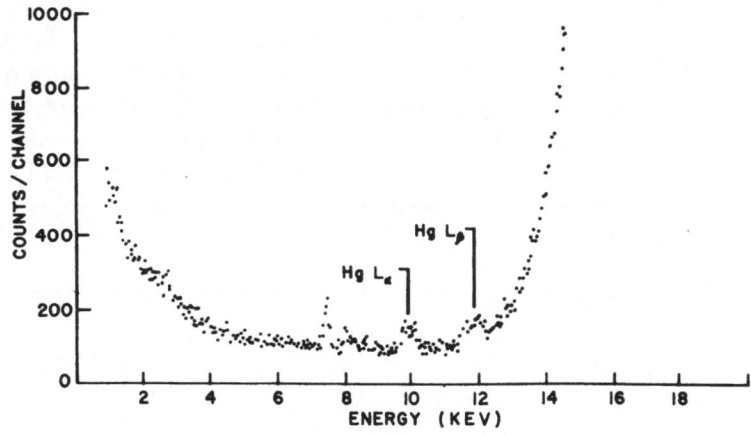

Figure 7.   Sprectrum of 10 ppm Hg in H_2O, 10-minute Count.

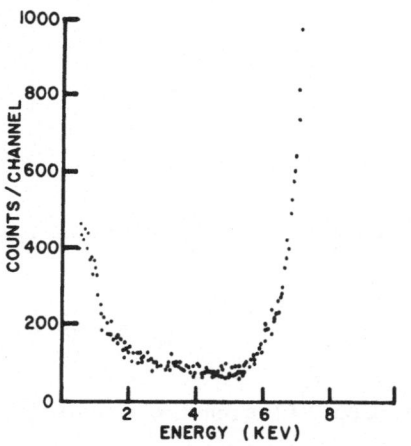

Figure 8a.   Demineralized Water
Sample, 10-minute Count.

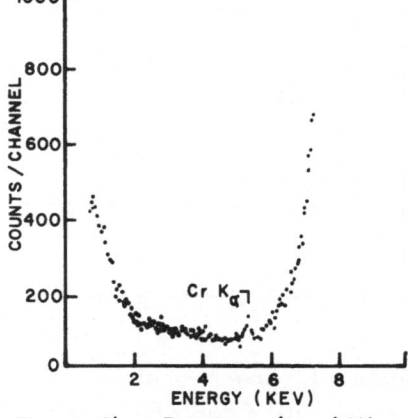

Figure 8b.   Demineralized Water
Sample with 10 ppm Cr Added,
10-minute Count.

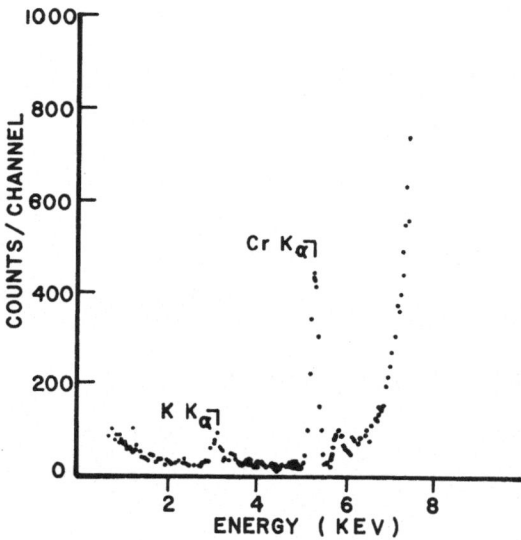

by weight, obtained with tungsten radiation filtered with a copper foil.

Small samples can be analyzed readily. The spectrum in Figure 9 is from a single crystal of $K_2Cr_2O_7$ (0.5 mg total weight) analyzed for ten minutes. Lines from both potassium and chromium are clearly shown.

Larger objects can give interesting results. Figure 10 shows the spectrum obtained from a glazed earthenware coffee cup in one minute's counting time. Traces of

Figure 9. Spectrum from a Single Crystal of $K_2Cr_2O_7$ (0.5 mg total weight).

potassium, calcium, and copper are present, along with large amounts of manganese, iron, and lead.

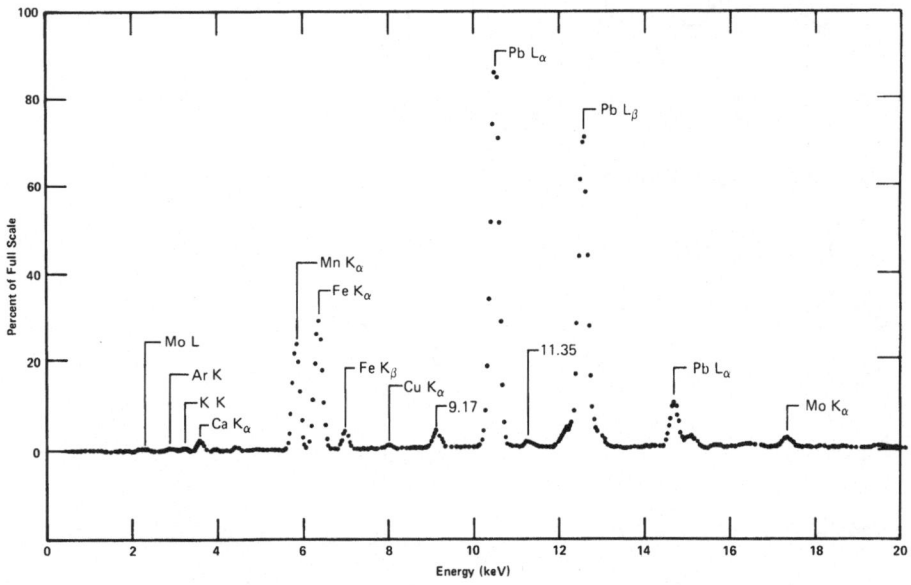

Figure 10. Spectrum of Glazed Coffee Cup. One-minute Counting Time. 10^4 Counts Full Scale.

Several other applications for the machine are not illustrated here. Workers at Vanderbilt University are presently evaluating the system as a tool for determining amounts of trace elements in blood and other biological samples. Analyses of percentage of sulfur in fuel oil, iron in engine oil, and silver in photographic film have been done with the present equipment.

In short, a powerful tool for rapid quantitative and accurate qualitative analysis of the range of elements from sodium through cesium in K radiation and through the highest Z elements in L radiation is now commercially available. The radioisotope version is portable, simple to operate, but still offers good sensitivity. The tube-excited version is versatile, available with a wide range of options, and offers sensitivity down to the order of 10 ppm for many elements in ten minutes' analyzing time.

ACKNOWLEDGMENTS

The authors wish to acknowledge the contributions of several workers to the analyzer project. Don Schechter did the development work on the x-ray tube, and Jim Arrington was responsible for mechanical design of the TEFA. George Presley did much of the electronics layout and construction for the x-ray generator power supply and controller.

REFERENCES

1. N. G. J. Moseley, "The High Frequency Spectra of the Elements," Phil. Mag. 26, 1024 (1913); 27, 703 (1914).

2. U. Fano, "Ionization Yield of Radiations. II. The Fluctuations of the Number of Ions," Phys. Rev. 72, 26 (1947).

3. Energy Dispersion X-Ray Analysis: X-Ray Probe and Electron Probe Analysis, J. C. Russ, coordinator, American Society for Testing and Materials, Special Technical Publication 485 (1971), contains more extensive information on solid-state spectrometers.

4. J. R. Rhodes, "Design and Application of X-Ray Emission Analyzers Using Radioisotope X-Ray or Gamma-Ray Sources," in J.C. Russ, coordinator, op. cit., especially pages 245-255, gives information on source geometry and radioisotope selection.

5. F. S. Goulding and J. M. Jaklevic, "Trace Element Analysis by X-Ray Fluorescence," UCRL-20625, UC-4 Chemistry, TID-4500 (57th ed.), reports a similar technique using a transmission anode tube.

RAPID RECORDING OF POWDER DIFFRACTION PATTERNS WITH Si(Li) X-RAY

ENERGY ANALYSIS SYSTEM: W AND Cu TARGETS AND ERROR ANALYSIS*

Cullie J. Sparks, Jr., Oak Ridge National Laboratory

Dale A. Gedcke, ORTEC, Inc.

Oak Ridge, Tennessee 37830

ABSTRACT

X-ray diffraction patterns using continuous radiation from copper and tungsten target x-ray tubes and detected with a Si(Li) energy analysis system are presented. Errors caused by a misaligned diffractometer and x-ray penetration into the sample are shown to be more difficult to correct and larger in magnitude than errors arising from energy calibration. All these errors can be minimized by mixing a standard with the unknown sample.

The energy resolution of the detector influences the breadth of the diffraction peaks more strongly than the standard slit systems available with commercial diffractometers. Thus, to reduce the recording time and maintain the same standard deviation for the data, one should increase the sizes of the front and receiving slits including the Soller slits. X-ray energy diffraction patterns can be recorded with standard deviations less than ±0.001 Å in the d spacing with only 200 sec measurement time using the standard diffractometer slit system. Copper targets are probably as useful as tungsten even though the continuous intensity is about three times less. Copper has fewer interfering characteristic lines, and its use permits convenient conversion to normal θ scanning diffractometer operation.

INTRODUCTION

X-ray powder diffraction patterns for phase identification and fluorescent analyses for elemental identification are among

---
*Research sponsored by the U.S. Atomic Energy Commission under contract with the Union Carbide Corporation.

240

the main industrial uses of x rays. The availability of solid
state x-ray detectors with energy resolution less than 200 eV
makes it possible to obtain accurate powder patterns in addition
to the fluorescent spectrum of the sample in minutes. The combination of diffraction and fluorescent analyses should permit a more
positive and rapid identification of the phases in the sample.
With an energy analysis system, the intensity of the diffracted
energy depends on the spectrum from the x-ray target, which changes
with use, and on the reflectivity of the diffracting planes as a
function of energy. Hence the relative intensities* are not as
useful as in the case of patterns measured with monochromatic
radiation. In the future, files of primary beam spectrums and
intensity ratios could be compiled. Acquisition of the fluorescent spectrum could more than compensate for the partial loss of
the intensity ratios, but only experience in actual laboratory
usage can show how valuable this compensation will be.

There is a very real advantage in an energy-analysis system
if one is interested in the detection of impurities. If the
impurity is crystalline, one can measure about one part in 100 by
diffraction analysis. It is possible, however, with fluorescence
to detect impurities to levels several orders of magnitude less.

The minimum energy breadth at full-width half maximum (FWHM)
of the diffraction peaks of about 200 to 400 eV is determined by
the resolution of the solid-state detector and associated electronics. However, typical diffractometers using fixed wavelength
give peak breadths (FWHM) of about $0.20°$ $\Delta 2\theta$ corresponding to from
about 30 to 180 eV over the entire 2θ range. Thus, peak overlap
will be more of a problem for complicated patterns using an energy-
analysis system, particularly if standards are mixed with the
sample.

The most obvious advantage of energy analysis systems is that
one records simultaneously the diffraction and fluorescent patterns
in very much less time than is necessary for a θ scan system with
monochromatic radiation which records only the diffraction pattern.
Since the early work of Giessen and Gordon (1), most of the research
in this field has been concerned with fluorescent analysis and the
characteristics of the solid-state detectors with its accompanying
electronics (2). We consider here the errors in the diffraction
pattern measured with an energy-analysis system. The results are
applied to diffraction patterns measured with copper and tungsten
continuous radiation from samples of silicon and stainless steel.
Errors related to the detector resolution, electronic nonlinearity,
energy scale calibration for the multichannel analyzer, diffractometer misalignment and beam penetration into the sample are
considered.

*As an aid in the reliable interpretation of the pattern, the
ASTM powder diffraction file lists the ratio of the intensity of
the various Bragg reflections to the most intense line for a single
wavelength.

EXPERIMENTAL ARRANGEMENT

An ORTEC solid state detector, having a lithium drifted silicon crystal with a 6-mm active diameter and 3.0-mm sensitive depth, was mounted on the detector arm of a General Electric XRD-5 diffractometer. The detector resolution at FWHM intensity was 210 eV at 6.4 keV and 310 eV at 25.2 keV for our mode of operation. Other associated electronics include an ORTEC 446 high-voltage power supply, ORTEC 716 linear amplifier, and ORTEC 730 single-channel analyzer. The General Electric XRD-5 high-voltage power supply provided full-wave rectified power to a G.E. CA-8S copper anode x-ray tube or to a CA-7 tungsten anode tube.

The XRD-5 diffractometer was aligned in the standard Bragg-Brentano para-focusing geometry with the standard slit systems. This geometry limits diffraction to those planes that are parallel or nearly so to the sample surface. However, some misalignment was deliberately retained to test the errors arising from this cause. Data were recorded in an ORTEC 1024 channel multichannel analyzer for a fixed time and the number of counts accumulated in each energy channel were printed and punched on tape.

ERROR ANALYSIS

Many authors have contributed to the understanding of the errors for the powder method of x-ray diffraction (3). The errors we treat are the same but require a slightly different geometrical construction because of the requirement of fixed scattering angle (2θ). Diffraction occurs from crystalline materials when Bragg's law is satisfied:

$$\lambda = 2d_{hk\ell} \sin \theta , \qquad (1)$$

where λ is the wavelength of the radiation diffracted and d the spacing between atomic planes identified by $hk\ell$. We have chosen to use the tables of x-ray wavelengths given by Bearden (4), which use the energy conversion constant, $\lambda E = 12.398105$ keV-Å, where E is the energy of the radiation in keV. Since our detector measures energy, we write Bragg's law as

$$d_{hk\ell} = 12.3981/2E \sin \theta , \qquad (2)$$

where $d_{hk\ell}$ is given in angstroms. We take the total differential of d to determine the variation of d with the variables E and θ. Thus,

$$\Delta d_{hk\ell}/d_{hk\ell} = - (\Delta\theta \cot\theta + \Delta E/E) . \qquad (3)$$

Since θ is fixed for this experiment, any error in θ caused by the misalignment of the diffractometer cannot be removed from the data by extrapolation unless several diffraction patterns are taken at different θ values. A sample with well-known d spacings could be used to remove all errors. Equation (3) shows clearly that the errors arising from a misaligned diffractometer are larger for small θ, and the errors in measurement of the energy diffracted by various hkl reflections will cause larger errors in d for smaller values of E. It is convenient to treat separately all diffractometer misalignment errors in terms of $\Delta\theta$ and all electronic errors in measurement of the energy scale in terms of ΔE.

Displacement Error

Figure 1 illustrates the geometry of the experiment with a source of x rays impinging on a powder sample and the diffracted ray making an angle of 2θ with the incident ray. The incident ray intersects the sample surface at the diffractometer axis. If the sample surface is displaced a distance D below the diffractometer axis and sin $2\Delta\theta$ is replaced by $2\Delta\theta$ and cos $2\Delta\theta$ by one, then one can show that the error in θ in radians is to a good approximation given by

$$\Delta\theta_D = D \cos \theta/R , \qquad\qquad (4)$$

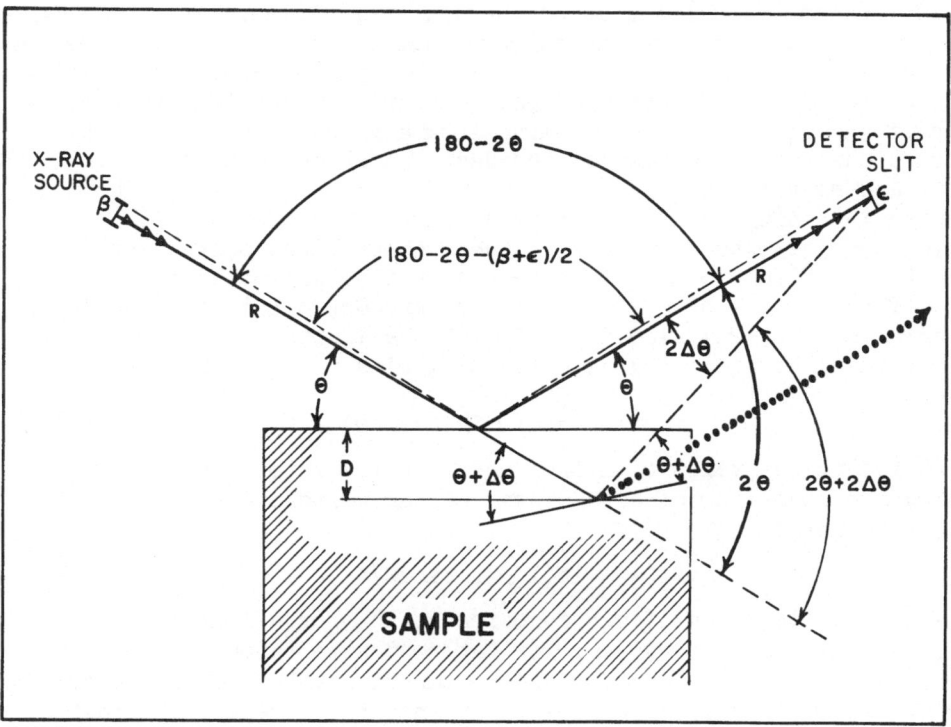

Figure 1. Geometry for the diffraction of radiation at fixed scattering angle, 2θ.

where R is the distance from the diffractometer axis to the detector slit. For diffraction with fixed wavelength while scanning θ, the detector would pick up the diffracted beam at the end of the dotted line with the same error in θ, as shown in Figure 1.

If we are to keep the displacement error to $0.01°\theta$ or about 0.001 in $\Delta d/d$, then D must be less than about 0.001 in. We cannot be sure that the sample surface, especially if it is a powder, is always placed to the exact center of the diffractometer. Other sources of angular error are due to flat sample, horizontal and vertical divergence of the incident and diffracted x-ray beam, and inherent mechanical defects in the diffractometer.

Beam Penetration Error

The x-ray beam may penetrate to some depth D below the surface before it is diffracted out and absorbed and cause an error similar to the displacement error (see Fig. 1). We must assign an energy to each peak to calculate the d spacing. Since the peaks have breadth, some consistent criteria such as midpoint at FWHM, centroid or intensity maximum must be used to designate its position. Without specifying which position is chosen, let f be that fraction of the peak intensity on the high energy side of the chosen center and (1-f) the intensity fraction on the low energy side. The beam penetration is a function of the energy since the attenuation of the beam in the sample is mostly photoelectric absorption. Since the diffraction process competes with the photoelectric absorption, the intensity of the diffracted beam coming from a given depth is directly proportional to the amount of incident radiation reaching that depth and emerging along the diffracted beam path length. The diffracted intensity from a layer at depth D is proportional to $\exp(-2\mu_\ell D/\sin\theta)$ where μ_ℓ is the linear absorption coefficient. The fraction f of the total diffracted intensity that comes from that part of the sample between the surface and depth D is obtained by integrating $\exp(-2\mu X/\sin\theta)dX/\sin\theta$ with the limits on X from 0 to D and dividing by the integral over the limits 0 to ∞. This result gives

$$f = 1 - \exp(- 2\mu_\ell D/\sin\theta) \ . \tag{5}$$

Substituting the value of D from equation (5) into equation (4) and taking the logarithm of both sides, we can write the error in θ from penetration as

$$\Delta\theta_P = - \ln(1 - f)\sin2\theta/4\mu_\ell R \ . \tag{6}$$

We know that $\ln\mu$ is a linear function of $\ln E$ for all practical purposes over the energy range of interest here (about 40 keV) except at absorption edges. Then μ_ℓ can be written as $\exp(-A\ln E+B)$, and the error in d from beam penetration into the samples is not a linear function of energy. The linear absorption coefficient of the powder sample must be known to calculate the penetration correction to the experimental values of $d_{hk\ell}$. This would require experimental measurement for unknown samples.

For theoretically dense iron, μ_ℓ = 62.3/cm, θ = 15° and f = 0.5, the penetration of 30 keV x rays into the sample will make an error of about 0.0013 Å in a d spacing of 3.5870 Å. We can expect more significant errors for atomic numbers below iron, for diffraction at an energy just below an absorption edge, and for low-density powders. We shall see later that beam penetration does account for a significant error in the lattice parameter of silicon powder at energies above 10 keV.

In addition, we can expect the higher energy diffraction peaks to have a low-energy tail. This skewness arises when the beam penetration is significant because $\sin(\theta+\Delta\theta)$ increases, so the energy must decrease correspondingly to satisfy equation (2) for a constant $d_{hk\ell}$. Examples of this effect can be seen in Figures 2(a) and 2(c) at higher energies. Wilson (5) has treated this absorption effect for fixed energy diffraction and shows skewness for the Bragg peaks in that case. All the d spacing information available for an unknown sample is obtained from equation (2). Thus, it is not possible to remove any errors by extrapolation.

Errors in Energy Measurement

Energy measurement errors will arise from the breadth and shape of the intensity profile and from calibration of the energy scale of the multichannel analyzer (including nonlinearity). If we have peaks of known energy near the diffraction peaks, it is an easy matter to calibrate all energy values. The integral nonlinearity of multichannel analyzers is about 0.05% of the maximum energy. The maximum error from nonlinearity for a multichannel analyzer set to receive from 0 to 40 keV is therefore about 20 eV. This error can be reduced to less than 10 eV by multiple point calibration with known sources. Thus, nonlinearity need not be a significant error.

Given the energy resolution of the system at FWHM and the total counts, N, in a Gaussian peak, the standard deviation, σ, in the measurement of the centroid (position) of the peak can be shown to be approximately

$$\sigma = \text{FWHM}/(2.35\, N^{\frac{1}{2}}). \qquad (7)$$

If FWHM is 235 eV and N is 100 counts then σ = 10 eV. The assumption is made that there are enough points to define the curve adequately (five or more points between FWHM).

Summary of Errors

From the above discussion, we know that the energy can be defined to about $\Delta E/E$ equal to 0.001 or less rather easily. To keep $\Delta\theta\cot\theta$ within this limit, the $\Delta\theta$ error can be as much as 0.01° if θ is at least 10°. This will keep either term from making a contribution to $\Delta d/d$ larger than 0.001. We conclude

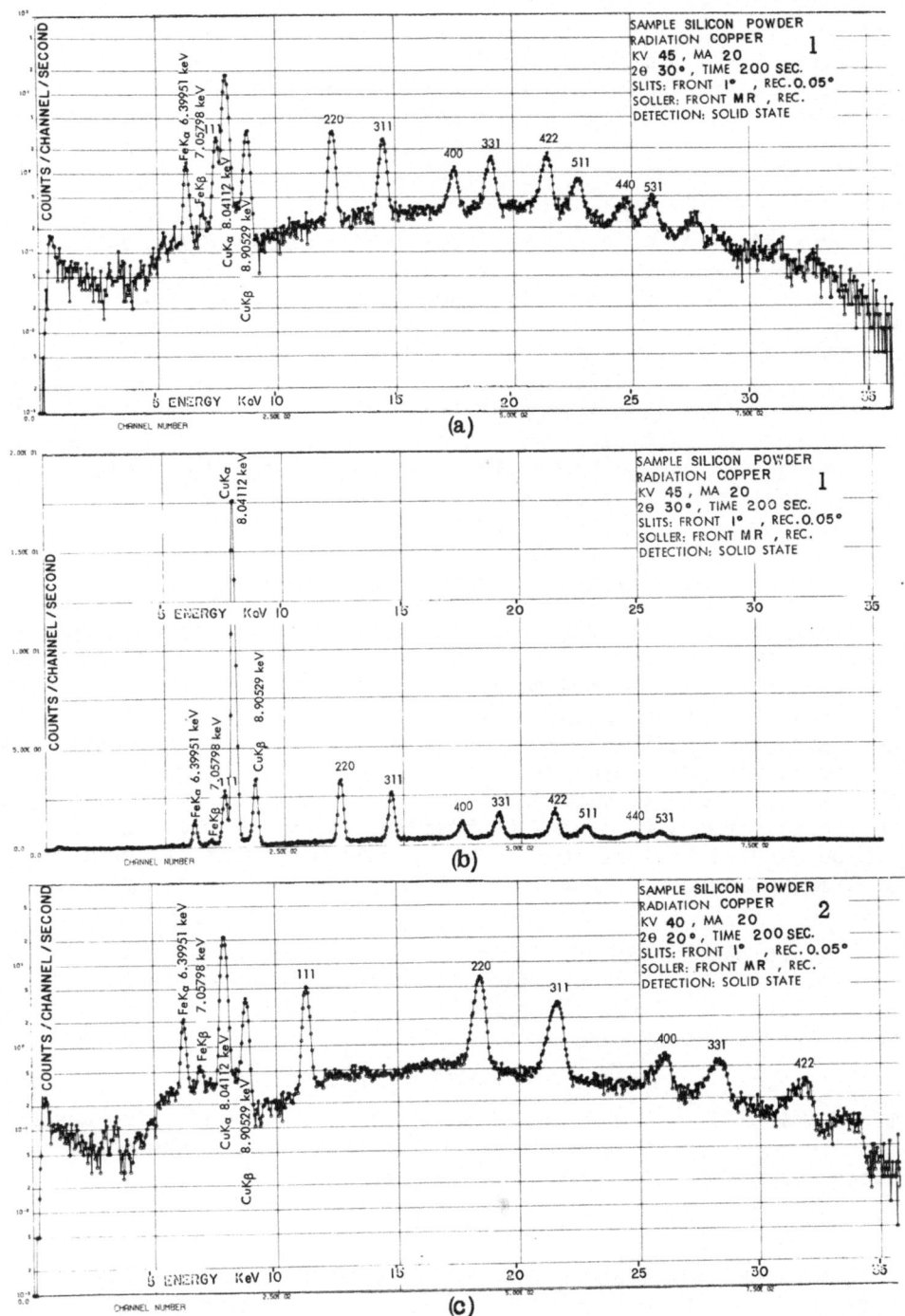

Figure 2.   Comparison of energy diffraction patterns from silicon powder using copper radiation.   (a) Semilog plot at 30° 2θ, (b) linear plot, and (c) semilog plot at 20° 2θ.

that the major source of errors need not arise from energy measurement and calibration but rather from diffractometer misalignment and beam penetration.

On substitution of the errors given in equations (4) and (6) into equation (3), the relative error in the determination of the interplanar spacings is given as

$$\Delta d_{hk\ell}/d_{hk\ell} = \cos^2\theta \ln(1-f)/2\mu_\ell R - D\cos\theta\cot\theta/R - \Delta\theta_0\cot\theta - \Delta E/E \ . \ (8)$$

The first term on the right-hand side is caused by penetration of the radiation into the sample before being diffracted. The second term is the displacement of the sample surface from the diffractometer axis due to misalignment or surface roughness of the sample. In the third term, $\Delta\theta_0$ is the error in setting the diffractometer at zero degrees 2θ. The fourth term contains all the errors in energy calibration of the multichannel analyzer. These errors can be kept sufficiently small compared to the other errors that they are of no consequence.

We have noted that there is no way to treat the data to remove any errors in θ or E without making measurements as a function of θ. However, energy diffraction patterns at two different θ angles are desirable to separate fluorescent lines from diffraction lines which shift in energy with changing θ. This provides a check on the alignment of the diffractometer, making it possible to adjust the sample position and zero setting until the two values of d agree for two different θ's. The highest energy d values would disagree the most for a significant beam penetration error because μ_ℓ is almost proportional to E cubed. Smooth, solid samples could be positioned accurately enough on the diffractometer to give negligible displacement error. Observation of the diffraction pattern to determine the presence of low-energy tails will be good evidence of penetration error.

RESULTS AND DISCUSSION

Typical energy diffraction patterns taken with copper continuum radiation are shown in Figure 2. All peaks identified by element and energy are fluorescent lines. These fluorescent peaks include the characteristic radiation of the target with iron contamination. All patterns were recorded in either 200- or 500-sec counting time. Semilog plots are used to bring out weak peaks. The midpoint at FWHM of the peak was recorded as the peak position. A linear plot is shown in Figure 2(b) for comparison. Comparison of Figure 2(a) and (c) for a silicon standard shows the same Bragg peaks occurring at higher energy, with a loss of the high-order peaks in the 20° pattern relative to the 30° 2θ pattern. The intensity decrease at higher energy occurs because the detector efficiency decreases above 20 keV and because the wavelength dependence of the intensity favors low energy. H. Cole (6) has made this latter point and has also given a useful graph showing d spacings as a function of 2θ and energy.

An example of the change due to beam penetration in the lattice parameter, a, as a function of energy for a silicon powder standard with a measured density of 1.39 g/cm³, θ = 15°, and f = 0.621 is plotted in Figure 3. The value of f is chosen to fit the data taken for condition 1 given in Figure 2(a). The a values plotted in Figure 3 as a solid line (f = 0.621) and dashed line (f = 0.486) are obtained from the first term of equation (8) and the relation a = a_{Si}(1 + Δd/d), where a_{Si} = 5.430 Å. More elaborate methods of choosing the peak position or center of gravity as given by Wilson (5) for fixed energy analysis could be followed. However, he finds for the center of gravity a value for ln(1 − f) within 3% of that used for the solid line in Figure 3. Taking f = 0.50 is equivalent to choosing as the center of the reflection that point where one-half of the intensity is on either side. All silicon diffraction data were corrected using these curves to remove the penetration effect.

In most conventional diffractometers operated in the fixed wavelength mode, the size of the x-ray source and the size of the receiving slit contribute greatly to the breadth of the Bragg reflections. Figure 1 gives the angle subtended by the size of the x-ray source and the receiving slit as β and ε, respectively. Thus θ varies by ±(β+ε)/2. For our measurements we view a line source about 1 mm wide at a takeoff angle of 4 to 6°. With a 0.1° Δ2θ receiving slit at 2θ = 30°, 2Δθ is about 0.12° or 40 to 120 eV at 10 to 30 keV. This makes a relatively small contribution to the FWHM of peaks already from 200 to 400 eV wide, assuming they are Gaussian in shape, and accounts for the fact that the principal effect in going from 0.05 to 0.1° slits is one of increased intensity. Another contribution to the breadth of the diffracted energy peak comes from the small particle size and strain in the sample which effectively broadens Δd. Because of the low absorption of silicon and the decreased density of the powder compact, the beam

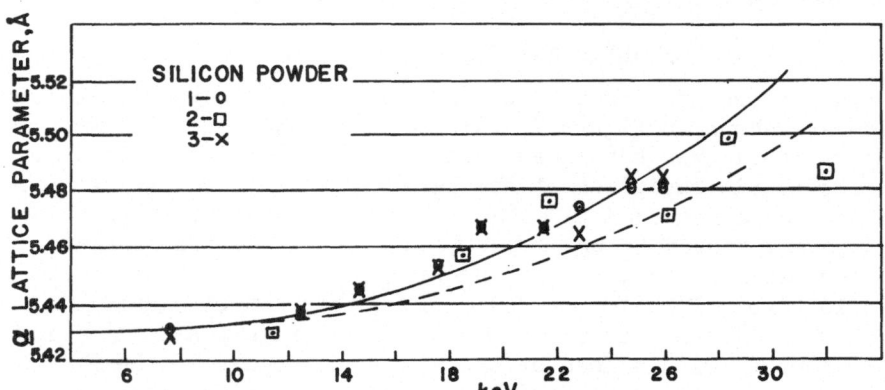

Figure 3. Effect of x-ray beam penetration on the lattice parameter of a silicon powder with a density of 1.39 g/cm³. The measured data are fitted by the solid (f = 0.621) or dashed (f = 0.486) line that gives the error in Δa as a function of energy.

penetration into the sample is the main cause of the large peak breadth above about 20 keV.

Figure 4 compares diffraction patterns recorded with (a) no Soller slit in front of a 0.1° receiving slit, and (b) a high-resolution Soller slit in front of a 0.05° receiving slit. In the figure legend, MR and HR stand for 4 and 2° resolution Soller slits, respectively. Note that Figure 4(a) has about five times the count rate of the pattern shown in Figure 4(b). But the latter shows a better peak-to-background ratio and about 20% better resolution for peaks at the highest energies. Least-squares fits to the lattice constant from these patterns gave a standard error of ±0.002 Å for both

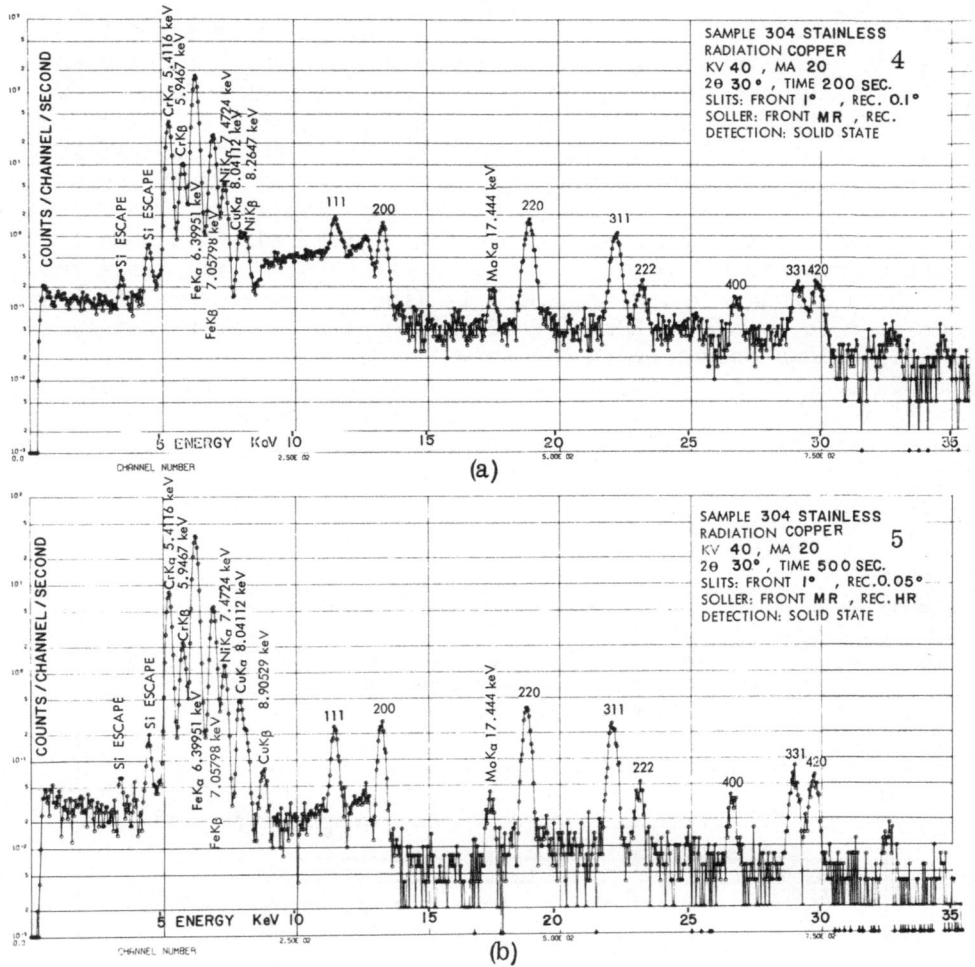

Figure 4. Energy diffraction pattern of type 304 stainless steel comparing the effect of (a) 0.1° receiving slit, no Soller, with (b) 0.05° receiving slit, high-resolution Soller.

(see Figure 5, plots 4 and 5). The pattern of Figure 4(a) has a total integrated intensity two times that of Figure 4(b); thus, the better resolution gained at the expense of the intensity and an increase of 2.5 in counting time did not affect the error. Equation (7) predicts this, since it shows that the standard deviation depends inversely on the square root of the total number of counts in the peak. Because the peak width was reduced only a small amount with the use of the higher resolution receiving slits, the intensity loss made it necessary to increase the counting time to reduce the standard error.

Energy diffraction patterns from a type 304 stainless steel sample, taken with tungsten continuum radiation at two different scattering angles, are shown in Figure 6. If we compare Figure 6 with Figure 4 (same slits and operating voltage but different x-ray targets) intensities of the Bragg peaks are about the same at 15 keV but are greater for tungsten radiation at ≥ 20 keV. We detected iron radiation from the tungsten tube; this contamination could account for our not observing a factor of three in intensity for the tungsten target over copper as measured by Gilfrich and Birks (7). Because Cu K radiation strongly fluoresces Fe $K\alpha$ and Cr $K\alpha$, these peaks are more intense for copper radiation than for tungsten radiation. More overlap with the pattern will arise from the three strong W $L\alpha$, W $L\beta$, and W $L\gamma$ lines and a weaker W $L\ell$ line than from Cu $K\alpha$ and Cu $K\beta$. One advantage of a copper target is that it allows operating in the normal mode of fixed wavelength with 2θ scan using the high resolution solid-state detector to discriminate against fluorescence from samples containing elements just below nickel in the periodic table. This consideration must be weighed against the factor of about 3 loss in intensity.

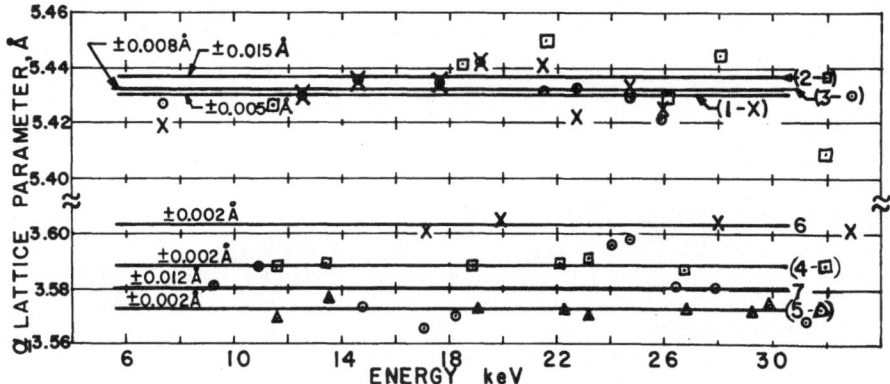

Figure 5. Least-squares fit of straight lines to the data taken by energy diffraction from a silicon powder (upper) and type 304L stainless steel sheet (lower). The standard errors are given.

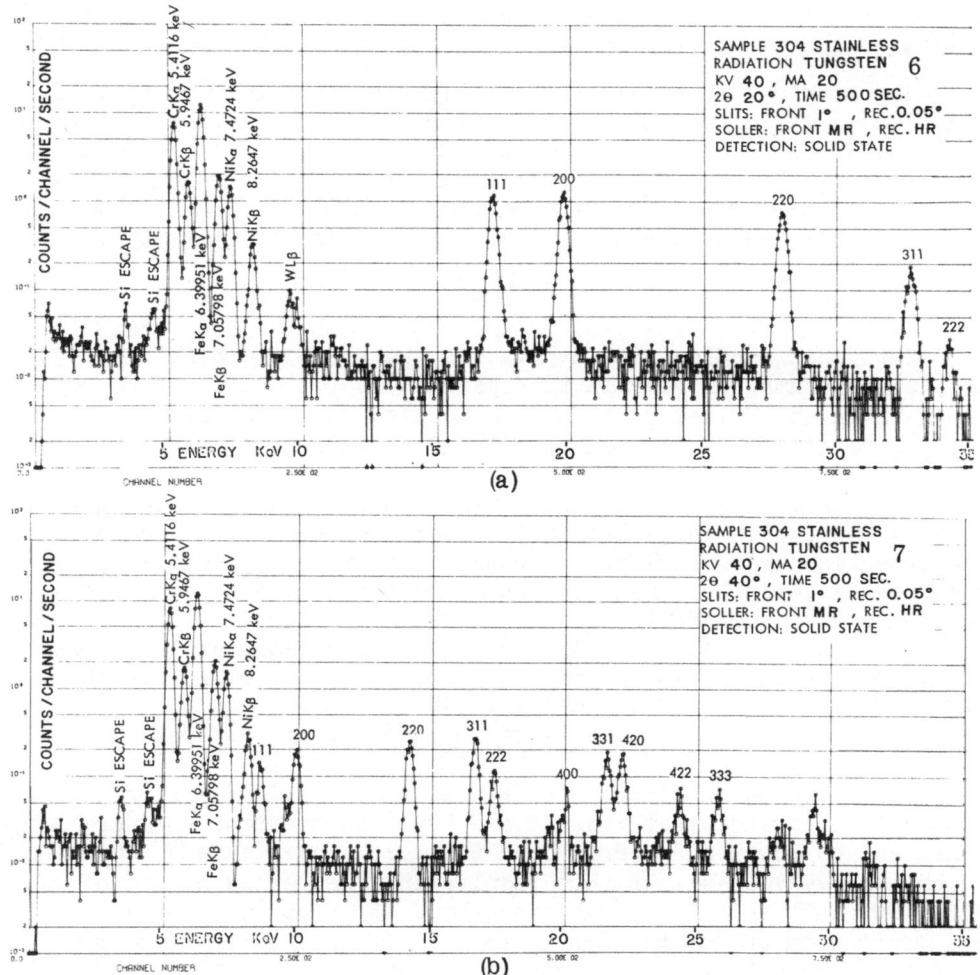

Figure 6. Energy diffraction pattern taken with radiation from tungsten x-ray target at scattering angles of (a) 20° 2θ and (b) 40° 2θ.

The data for type 304 stainless steel given in Figure 6 were fitted by a straight line using a least-squares procedure. These results appear in Figure 5, plots 6 and 7. Since the maximum penetration error at 35 keV was calculated to be about 0.002 Å in a for these data, no effort was made to correct for that effect. Because of the poor statistics from the low number of counts in Figure 6(b), the diffracted energy peaks were poorly defined, and even though there are about twice as many peaks in 6(b), the standard error is much larger than for the data from Figure 6(a). It is apparent that the low order hkℓ planes diffract more strongly than the higher order planes, and they occur at an energy at or near

the maximum in the continuum intensity from the tungsten target.
This explains the much higher intensities for Figure 6(a) over (b).

The spread in the values of the lattice parameter for type 304
stainless steel given in Figure 5 arises from the displacement error.
An error of ±0.01 Å in $\underline{a}$ corresponds to about ±0.003 in. displacement error. Though our interests here are in diffraction, we note
that the fluorescent lines have exactly the same intensity for both
scattering angles as predicted for samples with smooth surfaces.

CONCLUSIONS

X-ray diffraction with continuous radiation and Si(Li) energy-
analysis systems can give lattice parameters to ±0.002 Å from patterns taken in 200 sec with samples mounted on a standard commercial
x-ray diffraction system, such as the General Electric XRD-5.

Misalignment errors in the diffractometer are the more difficult to correct in comparison with errors in energy measurement in
the detection system. Energy errors can be made negligible by
adequate calibration. Sample displacement from the true diffractometer center and penetration of the harder radiation into the
sample are most likely the largest errors to be encountered. The
mixing of a standard known sample with the unknown will give the
highest accuracy in the d spacing as it will allow one to correct all errors from both misalignment and energy calibration.

Low energies and large θ's are favored for minimum error
from diffractometer misalignment and beam penetration. Higher
energies and lower θ's are favored for more accurate measurement
of the energy values, but this results in a decreased number of
diffraction lines. X-ray tubes operated at about 40 or 45 keV
and scattering angles between 20 and 50° 2θ will probably be most
useful. Though tungsten targets will give higher intensities,
copper targets will allow the usual diffractometer usage of scanning 2θ without changing x-ray targets.

The energy resolution of the standard diffractometer slits
is usually more than adequate for measurement of d spacings to
1 part in a thousand. More rapid data acquisition would require
use of lower resolution slits to gain intensity. The energy
resolution of the solid state detector will in general not be a
limiting factor in fast, accurate energy diffraction recording.
An exception may be in patterns where resolution is needed to
separate the lines.

A laboratory's needs will have to be the final criterion for
choosing the system of x-ray sources, slits, and detection
electronics.

ACKNOWLEDGMENTS

The authors wish to express appreciation to J. E. Horwedel for his assistance in processing the data and to L. A. Harris and R. W. Hendricks for some of the computations and plots.

REFERENCES

1. B. C. Giessen and G. E. Gordon, "X-Ray Diffraction: New High Speed Technique Based on X-ray Spectrography," Science 159, 973–975 (1968).

2. Energy Dispersion X-Ray Analysis: X-Ray and Electron Probe Analysis, Amer. Soc. Test. Mater. Spec. Tech. Publ. 485, American Society for Testing and Materials, Philadelphia (1971).

3. I. G. Edmunds, H. Lipson, and H. Steeple, "The Determination of Accurate Lattice Parameters," Chap. 15 in X-Ray Diffraction by Polycrystalline Materials, ed. by H. S. Peiser, H. P. Roaksby, and A.J.C. Wilson, The Institute of Physics, London (1955).

4. J. A. Bearden, X-Ray Wavelengths, NYO-10586, U.S. Atomic Energy Commission, Division of Technical Information (1964).

5. A.J.C. Wilson, "Geiger-Counter X-Ray Spectrometer — Influence of Size and Absorption Coefficient of Specimen on Position and Shape of Powder Diffraction Maxima," J. Sci. Instr. 27, 321–325 (1950).

6. H. Cole, "Bragg's Law and Energy Sensitive Detectors," J. Appl. Cryst. 3, 405 (1970).

7. J. V. Gilfrich and L. S. Birks, "Spectral Distribution of X-Ray Tubes for Quantitative X-Ray Fluorescence Analysis," Anal. Chem. 40, 1077–1080 (1968).

A COMPLETE INSTRUMENTAL SYSTEM FOR ENERGY DISPERSIVE

DIFFRACTOMETRY AND FLUORESCENCE ANALYSIS

G.W. Martin
Stanford Center for Materials Research
Stanford University, Stanford, California 94305

A.S. Klein
Nuclear Equipment Corporation
San Carlos, California 94070

ABSTRACT

A system has been designed and tested for rapid energy dispersive diffractometry and simultaneous fluorescence analysis. A
turntable composed of the sample chamber with attached air-cooled
x-ray tube allows the 2θ angle to be varied with respect to the
stationary Si(Li) detector. Data for most analyses can be obtained in one minute per sample. Results are stored in the memory
of a multichannel analyzer and are read out on a CRT, strip chart
recorder or tabulated in digital format by a computer.

INTRODUCTION

Use of semiconductor detectors for elemental analysis is widespread and several commercial units are available for this purpose.
In most cases, the x-ray spectrum of the sample is excited by
radioactive sources, but x-ray tubes are now beginning to replace
radioisotopes for this purpose (1).

The use of semiconductor detectors for x-ray diffraction was
described in 1968 but the technique is still largely confined to
laboratory studies using components from conventional diffractometry systems (2, 3). This method has been termed nondispersive
or energy dispersive diffraction analysis. A proposal was made to
refer to energy dispersive diffractometry as SPD (Spectrometric
Powder Diffractometry)(4).

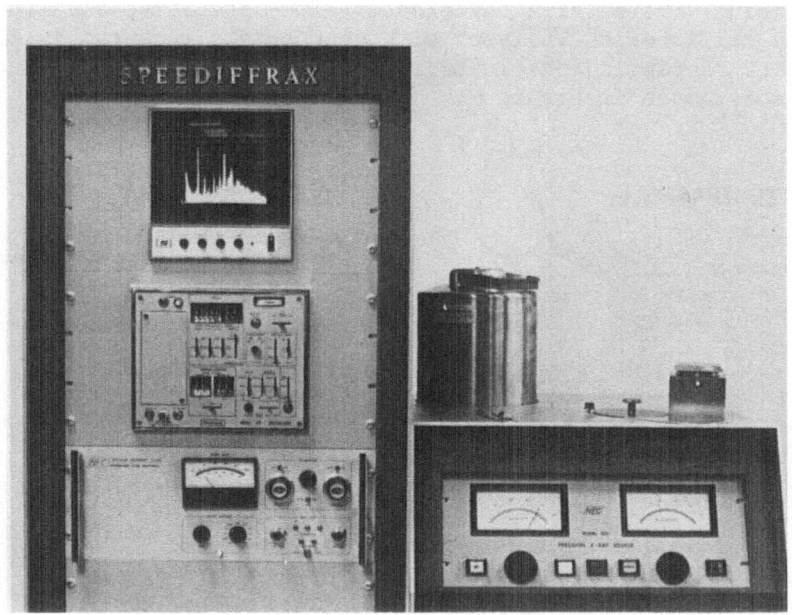

FIGURE 1: Basic SPEEDIFFRAX System

The SPEEDIFFRAX$^{(TM)}$ system described in this paper is the
first complete system manufactured for energy dispersive diffractometry and, as of this writing, is the only one commercially
available. It employs an air-cooled generator, x-ray tube and
sample chamber especially designed for simultaneous elemental and
compound analyses. A computer program developed for the system
can tabulate the data output in a format which is convenient for
interpretation.

DESIGN DESCRIPTION

Figure 1 is a photograph of the assembled instrument with a
200-channel multichannel analyzer and an oscilloscope read-out.
Teletype or other digital read-out is also available from the
MCA signal. The SPEEDIFFRAX is portable, requires no water cooling and only a standard 115V AC power input.

The essential components of the system are: 1) an x-ray
source which produces polychromatic radiation, 2) a sample chamber,

3) a lithium-drifted silicon detector with associated electronics, and 4) a multichannel analyzer with appropriate read-out instrumentation. A simplified block diagram of the system and its features is given in Figure 2.

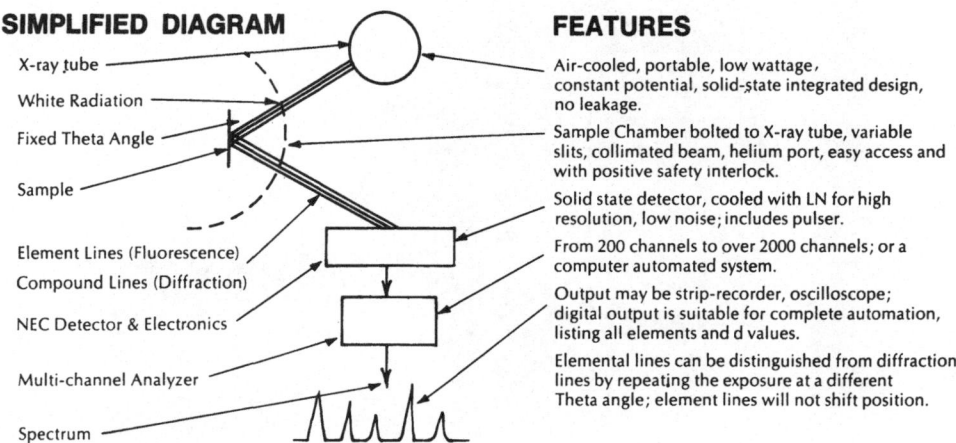

SIMPLIFIED DIAGRAM

X-ray tube
White Radiation
Fixed Theta Angle
Sample

Element Lines (Fluorescence)
Compound Lines (Diffraction)

NEC Detector & Electronics

Multi-channel Analyzer

Spectrum

FEATURES

Air-cooled, portable, low wattage, constant potential, solid-state integrated design, no leakage.

Sample Chamber bolted to X-ray tube, variable slits, collimated beam, helium port, easy access and with positive safety interlock.

Solid state detector, cooled with LN for high resolution, low noise; includes pulser.

From 200 channels to over 2000 channels; or a computer automated system.

Output may be strip-recorder, oscilloscope; digital output is suitable for complete automation, listing all elements and d values.

Elemental lines can be distinguished from diffraction lines by repeating the exposure at a different Theta angle; element lines will not shift position.

FIGURE 2:   Block Diagram of SPEEDIFFRAX System

X-ray Source

The generator produces an ultra-stable constant potential with voltage and current independently variable to a maximum of 50 KV and 5 ma. For diffractometry, the maximum voltage is used and the current set to an optimum level, usually 1 ma, as determined by the count rate meter. With elemental analyses, the x-ray tube voltage is set slightly above the excitation energy required by the particular element being analyzed.

In general, a tungsten or molybdenum x-ray tube is used. The tube shield contains dielectric oil which transfers heat from the tube to metal surfaces which in turn are air-cooled by a fan. A red warning light on the tube shield indicates when the x-ray tube is on. (Figure 3 shows the x-ray source control panel.)

Sample Chamber

A unique feature of the design is the intimate incorporation of the tube shield and x-ray exit port with the sample chamber to prevent any x-ray leaks. During operation of the system, no radiation is detectable around the x-ray tube shield or sample chamber. (The SPEEDIFFRAX has passed the California Radiation Code and has the same rating as a home television set.)

FIGURE 3:   Front Panel of the X-ray Power Supply

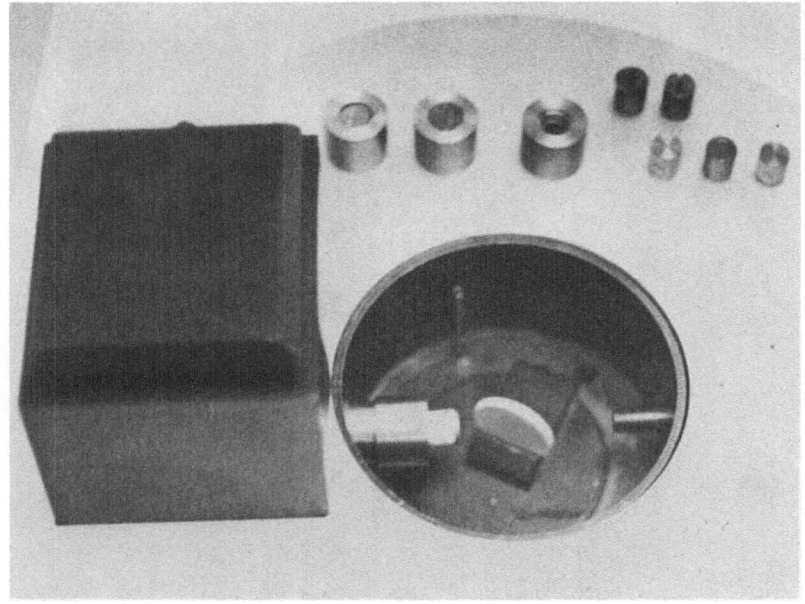

FIGURE 4:   Close-up of Sample Chamber and X-ray Source

Figure 4 shows a close-up of the sample chamber with the cover removed. When the cover is removed, a mechanical interlock deactivates power to the x-ray tube and a lead shield drops over the x-ray source port. The interior of the sample chamber is a cylinder of $2\frac{1}{2}$ inch radius and 2 inch height. Ports in the bottom of the chamber allow for an atmosphere of helium or other gases. (The helium inlet tube is shown at the back of the sample chamber.)

Various slits, filters and secondary targets are displayed in the background of Figure 4 (also foreground of Figure 5). These slits and metal foils may be placed over the beam collimator. Secondary targets are chosen for optimum fluorescence of particular elements in the sample providing for high signal and low background (1). In Figure 4, one of the interchangeable secondary targets may be seen on the beam collimator, protruding into the sample chamber from the left. A beam stop, seen protruding into the sample chamber from the right, is always in line with the main x-ray beam.

The receiving slit, in front of the detector, is continuously adjustable. It is controlled by the black knob shown in the lower left of Figure 5.

FIGURE 5: X-ray Source and Sample Chamber Turntable

The x-ray tube and sample chamber rotate in a horizontal plane around the sample holder (which is at the center of the diffracting circle). The Si(Li) detector remains stationary at all times; only the sample chamber assembly, with x-ray tube attached, moves when changing the 2θ angle. The 2θ scale is etched on the circumference of the sample chamber/x-ray tube turntable. The 2θ angle is variable to a maximum of 160°. High angles are desirable for elemental analysis. During the actual analysis, all parts are stationary.

The angle, omega, of the sample holder, with respect to the x-ray beam, is continuously adjustable and is independent of the 2θ angle. In Figure 5, the sample is placed in a transmission mode between mylar films whereas, in Figure 4, the sample is positioned for normal reflected diffraction.

Semiconductor Detector

The radiation detector is a lithium-drifted silicon wafer mounted in a cryostat attached to a 5-liter liquid nitrogen dewar. Description of these detectors is reviewed elsewhere (5, 6). The dewar need be refilled with liquid nitrogen only once every 4 or 5 days. The resolution of the system is 150 eV at 6 keV FWHM. The electronics associated with the detector contain two precision pulsers for calibration, a vacuum gauge and a count rate meter.

Data Output

The simple system, as pictured in Figure 1, contains a 200-channel analyzer, an oscilloscope output and a strip chart recorder (not shown). This is adequate for most qualitative elemental and compound analyses. Figure 6 shows a spectrum obtained from this system.

For more detailed analyses and/or greater convenience, a computer is desirable. A multiple iteration program produced the digital read-out shown in Figure 7. Figure 7 came from a 1000-channel analyzer using a 4K memory computer for digital output. A strip chart record of this spectrum is given in Figure 8.

Another program provides automatic calibration of the keV spectrum through use of a standard sample containing Ti (keV = 4.5) and Ag (keV = 22). A silver diffraction peak is used for calibrating the exact 2θ angle at which the analysis is performed. This calibration provides for greater accuracy in angle determination than is achieved by mechanical means.

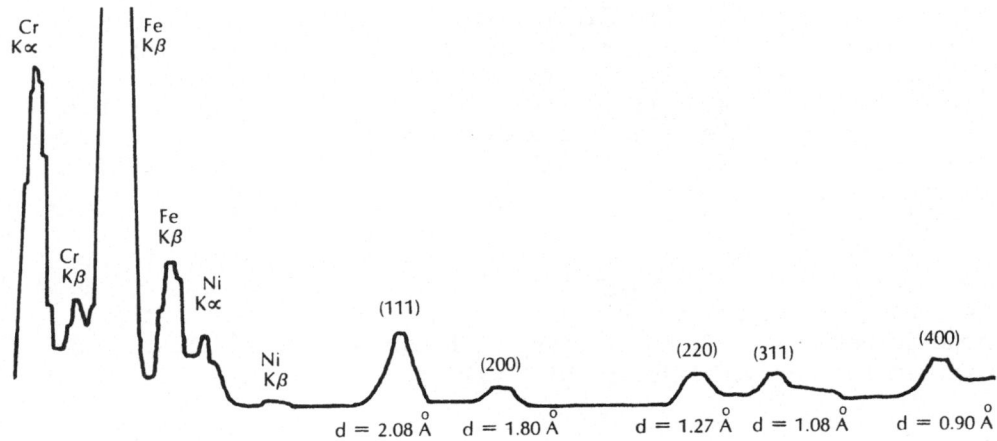

FIGURE 6: Spectrum of 303 Stainless Steel
This sample was analyzed in 17 seconds using
1 ma at 46 KV (tungsten tube).

1. SAMPLE: RHENIUM
2. TWO THETA ANGLE: 24.1 DEGREES
3. COUNTING TIME 10 MINUTES

CHANNEL	KEV	COUNTS	D-SPACING	Interpretation (by Analyst) Identity (HKL or element)	ASTM Value
217	8.616	223983	3.446	ReL α	
254	10.03	197451	2.961	ReL β	
297	11.69	13284.	2.538	ReL γ	
360	14.12	23964.	2.103	101	2.105A
465	18.22	5963.8	1.630	102	1.629A
552	21.56	10966.	1.377	110	1.380A
604	23.57	12952.	1.260	103	1.262A
653	25.47	14884.	1.165	112, 201	1.17, 1.15A
722	28.17	288.33	1.054	202	1.053A
756	29.46	222.32	1.008	104	1.009A
819	31.91	536.71	.9304	203	.9311A
860	33.53	1345.6	.8855	211	.8854A

FIGURE 7: Computer Read-out and ASTM Comparison

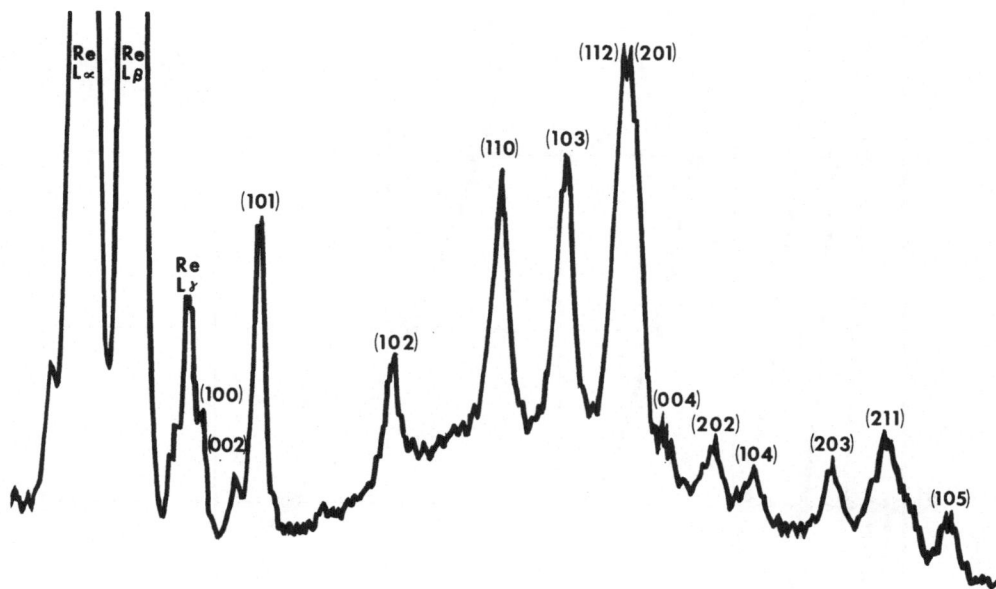

FIGURE 8: Spectrum of Rhenium Powder
10 minute count at 50 KV 3.5 ma
(tungsten tube)

DATA INTERPRETATION

Simultaneous elemental and compound analysis is the normal mode of operation in the SPEEDIFFRAX system. Both diffraction and fluorescence lines are usually present in the spectrum of a sample. In practice, two spectra at different 2θ positions are run on a sample. The fluorescence lines will not change position when the 2θ angle is changed; the diffraction lines will shift.

Tables for the energies of elemental lines are available (7). The d-values of the diffraction lines may be calculated from the formula $d = 6.2/keV(\sin \theta)$. When a computer is not used, it is convenient to select 2θ values of 24° or 36°; then the formula reduces to $d = 30/keV$ and $d = 20/keV$ respectively.

A graphical solution of the Bragg formula as adapted for energy dispersive analysis is shown in Figure 9. Given two of the three variables - 2θ, keV or d - the third may be calculated.

Several factors are involved in choosing a given 2θ value at which to do energy dispersive analysis. Figure 10 gives the "best"

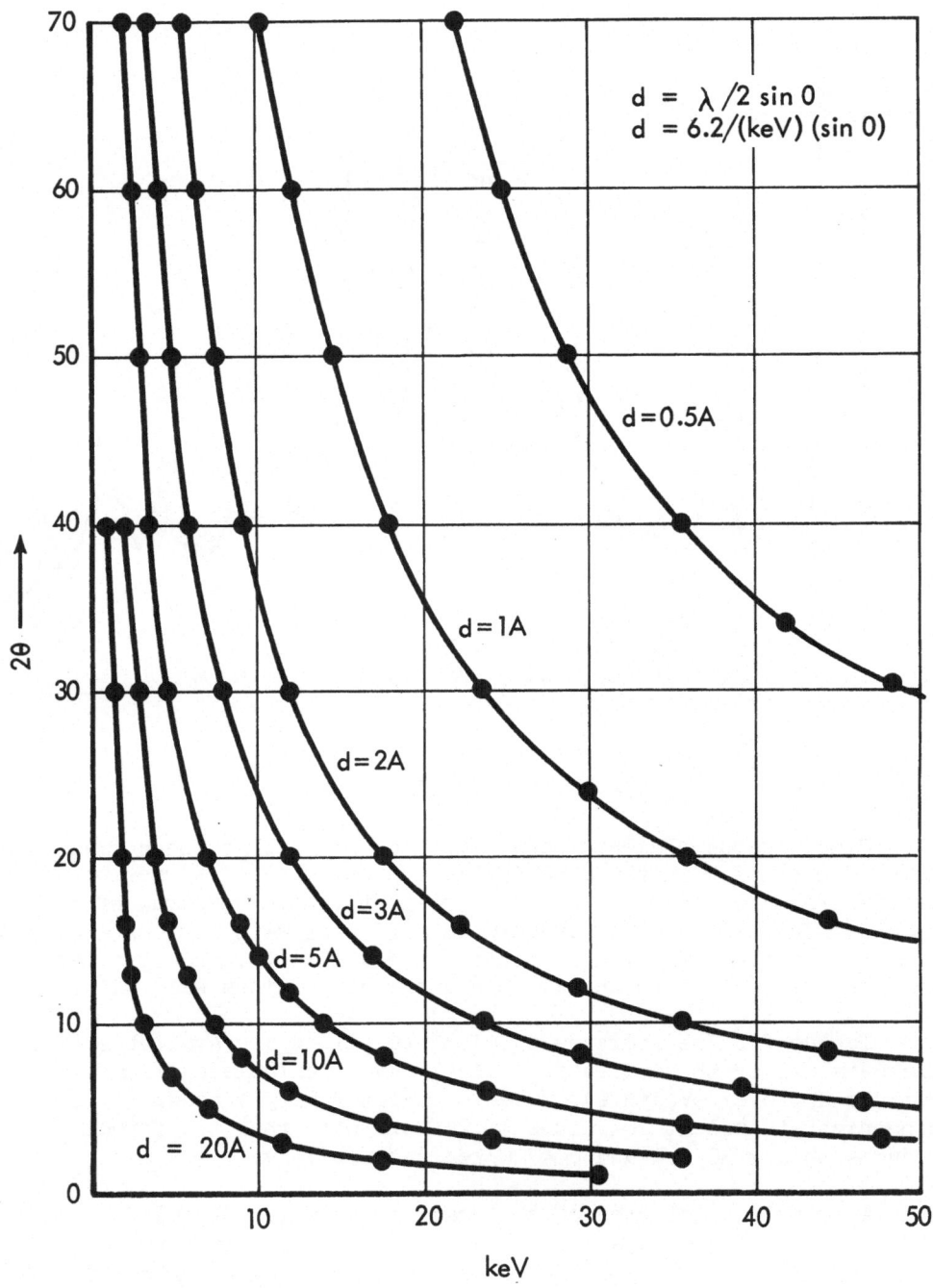

FIGURE 9:   Solution of Bragg Formula for Energy
                    Dispersive Diffractometry

2θ value for a particular d-spacing based upon an optimum value of
30 keV. Other factors, such as spectrum crowding, detector sensitivity, the pressure of fluorescent x-rays, etc., may dictate the
use of angles different than shown in Figure 10. For optimum elemental analysis, 2θ values above 120° are often used; no diffraction lines are probable in the spectrum taken above 90° 2θ (8).

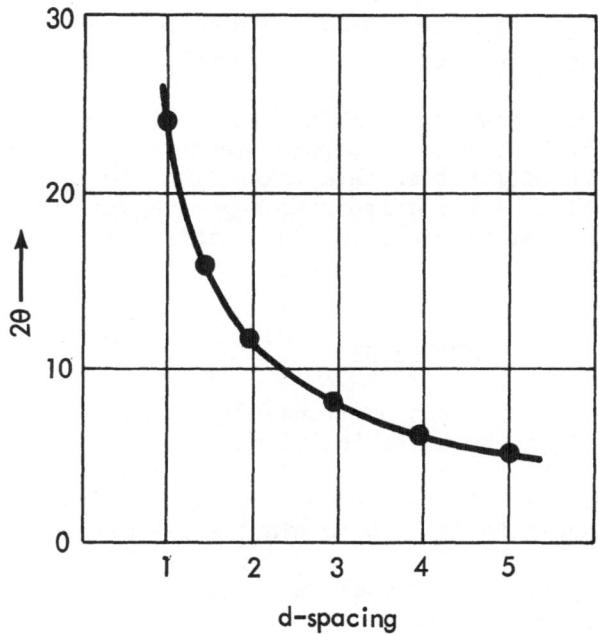

FIGURE 10: "Best" 2θ Angle for a Particular D-spacing

     Elemental analyses may be completed within a few seconds for
major constituents. Diffractometry may require several minutes.
With a helium atmosphere in the sample chamber, elements down to
Atomic Number 11 (sodium) may be detected. Elements over Atomic
Number 16 (sulfur) are readily detected in an air atmosphere.

     Since helium may be used for long wavelength diffraction, the
possibility of diffraction from thin films of 1,000 to 10,000 Angstroms is increased. One such case was investigated where a thin
film of SiO_2 occurred on a silicon wafer. By adjusting the SPEEDI-
FRAX system so that diffraction occurred at low energies (long
wavelengths), major lines of SiO_2 were observed in the spectrum,
whereas they were not obtained with a conventional diffractometer
using a copper x-ray tube.

The technique is also useful for the rapid determination of the orientation of single crystals, such as germanium or silicon wafers, since diffraction will occur over a wide range of 2θ values. The crystal "selects" that part of the incident white radiation to diffract which corresponds to its particular d-spacing. In this sense, the method is similar to the Laue photographic method where a stationary single crystal is irradiated with poly-chromatic radiation.

Evaluation of Diffraction Spectrum

The spectrum of a powdered sample of pure rhenium was obtained with the SPEEDIFFRAX system with computer read-out. Figure 7 shows this data along with diffraction data taken from the ASTM powder diffraction card file. A comparison shows good agreement in the d-values obtained. A strip chart record of this spectrum is given in Figure 8.

CONCLUSIONS

The SPEEDIFFRAX system has numerous advantages over more conventional x-ray diffraction units including:

1. Simultaneous element and compound analyses; normal time for analysis is 1 to 10 minutes per sample.

2. No moving parts during analysis; no complex goniometer for diffraction.

3. Air-cooled x-ray tube and single 115V AC power input makes system portable; entire generator and equipment is mounted on casters.

4. Simple operating procedure, automatic interlocks, radiation leakproof design, and low wattage requirements (about 1/10th that of conventional units) insure safe operation.

5. Digital output from analyzer allows for computer treatment of data. Computer tabulates energy (keV), and corresponding d-value for the given 2θ, and integrates counts for each peak.

REFERENCES

(1) F.S. Goulding and J.M. Jaklevic "Trace Element Analysis by X-ray Fluorescence" University of California Publication UCRL-202625 May 1971.

(2) B.C. Giessen and G.C. Gordon "X-ray Diffraction: New High
     Speed Technique based on X-ray Spectrophotograph" SCIENCE
     Volume 159 1968, Page 973.

(3) A.P. Langheinrich, J.W. Forster and W.M. Tuddenham "Recent
     Applications of Energy Dispersion X-ray Spectrometry"
     Paper presented at 1969 Northwest Regional Meeting of the
     American Chemical Society, Salt Lake City, Utah June 1969.

(4) B.C. Giessen and G.C. Gordon "Recent Developments in Spec-
     trometric Powder Diffractometry" Norelco Reporter Volume
     17, No. 2, 1970, Page 19.

(5) R. Fitzgerald and P. Gantzel "X-ray Energy Spectrometry in
     the 0.1-10 Å Range" American Society for Testing and Mater-
     ials, Special Technical Publication No. 485, 1971.

(6) W.B. Jones and R.A. Carpenter "Nondispersive X-ray Fluores-
     cence Spectrometer" Advances in X-ray Analysis, Volume 11
     Pages 214-229 edited by J.B. Newkirk, G.R. Mallett and J.
     Fay.

(7) G.G. Johnson, Jr. and E.W. White "X-ray Emission Wavelengths
     and keV Tables for Nondiffractive Analysis" ASTM Data Series
     D546, April 1970.

(8) A.P. Langheinrich, J.W. Forster, T.A. Linn, Jr. "Energy
     Dispersion X-ray (EDX) Analysis in the Non-Ferrous Mining
     Industry" Paper presented at the ISA AID National Symposium
     Houston, Texas, April 1971.

SMALL X-RAY TUBES FOR ENERGY DISPERSIVE ANALYSIS USING SEMI-

CONDUCTOR SPECTROMETERS

J. M. Jaklevic, R. D. Giauque, D. F. Malone
and W. L. Searles
Lawrence Berkeley Laboratory University of California

Berkeley, California 94720

ABSTRACT

Fast X-ray fluorescence analysis with radioisotope excitation requires intense sources to produce reasonable counting rates. The inconvenience of handling such sources and the small number of suitable radioisotopes places severe limitations on their use.

We have explored the possibility of using low-power X-ray tubes as exciting sources for energy-dispersive fluorescence analysis. The principal advantage to X-ray tubes is the ability to produce X-ray fluxes to three orders of magnitude higher than those obtained with convenient radioisotope sources while dissipating only a few watts in the tube. Furthermore, the variety of possible anode materials and range of currents in the tube make possible optimum choice of exciting energy and intensity for particular applications.

We have designed and tested such tubes in a variety of anode configurations suitable for fluorescence excitation. Using either X-ray filtering techniques or multiple fluorescence geometries it is possible to significantly reduce the Bremsstrahlung background relative to characteristic radiation.

As compared with normal radioisotope-target assemblies, excitation of a sample by the X-ray tube results in comparable sensitivity in only a tenth to one hundredth of the time.

INTRODUCTION

Maximum analytical sensitivity in energy dispersive X-ray analysis is obtained by the efficient excitation and detection of characteristic X-rays in the sample. However, this must be achieved in such a way as to minimize unwanted background in the detector; in most analytical applications using semiconductor spectrometers monoenergetic X-ray excitation is provided by either radioisotope sources or source-target assemblies (1). Although adequate in many applications, radioisotope excitation is limited to a small number of suitable isotopes. Their use for rapid trace analysis is precluded by the difficulties in the handling, storage and replenishment of intense radioactive sources.

We have explored the use of small X-ray tubes as an alternative means of fluorescence exciation with particular emphasis on applications to analysis of trace quantities (down to less than 1 ppm). Their principle advantage is the large characteristic X-ray flux which can be generated with a relatively small expenditure of power input to the tube. For example, an electron current of 1 μA at 27 keV will generate a total of 1.6×10^{10} K X-rays/sec when incident on a copper target (2). This is equivalent to 16 Curies/watt of power input. As the tube voltage, V, is increased the K X-ray intensity increases in proportion to $(V_O-1)^{1.67}$, where $V_O \equiv V/E_K$ and E_K is the K electron binding energy (2). The X-ray tube output can be varied in energy and intensity by changing the target material and electron current; and, unlike radioisotope sources, an electron beam can be turned off when not in use--an important safety consideration particularly in any large scale applications of the spectrometer systems outside carefully controlled laboratory environments. However, analytical sensitivity, using fluorescence excitation with normal X-ray tubes, is limited by the continuous Bremsstrahlung background generated by the deceleration of the electrons in the anode. Scattering of this spectrum from the sample to detector is unavoidable and a particularly serious limitation when using low-background semiconductor spectrometers (3). In the present work we describe techniques for generating nearly monoenergetic X-ray tube outputs with minimal background interference. Results are presented for both a filtered output X-ray tube and a secondary fluorescence tube.

X-RAY TRANSMISSION TUBE

The simplest method for reducing continuum background is to use suitable X-ray filters to transmit as much of the characteristic radiation as possible while attenuating all other energies. Since any given element is a good transmission filter for its own characteristic X-rays, the anode of the X-ray tube can be made of

an appropriate thickness to effectively filter the transmitted
X-ray spectrum. Although filtering the spectrum produced at back-
ward angles to the target would achieve the desired result, it
proves convenient to employ the transmission geometry since more
efficient anode to sample spacings are possible and the use of
multiple targets is facilitated. Further filtering can be included to selectively reduce the Kβ intensity and more closely approach
a monoenergetic X-ray beam.

Table 1 is a summary of relevant X-ray data for elements of
interest in this discussion. One point of immediate importance is
that the half-thickness for characteristic X-ray absorption is of
the order of 10^{-3} cm for all elements. This is a convenient value
since a filter thickness of 3 to 5 half-thicknesses provides both
adequate X-ray transmission and sufficient heat conduction to dissipate the incident power into the anode support.

A scale drawing of the X-ray tube is shown in Fig. 1. The
electron beam is obtained from a tungsten filament and the current
can be controlled either by the filament power or with the intermediate electrode voltage. The beam is focused on the target anode
which is held at the high positive voltage and supported by the
ceramic insulator. (Operation of the anode at positive voltage
instead of ground potential is dictated mostly by convenience of
design. There are possible advantages to operation at ground
potential since the exit window may then also be the transmission
target-filter; thus affecting higher geometric efficiency.) Optimum filter thickness is determined empirically but 3 to 5 half-
thicknesses of the target material are nearly optimum. A portion
of the total filter thickness is mounted external to the tube to
eliminate unwanted characteristic X-rays from the stainless steel
vacuum enclosure.

Most of the present data were obtained with molybdenum target
and total filter thickness of 0.012 cm. The tube was operated at
42 keV with variable currents up to 500 μA. The maximum beam
power of 25 watts was dissipated mostly by radiation from the anode
structure. Although increasing the beam voltage would result in
improved characteristic X-ray yield relative to Bremsstrahlung, a
practical limit is set by the transmission of the high energy continuum Bremsstrahlung X-ray through the filter. Using a voltage
regulated filament supply, the tube output was stable to within
± 10% over several hours; however, we anticipate providing feedback
stabilization in the near future.

TABLE I

Element	Z	K Absorption Edge (keV)*	Kα Energy (keV)*	Kβ Energy (keV)*	Half-Thickness at Kα Energy (cm)†
Al	13	1.560	1.486		6.1×10^{-4}
Ti	22	4.965	4.508	4.931	1.4×10^{-3}
V	23	5.464	4.949	5.426	1.2×10^{-3}
Fe	26	7.111	6.398	7.057	1.2×10^{-3}
Ni	28	8.332	7.471	8.263	1.3×10^{-3}
Cu	29	8.980	8.040	8.904	1.5×10^{-3}
Zr	40	17.999	15.744	17.704	5.3×10^{-3}
Mo	42	20.004	17.441	19.651	3.7×10^{-3}
Rh	45	23.220	20.165	22.777	3.5×10^{-3}

* See Ref. (6)
† See Ref. (7)

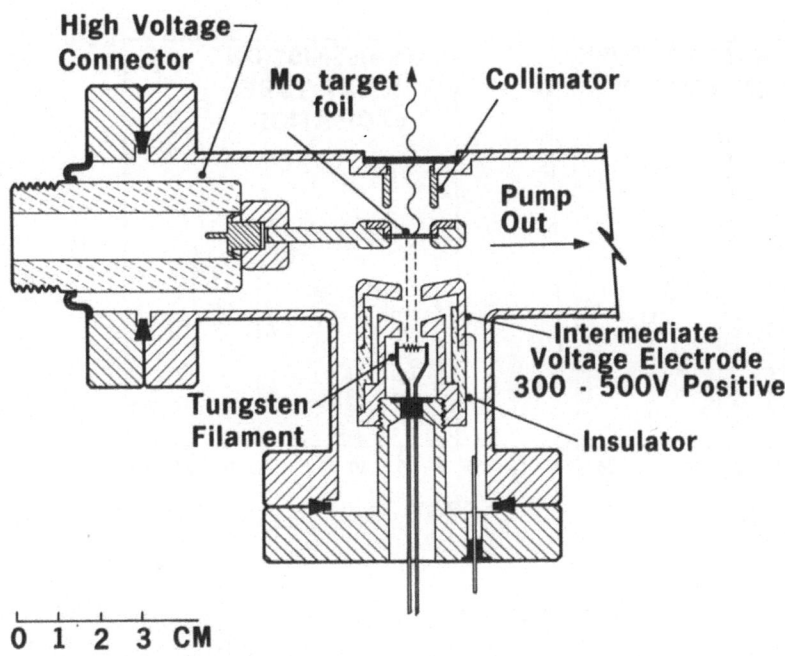

Figure 1.   Schematic of transmission X-ray tube.

Figure 2 is a comparison of fluorescence spectra obtained using radioisotope excitation and transmission X-ray tube excitation for the case of a biological sample. The spectra were obtained using a low-background guard ring reject system (3) and pulsed-light feedback electronics (4); only the high energy portions are shown since there is no significant difference in background at lower energies. The radioisotope source-target assembly employed a 125 millicurie ^{125}I source and molybdenum target with a separation of 1 cm between the source and molybdenum target, and 1 cm between the molybdenum target and the sample. The X-ray tube anode filter was 0.012 cm molybdenum and gave a ratio of 30:1 for the Kα X-ray peak height to background height just below the Kα peak in the tube output spectrum. With the X-ray tube operating at 100 µA and 42 keV the spectrum was acquired in approximately 15 minutes at a total counting rate of 4000 cts/sec. Trace elements concentrations for the 0.3 mm thick sample were obtained by the method described in Ref. (5). With a 500 µA beam current and the anode to sample distance of 8 cm as used in these measurements, the counting rate was 50 times that obtained with the radioisotope source.

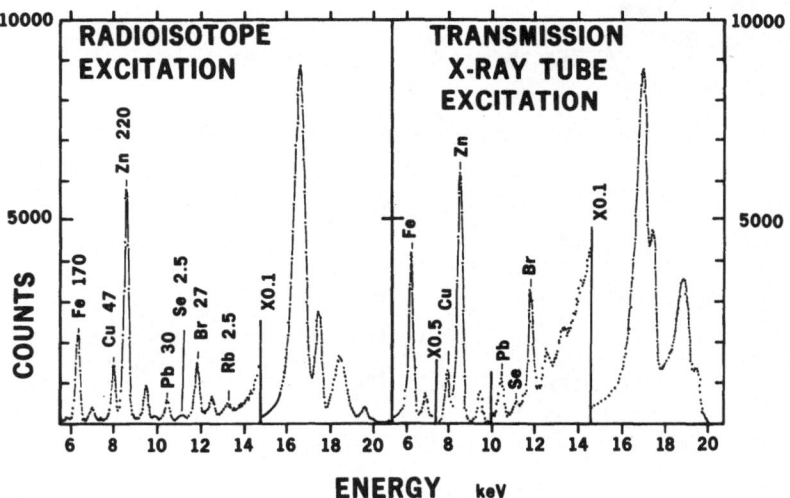

Figure 2. Comparison of excitation modes showing spectra obtained from biological specimen (lyophilized horse liver). Numbers on peaks indicate concentrations in ppm for the sample pellet.

A further improvement in excitation efficiency could easily
be realized by decreasing the anode-sample and sample-detector
geometry. It is apparent from Fig. 1, that relatively simple
mechanical modifications could be made to decrease these distances
to affect an efficiency improvement of at least 10 times. Looking
at the spectrum obtained with the transmission tube, one sees that
it should be possible to obtain sensitivities of substantially
better than 1 ppm in a few minutes at counting rates limited only
by the detector electronics. (For a more detailed discussion of
detection limits see Ref. (3).)

An interesting feature of the data is the difference in the
ratio of backscatter peak height to fluorescence X-ray intensity
between the two spectra. Although a part of this difference can
be accounted for by the lower peak to background in the X-ray tube
output spectrum, most of the difference is due to the difference
in source-sample-detector geometry. In the source-target assembly
the average scattering angle for the molybdenum X-rays is near 180°
whereas the X-ray tube is mounted horizontally with a total scattering angle of approximately 90°. These angles are the respective
maximum and minimum in the incoherent scattering differential cross
section. These conclusions are supported by data obtained with
the secondary fluorescence tube operated in a similar geometric
configuration.

SECONDARY FLUORESCENCE X-RAY TUBE

A substantial further reduction in continuum background can
be realized by using a secondary excitation mode analogous to the
source-target arrangement. The X-rays generated in the anode by
the electron beam are made to strike a secondary target--the characteristic X-ray from this target then are used to excite X-rays
from the sample. Although this conversion process results in a
reduction in X-ray output for a given electron beam power, it virtually eliminates the continuous background in the output spectrum.
Since the secondary target need not be an electrical conductor or
dissipate large amounts of power, it is possible to generate characteristic X-rays from elements not available as metallic foils.

The choice of anode-secondary target combination is determined
by choosing an anode material with Kα X-ray energy as near as possible to the maximum photoelectric cross section for the secondary
target. Referring to the data in Table 1, it is apparent that the
rhodium to molybdenum anode target combination is especially fortu-
itous in this regard. In fact the total conversion efficiency of
rhodium to molybdenum X-ray is 0.52--i.e. for every rhodium Kα
X-ray incident on a thick molybdenum target, 0.52 molybdenum K

X-rays are generated isotropically. The total efficiency for secondary fluorescence excitation is then determined principally by the solid angle subtended by the secondary fluorescence target relative to the anode.

In Fig. 3 is shown a schematic of an experimental tube used for secondary fluorescence target excitation. Electrons emitted from the filament are accelerated into the annular rhodium anode through a series of openings in the molybdenum secondary target. The collimator shields the external sample from the rhodium anode but not the molybdenum target. The geometric efficiency of the rhodium to molybdenum conversion is approximately 3 to 5% yielding an output rate approximately 0.1 times that of the transmission X-ray tube for a given beam power. (The reason that the difference is not larger is due to the less than 10% X-ray transmission through the 0.012 cm filter in the transmission tube.) It is possible to increase the rate further by operating at a higher anode potential since transmission of Bremsstrahlung through the filter is no longer a problem--in the present series of measurements the potential was limited to 45 keV only by the high-voltage connector rating.

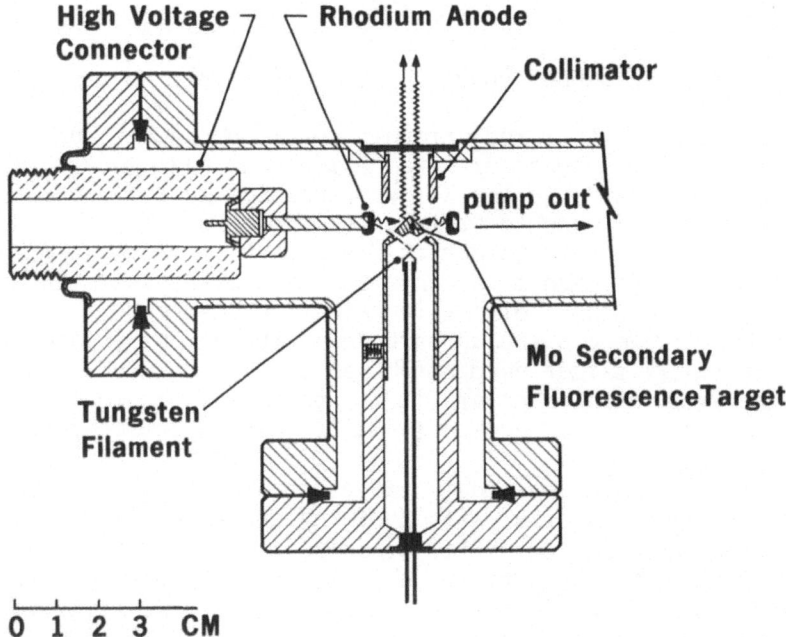

Figure 3. Schematic of secondary fluorescence tube with rhodium anode and molybdenum secondary target.

Direct measurement of the output spectrum show a ratio of peak-to-background height of greater than 300:1. Figure 4 is a comparison taken with radioisotope excitation similar to that shown in Fig. 2. The improved peak-to-background of the bromine peaks compared to the transmission tube reflects the improved quality of the secondary fluorescence input spectrum. The differences between the transmission tube and the radioisotope excitation are small. The greatly enhanced ratio of coherent to incoherent scattering intensity in the spectrum obtained with the X-ray tube has been attributed to the fact that the beam was more widely divergent at the sample due to a closer geometry employed with the particular tube used. This allowed a larger number of small angle scattering events to reach the detector thus enhancing the coherent contribution.

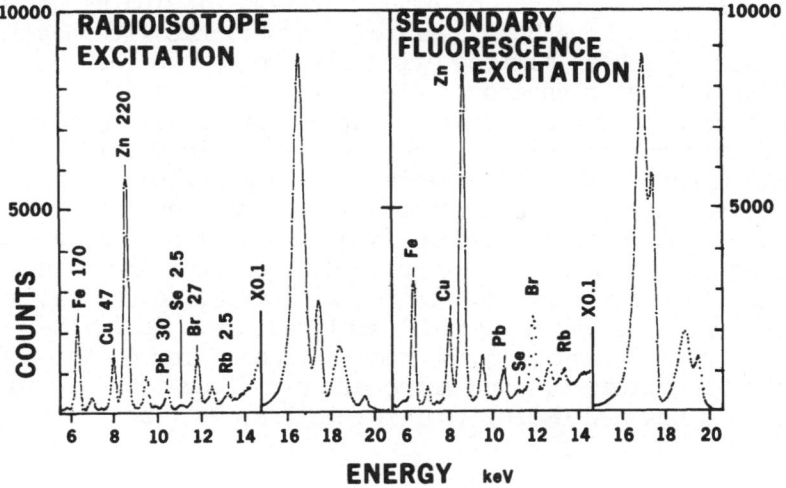

Figure 4. Comparison of excitation modes using same sample as Fig. 2.

SUMMARY AND CONCLUSIONS

Measurements performed with electron beam excited X-ray sources have demonstrated their greatly increased counting rate capability relative to radioisotope excitation; techniques for eliminating continuum background have minimized degradation in analytical sensitivity when using X-ray tube excitation. Although the peak-to-background ratio with the transmission X-ray tube was considerably worse than that possible with secondary fluorescence, it has the advantages of higher X-ray output flux and simplicity of design. Other potential advantages not exploited in the present work are improved geometrical efficiency with the anode held at ground potential and possible multiple anode operation with electronic switching between targets. The secondary fluorescence tube is characterized by greater peak to background, but a lower efficiency for X-ray production; an additional advantage being the potentially greater flexibility in secondary target material.

Although we have emphasized the example of molybdenum X-ray sources, tubes with a variety of other materials including copper, nickel and zirconium have been operated. Experience with heavier element targets has been limited primarily due to the lack of suitable high voltage connectors. In addition to extending the measurements to higher energy X-ray sources, it is also possible to conceive of a variety of different mechanical improvements which could improve geometric efficiency. In particular one can conceive of secondary fluorescence tubes with an annular target similar to the source-target assembles or small diameter transmission tubes for improved geometries. However, our experience has indicated that count rates obtained with existing designs are adequate for most present analytical applications. As further discussed in Ref. (3) and (5), the combination of low-background semiconductor spectrometers and monoenergetic X-ray tube excitation provides a new and useful tool for multielement analysis.

ACKNOWLEDGMENTS

The authors wish to express their appreciation to F. S. Goulding for his many contributions to this work and to other members of the Nuclear Chemistry Instrumentation Group for their assistance. We would like to acknowledge J. Anderson for his assistance in the fabrication of the X-ray tubes.

This work is part of the program of the Nuclear Chemistry Div. of the Lawrence Berkeley Laboratory, and was supported by the United States Atomic Energy Commission.

REFERENCES

1. R. D. Giauque, "A Radioisotope Source-Target Assembly for X-Ray Spectrometry", Anal. Chem., **40** (1968) 2075.

2. M. Green and V. E. Cosslett, "The Efficiency of Production of Characteristic X-Radiation in Thick Targets of a Pure Element", Proc. Phys. Soc. (London) **78** (1961) 1206.

3. F. S. Goulding J. M. Jaklevic, B. V. Jarrett and D. A. Landis, "Detector Background and Sensitivity of X-ray Fluorescence Spectrometers", to be presented at the 20th Annual Denver X-ray Conference, August 11-13 1971 LBL-9 Lawrence Berkeley Lab.

4. D. A. Landis, F. S. Goulding and R. H. Pehl, "Pulsed Feedback Techniques for Semiconductor Detector Radiation Spectrometers", IEEE Trans. Nuc. Sci., NS-18, No. 1 (1971) pg. 115-124.

5. R. D. Giauque and J. M. Jaklevic, "Rapid Quantitative Analysis by X-Ray Spectrometry", to be presented at the 20th Annual Denver X-ray Conference, August 11-13 1971.

6. J. A. Bearden, "X-Ray Wavelengths", Rev. Mod. Phys., **39** (1967) 78.

7. W. H. McMaster, N. K. Del Grande, J. H. Mallett and J. H. Hubbell "Compilation of X-ray Cross Sections", UCRL-50174 Section II Lawrence Radiation Laboratory, Livermore, California

DISCUSSION

R. JENKINS (Philips Electronic Instruments): Do you have any idea of the contamination rates of these tubes?

J. M. JAKLEVIC: We have not observed any contamination of the anode over the limited observation period available. However, the fact that the output spectrum is filtered may eliminate many contaminant lines.

RAPID ANALYSIS OF Mn IN PLAIN CARBON STEELS BY NONDISPERSIVE

X-RAY FLUORESCENCE SPECTROSCOPY

R. J. Gehrke and M. S. Cole

Aerojet Nuclear Company, National Reactor Testing Station

Idaho Falls, Idaho

W. A. Ryder

Allied Chemical Corporation

Idaho Falls, Idaho

ABSTRACT

Plain carbon steels are primarily composed of iron ($\sim$ 97%), but generally have small quantities of carbon, manganese, sulfur, phosphorous and silicon also present. Lead or copper may also be present. The steel industry is in need of an on-line technique of analysis for manganese in these steels. The manganese concentration of these steels varies from 0.3 to 1.5%. A technique is presented for the rapid analysis of manganese in carbon steels using energy-dispersive x-ray fluorescence spectroscopy. It is capable of determining the manganese content of a carbon steel in less than 30 sec with an uncertainty of less than 0.05% manganese. Because this method can analyze a steel from a distance of two feet, it should be possible to adequately protect the x-ray fluorescence spectrometer from the environment even when analyses are made of hot steel ingots at temperatures ranging up to 2400°F.

INTRODUCTION

Plain carbon steels account for about 90% of all steel made, and a large steel plant is capable of producing in excess of two million tons of steel annually. As a result of the large tonnage

of steel manufactured in a carbon steel plant, analyses of the
different grades of carbon steel are essential. However, the present carbon steel analysis techniques are not easily adopted for the
on-line analysis of manganese. Because of the need for an on-line
manganese analysis of carbon steel, this study was undertaken to
investigate the feasibility of using a high resolution Si(Li) x-ray
fluorescence spectrometer to provide this analysis. The precision
desired by the steel industry of an on-line manganese analysis is
0.05% manganese. Because of the inherent versatility of the Si(Li)
x-ray fluorescence spectrometer, the elements of titanium, vanadium
and chromium can simultaneously be determined along with manganese.
By equipping this x-ray fluorescence spectrometer with another
higher energy (> 20 keV) x-ray fluorescence source, other elements
including nickel, copper, zirconium, niobium, molybdenum, tungsten
and lead could also be determined in an additional measurement.
Analysis of these elements could be valuable in the on-line analysis
of alloy steels.

EXPERIMENTAL MEASUREMENTS

Several techniques are described in the literature(1-3) for
the identification of high alloy steels using x-ray fluorescence
spectrometry. However, none of these studies are concerned with
the analysis of manganese in plain carbon steels. Due to the lack
of energy resolution, x-ray fluorescence spectrometers using a
NaI(Tl) scintillation detector are unable to quantitatively determine the manganese content of carbon steels. Present state-of-the-
art Si(Li) spectrometers can resolve the Mn K_α from the Fe K_α x ray
when their peak intensities are about equal. But, when the Fe K_α
peak is 100 times stronger than the Mn K_α peak, these x rays are no
longer resolved. This is the case in an x-ray fluorescence spectrum
of carbon steel excited by K x rays from nickel or heavier elements.
However, by preferentially exciting the manganese over the iron
using Co K x rays the Fe K_α peak in the x-ray fluorescence spectrum
is sufficiently reduced relative to the Mn K_α peak so that these
K_α x rays are resolved with a high resolution Si(Li) spectrometer
(< 300 eV FWHM at 6.4 keV). An example of a Si(Li) x-ray fluorescence spectrum from a carbon steel containing 0.84% manganese is
illustrated in Figure 1.

Using Co K x rays for excitation carbon steel samples of known
manganese content were analyzed with a high resolution Si(Li) x-ray
fluorescence spectrometer in counting times varying from 20 to 45
sec. The steel samples were excited either by secondary fluorescence (i.e., a primary radiation from a copper x-ray generator producing Co K x rays in a cobalt metal target) or by direct fluorescence (i.e., a cobalt x-ray generator). In direct fluorescence,
the exciting x rays, although very intense, contain a large undesirable bremsstrahlung component. This bremsstrahlung is undesirable

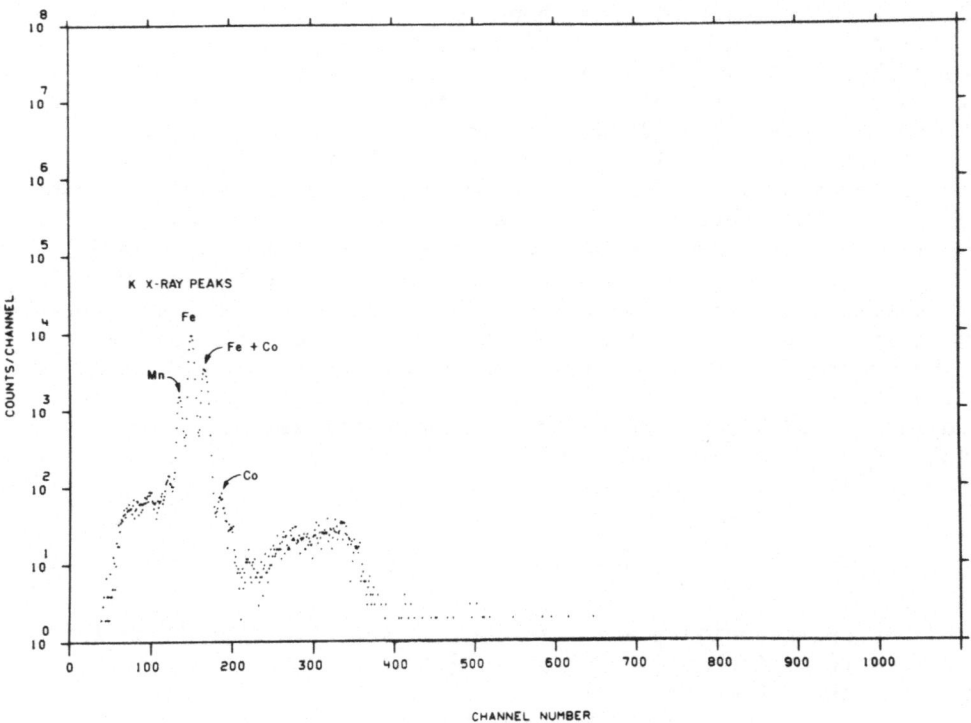

Fig. 1. X-ray fluorescence spectrum from carbon steel containing 0.84% manganese. Co K x rays used for fluorescence of sample.

because it does not preferentially excite the manganese in the steel. On the other hand in the case of secondary fluorescence, the exciting radiation, although relatively free of bremsstrahlung, is reduced several orders of magnitude in intensity through the secondary excitation. Figure 2 illustrates the x-ray fluorescence spectra from a carbon steel sample excited by the two different methods. The manganese K_α peaks are normalized to the same height and the effect of each exciting source can be observed from the relative intensities of the iron K_α x rays. Even though the manganese K_α peak is well resolved from the iron K_α peak in each case, the manganese K_α to iron K_α intensity ratio for direct fluorescence is about half that for secondary fluorescence. However, due to the large intensity loss in secondary fluorescence, the best choice of exciting source depends on the experimental conditions. Because the iron content in carbon steels does not vary significantly from sample to sample, the Fe K_α x-ray peak serves as an internal standard.

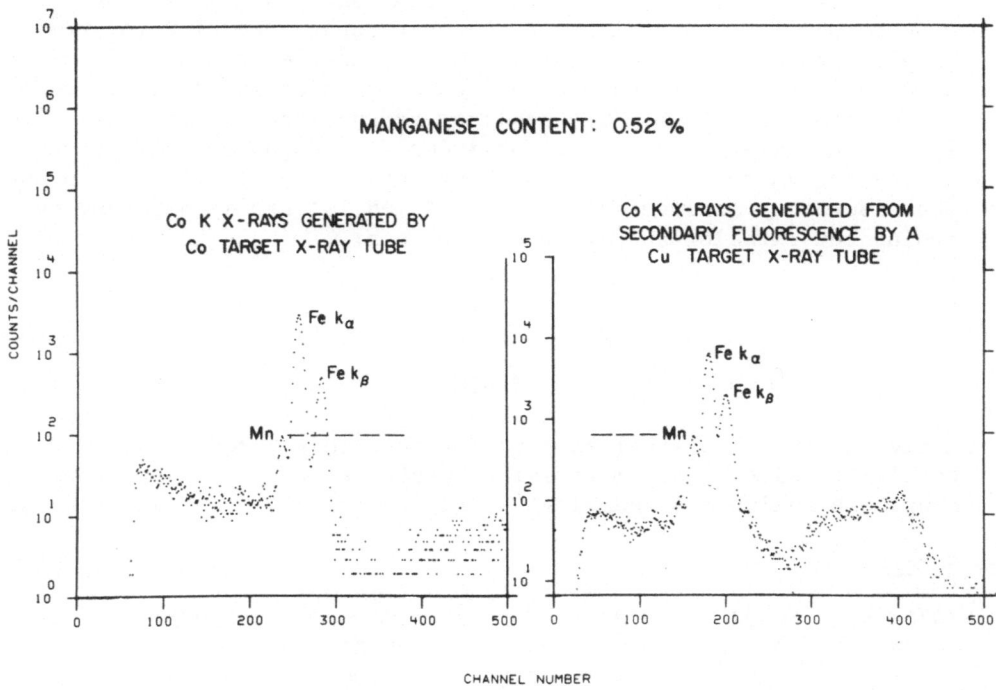

Fig. 2. Comparison of the x-ray fluorescence spectra from a carbon steel sample excited by two different methods.

The x-ray generator used to directly excite the steel samples is an x-ray diffraction unit equipped with a cobalt target x-ray tube. This x-ray tube was operated at 35 KV potential and with a tube current of 4 ma. With this system data were taken at source-sample and sample-detector distances of 2 ft and 9 in. respectively. Sets of data on the steel samples of known manganese content were taken for count times of 20 sec and 40 sec.

The x-ray generator used for secondary fluorescence of the steel samples is an x-ray diffraction unit equipped with a copper target x-ray tube. This x-ray tube was operated at 30 KV potential and with a tube current of 20 ma. The copper radiation emitted from this tube is used to fluoresce cobalt x rays from a cobalt metal target. These cobalt x rays are used to excite the steel samples. The source-sample and sample-detector distances were 2 in. and 4 in. respectively. Analyses of the steel samples of known manganese content were made with count times of 45 sec. In all measurements a 30 mm^2 x 3 mm Si(Li) detector with a resolution of 210 eV FWHM at 6.4 keV was used to acquire the data.

The steel samples containing known manganese concentrations were obtained from the Homer Research Laboratories of Bethlehem Steel Corporation. These samples have a course ground surface and were analyzed by Bethlehem on a vacuum emission spectrometer under production conditions. The analyses of the steel samples are listed in Table I. Laboratory analyses for manganese were performed on samples 4A and 10A using the "wet" arsenite titration technique. The laboratory analyses of samples 4A and 10A are in parentheses behind the emission spectrometer analyses.

Table I

Analyses of carbon steel samples obtained under production conditions with a vacuum emission spectrometer. The laboratory manganese analyses of samples 4A and 10A are given in parentheses.

Sample	C	Mn		P	S	Si
476	0.22	0.89		0.012	0.025	0.32
3A	0.22	0.52		0.010	0.032	0.07
4A	0.21	0.72	(0.746)	0.008	0.028	0.054
5A	0.20	0.90		0.009	0.035	0.10
7A	0.24	0.84		0.009	0.035	0.26
10A	0.41	1.50	(1.54)	0.009	0.027	0.22

Because this x-ray fluorescence technique uses the iron content of the steel as an internal standard, we investigated the possibility of making an analysis without removing the oxide coating normally covering hot steel. Steel samples with oxide coatings on all but one side were analyzed. The analysis obtained from the oxide side was about 0.05% manganese less than that obtained from the uncoated side.

DATA

The manganese K_α and iron K_α x-ray peak areas were determined by fitting a Gaussian function to the data points comprising each peak. The ratio of the areas of the manganese K_α to the iron K_α peaks ($R = I_{Mn}/I_{Fe}$) were determined for all the calibration samples and plotted on a linear scale as a function of the manganese concentration. The data were analyzed on-line with a small digital processor (4 K memory) coupled to the multichannel analyzer of the Si(Li) spectrometer and off-line by the large computer at the

National Reactor Testing Station. The results obtained by each
processor were essentially the same.

An algorithm which provides an even faster on-line analysis
is based on determining the manganese K_α and iron K_α peak heights
without determining the peak areas. This algorithm does not make
use of all the information in the data comprising each peak and its
use results in some loss of precision. However, by sacrificing
precision for a greatly simplified analysis program, a faster
analysis routine requiring much less computer memory can be written.
This savings of computer analysis time and memory is advantageous
particularly when speed and economy are the decisive factors.

RESULTS

Figures 3, 4 and 5 illustrate the results of the calibration
runs. A straight line has been fit through the data. The legend
associated with each plot describes the method of excitation, count
time and source-sample and sample-detector distances. The standard
deviation of each ratio is also given on the plots. The data obtained by secondary fluorescence of the steel samples for counting
times of 45 sec has an average deviation from the straight line fit

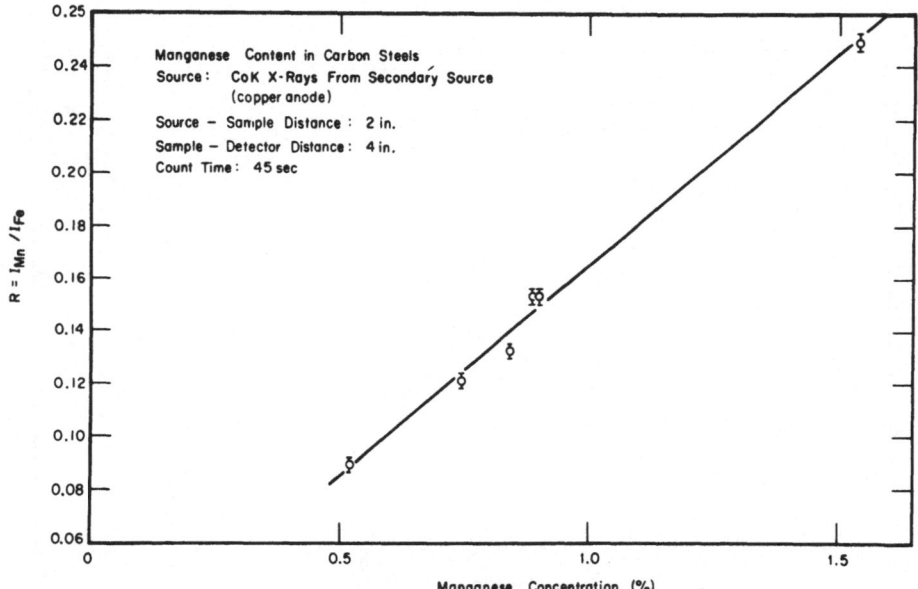

Fig. 3. Calibration curve for x-ray fluorescence of manganese
in plain carbon steel using Co K x rays generated by secondary
emission. Counting time: 45 sec.

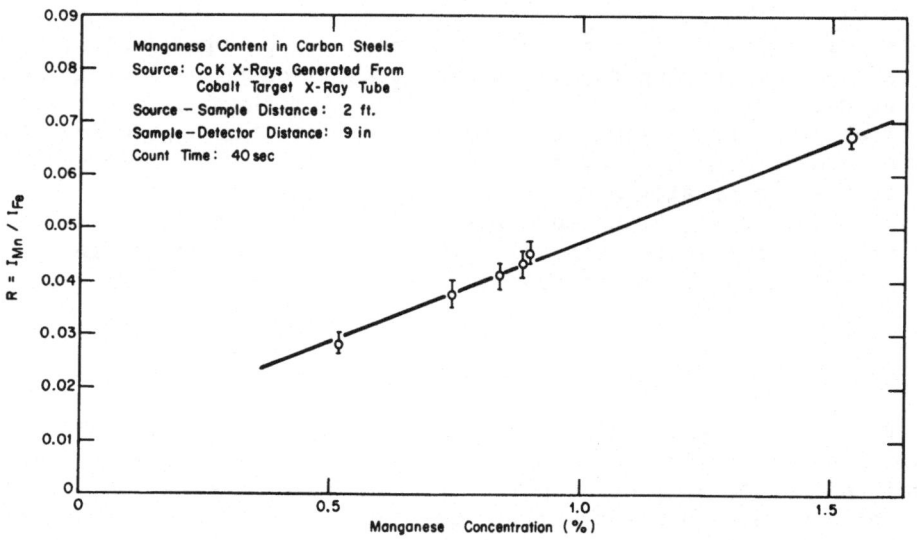

Fig. 4. Calibration curve for x-ray fluorescence of manganese in plain carbon steel using Co K x rays produced by cobalt target x-ray generator. Counting time: 40 sec.

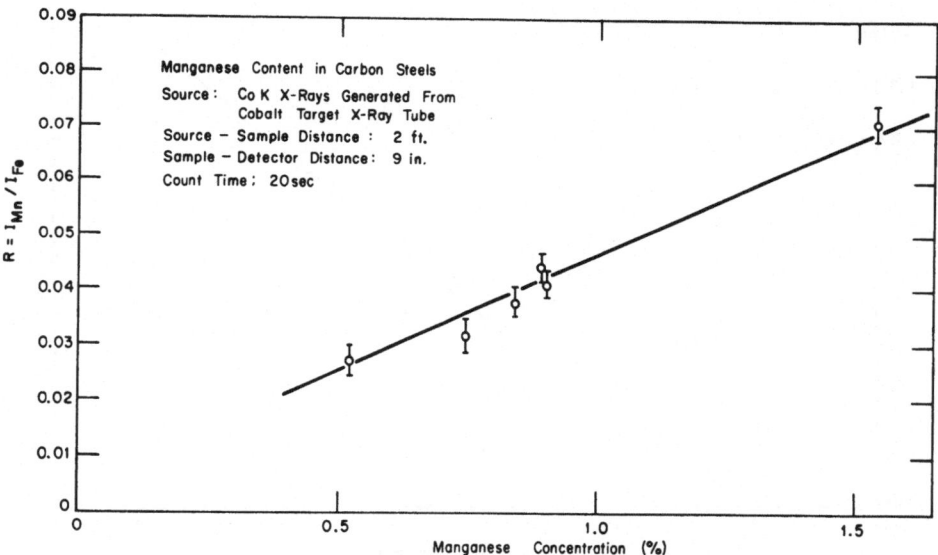

Fig. 5. Calibration curve for x-ray fluorescence of manganese in plain carbon steel using Co K x rays produced by cobalt target x-ray generator. Counting time: 20 sec.

corresponding to 0.02% manganese. The data obtained by direct
fluorescence of the steel samples for counting times of 20 sec has
an average deviation from the straight line fit corresponding to
0.07% manganese and that for counting times of 40 sec has an average deviation corresponding to 0.02% manganese.

DISCUSSION

The use of energy-dispersive Si(Li) x-ray fluorescence
spectroscopy for determining the manganese content in carbon steels
in on-line process control appears promising. With source-sample
and sample-detector distances of approximately 2 ft, it should be
possible to adequately protect the x-ray fluorescence spectrometer
from the environment when analyses are made of hot steel samples
(i.e., ingots, blooms, billets, etc.). Cobalt target x-ray generators capable of continuous operation at tube currents of 50 ma
are commercially available as stock items making it possible to
reduce count times to less than 10 sec for manganese analyses and
still perform the analysis with a precision of 0.05% manganese.
Thin cobalt foils could be used to filter out a large amount of the
bremsstrahlung radiation emitted from the cobalt x-ray generator
to further improve the quality of the fluorescing radiation.
Because the iron in the steel sample is used as an internal standard, drift in the x-ray tube current has a negligible effect on
the analysis.

ACKNOWLEDGMENTS

This work was supported under auspices of the U. S. Atomic
Energy Commission under contract with Aerojet Nuclear Company
(formerly Idaho Nuclear Corporation).

We wish to acknowledge the assistance of Bethlehem Steel
Corporation in providing the steel samples. In particular we wish
to thank Dean Flinchbaugh of Bethlehem Steel for several informative discussions relating the viewpoint of the steel industry to
the problems of on-line manganese analysis. His helpful suggestions
and comments are deeply appreciated. We are also grateful to
J. E. Cline for his critical reading of the manuscript.

REFERENCES

1. J. R. Rhodes, C. B. Hunter, D. L. Kellog, R. D. Sieberg and
    T. Furuta, "Application of a Computer-Coupled Radioisotope
    X-Ray Spectrometer to Analysis of Steels", in C. S. Barrett,
    J. B. Newkirk and C. O. Ruud, Editors, Advances in X-Ray
    Analysis, Vol. 14, Plenum Press, New York, pp. 127-138 (1971).

2. B. Sellers and J. Brinkerhoff, "Signature Comparison Technique for Rapid Alloy Sorting with a Radioisotope Excited X-Ray Analyzer", Materials Research and Standards, Vol. 10, No. 11, pp. 16-18 (1970).

3. "Radioisotope X-Ray Fluorescence Spectrometry", IAEA Technical Report Series No. 115 Vienna (1970), IAEA STI/DOC/10/115.

THE USE OF FIELD EMISSION TUBES IN X-RAY ANALYSIS

J. H. McCrary* and Ted Van Vorous

Vacuum Technology Associates

Broomfield, Colorado 80020

ABSTRACT

Recently developed, miniature, steady state, field emission
tubes are finding application in several areas of x-ray analysis.
These tubes require only a high voltage, low current power supply
to produce relatively intense beams of x-rays. Since anodes can
be fabricated from almost any element, and since the tubes can be
operated at potentials up to about 70 kV, many different output
x-ray spectra are available. Miniaturized battery operated x-ray
sources of this type, occupying a volume of about one liter, have
several advantages over radioisotope sources. These include cost,
safety, and controllable output spectra and intensity. X-ray
sources for energy dispersive fluorescence analyzers are designed
so that no scattered characteristic radiations will interfer with
the analysis of the sample fluorescence. Sources which are essentially monoenergetic can be fabricated for use in non-dispersive
x-ray fluorescence analyzers. Because of the intensity and safety
of the field emission tubes, such analyzers can be made which are
sensitive while compact, portable, and inexpensive. In x-ray
absorption analysis the measurement of absorption edge jump ratios
provides a quantitative measure of sample impurities. Field
emission tubes whose output spectra consist primarily of bremsstrahlung are particularly well suited to such measurements. The
techniques involved in using these tubes in x-ray analysis are
described.

* Consultant

INTRODUCTION

Miniature steady state x-ray machines (1) utilizing field
emission tubes (abbreviated FET) and portable power supplies are
commercially available and are finding application in several
areas of x-ray analysis. Figure 1 is a sketch of a typical FET.
The essential features of the tube are the cold, field emission
cathode (made from an ordinary sewing needle), the hemispherical
anode, the thin exit window, the high voltage insulator, and the
high vacuum ($\sim 10^{-7}$ Torr). The radius of the cathode is <0.0005
inch. Anodes are fabricated from any material which can be machined
or plated. Many different output spectral shapes are thus available. Exit windows typically are made of 0.005 inch thick beryllium. Permanent filters can be used in lieu of, or plated on, the
beryllium window. Typical high voltage insulators, capable of
withstanding 30 kV, permit an overall tube length of 2 inches.
This length can be varied to meet other high voltage requirements.
Since the power dissipated in the anode is approximately one watt,
no auxilliary anode cooling is required. The diameter of the x-ray
source area on the anode surface is less than 0.005 inch.

The application of high voltage to the electrodes of the FET
causes electrons to be emitted from the cathode and to be accelerated to the anode where their kinetic energy is converted into
heat and x-rays. Power supply requirements are not stringent.
Either A.C. or D.C. supplies can be used. In general, highly
regulated D.C. power supplies capable of providing $\sim$ 30 µA at
30 kV are used to extract maximum and constant x-ray output from
the field emission tubes. Supplies of this type are miniaturized
($\sim$ 1 liter, 1 kg) and are battery operated for use in remote
locations. A high voltage control provides an adjustment for the
x-ray intensity. Figure 2 shows the x-ray beam intensity as a
function of high voltage for a copper anode FET. Anodes made
from heavier elements produce more intense x-ray beams.

Figure 3 is a graph of the output spectrum of the copper tube
operated at 30 kV as measured with a Si(Li) spectrometer. Approximately 70 percent of the x-ray quanta are contained within the
characteristic copper K_α and K_β peaks. By selecting suitable
anode material, accelerating potential, and filtration, the output
spectrum from an FET can be made essentially monoenergetic or all
bremsstrahlung. Figure 4 is a plot of the spectrum from a copper
anode FET operated at 16 kV with a 0.001 inch copper filter.
Eighty percent of the x-rays in this spectrum (eighty-five percent
of the energy flux) are in the copper K_α and K_β peaks. Conversely,
a tube containing a silver anode and a thin aluminum window operated at 25 kV would emit an x-ray spectrum which consists entirely
of bremmstrahlung. Silver K radiation is not excited, and the 3
keV silver L x-rays are absorbed by the aluminum window.

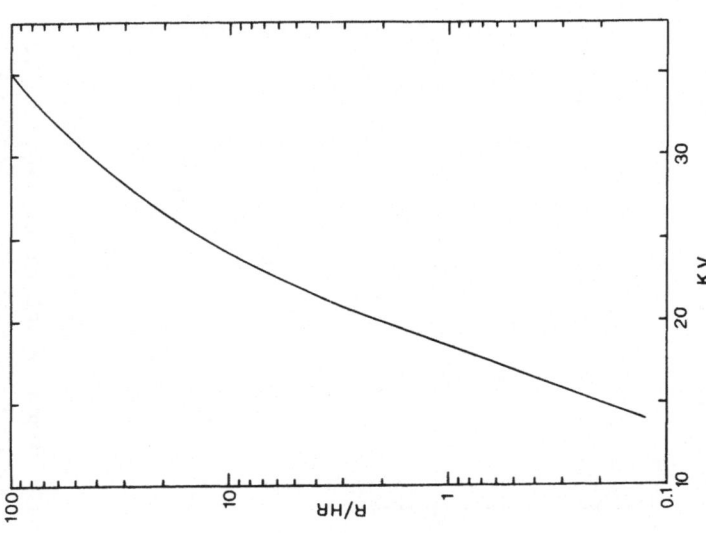

Figure 2. X-ray intensity at 30 cm (Roentgens/hour) vs. applied high voltage (kilovolts, constant potential) for a typical copper anode FET.

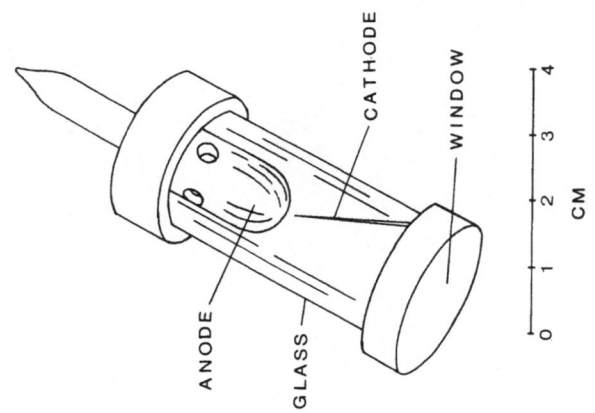

Figure 1. Sketch of typical miniature field emission x-ray tube.

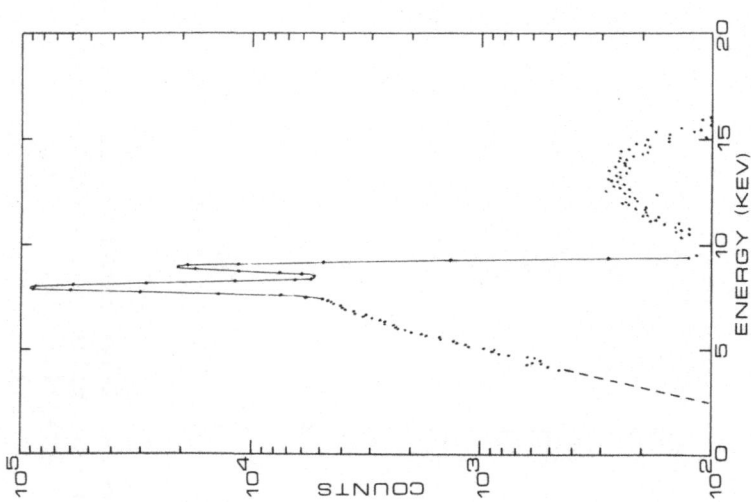

Figure 4. Output spectrum of copper FET operated at 16 kVCP with 0.001 inch copper filter.

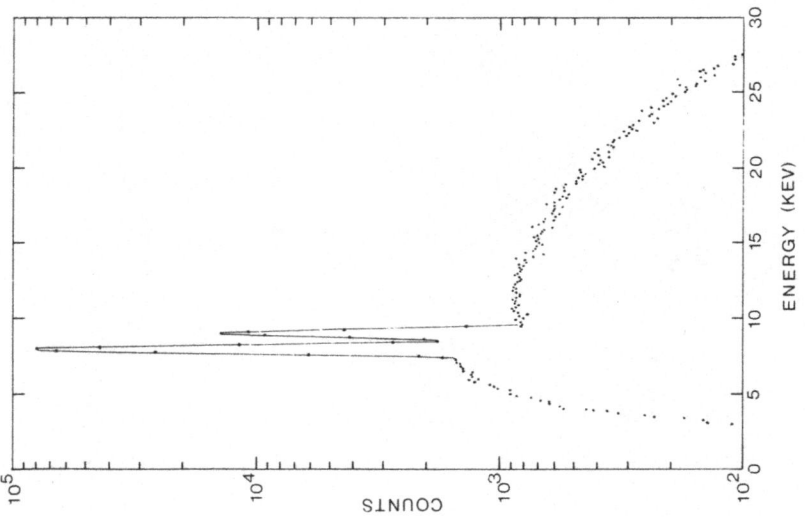

Figure 3. Copper FET output spectrum as measured with a Si(Li) spectrometer. Applied voltage = 30 kV (constant potential).

The FET is used to replace radioisotope photon sources in
many applications. The inherent advantages of the FET include
the following: (1) Since anodes are made from most elements,
many characteristic x-ray energies are available. (2) Sources
can be fabricated such that monoenergetic x-rays or bremsstrahlung
dominate the output spectrum. (3) The FET is effectively a point
source of radiation. (4) The x-ray intensity from the FET can be
varied by adjusting the high voltage. (5) FET's operated at maximum voltage are equivalent in intensity to several Curies of a
comparable radioisotope. (6) Since FET x-ray sources can be
turned off, they are safer than radioisotopes and installation
is simplified in some applications. (7) FET sources are in general
lower priced than radioisotope sources. Costs are further reduced
by the interchangeability of tubes and filters in FET systems.

X-RAY FLUORESCENCE ANALYSIS

Energy dispersive x-ray fluorescence analyzers utilize a
photon source to excite fluorescence within a sample, a Si(Li) or
Ge(Li) spectrometer to detect and analyze the energy of the fluorescent x-rays, and a multichannel analyzer which accumulates, displays, and permits access to the spectral data. The phrase "non-
dispersive x-ray fluorescence analyzers" will be used here to
denote those instruments which make use of photon sources, filters,
and low resolution counters to selectively excite and detect fluorescence from a given sample element without requiring the use of
an x-ray spectrometer and a multichannel analyzer. These non-
dispersive devices can in general be used only to determine the
presence and abundance of a single element within a sample. In
some applications, however, this limitation is offset by the
simplicity, portability, and low cost of these instruments.

The field emission tube is a direct replacement for radioisotope sources in energy dispersive systems. Its main advantages
are its high and variable intensity and the interchangeability of
tubes. Although either bremsstrahlung or monoenergetic FET sources
can be used, the bremsstrahlung sources appear to be the most promising since they emit no characteristic radiations to be scattered
by the sample and thus interfer with its analysis. Silver or tin
anode FET's operated at 20 to 30 kV are available for use with
Si(Li) spectrometers; tungsten anode tubes operated in the 50 to
70 kV range are recommended for use in Ge(Li) systems. The source-
sample-detector geometry used in the FET excited energy dispersive
analyzer is identical with that used in radioisotope systems.

There are several types of non-dispersive fluorescence x-ray
analyzers which make use of FET sources. These devices do not contain a sophisticated x-ray spectrometer and can thus be fabricated
in a portable, battery operated configuration. One such analyzer

is similar in principle to that described by Rhodes (2) in 1967.
However, the use of the FET source removes many of the limitations
and disadvantages inherent to that device. A schematic of the
present analyzer is shown in Figure 5. Basically, it consists
of a monoenergetic FET source and two shielded NaI(Tℓ) detectors
which view the fluorescing sample from identical geometry through
balanced filters whose absorption edges bracket the characteristic
radiation from the sample element of interest. The energy of the
source radiation ideally should be about twice that of the sample
fluorescence so that scattered radiation can be discriminated
against electronically. Because of the high intensity of the FET
source, small NaI detectors can be used, and measurements can be
made quickly and continuously. Since the two counters are operated simultaneously, analyses performed with this instrument are
not sensitive to variations in x-ray tube output. The long counting times associated with radioisotope excited analyzers are not
required.

Consider an analyzer which was designed to detect and measure
the copper content of ores. Referring to Figure 5, the radiation
source is an FET with a zirconium anode and zirconium filter
(to absorb most of the bremsstrahlung) operated at 30 kV. The
E- detector is covered with a cobalt filter sufficiently thick
to absorb most of the copper fluorescent x-rays while the nickel
filter over the E+ detector passes most of this characteristic
radiation. The two filters are balanced in thickness so that they
both absorb the same quantity of those x-rays whose energies are
not between the cobalt and nickel K edges. Amplified pulses from
both detectors pass through dual discriminators, which remove electrical noise and pulses from high energy (15 keV) scattered x-rays,
into a count rate ratio meter. The ratio of the count rates indicated on this meter is directly related to the relative abundance
of copper in the ore sample.

Two variations of the analyzer described above will be mentioned. The first of these employs two monoenergetic FET sources
whose x-ray energies, which are slightly different, bracket the
absorption edge (K or L) of the element being analyzed. A NaI
counter measures the intensity of the secondary radiation from
the separate areas of the sample illuminated by the sources.
The ratio of the count rates indicates the abundance of the
element in question. This analyzer is most adaptable to the
analysis of uniform samples. Another variation consists of a
single monoenergetic source whose energy is slightly higher than
the absorption edge of the element being analyzed. A single
counter detects its fluorescent x-rays. This type of device
is applicable to the analysis of elements whose atomic number
differs appreciably from those of the matrix elements.

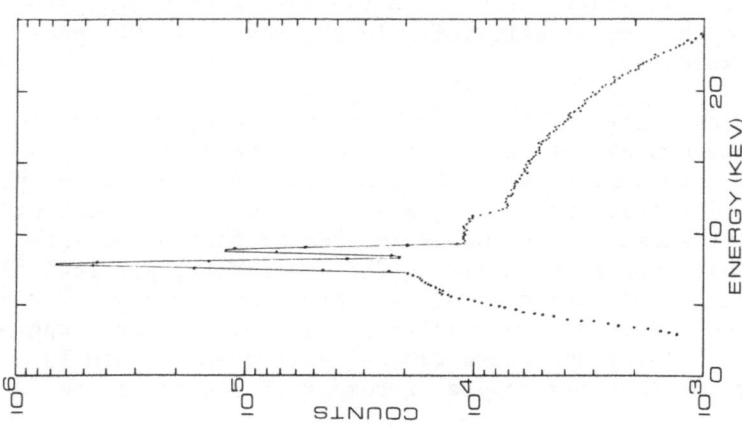

Figure 6. Spectrum from copper FET transmitted by a filter containing 1.9 mg/cm^2 of germanium.

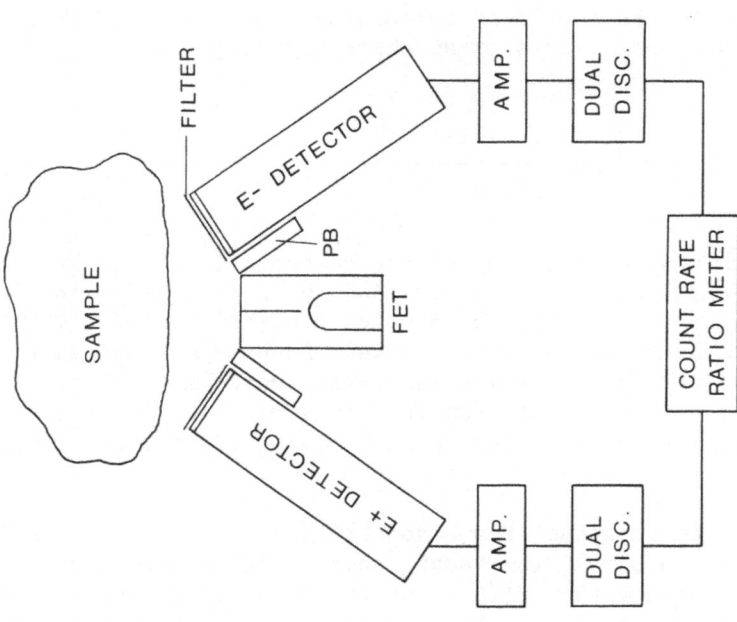

Figure 5. Schematic diagram of a typical non-dispersive x-ray fluorescence analyzer.

The sensitivity of these non-dispersive x-ray analyzers is strongly dependent upon the energy of the fluorescent x-rays and upon the absorption properties of the sample matrix. In the analysis of most metals in their ores, where the matrix consists primarily of light elements, sensitivities in the range of <0.1 percent are to be expected.

Since tubes and filters are interchangeable in all of these instruments, a given analyzer is not limited to the detection of a single element. By employing the L absorption edges and L x-rays for the heavier elements, analyzers can be built which will detect almost any element whose atomic number is greater than about fifteen. Non-portable analyzers utilizing FET sources are particularly adapted to industrial process or quality control applications where rapid or continuous readings are required. Portable, battery operated non-dispersive x-ray analyzers can be fabricated for use in the field or for use in other remote or temporary installations.

X-RAY ABSORPTION ANALYSIS

Absorption edge analysis, which has been in use for nearly fifty years (3,4), provides a quick, direct method for quantitatively analyzing thin samples. The technique consists of irradiating the sample with continuum x-rays and analyzing in intensity the energy of the transmitted beam. The thickness of a given element contained in the sample is calculated from the transmitted x-ray intensities at both sides of an absorption edge of the element from

$$G = \frac{\log I_1/I_2}{\mu_2 - \mu_1} \qquad (1)$$

where G is the thickness in gm/cm^2 of the element, I_1 and I_2 are the transmitted x-ray intensities below and above the absorption edge, and μ_1 and μ_2 are the mass absorption coefficients of the element in cm^2/gm below and above the edge. Absorption edge analyses are generally made with a machine x-ray source and a Bragg diffraction spectrometer. Measurements with similar accuracy can be made more rapidly and at a lower cost by using a bremsstrahlung FET source and a Si(Li) spectrometer.

Figure 6 is the spectrum from a copper FET (operated at 25 kV) transmitted through a beryllium sample coated with a thin layer of germanium, as measured with a Si(Li) spectrometer. The spectrometer resolution was 220 eV at 6 keV, the source-detector distance was 30 cm, and the counting time was 1,000 seconds. Clearly visible in the spectrum is the change in intensity at the germanium K edge (11.1 keV).

The intensities were extrapolated over the spectrometer resolution half-width to the K edge. From these intensities and from published values of mass absorption coefficients (5) the thickness of the germanium was calculated to be 1.9 mg/cm^2. By counting for longer times and by comparing transmitted spectra with incident spectra, impurity thicknesses in the range of 10 to 20 μg/cm^2 can be detected.

Other areas of application for the FET in x-ray absorption analysis include radiography and the use of x-ray transmissivity in the monitoring of thin material production. Portable, hand-held, light-duty radiography units are being built for use in remote or difficult locations, replacing radioisotopes in many applications. Tubes operated at 30 to 40 kV produce field intensities of several R/sec at a distance of an inch from the window. The extremely small (<0.005 inch diameter) source size permits high resolution radiographs to be made at very short source-film distances. A medical radiography unit, presently being developed, will be operated at 75 kV and will produce dental and other light medical radiographs with exposures of a few tenths of a second. Its cost and size will both be smaller than those for comparable conventional units now in use. FET's with low Z anodes operated at ∿10 kV are useful in the field of micro-radiography. The soft x-rays provided by these point sources produce high resolution, high contrast radiographs of thin sections of organic materials.

In many areas of x-ray analysis the miniature field emission x-ray tube offers important advantages over conventional radiation sources. Its unique features will permit an increased level of versatility, sensitivity, and safety in the design and application of instrumentation for x-ray analysis.

REFERENCES

1. J.H. McCrary and L.D. Looney, "Miniature Field Emission X-Ray Tube," Rev. Sci. Instr. 41, 1095-1096 (1970). U.S. Patent application filed September 1970.

2. J.R. Rhodes, "Some Examples of Ore and Alloy Analysis Using A Multipurpose Portable Analyzer," Proceedings of Second Symposium on Low Energy X- and Gamma Sources and Applications, ORNL-IIC-10, 843-869 (1967).

3. R. Glocker and W. Frohnmayer, "Uber die Roentgenspektroskopische Bestimmung des Gewichtsanteiles Eines Elementes in Gemengen und Verbindungen," Ann. Physik 76, 369-395 (1925).

4.   R. E. Barieau, "X-Ray Absorption Edge Spectrometry as an
     Analytical Tool," Analytical Chem. 29, 348-352 (1957).

5.   E. Storm and H. I. Israel, "Photon Cross Sections from 1 keV
     to 100 MeV for Elements Z = 1 to Z = 100," Nuclear Data
     Tables, Section A, 7, 565-681 (1970).

OLD ERRORS AND NEW CORRECTIONS IN X-RAY LINE PROFILE ANALYSIS

A. Kidron and R.J. De Angelis
Department of Metallurgical Engineering and
Materials Science
University of Kentucky
Lexington, Kentucky   40506

ABSTRACT

In recent years there has been an increasing awareness of the errors involved in X-Ray line profile analysis which is used to calculate coherently diffracting particle size, local strains and stacking fault probabilities in materials. Here a modified version of a Least Squares Analysis (L.S.A.) of the Fourier coefficients is presented. This method gives a possibility of making new corrections in a more natural and accurate way to the errors in the line profile analysis.

The first error involves the fact that experimentally one accumulates data as a function of θ – the Bragg angle, whereas the Fourier analysis is made in terms of equidistant steps in $\sin \theta$. In the L.S.A. the data does not have to be given in equidistant steps of $\sin \theta$ in order to be analyzed, so that this type of error does not exist at all.

In the L.S.A. there is also no need for the explicit separation of the $K_{\alpha 1}$ component from the total doublet line profile. This type of analysis is taken here one step further and it is shown that one can calculate the Fourier coefficients of each one of two merging lines. Each of these lines may be doublets in themselves and the two series of Fourier coefficients are calculated around the centers of gravity of their $K_{\alpha 1}$ components respectively.

The existence of an extra background which usually affects the Fourier coefficients when calculated by the Fourier transform method does not affect the L.S.A. except for a normalization factor. Here it is shown that this gives the possibility of calculating the strains without errors associated with the extra background.

295

INTRODUCTION

The observed intensity of a diffraction line from a cold worked material is given (1) by:

$$I(h_3) = \sum_{\ell=-\infty}^{+\infty} (A_\ell \cos 2\pi\ell h_3 + B_\ell \sin 2\pi\ell h_3) \qquad (1)$$

where $h_3 = 2a_3 \sin\theta/\lambda$, a_3 - the interplanar distance, θ - the Bragg angle, λ - the wavelength of the radiation and ℓ is the harmonic number. A_ℓ, B_ℓ are the Fourier coefficients from which one can determine the coherently diffracting particle sizes, the microstrains and the faulting parameters (1).

By changing h_3 and a_3 to the fictitious variables h_3' and a_3' such that $h_3' = 2a_3' \sin\theta/\lambda$ and by normalizing the total span of h_3' (including all non zero intensity points of the line) so that $\Delta h_3' \equiv 1$, equation (1) can be written (2) in the more useful form:

$$I(S-So) = Ao + \sum_{\ell=1}^{t} \left(2A_\ell \cos \frac{2\pi\ell(S-So)}{\Delta S} + 2B_\ell \sin \frac{2\pi\ell(S-So)}{\Delta S}\right) \qquad (2)$$

Here $S \equiv 2 \sin\theta/\lambda$ and $So \equiv 2 \sin\theta o/\lambda$, where θ_o is the Bragg angle at the peak position (or at the centroid of the profile) and $\Delta S = 2 \sin\theta_2/\lambda - 2 \sin\theta_1/\lambda$, where (θ_1,θ_2) is the interval of the Bragg angle which includes all the intensity points of the diffraction line. The harmonic number ℓ corresponds now to a real distance normal to the diffraction plane $L = \frac{\ell}{\Delta S}$. t is the number of significant values of A_ℓ or B_ℓ, whichever is larger.

If one takes the whole profile and divides it into equal intervals in θ, then it can be shown (2) that for $\theta \leq 60°$ we will have

$$\frac{S-So}{\Delta S} \simeq \frac{j}{q} \qquad (3)$$

where q is the number of divisions and $j = 0, \pm1, \pm 2, --, \pm \frac{q}{2}$. Here $j = 0$ is the peak position (or centroid) of the line and the positive integers +1, +2, etc. give the positions of the divisions at the high angle side of the line and similarly with the negative numbers -1, -2, etc. for the other side of the line. In this way equation (2) can be written as:

$$I(S-S_o) = A_o + \sum_{\ell=1}^{t} \left(2A_\ell \cos \frac{2\pi\ell j}{q} + 2B_\ell \sin \frac{2\pi\ell j}{q}\right) \qquad (4)$$

The parameters of interest here are A_ℓ and B_ℓ. The usual way to calculate them (1,2) is by a Fourier Inversion of equation (4).

Another way of calculating A_ℓ and B_ℓ is by looking at eq. (4) as a set of linear equations with A_ℓ and B_ℓ as unknowns, and calculating the latter by a Least Squares Analysis (L.S.A.) of equations (4). This was done recently (3), where it was shown that the L.S.A. method has a few important advantages over the usual Fourier Transform method. The L.S.A. of equations (4) makes it also possible to separate the K_α doublet into its components K_{α_1} and K_{α_2} and one can even calculate directly the Fourier coefficients of the K_{α_1} component from the total doublet intensity (4).

In the present paper we will show that one has some other advantages in using the L.S.A. method and that this method can help correct partly or totally some of the common errors encountered in the Fourier Analysis of line profiles.

EQUIDISTANT STEPS IN THE BRAGG ANGLE

The first important error in the Fourier analysis of X-Ray diffraction profiles arises from the fact that the approximation given in equation (3) is not good for lines where $\theta > 60°$ or even for lower angle lines which are very broad. Usually what one does (2) is translate the θ values given experimentally in equal steps, into $\sin \theta$ values and then interpolate the experimental intensities and have them for values of equal steps in $\sin \theta$. It is not easy to make such a translation accurately specially for sharp lines at high angles.

In the L.S.A. of line profiles it is not necessary to have the experimental intensity in equal steps of $\sin \theta$. Indeed, the equations (2) can be written for any set of values of S as long as the latter are well distributed along the profile. What we did then was to measure the intensities in equal steps of θ (by step-scanning) then change θ to S and feed the values of S and So to equations (2). A L.S.A. of the latter gave us the required Fourier coefficients A_ℓ, B_ℓ.

OVERLAPPING DIFFRACTION LINES

Another error associated with the Fourier transform method is the case when some diffraction lines merge one into another. One way out of this error is by separating the lines by some kind of "intuition", but this may be very treacherous. Another way is by using only lines which are well separated and this usually means the use of a single line (without a second order line) to calculate the effective diffraction particle size, the local microstrains, and the stacking fault probabilities (5). We will discuss below the errors associated with such solutions of the problem, but first the solution to the Fourier Analysis of overlapping diffraction lines using L.S.A. will be given.

In the L.S.A. we have a set of linear equations of the form:

$$I(n) = \sum_{k=1}^{v} F_k \cdot C_{n,k} \qquad (5)$$

By taking $F_k = A_0, 2A_1, ---, 2A_t, 2B_1, ---, 2B_t$; $v = 2t + 1$ and
$C_{n,k} = 1, \cos \dfrac{(2\pi \cdot 1 \cdot (S-S_0))}{\Delta S}, \cos \dfrac{(2\pi \cdot 2 \cdot (S-S_0))}{\Delta S}, ---, \cos \dfrac{(2\pi \cdot t \cdot (S-S_0))}{\Delta S}$

$\sin \dfrac{(2\pi \cdot 1 \cdot (S-S_0))}{\Delta S}, \sin \dfrac{(2\pi \cdot 2 \cdot (S-S_0))}{\Delta S}, ---, \sin \dfrac{(2\pi \cdot t \cdot (S-S_0))}{\Delta S}$

and by writing this for all n, each one of them corresponding to a
different value of $(S-S_0)$ one can evaluate the set of equations (5)
by a L.S.A. and obtain all F_k, i.e. all the coefficients A_ℓ and B_ℓ
in (2).

This procedure is good for a singlet diffraction line. When
we have a diffraction line composed of a doublet then one can still
use equations (5) except that now (4) we have $C'_{n,k}$ instead of $C_{n,k}$
with:

$$C'_{n,k} = R \, C_{n,k} + (1-R) C_{n-s,k} \qquad (6)$$

where R is the ratio of the total intensity of the $K_{\alpha 1}$ component
to the total intensity of the doublet; and S is the separation between the $K_{\alpha 1}$ and $K_{\alpha 2}$ components.

Now consider two diffraction lines, each one of them a doublet
in itself and neither one having a necessarily symmetrical $K_{\alpha 1}$ component (see Fig. 1).

Now the intensity of the first doublet can be written according
to equations (5) and (6) as:

$$I_1(n) = \sum_{k=1}^{v} F_k(1) \, C'_{n,k}(1) \qquad (7)$$

where $F_k(1)$ are the Fourier coefficients of the $K_{\alpha 1}$ component as
calculated around its peak position (or centroid) which we call
$S_0(1)$. Similarly for the second doublet we will have:

$$I_2(n) = \sum_{k=1}^{w} F_k(2) \, C'_{n,k}(2) \qquad (8)$$

where $F_k(2)$ are the Fourier coefficients of the $K_{\alpha 1}$ component of the
second doublet as calculated around its peak position (or centroid)
$S_0(2)$. In equations (7) and (8) we used the same n (i.e. no shift
between them). We could do this by adding zero intensity points to
each one of the doublets so that the spans (in S or in n) of both
are congruent.

The total intensity of the two doublets together will be given by: $I(n) = I_1(n) + I_2(n)$

$$= P \cdot \sum_{k=1}^{v} F_k(1) \cdot C'_{n,k}(1) + (1-P) \sum_{k=1}^{w} F_k(2) \cdot C'_{n,k}(2). \qquad (9)$$

where P is the ratio of the integrated intensity of I_1 to the integrated intensity of the two doublets together. If we know $So(1)$ and $So(2)$, we can calculate the coefficients $C'_{n,k}(1)$ and $C'_{n,k}(2)$ in (9) and then this set of equations can be L.S. analyzed to yield $F_k(1)$ and $F_k(2)$. But usually this procedure does not work because the solution of the equations (9) is not necessarily unique and usually a non converging solution is obtained.

We solve the problem in two steps. First we take the intensity points of the first doublet from $n = 1$ to some $N = J$ where the total (two doublets together) intensity is minimum (see Fig. 1). This profile is then L.S. Analyzed using equations (7). This is permissible, because in the L.S.A. one can drop part of the data and still obtain reasonable results (3). In our case the results will not be very good unless the merging of the two doublets is small. (In general dropping about 10% of the non zero intensity points on one side of a line will not change significantly the resulting Fourier coefficients (1)). Then we L.S. Analyze the second doublet taking the intensity points from $N = J$ to the high angle end of the lines, and using equations (8). This calculation and the one above give us the approximate solutions of $F_k(1)$ and of $F_k(2)$. These solutions are then plugged into the set of equations (9) as a first approximation. The set (9) can then be solved by a computer program which uses a successive approximations method in the Least Squares sense.

The separation $So(2)-So(1)$ between the two lines can be calculated easily from the lattice parameter. On the other hand the ratio of the intensity of the two lines $P/(1-P)$ will not be given usually by the structure factors and the multiplicities because in real cases there will usually be some extinction and preferred orientation effects. The way to overcome this is by taking the area of the first line up to the point where the intensity is minimum (i.e. $N = J$) and divide it by the total area of the two lines. This will give an approximate value of P (eq. 9) which can be changed slowly until the back calculated curve fits best the observed intensity.

The procedure described above was carried out on two Gaussian doublets which were generated for this purpose. The first doublet was made out of two asymmetric components ($K_{\alpha 1}$ and $K_{\alpha 2}$). Each one had a left (low angle) side according to exp $- (0.10x)^2$ and a right side according to exp $- (0.08x)^2$. The intensity of the components was 2 to 1. Similarly the second doublet was made out of two

A. Kidron and R. J. De Angelis

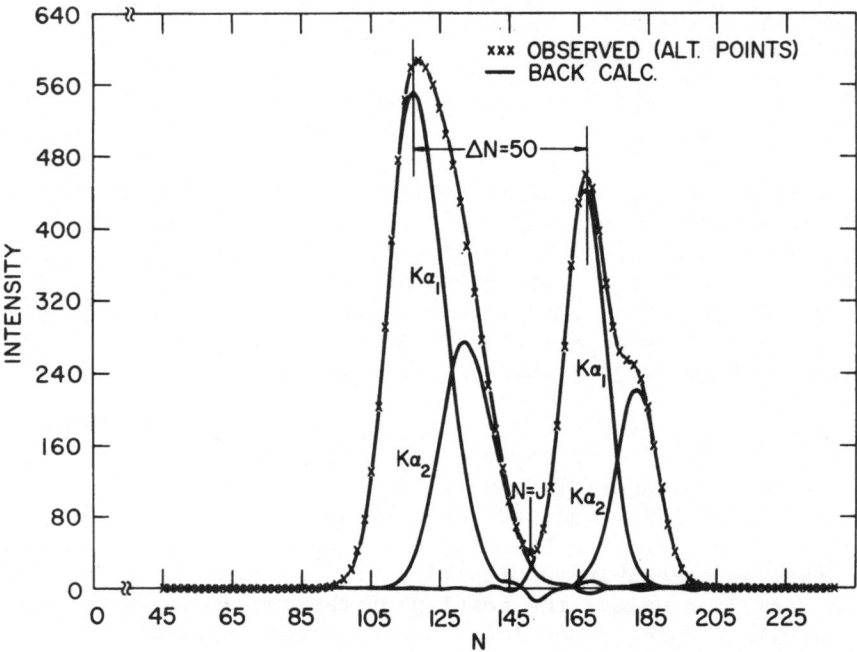

Fig. 1. The intensity of two overlapping doublets versus the station number. The "observed" intensity is given for alternate points.

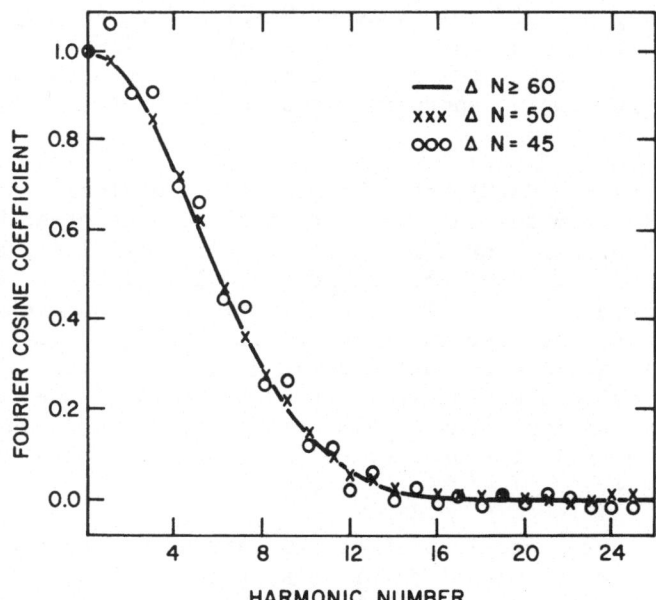

Fig. 2. The Fourier cosine coefficients of the first $K_{\alpha 1}$ line of Figure 1 as calculated around its peak position versus the harmonic number.

components with the same intensity ratio except that each one of the components was now symmetric with exp $- (0.12x)^2$ on both sides. Figure 1 gives the plot of the "observed" intensities obtained in this way. The separation between the peaks of the $K_{\alpha 1}$ components was taken as $\Delta N = 50$ divisions. The total number of divisions was 252 and the number of non zero (greater than 10^{-4}) intensity points was 136. The ratio of total intensity of the first doublet to the total intensity of both doublets was $P = 0.623$. The same calculations were also done for similar lines but for separations $\Delta N = 40, 45, 55, 60, 65, 70$. For $\Delta N = 40$ the whole procedure broke down, resulting in non converging high value ($\sim 10^3$) Fourier coefficients. As can be seen in figures 2,3 the resulting cosine Fourier coefficients were reasonable for $\Delta N = 45$, good for $\Delta N \geq 50$ and "very good" for $\Delta N \geq 60$. On the other hand the sine Fourier coefficients (figure 4) were good only for $\Delta N \geq 60$ (note the difference in the scale of the sine coefficients).

In Figure 1 plotsof the back calculated total intensity as well as the individual $K_{\alpha 1}$ and $K_{\alpha 2}$ components are given. The fit between the observed and back calculated intensities is very good.

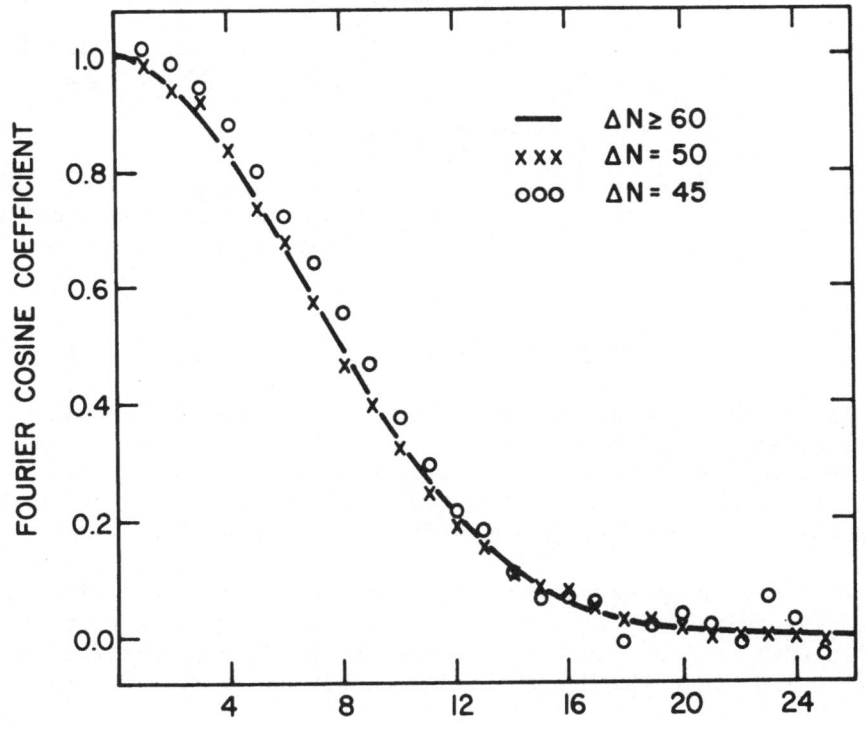

Figure 3. The Fourier cosine coefficients of the second $K_{\alpha 1}$ line of Figure 1 as calculated around its peak position versus the harmonic number.

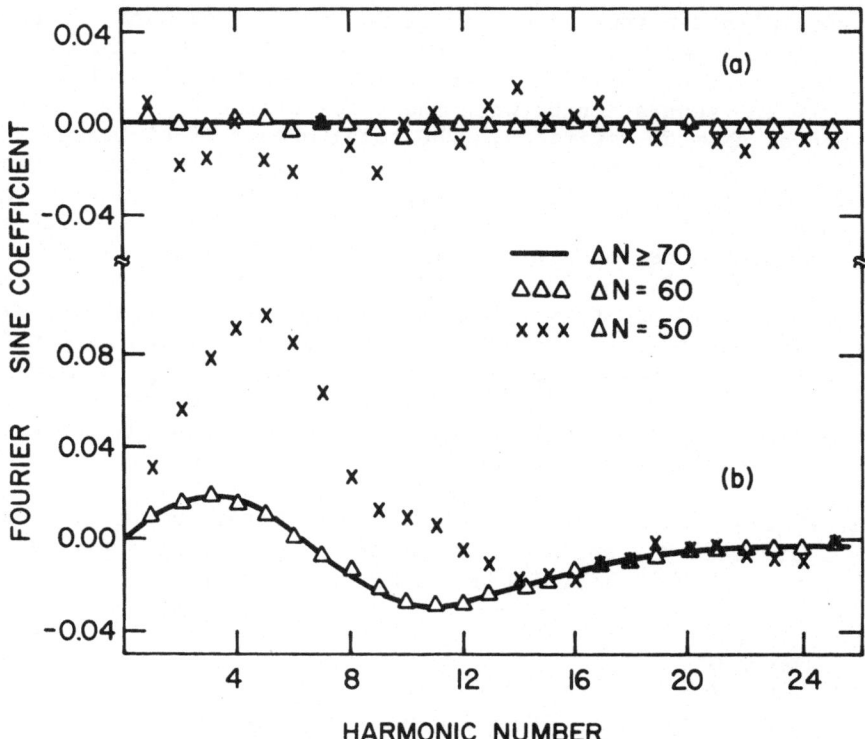

Figure 4.   The Fourier sine coefficients of the $K_{\alpha 1}$ lines of Figure
1.   (a) For the second $K_{\alpha 1}$ line around its peak position.
(b) For the first $K_{\alpha 1}$ line around its peak position.

EXTRA BACKGROUND

The proper selection of background is the most difficult problem associated with the analysis of X-Ray Diffraction profiles.   The
incorrect choice of background leads to an error in the computed
coefficients which may be propagated to the coefficients of the true
diffraction profile through the Stokes correction (1).

If there is an extra background associated with the line, i.e.
if $I_{obs} = I_{true} + C$ and if we normalize the resulting Fourier
coefficients so that the first cosine (Real) Fourier coefficient is
equal to 1.0, then the other Fourier coefficients will be given by:

$$G_r(L) = C_1\, G_r'(L) \quad , \quad G_i(L) = C_1\, G_i'(L)$$
$$H_r(L) = C_2\, H_r'(L) \quad , \quad H_i(L) = C_2\, H_i'(L) \quad \Big\} \quad L=2,3,\ldots,t \quad (10)$$

where $G_r(L)$ and $G_i(L)$ are the cosine (real) and sine (imaginary)
Fourier coefficients for the "annealed" peak respectively and $H_r(L)$

and $H_i(L)$ are the similar coefficients for the "broadened" peak. Here the primes refer to the error free quantities and generally the constant C_1 will be different from C_2.

The true cosine Fourier coefficients are given (1) by:

$$A_L = \frac{H_r(L) \cdot G_r(L) + H_i(L) \cdot G_i(L)}{G_r^2(L) + G_i^2(L)}$$

Plugging equations (10) into the last equation will give for the calculated true cosine coefficients:

$$A_L = A_L' \cdot \frac{C_2}{C_1} = A_L' \cdot C \tag{11}$$

where $C \equiv C_2/C_1$ and A_L' are the error free coefficients.

In order to separate the particle size from the microstrains we use the Warren-Averbach procedure (1). Accordingly we write

$$\ln A_L' = \ln A_L^S - 2\pi^2 <\varepsilon_L^2> L^2 d^2 \tag{12}$$

where A_L^S are the particle size coefficients, $<\varepsilon_L^2>$ are the Mean Square strains and d is the interplanar spacing. From (11) and (12) we get:

$$\ln A_L = \ln A_L' + \ln C = \ln A_L^S + \ln C - 2\pi^2 <\varepsilon_L^2> L^2 d^2 \tag{13}$$

If there is no background error then $\ln C = 0$ and a plot of $\ln A_L$ versus d^2 (for each value of L) gives us the parameters A_L^S from which one can find the particle size and also we get $<\varepsilon_L^2>$, i.e. the Mean Square Strains versus L.

But if $\ln C \neq 0$ then things become a little more complicated. To facilitate the discussion let us say that we have a cubic crystal and that we are looking at a given (h,k,ℓ) line with $d_1^2 = ho^2(1)/a^2$ where $ho^2 = h^2 + k^2 + \ell^2$ and a is the lattice parameter, and at its second order line $(2h, 2k, 2\ell)$ with $d_2^2 = ho^2(2)/a^2$. For the first line equation (13) will be:

$$\ln A_L(1) = \ln A_L^S + \ln K_1 - 2\pi^2 <\varepsilon_L^2> L^2 ho^2(1)/a^2 \tag{14}$$

and for the second line:

$$\ln A_L(2) = \ln A_L^S + \ln K_2 - 2\pi^2 <\varepsilon_L^2> L^2 ho^2(2)/a^2 \tag{15}$$

where K_1 and K_2 are constants.

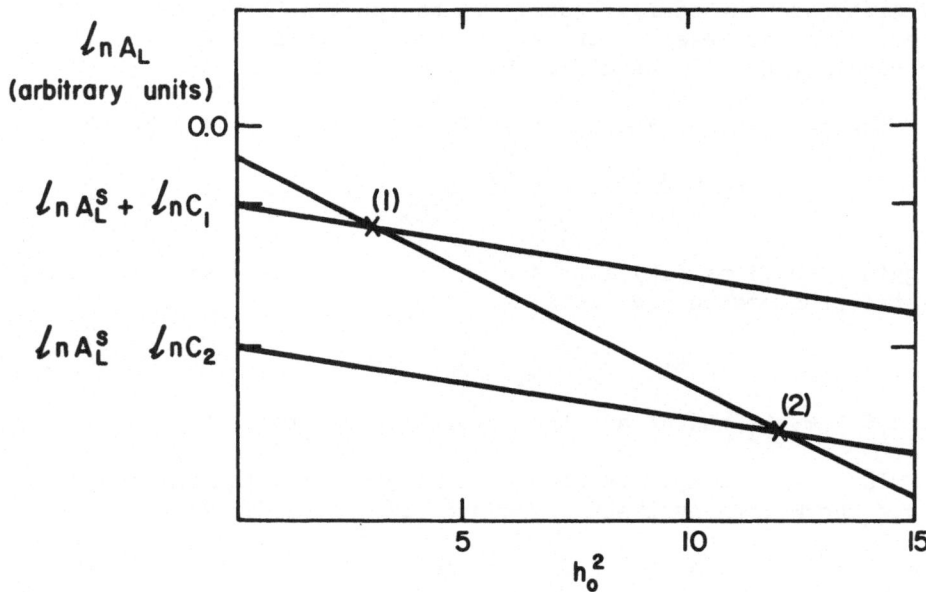

Figure 5.   Illustration of the errors involved when an extra back-
            ground is added to the intensity (see text).

Referring to figure 5 let us plot $\ln A_L$ versus h_o^2 for a given
L. On this plot the point given by equation (14) is denoted (1)
and the point given by equation (15) is denoted (2). If we drew a
line through point (1) with slope equal to $-2\pi^2 \langle \varepsilon_L^2 \rangle L^2/a^2$ it would
intercept the ordinate at $(\ln A_L^S + \ln K_1)$. A second line passing
through the point (2) with a slope of $-2\pi^2 \langle \varepsilon_L^2 \rangle L^2/a^2$ would intercept the ordinate at $(\ln A_L^S + \ln K_2)$. Generally $K_1 \neq K_2$, so that
the two intercepts are different. In the Warren-Averbach procedure
we draw a line through the two points (1) and (2). This line has a
slope different from $-2\pi^2 \langle \varepsilon_L^2 \rangle L^2/a^2$ and consequently the strains
$\langle \varepsilon_L^2 \rangle$ calculated from them would be wrong. Also the intercept of
this line with the ordinate would be different from $\ln A_L^S$ leading
to an erroneous value of the particle size.

By the following procedure we can find the correct values of
the strains and particle sizes.

Let us denote by S(L) the slope that we actually measure, i.e.

$$S(L) = \frac{\ln A_L(2) - \ln A_L(1)}{ho^2(2) - ho^2(1)}$$

which by virtue of (14) and (15) becomes:

$$S(L) = \frac{\ln (K_2/K_1)}{ho^2(2) - ho^2(1)} - \frac{2\pi^2 <\varepsilon_L^2> L^2}{a^2} \tag{16}$$

A plot of S(L) versus (L) will not necessarily be a straight line. But if one calculates the Fourier coefficients for values of L close enough one to another (by adding zero intensity points to both sides of the line profile and thus making the span of the line as wide as necessary) then it is possible to extrapolate the curve given by equation (16) to L = 0 and thus obtain the value of $\ln(K_2/K_1)$. Once this value is known equation (16) will give the mean square strains $<\varepsilon_L^2>$ for each value of L.

It has been shown (5) that if one assumes $\varepsilon \propto \frac{1}{r}$ where r is the distance from a dislocation then for low values of L we will have

$$<\varepsilon_L^2> = G^2(hk\ell)/L \tag{17}$$

where G^2 is a constant not depending on L. Accordingly equation (13) can be written as

$$\ln A_L = \ln A_L^S + \ln C - \frac{2\pi^2 ho^2 G^2(hk\ell)}{a^2} L$$

also for small L we have

$$\ln A_L^S = - \frac{L}{D_{eff}(hk\ell)} \tag{18}$$

where D_{eff} is the effective particle size so that

$$\ln A_L = \ln C - (\frac{1}{D_{eff}} + \frac{2\pi^2 ho^2 G^2(hk\ell)}{a^2}) L \tag{19}$$

This last equation has been obtained by Rothman & Cohen (5) (except for the term $\ln$ C) for error free measurements. We want only to stress that the relationship (17) does not hold for very small L and that (18) holds only for small L. So that equation (19) will hold for small but not very small L, e.g. between 10 to 100 A° in most common cases. This means that generally the plot of $\ln A_L$ versus L will have a curvature (usually negative) for very small (L) and consequently the straight line that one obtains at higher L where equation (19) holds will not extrapolate to $\ln$ C (or to zero in the error free measurement).

For the portion of the plot of $\ln A_L$ versus L where equation (19) holds we will have a straight line which will give us the quantity

$$\frac{1}{D_{eff}} + \frac{2\pi^2 ho^2 \; G^2(hk\ell)}{a^2} \; .$$

Having obtained already the quantity $2\pi^2 <\varepsilon_L^2 > L^2/a^2 = 2\pi^2 G^2(hk\ell)L/a^2$ through equation (19) we can now evaluate D_{eff} − the effective particle size.

ACKNOWLEDGEMENT

The authors wish to thank Mrs. A. Leigh for her help in the use of the University of Kentucky Numerical Library computer programs.

Research was sponsored partly by the Office of Aerospace Research, United States Air Force, under Contract F33-615-69-C1027 and partly by the University of Kentucky Research Foundation.

REFERENCES

1.   B.E. Warren, X-Ray Diffraction, p. 251, Addison Wesley Publ. Co. (1969).

2.   C.N.J. Wagner, "Analysis of the Broadening and Changes in Position of Peaks in an X-Ray Powder Pattern," in J.B. Cohen and J.E. Hilliard, Editors, Local Atomic Arrangements Studied by X-Ray Diffraction, ASM Conference, Vol.36, p. 219-269, Gordon and Breach, Sc. Publ. (1965).

3.   A. Kidron and R.J. De Angelis, "Least Squares Analysis of Fourier Coefficients and its Application to X-Ray Diffraction Profiles," in G.M.L. Gladwell, Editor, Symposium on Computer Aided Engineering, p. 285-297, University of Waterloo Press (1971).

4.   A.Kidron and R.J. De Angelis, "Direct Evaluation of $K_{\alpha 1}$ Fourier Coefficients in X-Ray Profile Analysis," Acta Cryst. A 27, (1971), in print.

5.   R.L. Rothman and J.B. Cohen, "A New Method for Fourier Analysis of Shapes of X-Ray Peaks and its Application to Line Broadening and Integrated Intensities Measurements," in C.S. Barrett, J.B. Newkirk and G.R. Mallet, Editors, Advances in X-Ray Analysis, Vol. 12, p. 208-235, Plenum Press (1969).

THE EFFECTS OF SELF-IRRADIATION ON THE LATTICE OF $238(80\%)PuO_2$ *

R. B. Roof, Jr.

Los Alamos Scientific Laboratory

Los Alamos, New Mexico 87544

ABSTRACT

As a function of self-irradiation, the crystalline lattice of $238(80\%)PuO_2$ is gradually altered. The technique of x-ray line broadening was used to search for changes in the crystallite size and strain in the lattice following an initial annealing treatment. The integral breadth, Fourier coefficient, and variance methods were used to analyze the broadening of the x-ray powder pattern lines. The mechanism of alteration appears to be one of retained strain as no evidence was forthcoming from these techniques to indicate that the material undergoes significant crystallite size change. The strain in the lattice increased from zero at time zero to approximately 0.2% during a 2-year period.

INTRODUCTION

This paper presents the results of a study of the changes in crystallite size and strain in the lattice of $238(80\%)PuO_2$ as a function of time of self-irradiation. The temperature of the sample was somewhat above room temperature due to radioactive self-heating. The experimental examining technique employed was x-ray line broadening analysis.

A material that has a large crystallite size and no lattice strain or other faults will, in general, yield a quite sharp x-ray diffraction powder pattern line. (See Fig. 1.) A broadened x-ray

---

* Work done under the auspices of the U. S. Atomic Energy Commission.

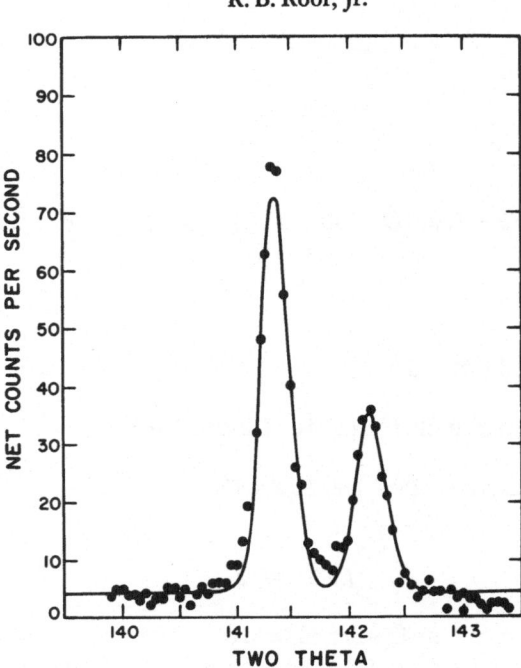

Fig. 1.   The 622 reflection of 238(80%)PuO$_2$ examined with copper
          Kα x-ray radiation. Plotted points are the average of
          data taken 7, 8, and 9 weeks after an initial annealing
          treatment of the sample.

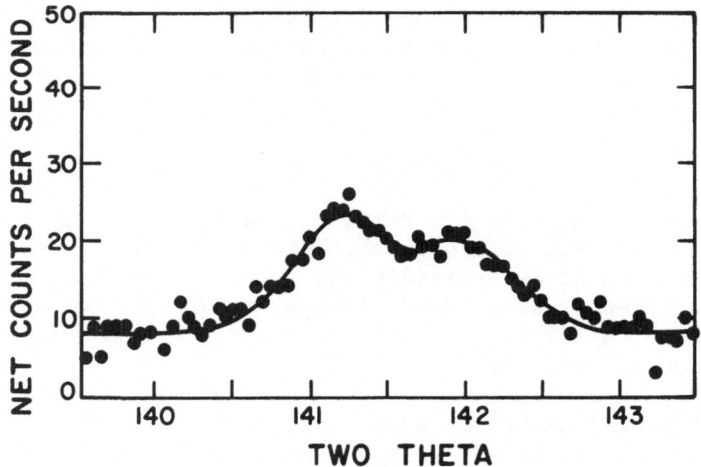

Fig. 2.   The 622 reflection of 238(80%)PuO$_2$ examined with copper
          Kα x-ray radiation. Plotted points are the average of
          data taken 95, 96, and 97 weeks after an initial anneal-
          ing treatment of the sample.

powder pattern line occurs when some physical operation decreases
the crystallite size and/or introduces large strains into the
lattice. (See Fig. 2.) By comparing the broadened x-ray line
with the sharp line through the use of the mathematical procedure
of "deconvolution", estimates of the size of the crystallites and
the amount of strain can be made.

SAMPLE PREPARATION

The sample consisted of a disk of 238(80%)PuO_2 approximately
1/2 in. dia by 1/8 in. thick which had been sintered at 1625°C.
Calculated density, determined from size and weight, was 10.7 g/cc.
One side of the disk was metallographically polished to provide a
flat surface. The disk was then placed in a brass sample holder
containing an oversized hole filled with crumpled aluminum foil.
A glass plate was used to press the disk into the foil insuring
that the disk was held firmly and that its surface was coincident
with the surface of the sample holder. This combination was encased
in 0.00025-in.-thick aluminum foil, the foil being sealed with lead
tape to complete the assembly. The package was inserted into the
sample holder of a General Electric XRD-3 x-ray diffraction instrument. Since preliminary examination had indicated that the sample
contained preferred orientation effects, the azimuthal rotation of
the sample was adjusted so that the maximum intensity and greatest
resolution were obtained for a high-order back-reflection line.
The sample remained at its equilibrium temperature, in this orientation, in the sample holder on the diffractometer unit for the
duration of the experiment.

DATA COLLECTION

Experimental data consisted of intensity counts (usually for
10 sec or more) made at equal intervals of arbitrary length across
a line profile. Copper $K\alpha$ radiation was employed; $\lambda K\alpha = 1.54178$,
$\lambda K\alpha_1 = 1.54051$, and $\lambda K\alpha_2 = 1.54433$ A. Pulse height discrimination
was used to aid in separating the desired x-rays from the general
intense radioactive background. Data were taken on the 111, 200,
220, 311, 333(511), and 622 diffraction lines by step scanning in
units of 0.05 °2θ over a total range of 3.5 to 5.0 °2θ. The
initial intensity measured included both the copper $K\alpha$ x-rays and
the radioactive background from the sample. After being step
scanned across the line profile, the sample was shielded from the
x-rays and the step scan was repeated to determine the radioactive
background. The radioactive background was then substracted, step
by step, from the initial intensity to yield intensity in units of
net counts per second.

Data were collected once a week during the early stages of the experiment. However, since changes in the line broadening occurred at such a slow rate (as will subsequently be shown), the data were later averaged in 3-wk unit blocks in order to improve the representation of the line profile.

METHODS OF ANALYSIS

In general, there are three methods used in the determination of crystallite size and lattice strain: the obtaining of the integral breadth, the examination of the Fourier coefficients describing the shape of the line, and the evaluation of the second moment of the line profile about its centroid. Since these techniques have been extensively described in the literature (1)(2)(3), they will not be discussed here.

A computer program UNFOLD was written to assist in the analysis of the experimental data. All three of the general methods for determining crystallite size and strain were incorporated into the program. A unique feature of the program is the inclusion of Wilson's equations (4)(5), for determining standard deviations of the integral breadth, the Fourier coefficients, and the variance of the line profile as a function of intensity data in units of x-ray counts per second.

ANALYSIS OF EXPERIMENTAL PEAKS

The experimental intensity data from the six reflections 111, 200, 220, 311, 333(511), and 622 were averaged in 3-wk unit blocks and subjected to least-squares fits of an equation consisting of the summation of Gaussian curves having a polynomial background. That is,

$$y = P_1 + P_2 x + P_3 x^2 + \sum_{j=2}^{\leq 13} \frac{P_{3j-2}}{\sqrt{2\pi} \cdot P_{3j-1}} \cdot \exp{-1/2 \left(\frac{x-P_{3j}}{P_{3j-1}}\right)^2}, \quad (1)$$

where P_1, P_2, P_3 are the coefficients of the polynomial background,

P_{3j-2} is the area under the profile,

P_{3j-1} is the Gaussian half-width parameter σ of the peak,

P_{3j} is the 2θ position of the peak, and

x   is the 2θ position at which the

y   counts per second intensity data are taken.

The Gaussian half-width parameter σ was used to determine whether significant line broadening had occurred as a function of time. As Fig. 3 indicates, no significant broadening had occurred after the passage of approximately 30 weeks. The 622 and 333(511) reflections consist of resolvable $K\alpha_1$ and $K\alpha_2$ peaks. In Fig. 3 the sum of the half-width parameter σ for each peak is plotted for times through 33 weeks.

The lack of data for a subsequent period of 60 weeks reflects the state of suspended animation of the experiment while the author was on leave at the University of Western Australia. Upon his return the experiment was reactivated and significant broadening was noted, particularly in the 622 reflection but also in the 333(511) peak. The half-width parameter σ plotted for times in excess of 90 weeks is for a single Gaussian curve through the broadened peak.

Integral Breadth Method

Observed diffraction profiles consist of a pure diffraction profile that is broadened by various instrumental effects. Removal of the instrumental aberrations from the data was accomplished by subtracting a sharp reference line from the broadened observed line by utilizing the method of deconvolution. The diffraction profile due only to the effects of crystallite size and lattice strain may then be resynthesized by the summation of a Fourier series.

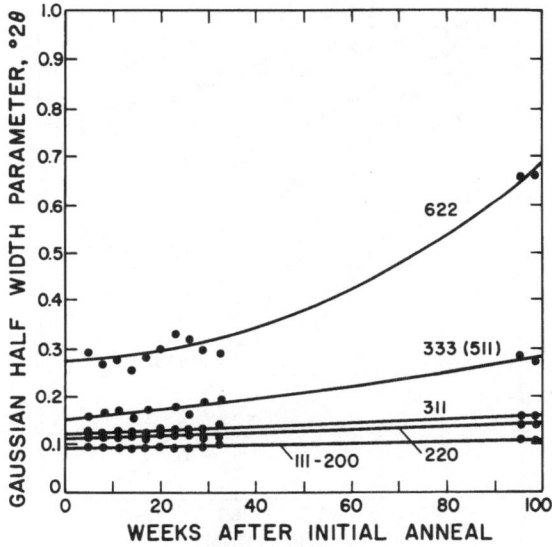

Fig. 3. The change in Gaussian half-width as a function of time for six reflections from $238(80\%)PuO_2$.

Input to the computer program UNFOLD consisted of the experimental 7, 8, 9-week data as the sharp reference line, and the experimental 95, 96, 97-week data as the broadened observed line. The normalized unfolded Fourier coefficients $C = (A^2 + B^2)^{1/2}$ are plotted in Fig. 4.

An example of the pure diffraction profiles obtained by using the unfolded Fourier coefficients to resynthesize the curves is given in Fig. 5. The integral breadth $\beta(s)$ was obtained from these profiles either by fitting Eq. (1) to the data ($\beta(2\theta)\,G$) or from the summation of the Fourier coefficients ($\beta\,(2\theta)\Sigma$); and then multiplying by $(2\pi/360)*(\cos\theta/\lambda)$. The integral breadths calculated by this procedure are listed in Table 1.

Table 1. Experimental Integral Breadths, $\beta(s)$, 238(80%)PuO_2.

hkℓ	β (s)G, $\overset{\circ}{A}{}^{-1}$	β (s)Σ, $\overset{\circ}{A}{}^{-1}$
111	1.41 ± 6 * 10^{-3}	1.39 ± 1 * 10^{-3}
200	1.65 6	1.59 2
220	1.98 6	2.03 2
311	2.18 6	2.44 3
333(511)	3.60 6	3.65 9
622	4.65 6	4.75 15

The data of Table 1 are plotted in Fig. 6 as $\beta^2(s)$ vs $\sin^2\theta/\lambda^2$. The straight line was fitted by least-squares techniques.

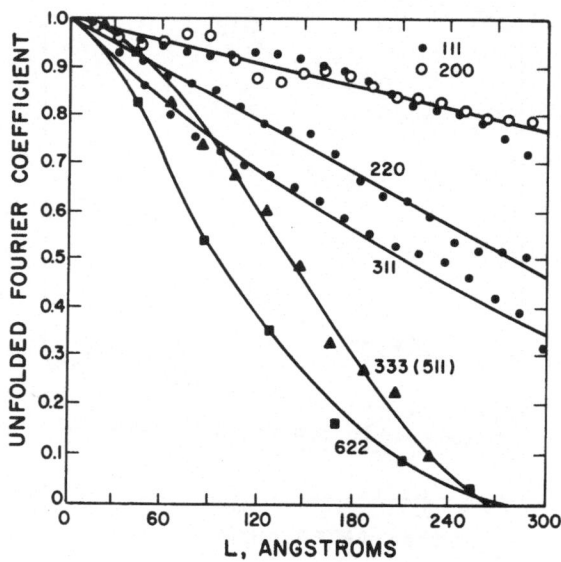

Fig. 4. Normalized unfolded Fourier coefficients for six reflections from 238(80%)PuO_2.

The crystallite size D and average strain ϵ_I, determined from the intercept and slope, respectively, of the straight line, are: D = 1820 ± 595 Å and ϵ_I = 0.00184 ± 2.

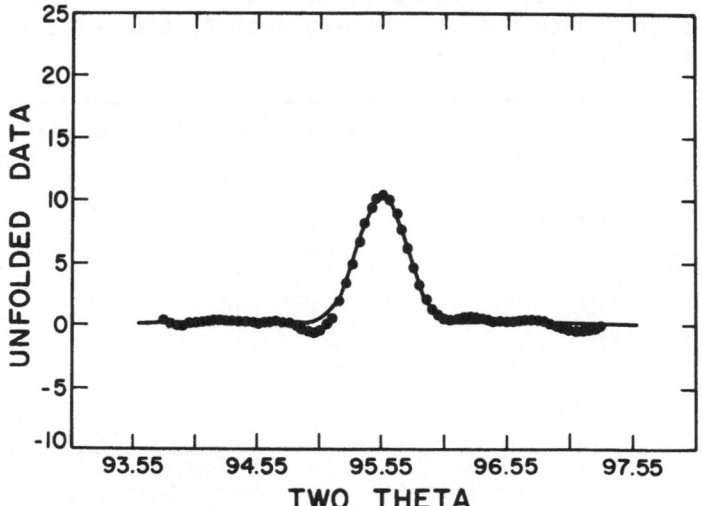

Fig. 5.   The pure diffraction profile for the 333(511) reflection from 238(80%)PuO_2. Resynthesized by using the Fourier coefficients of Fig. 4.

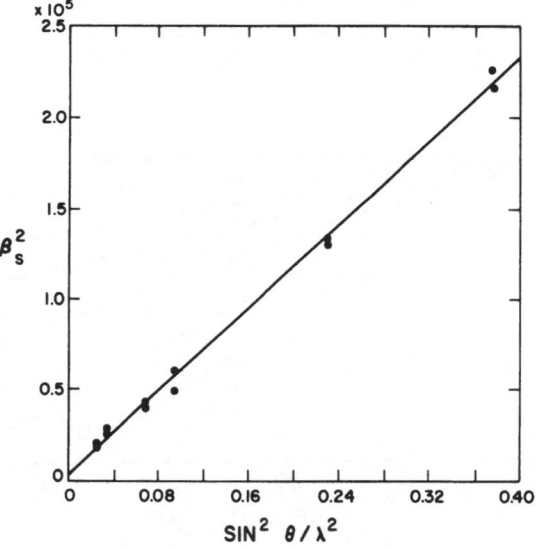

Fig. 6.   The square of the integral breadth as a function of sin2θ/λ^2 for six reflections from 238(80%)PuO_2.

Fourier Coefficient Method

The Fourier coefficients in Fig. 4 are plotted in Fig. 7 as a function of selected values of L. Table 2 lists the intercepts and the slopes obtained by least-squares fitting of the data to the general exponential equation y = a exp bx.

Table 2. Intercepts, slopes, and strains from Fig. 7.

L, Å	Intercept	Slope	Strain, ϵ_F	
30	0.97 ± 1	-0.15 ± 6 * 10^{-2}	0.00158	± 29
60	0.95 3	-0.67 16	0.00166	20
90	0.96 4	-1.45 26	0.00164	15
120	0.95 4	-2.12 33	0.00148	12
150	1.00 3	-3.29 31	0.00147	7
180	1.03 3	-4.93 38	0.00150	6
210	1.03 3	-6.08 39	0.00144	5
240	1.04 4	-7.25 65	0.00137	6

If the Fourier crystallite size coefficients, the intercepts of Table 2, are plotted as a function of L, it can be shown that the intersection on the L axis of a line drawn through the points will be a large and relatively meaningless number. Thus, the crystallite sizes determined from examination of the Fourier coefficients and from integral breadths are in general agreement. If the strain determined by the Fourier coefficient method ϵ_F, as listed in Table 2, is plotted as a function of L and extrapolated to L = 0, the strain is approximately 0.00170, which is in good agreement with the strain determined from integral breadths.

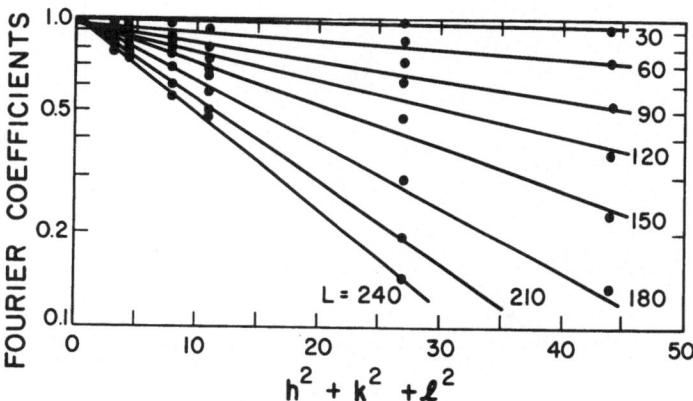

Fig. 7. The normalized Fourier coefficients of Fig. 4 plotted as a function of L for $h^2+k^2+\ell^2$ values of reflections from 238(80%)PuO_2.

Variance Method

The variance W of the unfolded diffraction profile is obtained by subtracting the second moment of the sharp line from the second moment of the broadened line. That is,

$$W = 2nd \ M_c^B - 2nd \ M_c^S . \tag{2}$$

W is dependent on the range of ΔS over which the line profiles are examined. An example of the variance obtained as a function of ΔS is given in Fig. 8 for the 311 reflection. As Halder and Wagner (3) note, the experimental determination of the true value of ΔS can be uncertain, especially if the experimental background varies appreciably from point to point. (See Fig. 2.) The curves of W vs ΔS for the reflections contain many breaks, and the selection of which break represents ΔS for a given hkℓ reflection is difficult without auxiliary information. Using the integral breadth and Fourier coefficient information as a guide, the author selected the values of W and ΔS listed in Table 3.

Table 3. Values of W and ΔS for hkℓ reflections of 238(80%)PuO_2.

hkℓ	S	S2	ΔS, $\overset{\circ}{A}^{-1}$	W, $\overset{\circ}{A}^{-2}$	
111	0.320	0.102	0.0180	1.10 ± 16	* 10^{-6}
200	0.370	0.136	0.0120	0.40	16
220	0.524	0.274	0.0145	1.43	15
311	0.612	0.374	0.0160	3.22	20
333(511)	0.960	0.920	0.0190	2.80	59
622	1.226	1.504	0.0178	5.56	62

In Fig. 9, $W/\Delta S$ is plotted as a function of $S^2/\Delta S$. The crystallite size obtained from the intercept is 1250 ± 1075 $\overset{\circ}{A}$, and the strain ϵ_W obtained from the slope is 0.00177 ± 23.

The variance method, as well as the integral breadth and Fourier coefficient techniques, indicates a large crystallite size for the material under examination. The strain in the lattice, as determined by the three different methods, is in quite good agreement. The integral breadth yields ϵ_I = 0.00184, the Fourier coefficients yield ϵ_F = 0.00170, and the variance yields ϵ_W = 0.00177.

CONCLUSIONS

Three different methods of examining x-ray line broadening have been applied to data obtained from a sample of 238(80%)PuO_2. The techniques are complementary and all yield essentially identical results.

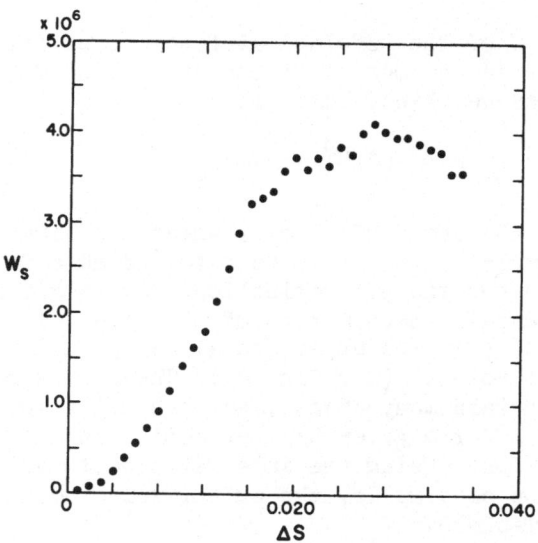

Fig. 8.   The variance W as a function of ΔS for the 311 reflection
          from 238(80%)PuO_2. $S = 2 \sin\theta/\lambda$, $\Delta S = S_2 - S_1$.

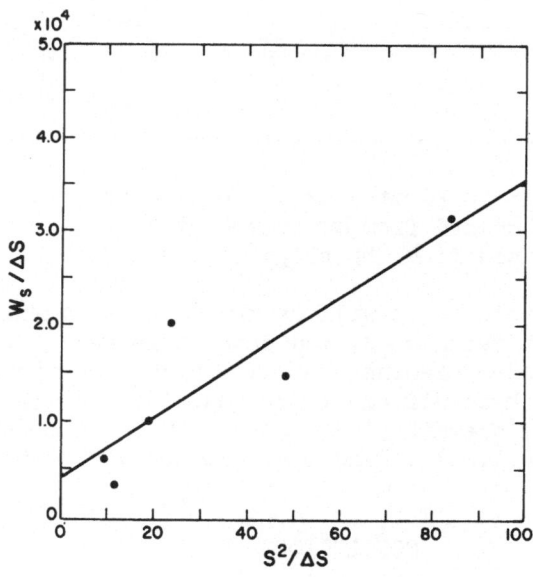

Fig. 9. $W/\Delta S$ vs $S^2/\Delta S$ for six reflections from the material
          238(80%)PuO_2.

The sample was given an initial annealing treatment which was presumed to have strain relieved the material and promoted the growth of the crystallites. The analysis of the data indicates that, during a period of 2 years under the influence of self-irradiation and at equilibrium temperature, this material undergoes no significant reduction in crystallite size but does sustain a strain of approximately 0.2%. The source of the strain is most likely the entrapment in the lattice of helium that results from the radioactive disintegration of the 238 Pu atoms, although the collection of vacancies, the insertion of interstitals, and the introduction of more complex displacement-generated defects may also contribute to the strain. The internal heating of the sample, however, would tend to mitigate some of these effects through thermal diffusion of the helium atoms and recombination or annealing of some of the defects.

The indicated conslusion is that this material is quite resist-ant to the damaging effects of self-irradiation. Metamictization, the pseudo phase-change from a crystalline to a quasi-amorphous state, which occurs in natural minerals containing alpha-emitting radioactivity, has not yet been observed. However, natural minerals are examined by man after geologic time periods have passed, whereas the current material sample is only 2 years old.

Extrapolation to projected future behavior indicates that this material will retain its structural integrity. Two states of even-tual ultimate behavior can be envisioned. Either the strain will gradually increase, probably asymptotically, to reach some equilib-rium value, or it will increase to a point where the lattice will rupture, thereby relieving the stress causing the strain and result-ing in a noticeable reduction in the crystallite size. This second process might be defined as incipient metamictization. Continued observation may indicate which path will be followed. It is doubt-ful, however, that either process will substantially alter the structural integrity of the material.

REFERENCES

1. H. D. Klug and L. E. Alexander, X-ray Diffraction Procedures, p. 491-538 John Wiley & Sons (1958).

2. C. N. J. Wagner, "Analysis of the Broadening and Changes in Position of Peaks in an X-ray Powder Pattern," in J. B. Cohen and J. E. Hilliard, Editors, Local Atomic Arrangements Studied by X-ray Diffraction, p. 217-269 Gordon & Breach Science Publishers, Inc. (1966).

3. N. C. Halder and C. N. J. Wagner, "Analysis of the Broadening
    of Powder Pattern Peaks using Variance, Integral Breadth, and
    Fourier Coefficients of the Line Profile," in W. M. Mueller,
    G. R. Mallett, and M. Fay, Editors, <u>Advances in X-ray Analysis</u>,
    Vol. 9, p. 91-101 Plenum Press (1966).

4. A. J. C. Wilson, "Statistical Variance of Line-Profile Para-
    meters, Measures of Intensity, Location and Dispersion," Acta
    Cryst. <u>23</u>, 888-898 (1967).

5. A. J. C. Wilson, "Statistical Variance of Line-Profile Para-
    meters: Addendum.," Acta Cryst. <u>A25</u>, 584-585 (1969).

THE DISORDER-ORDER TRANSFORMATION IN Ni_4Mo

Fu-Wen Ling and E. A. Starke, Jr.

Georgia Institute of Technology

Atlanta, Georgia 30332

ABSTRACT

The progressive ordering of a single crystal of Ni_4Mo by isothermal ageing at $650°C$ (transformation temperature = $868°C$) has
been studied by x-ray line broadening techniques using the Warren-
Averbach method employing computer techniques. The long-range-order
parameter, antiphase domain size, and internal strains were measured
as a function of ordering time and compared with those previously
obtained at $700°C$. The activation energies for domain growth and
ordering were found to be 91 kcal/mole and 44.5 kcal/mole respectively. The rms strain developed during ordering was found to be
dependent on the degree of tetragonality of the structure.

INTRODUCTION

Alloys which develop long-range order are of considerable interest since the atomic arrangements have a marked influence on most
properties. Cohen (1) has recently reviewed the effect of order on
various physical and mechanical properties. Quantitative information on the degree of long-range order can be deduced from an analysis of the integrated intensities of fundamental and superlattice
reflections. However, in order to obtain information on domain size,
and internal strains produced by the transformation (parameters important in understanding the kinetics of the ordering process) an
analysis of the peaks shapes is necessary. This may be extremely
difficult or impossible in those polycrystalline materials where
there is a superposition and/or overlapping of the superlattice and
fundamental reflections. The problem may be overcome by using a
single crystal for the ordering studies. However, even for a single

crystal problems may arise during ordering if the domain structure
is such that overlapping of the superlattice reflections occur.
Ni_4Mo is conducive to analysis of peak shapes of single crystals
since the antiphase domains developed on ordering exhibit rotational relationships which result in a separation of the superlattice
reflections.

A previous study of isothermal ordering a single crystal of
Ni_4Mo at $700^{\circ}C$ indicated that ordering occurred homogenously by
short range diffusion in the early stage (2). The late stage of
ordering, i.e., for $S \geq 0.8$, was described by the elimination of
domain boundaries due to domain growth. This was suggested since
the increase in order and domain growth had the same time dependence. In addition, domain growth satisfied a rate relationship
analogous to grain growth. The internal strains developed during
ordering were correlated with the degree of order and domain size.

This paper describes the results of a study of the disorder-
order transformation on a single crystal of Ni_4Mo at $650^{\circ}C$ and
compares them with the results obtained previously (2) at $700^{\circ}C$.
The objectives of this study were to establish the temperature
dependence of ordering kinetics, domain growth, and internal strains
produced during ordering. In addition, the data obtained at $650^{\circ}C$
allows the calculation of the activation energy for domain growth
and ordering in Ni_4Mo.

EXPERIMENTAL PROCEDURE

Sample preparation and x-ray methods used in this study have
been previously described (2) and will not be repeated in detail.
The domain size and microstrains developed during progressive ordering at $650^{\circ}C$ were measured by the x-ray line broadening method
developed by Warren and Averbach and described by Warren (3). The
$(110)_T$, $(220)_T$, $(330)_T$, $(440)_T$, $(200)_T$, $(400)_T$, $(600)_T$, $(011)_T$,
$(022)_T$, and $(0\bar{3}3)_T$ reflections were examined after various ordering
times. The sub-T designates the use of tetragonal indices whereas
the sub-C will designate cubic indices. The intensity profiles were
transformed into a Fourier series, corrected for $\alpha_1\alpha_2$ doublet by the
Rachinger (4) method, and for instrumental effects by the method of
Stokes (5). The (200) and (400) reflections of a LiF single crystal,
$2\Theta \approx 45^{\circ}$ and 100° for CuKα radiation, were used for the instrumental
corrections of Ni_4Mo reflections with Bragg angles of $2\Theta < 80^{\circ}$ and
of $2\Theta > 80^{\circ}$ respectively. The logarithm of the corrected Fourier
coefficients of the cosine terms were plotted versus $1/d^2$, and the
rms strain values were calculated from the slopes which are equal
to $-2\pi^2L^2<\varepsilon_L^2>$. Typical curves are given in Figure 1 which represents the coefficients obtained from the $(011)_T$, $(022)_T$, and $(033)_T$
reflections. The intercepts at $1/d^2 = 0$ give the values of the size
coefficients A_L^s. The domain size for the various directions was

determined from the intercept on the abscissa of the extrapolated value of the A^s versus L plot, Figure 2.

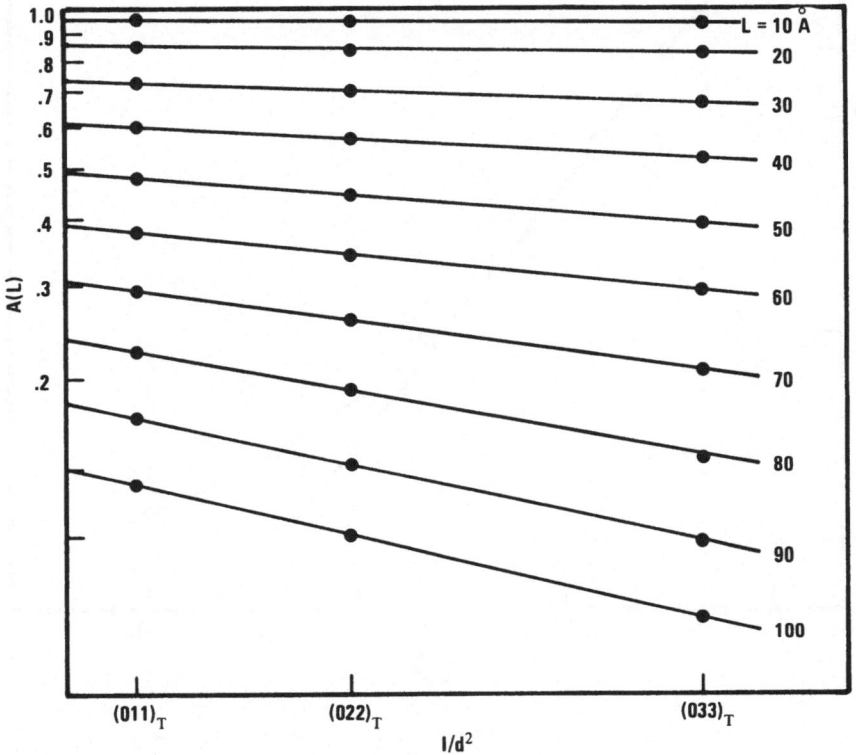

Figure 1. Fourier coefficients of various orders of reflection of a $(011)_T$ plane for different L values for Ni_4Mo ordered at $700°C$ for 350 minutes.

The long-range-order parameter, S, was calculated from a comparison of the integrated intensities of the superlattice and fundamental reflections using the relation

$$S^2 = \frac{E_i^s / R_i^s}{E_j^f / R_j^f}$$

where E_i^s and E_j^f are the measured intensities of the superlattice and fundamental reflections respectively and R_i^s and R_j^f are their corresponding calculated intensities. They are determined by $R_i^s = v_i P_i F_i^2 (LP)_i \exp(-2M_i)$ and $R_j^f = P_j F_j^2 (LP)_j \exp(-2M_j)$. P is the multiplicity factor, and is one for a single crystal, F is the structure factor, and LP is the Lorentz-polarization factor. The value of v_i

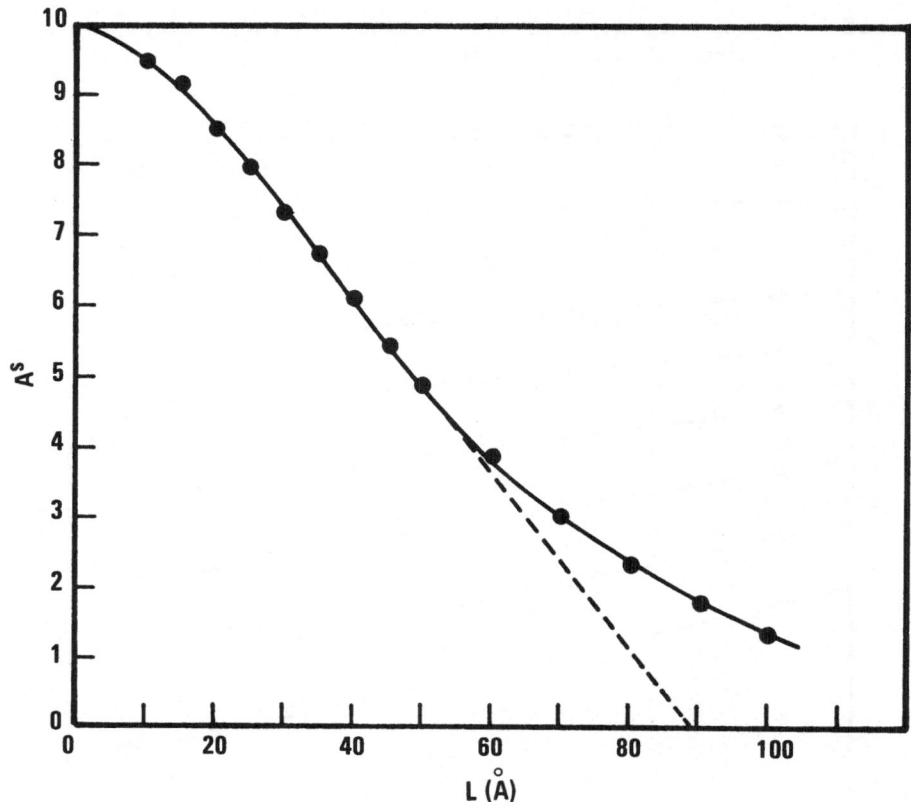

Figure 2. Fourier size coefficients in the <011> direction versus L for Ni_4Mo ordered at $700^{\circ}C$ for 350 minutes.

in the calculation represents the probability of having a particular orientation of the ordered domain. Since there are six unique ways (6) in which the ordered lattice can develop with respect to the disordered lattice, this factor is 1/6 if all orientations have equal probability of forming.

Previous work (2) at $700^{\circ}C$ has shown that the domains which have their C-axis close to the sample surface normal predominate in the late ordering stages since the contraction is more easily accommodated perpendicular than parallel to the surface. The amount of this preferred orientation depends on the strain and can be estimated by comparing the ratio of the integrated intensities of the different superlattice reflections which result from different domain orientations. Unlike the $700^{\circ}C$ study, no preferred orientation was observed until 8000 minutes of ordering at $650^{\circ}C$ at which time S = 0.94. Up to this time, the S was calculated from the integrated intensities of the parallel $(440)_T$ superlattice and $(420)_C$ fundamental reflections.

RESULTS

The results of measurements of the LRO parameter after various ordering times at 650°C are given in Figure 3 along with the data obtained previously (2) at 700°C. The logarithm of S versus the logarithm of time is approximately linear for the 650°C data, however the slope is different from that observed in the late ordering stage at 700°C. The slopes of the logarithm of the domain size versus the logarithm of time for the 650°C and 700°C data are the same in the late ordering stage, Figure 4. The deviation from a straight line for early times is due to the neglection of D_0, the initial microdomain size. Domain growth has been shown to follow the relationship $D^n - D_0^n = kt$ where n is a constant which mostly depends on the impurity content of the alloy. If the domain size curves for the two ordering temperatures are extrapolated toward zero time, they appear to converge to ~$20\overset{\circ}{A}$. This is the value for the microdomain size suggested for Ni_4Mo (7,8,9). If we assume an Arrhenius equation for the domain growth, i.e., $D^n = kt \exp(-Q/RT)$, the activation energy can be calculated from the slope of the logarithm of time versus the reciprocal of the temperature. The value obtained was 91 kcal/mole.

The time-temperature-transformation curves for the ordering reaction in Ni_4Mo have been previously determined by Guthrie and Stansbury (10). We have added to their curves the end points of the transformation, defined as $S = 1$, for 700°C and 650°C isothermal ordering. The data of Guthrie and Stansbury is replotted along with the results of this experiment in Figure 5.

The rms strain, measured in three crystallographic directions, are plotted as a function of ordering time in Figure 6. As for domain growth and ordering, the rms strain increased at a much slower rate at 650°C than at 700°C. The maximum rms strain obtained for our 700°C study was larger than that obtained at 650°C for the times studied.

DISCUSSION

The results obtained on the change in long-range-order parameter with time at 650°C, depicted in Figure 3, show that although the logarithm of S versus the logarithm of time relationship is linear, the slope is different from that obtained at the higher ordering temperature. This is in contrast to the logarithm of the domain size versus the logarithm of time relationship which had the same slope at both temperatures, Figure 4. Consequently, our previour conclusion (2) concerning the relationship between order and domain growth is questionable.

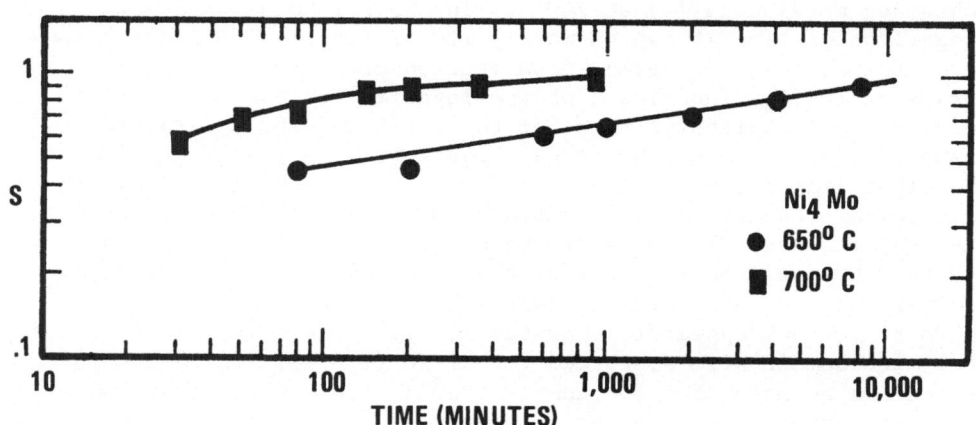

Figure 3.   The long-range-order parameter versus ordering time for Ni$_4$Mo.

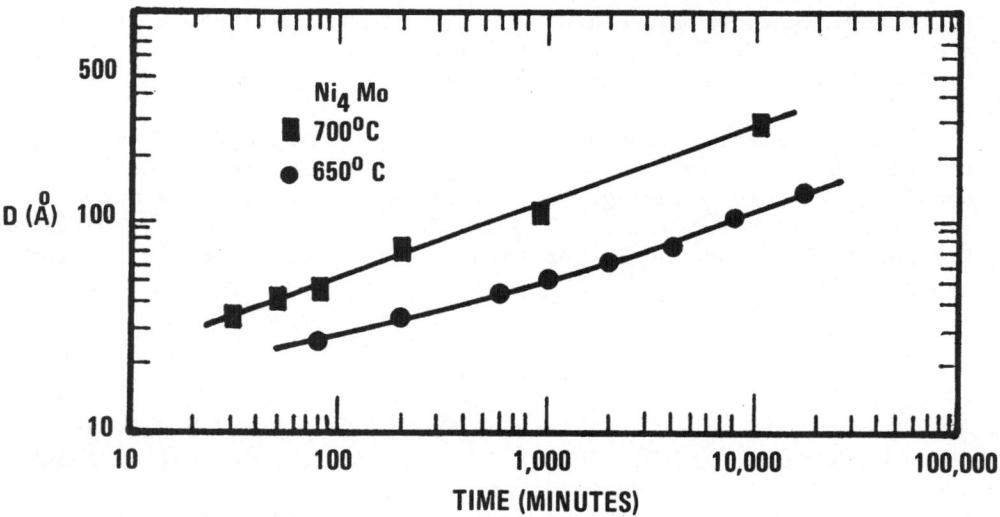

Figure 4.   The domain size versus ordering time for Ni$_4$Mo.

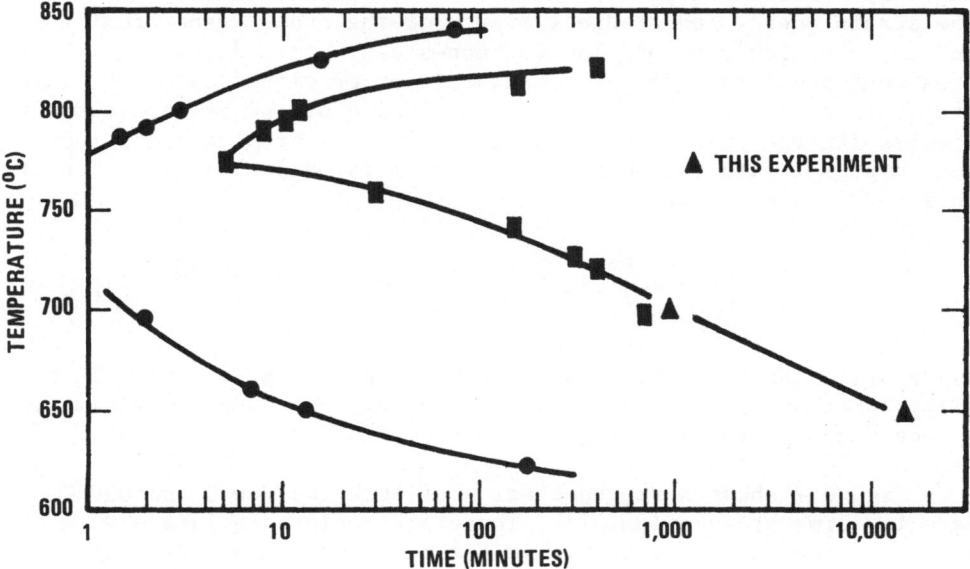

Figure 5. The time-temperature-transformation curves for the ordering reaction of Ni_4Mo. All but the triangular points are from Guthrie and Stansbury (10).

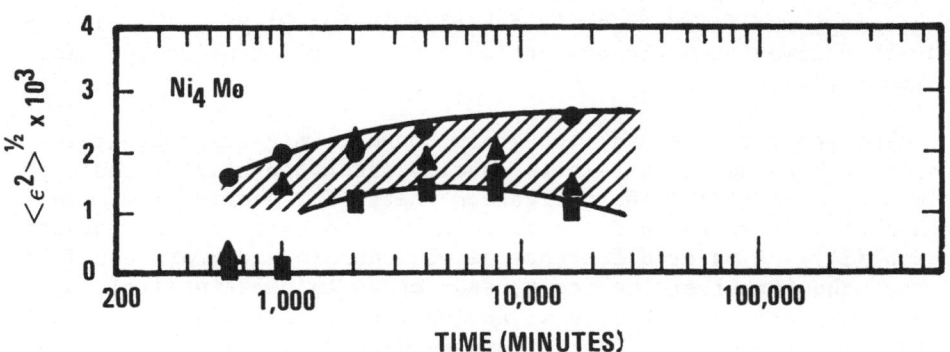

Figure 6. The rms strain versus ordering time at $650^{\circ}C$ for Ni_4Mo.

We have recently developed a theory for the kinetics of ordering (11) which is temperature consistent. The new theory considers
the strain energy and gradient energy of the order transformation
and employs Fick's second law for non-steady state diffusion. This
treatment predicts that the increase in order with time in the late
stages is controlled by long-range diffusion of one of the atomic
species (in the present case molybdenum atoms) from the center of
the domains toward the boundaries. The mathematical description of
this model is given by

$$\ln (1 - S) = -G_1(T) t^{(n - 2)/n} - G_2(T) t^{(n - 4)/n} \qquad [1]$$

where $G_1(T)$ and $G_2(T)$ are functions of temperature, T, and n is an
alloy dependent parameter greater than two. For long order times,
t, the second term is negligible in comparison with the first.

Figure 7 shows a representation of both the 700°C and 650°C
data in terms of equation [1]. These ln-ln plots of $\ln(1 - S)$ and
time have the same slopes. If we assume a temperature dependence
of the form

$$G_1(T) = G_o \exp(-Q/RT) \qquad [2]$$

we may calculate an activation energy for ordering from the logarithm
of $\ln(1 - S)$ versus the reciprocal of the temperature. The activation energy was found to be 44.5 kcal/mole for Ni_4Mo. This is considerably lower than the activation energy for diffusion in this
system (2,12).

The activation energy for grain boundary migration or grain
growth in high purity metals is usually about one half of that for
bulk diffusion (13). The activation energy for domain growth in
high purity Cu_3Au has been reported to be the same as that for diffusion (14). A major difference between domain and grain boundaries
is that the former are coherent, and for Cu_3Au contain little or no
coherency strains. There is no activation energy data for diffusion
available for ordered Ni_4Mo to compare with that found for domain
growth, i.e., Q = 91 kcal/mole. However, there are two factors
which may influence the value of Q: (i) the impurity content may
affect the measured value similar to the effect found for grain
boundary migration (13,15), (ii) the strain associated with the coherency may lower the activation energy for domain boundary migration.

The rms strain developed during ordering is dependent on at
least three factors: (i) degree of order, (ii) domain size, and
(iii) thermal fluctuations. The tetragonality of the ordered

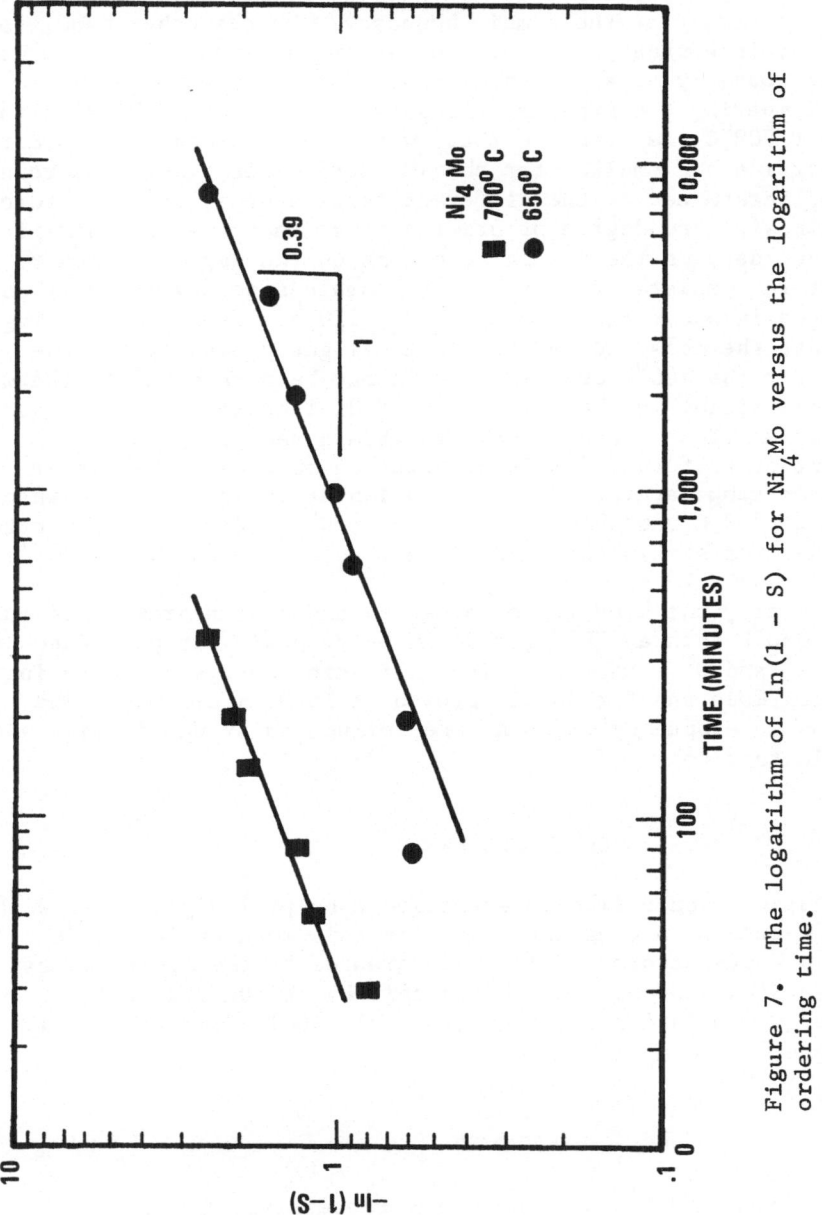

Figure 7. The logarithm of ln(1 − S) for Ni₄Mo versus the logarithm of ordering time.

structure does not only depend on the degree of order but also on the domain size. For the same degree of order, the tetragonality is more easily formed when the domain size is large. Consequently, for the same degree of order the larger the domain size the higher the coherent strain at the domain boundary. On the other hand, for the same domain boundary strain, the larger the domain size the lower the domain boundary area per unit volume and the lower the rms strain. Comparing the experimental data obtained at 650°C with that obtained at 700°C one observes that, with the same degree of order, the rms strains are smaller for the smaller domain size. The relationships, determined at the different temperatures, of rms strain with domain size and degree of order suggest that the rms strain is more dependent on the domain size than on the degree of order. When order is complete, i.e. $S = 1$, the domain size developed at both temperatures is about the same, i.e. $D = 110$ Å. It is interesting to note that the measured rms strain is slightly smaller for the 650° than for the 700°C ordering. This may be attributed to thermal fluctuations aiding in the generation of dislocations at the coherent boundaries (2). The critical rms strain necessary for the development of preferred domain orientation appears to be higher at the lower temperature. Preferred orientation was observed when $S = 0.74$ for 700°C and when $S = 0.94$ for 650°C ordering. The latter has a higher rms strain than the former.

In summary, the kinetics of ordering and domain growth are much slower at 650°C than at 700°C, and can be respectively predicted by equation [1] and $D^n - D_0^n = kt$. The activation energy for ordering is 44.5 kcal/mole and for domain growth it is 91 kcal/mole. The rms strains developed by ordering are influenced by domain size and thermal fluctuations.

ACKNOWLEDGEMENTS

The authors would like to acknowledge helpful discussions with Dr. B. G. LeFevre. During the course of this work one of us (Ling) was under a Post-Doctorate Fellowship granted by the National Institute of Dental Research. The Sponsorship by the United States Atomic Energy Commission under contract AT-(40-1)-3908 is greatly acknowledged.

REFERENCES

1. J. B. Cohen, "A Brief Review of the Properties of Ordered Alloys," J. Mater. Sci. 4:1012, 1969.

2. Fu-Wen Ling and E. A. Starke, Jr., "The Development of Long-Range Order and the Resulting Strengthening Effects in Ni_4Mo," Acta Met. (in press).

3. B. E. Warren, X-Ray Diffraction, Addison-Wesley Publ. Co., Reading, Mass., 1969.

4. W. A. Rachinger, "A Correction for the $\alpha_1\alpha_2$ Doublet in the Measurement of Widths of X-Ray Diffraction Lines," J. Sci. Inst. 25:254, 1948.

5. A. R. Stokes, "A Numerical Fourier-Analysis Method for the Correction of Widths and Shapes of Lines on X-Ray Powder Photographs," Proceedings of the Physical Society 61:382, 1948.

6. Fu-Wen Ling, "The Determination of Microstrains and Antiphase Domain Size Produced During Ordering of a Ni_4Mo Single Crystal," Ph.D. Thesis, Georgia Institute of Technology, Atlanta, Georgia, December, 1970.

7. J. E. Spruiell and E. E. Stansbury, "X-Ray Study of Short-Range Order in Ni Alloy Containing 10.7 & 20.0 At.% Mo," J. Phys. Chem. Solids 26:811, 1965.

8. E. Ruedl, P. Delavignette, and S. Amelinckx, "Electron Diffraction and Electron Microscopic Study of Long- and Short-Range Order in Ni_4Mo and of Substructure Resulting from Ordering," Phys. Stat. Solidi 28:305, 1968.

9. Paul R. Okamoto, "Initial Stages of Ordering in Ni_4Mo," Ph.D. Thesis, University of California, Berkeley, California, March, 1970.

10. P. V. Guthrie and E. E. Stansbury, "X-Ray and Metallographic Study of the Nickel-Rish Alloys of the Nickel-Molybdenum System II," USAEC Report ORNL-3078, Oak Ridge National Laboratory, Tenn., July, 1961.

11. Fu-Wen Ling and E. A. Starke, Jr., "The Kinetics of Disorder-Order Transformation," to be published.

12. Y. Adda and J. Philibert, "La Diffusion dens Res Solides," Tome II, Presses Universitaires de France, Paris, 1966.

13. Paul Gordon and T. A. El-Bassyouni, "The Effect of Purity on Grain Growth in Aluminum," Trans. AIME 233:391, 1965.

14. G. E. Poquette and D. E. Mikkola, "Antiphase Domain Growth in Cu_3Au," Trans. AIME 245:743, 1969.

15. Paul Gordon and R. A. Vandermeer, "The Mechanism of Boundary Migration in Recrystallization," Trans. AIME 224:917, 1962.

A STRATEGY FOR RAPID AND ACCURATE (p.p.m.) MEASUREMENT OF

LATTICE PARAMETERS OF SINGLE CRYSTALS BY BOND'S METHOD

R. L. Barns

Bell Telephone Laboratories, Incorporated

Murray Hill, New Jersey   07974

ABSTRACT

All published accounts of the use of Bond's method for lattice
parameter measurements have used step-scanning (at equal angle
increments) of the diffraction peaks, followed by graphical or
computer analysis of the data to locate the peak positions. It
has been found that the peak angles can be determined with little
loss in accuracy or precision by manually setting the crystal angle
to give a counting rate (observed on a rate-meter) equal to 1/2 the
peak rate and defining the peak angle as the average of the angles
on the two sides of the peak. Because of the asymmetry of the
spectral line, defining the peak in this manner results in a shift
of the peak angle from that determined by the mid-chord peak method.
This shift can be compensated by determining an effective value of
the wavelength based on a silicon standard. Using the method described, a lattice parameter measurement, including mounting and
orienting the sample, taking the data and computing the result
using a time-sharing computer terminal, can be made in less than
20 mins.

INTRODUCTION

The difficulties of measuring accurate or precise lattice
parameters by powder methods is well documented (1)(2)(3) and
agreement better than 100 p.p.m. between different laboratories
is, in general, not to be expected. Furthermore, several man-
hours are required for each measurement.

Bond's method (4) of measuring lattice parameters on a single crystal reduces the uncertainty to about 1 p.p.m. (5)(6)(7) and, as shown below, requires only about 20 mins. for a complete measurement (even by manual methods with simple equipment), including data processing. This uncertainty is ignoring the standard deviation (σ) in the absolute value of the X-ray wavelengths, recently reduced to about 5 p.p.m. by Bearden (8) and which should shortly be reduced even further by Deslattes (9).

Measurement of the lattice parameter of undoped silicon using Bond's method has been reported by six authors. The results are summarized in Table 1. All of these agree within the stated uncertainties except (10) (to be discussed below). Hence, it appears that Bond's method is capable of giving lattice parameters with realistic accuracy estimates based solely on replicate determinations provided that a consistent method is used to define the peak of a diffraction profile.

TABLE 1

Measurements of High Purity, High Perfection
Silicon Corrected to 25°C $\lambda = 1.540562\text{Å}$ (8)

Ref.	θ (444)	a_0	θ (333)	a_0
(4)	79°18'43.8±1.2"	5.430930 ±0.000006	47°28'33.5±0.4"	5.430935 ±0.000033
(7)	41.2±0.2	5.430943 ±0.000001		
(10)	37.7±0.11	5.430960 ±0.0000005		
(11)	41.2±2.6	5.430943 ±0.000013	33.2±1.1	5.430923 ±0.000091
(12)-(13)			32.3±0.2	5.430945 ±0.000011
(14)		5.430945 ±0.000002		

LOCATION OF THE DIFFRACTION PEAK

Many different ways of defining the peak of a diffraction curve have been used and a good review is given by Thomsen and Yap (15). These methods require step-scanning of the diffraction peak followed by computer or graphical analysis. In the Bond method, the stepping process is the most time-consuming part of a measurement and this work was undertaken to develop a method of reducing the effort required without complicating the apparatus, such as by adding dedicated computers and automatic control (10)(14).

It is clear that, for a given total counting time, the peak can be located most accurately by taking data where the slopes of the curve are maximum (16). Donnay and Donnay (17) used the center of a chord at 2/3 peak height to determine lattice parameters from powder diffractometer charts, but Backovsky (16) shows that any location on the curve from 1/2 peak intensity to 3/4 peak is almost equally suitable. However, as shown below, consistency in the use of a particular intensity location on the curve is necessary.

MEASUREMENT METHOD

This process will be called the half-peak midchord method (HPM).

The apparatus described in (7) was used. Samples are prepared, their orientation adjusted, and the temperature controlled as described in (4) and (7). (Several recent papers deal with orientation errors (18), (19), (20), (21).)

The diffracted intensity is observed on a ratemeter, preferably displayed on a strip-chart recorder. The time constant chosen for the ratemeter is not critical and is usually set so that, at the peak, the random fluctuations amount to 1-3% of full-scale.

The crystal is first turned to within a degree or so of its diffracting position in one counter, say the right. Background is subtracted by suppressing the zero on the recorder. The peak intensity is then observed by slowly scanning through it manually. The value of 1/2 peak intensity is recorded for this counter.

The crystal angle is then adjusted to give a chart indication of 1/2 peak intensity (for this counter) on one side of the peak and the angle is accurately read and recorded. The crystal angle is then adjusted to give 1/2 peak intensity on the other side of the peak and this setting is read and recorded.

This procedure is repeated for the other counter, say the left. Usually, the peak intensity in the two counters is unequal (due principally to unequal sensitivity of the scintillation counters, but this is unimportant.

The observations are then repeated in the right and then the
left counters, omitting the observation of the peak intensity.
This repetition is desirable to avoid gross angle reading errors,
to give a replicate determination from which an error estimate can
be made and to detect any instrumental instabilities such as
slippage of the sample holder in the circle chuck or drift in the
crystal mounting adhesive or clamp. This gives eight readings
which are paired three ways for a triplicate determination of theta.

A time-sharing computer program, written in BASIC (for convenience in entering the data), is then used to calculate the
results. The program: (a) applies interpolator and circle-error
corrections (from a Fourier equation derived from the circle calibration (7)), (b) calculates the peak angles from the average of
the angles at 1/2 peak intensity, (c) calculates theta from these
peak angles(4), (d) calculates the full width at 1/2 max, for all
the peaks, (e) applies all the corrections(4) to give the true
d-spacing, (f) calculates the lattice parameter and various error
estimates.

EFFECTIVE WAVELENGTH DETERMINATION

The longer wavelength characteristic spectral X-ray lines,
e.g., copper and chromium Kα) are asymmetric and the mid-chord
line is not a straight line. Hence, using the average of the chord
between 1/2 peak points or any other selected points will not give
the correct d-spacing, except fortuitously.

For comparative lattice parameter measurements using reflections with nearly equal Bragg angles (thetas), the "true" wavelengths (8) can be used. But, to compare different materials or
different reflections, a means of compensating this error is
needed· This can be done simply by determining a value for the
effective wavelength using the same definition of the peak to be
used for unknown samples and a crystal of known d-spacing. It was
shown above that present-day undoped, high-quality silicon is
satisfactory as a d-spacing standard at least to a few p.p.m.
uncertainty.

The Bragg angle was measured using the 111, 333 and 444 reflections from a sample of heavily etched, undoped, high-quality
silicon (Texas Instruments Vac. Float Zone p-type, 10-100 Ω cm.,
<10 disloc/cm^2). Each reflection was measured at 4 locations
around the divided circle about 90° apart to further reduce circle
errors (4). The data and the calculation of the effective wavelength appropriate for the HPM method are shown in Table 2.

TABLE 2

Data at 25°C for Si Sample by HPM Method and the Calculation of the Effective Wavelength

$a_o = 5.430943\text{Å}$ ($\lambda CuK\alpha_1 = 1.540562\text{Å}$) at 25°C σ = 0.6 p.p.m.

		444	333	111
(1)	$\sqrt{h^2+k^2+l^2}$	6.928203	5.196152	1.732051
(2)	$d_o = a_o/(1)$	0.7838891	1.0451855	3.135560
(3)	θ_{obs} (HPM method)	79°18'52.37±.7"(a)	47°28'37.20±.14"	14°13'24.09±.3"
		51.71±.5	35.79±.4	24.12±.2
		52.37±.6	38.45±.3	24.45±.2
		52.21±.3	36.38±.8	23.78±.0
			36.43±.2	
			37.27±.2	
			36.21±.14	
			35.69±.4	
		AVG = 52.16"	AVG = 36.68"	AVG = 24.12"
		σ = 0.15"(b)	σ = .35"	σ = 0.15"
		= 0.14 p.p.m.	= 16 p.p.m.	= 28 p.p.m.
(4)	$\Delta\theta = (\theta_{obs}-\theta_{corr})$(c)	12.325"	3.889"	6.638"
(5)	$\theta_{corr} = \theta_{obs} - \Delta\theta$	79°18'39.84"	47°28'32.79"	14°13'17.48"
(6)	wt. = tan θ·(1/σθ)	35	2.86	1.7
(7)	Σ wts.	39.56		
(8)	λ from (2) and (5)	1.5405751	1.5405859	1.5406356

(9) λ HPM = $\dfrac{35 \cdot 1.5405751 + 2.86 \cdot 1.5405859 + 1.7 \cdot 1.5406356}{39.56}$ = 1.540575   σ = 0.6 p.p.m.

{(a)} This σ from the triplicate measurement.
{(b)} σ for small samples calculated as range/n (22)
{(c)} θcorr. is corrected for refraction and axial divergence.

As can be seen, the measurements of the (444) reflection completely dominate the final value because its theta is so large
(tan θ is large) and because angular measuring errors are the
dominant errors. Hence, for future extensions of the method, e.g.,
to use the 3/4 max. points (which might be less susceptible to
effects of α2 overlap for samples having broad maxima), determination of the effective wavelength from the (444) reflection alone
would probably suffice.

Comparison of Tables 1 and 2 show that the difference in θ
for the (444) is about 11±0.8". Inspection of step-scanned peaks
(7) shows the asymmetry should give about 8.5±1". The θ found by
Baker, et al. (10) by the mid-chord peak method between 84% max.
points differs from (4), (7), (11) and (13) by 3.5" and hence
could be due simply to Baker's implicit redefinition of λ.

ERROR ANALYSIS

The uncertainties in measurement by the HPM can be evaluated
as follows. The calibration supplied with the Hilger-Watts cli-
nometer (used as the divided circle in these measurements) shows
the error at 5° intervals. A Fourier analysis shows strong second
and fourth harmonics with weaker first, third, fifth, and sixth.
The equation from this analysis is included in the BASIC program
to correct angle readings. The residuals, from this equation and
individual calibration values, have 95% of their values between
+2" and -1.7". Hence, 2σ for any circle mark is taken as ±1.8"
and σ is ∿ ±1".

An estimate of the total uncertainty due to counting statistics (i.e., the uncertainty is setting to 1/2 peak intensity) and
to circle reading errors was evaluated by inspecting the standard
deviation of the average of 170 triplicate lattice parameter
measurements made on a variety of samples and thetas which thus
includes a large range in observed line widths. The average
uncertainty in the average theta was found to be 0.21% of FWHM
with 95% of the values lying between 0 and 0.55%.

This seems reasonable as shown by the following analysis.
For the Lorentzian $I = 1/(1+\phi^2/w)$ where I represents the intensity,
φ the crystal angle and w is the full width at half-max., a 1%
change in I is produced by a change in crystal angle of 1% of the
FWHM, near the 50% I points.

Since eight settings of the angle of 1/2 max. are involved in
each triplicate measurement, the uncertainty for each setting is
observed as .21· $\sqrt{8}$ = .51% of FWHM, or, for a "typical" value of
theta of 60° where FWHM is 150", a Δθ of 0.76". The σ of reading

each angle is estimated to be 1.5" or 0.6" for 6 settings which leaves .76-0.6 = 0.16" as the σ due to the uncertainty from counting statistics. At 60°, a Δφ of 0.16" corresponds to a change in intensity of 0.11% of peak intensity for the triplicate or 0.11 √6 = .27% for each setting. For a typical peak intensity of 3,000 counts per second 0.27% is an uncertainty of 810 counts which implies that $810^2 = 6.5 \cdot 10^5$ counts are observed or that a 200 sec. counting interval is included in the 6 settings.

To summarize, the standard deviation inherent in the method is approximately 0.16" and the additional σ due to errors in the circle used were 1.2" for a total σ of 1.2". Thus, for a reflection of θ = 60° or more, the σ of the d-spacing is less than 3.4 p.p.m. plus the uncertainty in the wavelength determined here of 0.6 p.p.m. plus the uncertainty of Bearden's wavelength of 5 p.p.m.

CONCLUSIONS

Lattice parameter measurement by Bond's method (using step-scanning of the peaks and the mid-chord peak definition of peak position) requires about 2 hours and gives a standard deviation of about 0.59 arc sec. (i.e., 0.55 p.p.m. for θ = 79°) (7).

Measurement by the half-peak mid-chord (HPM) method described here requires about 20 mins. and gives a standard deviation of about 1.2" (i.e., 1.1 p.p.m. for θ = 79°) and uses simple and relatively inexpensive apparatus. This method of defining the peak of a curve might profitably be used for powder diffraction, optical spectra, etc.

ACKNOWLEDGMENTS

I am grateful to R. A. Laudise for support and encouragement throughout this work and to S. C. Abrahams for useful comments on the manuscript.

REFERENCES

1. W. Parrish, "Results of the I.U.Cr. Precision Lattice Parameter Project", Acta Cryst. 13, 838-850 (1960).

2. L. F. Vassamillet and H. W. King, "Precision X-Ray Diffractometry Using Powder Specimens", in W. M. Mueller and M. Fay (Eds.) Advances in X-Ray Analysis, Vol. 6 p. 142-157, Plenum Press (1963).

3. G. Boom, "Accurate Lattice Parameters and the LPC Method",
    Thesis, University of Groningen (1966).

4. W. L. Bond, "Precision Lattice Constant Determination", Acta
    Cryst. 13, p. 814-818 (1960).

5. K. E. Beu, F. J. Musil and D. R. Whitney, "Precise and
    Accurate Lattice Parameters by Film Powder Methods. I. The
    Likelihood Ratio Method", Acta Cryst. 15, p. 1292-1301 (1962).

6. K. E. Beu, "Further Developments in a Likelihood Ratio Method
    for the Precise and Accurate Determination of Lattice Param-
    eters", Acta Cryst. 22, p. 932-933 (1967).

7. R. L. Barns, "A Survey of Precision Lattice Parameter Measure-
    ments as a Tool for the Characterization of Single-Crystal
    Materials", Mat. Res. Bull. 2, p. 273-282 (1967).

8. J. A. Bearden, "X-Ray Wavelengths", p. 10, U.S. Atomic Energy
    Comm. (1964).

9. R. D. Deslattes, "Optical and X-Ray Interferometry of a
    Silicon Lattice Spacing", App. Phys. Lett. 15, p. 386-388
    (1969).

10. T. W. Baker, J. D. George, B. A. Bellamy and R. Causer, "Fully
    Automated High Precision X-Ray Diffraction", in J. B. Newkirk,
    G. R. Mallett and H. G. Pfeiffer (Eds.), Advances in X-Ray
    Analysis, Vol. 11, p. 359-375 (1968).

11. V. I. Lisiovan and R. R. Dikovskaya, "Local Precision Deter-
    mination of Lattice Constants of a Single Crystal",
    Instruments and Exper. Tech. (Eng. transl.), 4, p. 992-994
    (1969).

12. I. Henins and J. A. Bearden, "Silicon-Crystal Determination
    of the Absolute Scale of X-Ray Wavelengths", Phys. Rev. 135,
    p. A890-A898 (1964).

13. R. D. Deslattes, H. S. Peiser, J. A. Bearden and J. S. Thomsen,
    "Potential Appl. of the X-Ray/Density Method for Comparison
    of Atomic-Weight Values", Metrologia, 2, p. 104-111 (1966).

14. A. Segmuller, "Automated Lattice Parameter Determination on
    Single Crystals", in B. L. Henke, J. B. Newkirk and G. R.
    Mallett (Eds.), Advances in X-Ray Analysis, Vol. 13, p. 455-
    467 (1970).

15. J. S. Thomsen and F. Y. Yap, "Effect of Statistical Counting
     Errors on Wavelength Criteria for X-Ray Spectra", J. of
     Research, NBS-A, 72A, p. 187-205 (1968).

16. J. Backovsky, "On the Most Accurate Measurements of the Wave-
     lengths of X-Ray Spectral Lines", Czech. J. Phys. 15, p. 752-
     759 (1965).

17. G. Donnay and J. D. H. Donnay, "The Symmetry Change in the
     High-Temperature Alkali Feldspar Series", Am. J. Science,
     Bowen Vol. (Pt. 1), p. 115-132 (1952).

18. J. Burke and M. V. Tomkeieff, "Specimen and Beam Tilt Errors
     in Bond's Method of Lattice Parameter Determination", Acta
     Cryst., A24, p. 683-685 (1968).

19. J. Burke and M. V. Tomkeieff, "Errors in the Bond Method of
     Lattice Parameter Determinations - Further Considerations",
     J. Appl. Cryst. 2, p. 247-248 (1969).

20. E. E. Gruber and R. E. Black, "Analysis of the Axial Mis-
     alignment Error in Precision Lattice Parameter Measurement
     by the Bond Technique", J. Appl. Cryst. 3, p. 354-357 (1970).

21. M. A. G. Halliwell, "Measurement of Specimen Tilt and Beam
     Tilt in the Bond Method", J. Appl. Cryst. 3, p. 418-419 (1970).

22. J. Mandel, "The Statistical Analysis of Experimental Data",
     p. 111, Interscience Publ., John Wiley and Sons, New York,
     1964.

X-RAY SPECTRAL DISTRIBUTIONS FROM THICK TUNGSTEN TARGETS IN THE

ENERGY RANGE 12 TO 300 kV*

Ellery Storm, Harvey I. Israel, and Douglas W. Lier

Los Alamos Scientific Laboratory,University of California

Los Alamos, New Mexico   87544

ABSTRACT

Bremsstrahlung emission from four x-ray tubes operating at 12
to 300 kV was measured. Spectral distributions are given in terms
of absolute-photon and energy fluxes. Silicon and germanium semiconductors, sodium iodide scintillators, and a xenon proportional
counter were used to measure the spectra. Detector distortions were
corrected by assuming an undistorted spectrum, distorting the spectrum with a Monte Carlo computer program, and comparing the results
with measurement. The assumed undistorted spectrum was revised until
the Monte Carlo calculation gave satisfactory agreement with spectral
measurements from all four types of detectors. The effects on the
spectra of varying the tube potential, tube current, target aperture,
and detector aperture were investigated. Corrections for the intervening material and the solid angle subtended by the detector were
applied to obtain the absolute-photon and energy flux from the target. Total fluxes in the L lines,K lines, and continuum are given.
X-ray-production efficiencies varied from 0.026% at 12 kV to 1% at
300 kV. The constant of proportionality varied in the range $(0.23$
to $0.73) \times 10^{-6}$ kv^{-1}.

INTRODUCTION

This paper presents the results of an experimental study of
thick-target bremsstrahlung emission from four commercial x-ray
units, operating at 12 to 300 kV. Commercial x-ray units are widely
used in radiation therapy, diagnostic radiography, and material
analysis, where a knowledge of the spectral distribution in terms
of absolute intensity is desirable. Previous spectral measurements

in this energy region are summarized in Refs. 1 and 2. In most
cases, data are given in terms of relative photon intensities.
Dyson(3) and Placious(4) have measured the absolute bremsstrahlung
emission from experimental electron accelerators. The only absolute
spectral measurements of commercial x-ray units appear to be those
of Ehrlich,(5) Hettinger and Starfelt,(6) and Unsworth and Greening.
(7) Sodium iodide and proportional counters were used as detectors
in these measurements.

The present measurements were undertaken to provide more experimental data on the absolute bremsstrahlung emission of commercial
x-ray units. The semiconductors, with their improved resolution,
were used in these measurements,together with sodium iodide scintillators and a proportional counter. A Monte Carlo computer program
was utilized to correct for detector distortions, to separate the
characteristic lines from the continuum, and to resolve the K-edge
discontinuity in the continuum. Because the absolute intensity depends upon the tube potential, tube current, inherent filtration,
wave form, target material, target angle, detector angle, target aperture, and detector aperture, each of these parameters has been
specified and, where possible, the effect of varying them has been
determined.

EXPERIMENTAL DETAILS

The spectra from four x-ray units, a 300-kV Norelco, a 275-kV
Triplett-Barton, a 100-kV Picker, and a 60-kV Picker were studied.
The more important features of these units are summarized in Table I.

TABLE I. DESCRIPTION OF X-RAY UNITS

X-Ray Unit	X-Ray Tube[a]	Voltage Range (kV)	Current Range (mA)	Wave Form	Target Angle (°)	Inherent Filtration (g/cm^2)
No. American Philips Co. (Norelco)	Müller MÖ 301/10	100-300	2-10	Constant-potential	22.5	0.65 Pyrex glass 0.71 Araldit plastic 0.44 beryllium 1.4 oil
Triplett & Barton, Inc.	Machlett EG-252-C	60-200	1-10	Self-rectified	20	0.425 beryllium
Picker X-Ray Corp.	Machlett MR-100	40-100	1-20	Self-rectified	20	0.28 Pyrex glass
Picker X-Ray Corp.	Machlett OEG-60G	12- 60	2-50	Constant-potential	45	0.046 beryllium

[a]All four x-ray tubes had tungsten targets.

Except for the exit window, the x-ray tubes were completely encased
in lead shielding to prevent radiation leakage. To reduce the
counting rate to a measurable level, the detectors were placed approximately 50 m from the target. An evacuated pipe was placed between the x-ray tube and the detectors to reduce air attenuation.
The pipe was found to introduce a slight shift in the spectra at the
higher energies owing to small-angle scatter.

The spectral measurements were made with four types of detectors:
a germanium semiconductor, a silicon semiconductor, a sodium iodide
scintillator, and a xenon proportional counter. The detector angle
relative to the electron beam direction was 90°. Fluorescent x-ray
sources and isotope gamma rays were used to calibrate the detectors.
The measurements were corrected for Gaussian broadening, photopeak
efficiency, escape-peak losses, Compton scattering, energy linearity,
attenuation of intervening material, and the solid angle subtended
by the detector. For further information about the apparatus and
the experimental method, see Refs. 8 and 9.

PARAMETER VARIATIONS

Tube Current

The shape of the x-ray spectrum was not significantly affected
when the current was varied, but changes in intensity per mA were
observed. This may be seen in Fig. 1, where the average ratio of
the count obtained at a given energy and current to the count obtained at the same energy and lowest current (1 or 2 mA) is plotted
as a function of tube current. The self-rectified Triplett-Barton
and 100-kV Picker units show the same tendency for the intensity per
mA to decrease nearly linearly as the current is increased. Our
final spectral measurements were made at the lowest currents.

Target Aperture

The target aperture appeared to have a small effect on the spectral shape, but the absolute intensity was quite sensitive to the
aperture area. This may be seen from Fig. 2, where the average ratio
of the count obtained at a given energy with a given aperture to the
count obtained at the same energy with the largest aperture is plotted as a function of the aperture area. The count ratio increases
rapidly at first, then levels off and becomes constant above an aperture area of 1.5 cm^2. The rest of our spectral measurements were
made with target aperture areas greater than 1.5 cm^2.

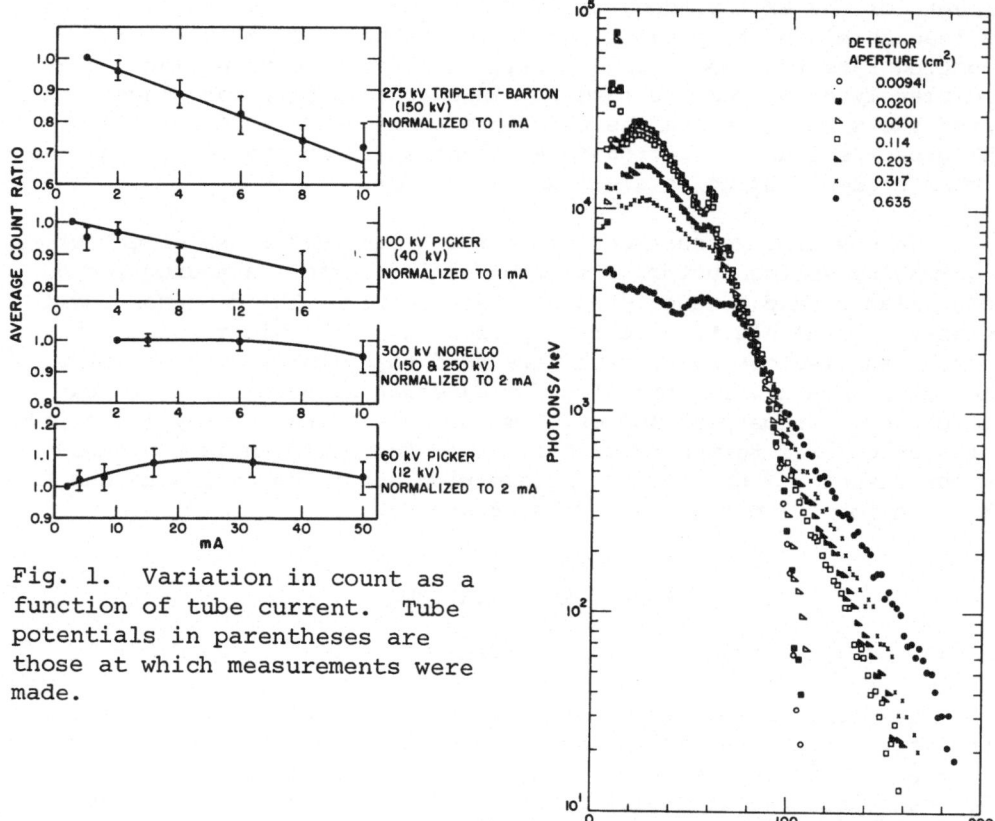

Fig. 1. Variation in count as a function of tube current. Tube potentials in parentheses are those at which measurements were made.

Fig. 3. Triplett–Barton 100-kV spectrum measured with various detector apertures in front of the germanium detector.

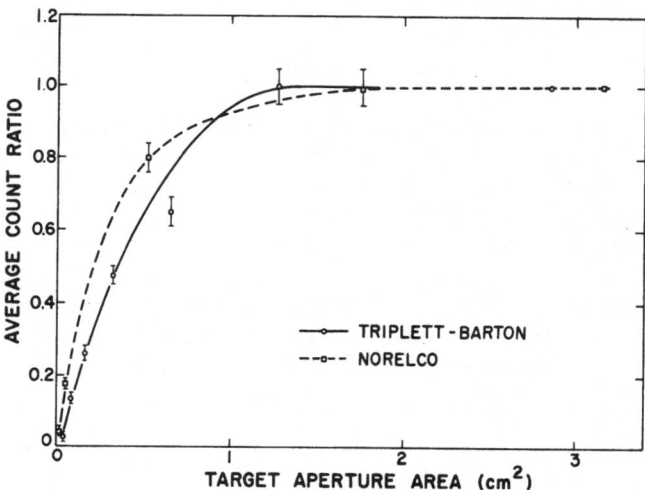

Fig. 2. Variation in count as a function of target aperture area.

Detector Aperture

The intensity was found to vary directly with the detector aperture area when the counting rate was low. However, if the detector aperture area was too large, the high counting rate overloaded the detector and the spectra were distorted. Typical results are given in Fig. 3, which shows the 100-kV Triplett-Barton spectrum measured with various apertures in front of the germanium detector. For ease of comparison, the counting times were adjusted to give the same cm^2-min. The points measured with the three smallest apertures are very close to one another, indicating that the intensity varied directly with the aperture area up to a counting rate of 1.2×10^4 counts/sec. As the area increased, more and more counts were lost in the lower energies and recorded at the higher energies owing to coincidence effects. At a rate of 3×10^6 counts/sec, all trace of the L- and K-line structure was lost and counts were recorded at energies as high as 200 keV, although the exciting potential was only 100 kV.

RESULTS

The Measured Bremsstrahlung Spectra

X-ray spectra measured with the silicon, germanium, sodium iodide, and xenon detectors were obtained with the Norelco unit operating at 100, 150, 200, 250, and 300 kV; the Triplett-Barton unit operating at 60, 80, 100, 150, and 200 kV; the 100-kV Picker unit operating at 40, 60, 80, and 100 kV; and the 60-kV Picker unit operating at 12, 20, 40, and 60 kV. A complete set of figures showing these measurements may be found in Los Alamos Scientific Laboratory report LA-4624.(9) Typical of the results obtained are the Triplett-Barton measurements reproduced in Figs. 4 to 7.

Each detector distorts the spectra differently. The method of deducing the undistorted incident spectra has been described in detail in Ref. 8. Briefly, a hand calculation partially corrects the distortion and gives an estimate of the undistorted spectrum. The estimate, together with the detector resolution and geometry, is used as input to a computer program that utilizes Monte Carlo techniques to distort the spectrum. The calculated distorted spectrum is compared to the measurements, and the estimate is revised until satisfactory agreement is obtained with all four types of detectors.

The Monte Carlo calculations of the distorted Triplett-Barton spectra are given by the solid lines in Figs. 4 to 7, and the undistorted input spectra are shown in Fig. 8. Final estimates of the 60-kV Picker and 100-kV Picker undistorted spectra for each measured tube potential are given in Figs. 9 and 10, respectively. The undistorted Norelco spectra for measured potentials may be found in Ref. 9. Experimental conditions for Figs. 8 to 10 are given in Table II.

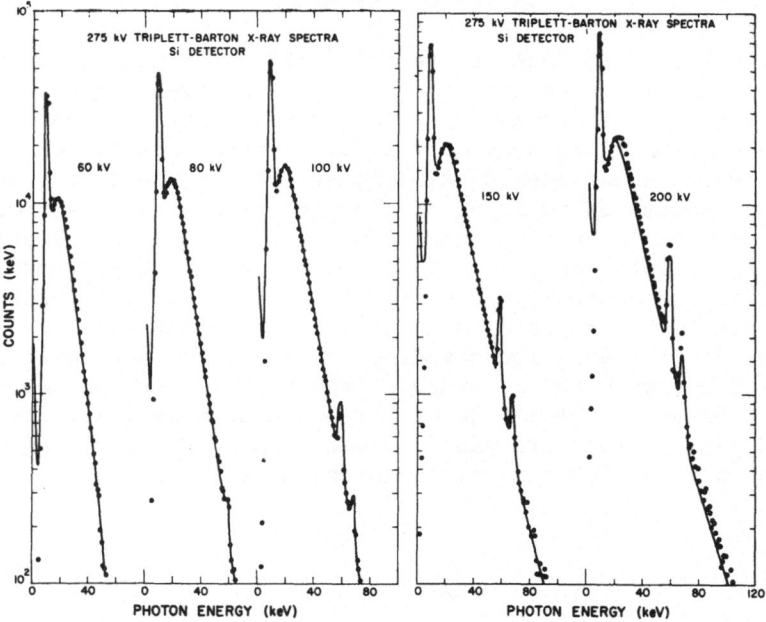

Fig. 4.   Silicon detector measurements of the Triplett-Barton x-ray spectra.   In Figs. 4 through 7, the points are measured values and the solid lines are Monte Carlo calculations.

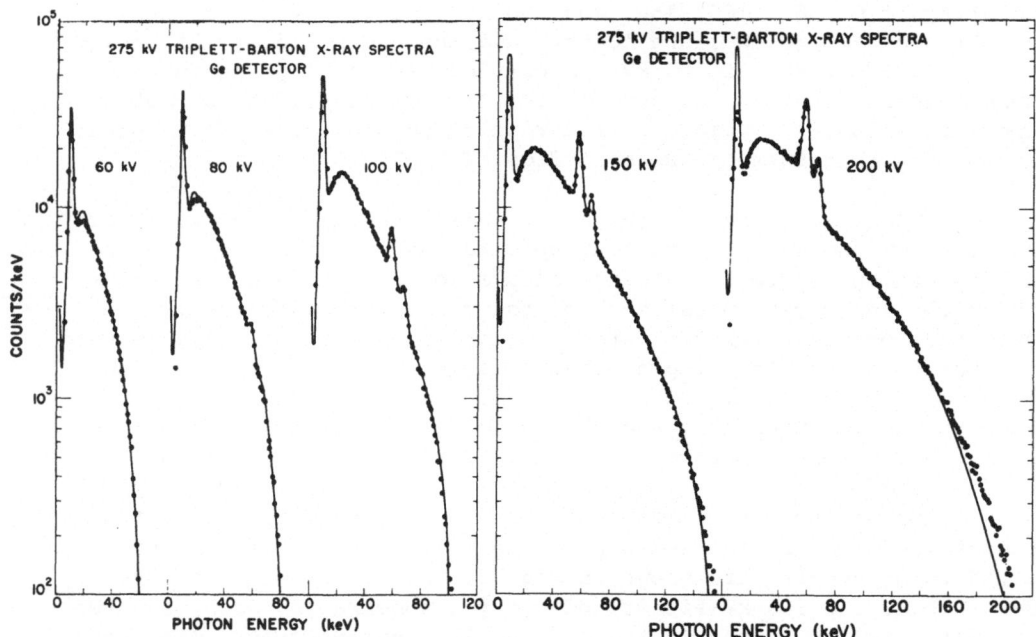

Fig. 5.   Germanium detector measurements of the Triplett-Barton x-ray spectra.

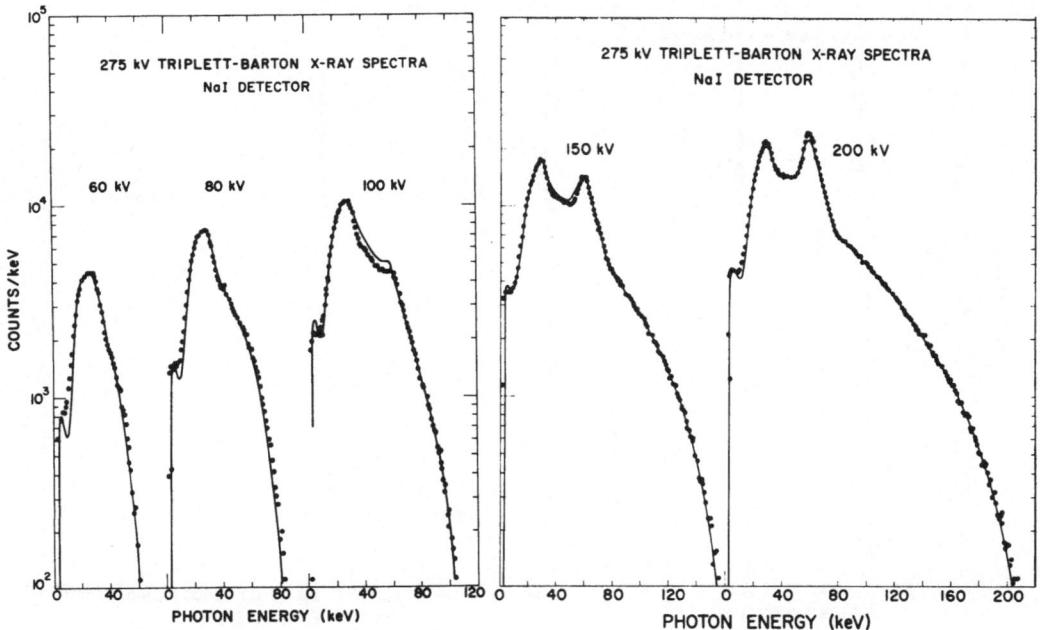

Fig. 6. Sodium iodide detector measurements of the Triplett-Barton x-ray spectra.

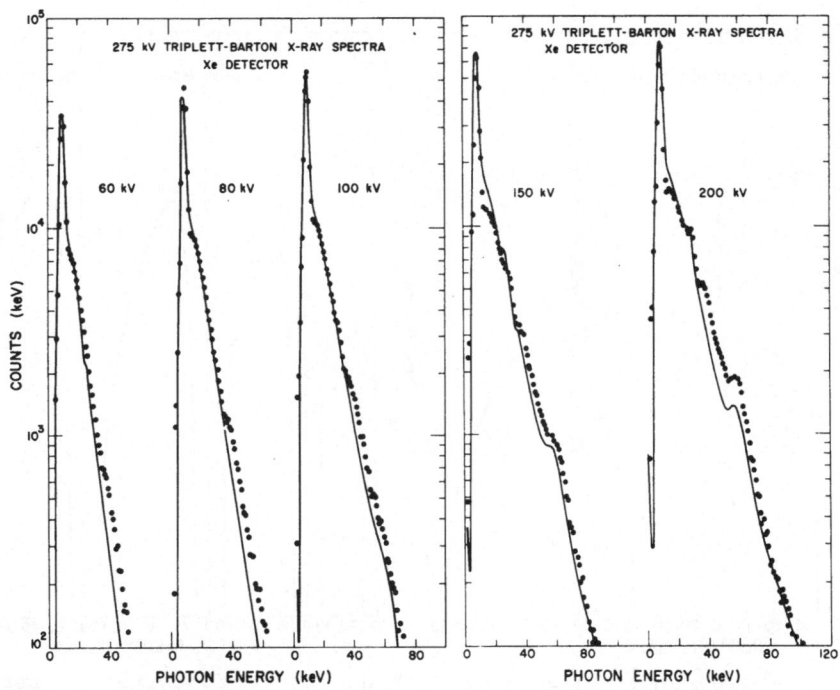

Fig. 7. Xenon detector measurements of the Triplett-Barton x-ray spectra.

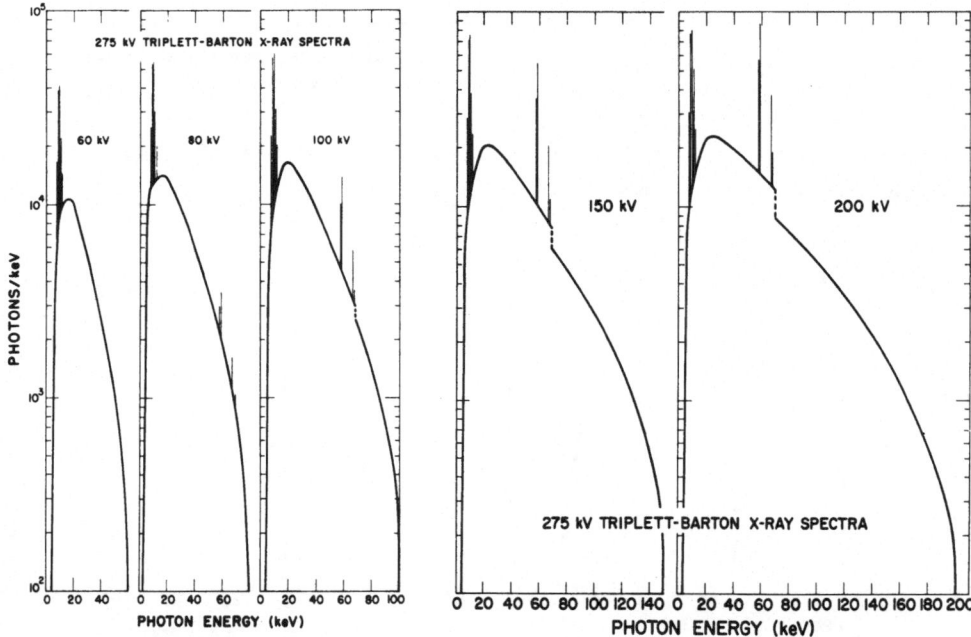

Fig. 8.   Undistorted spectra from the Triplett-Barton x-ray unit.

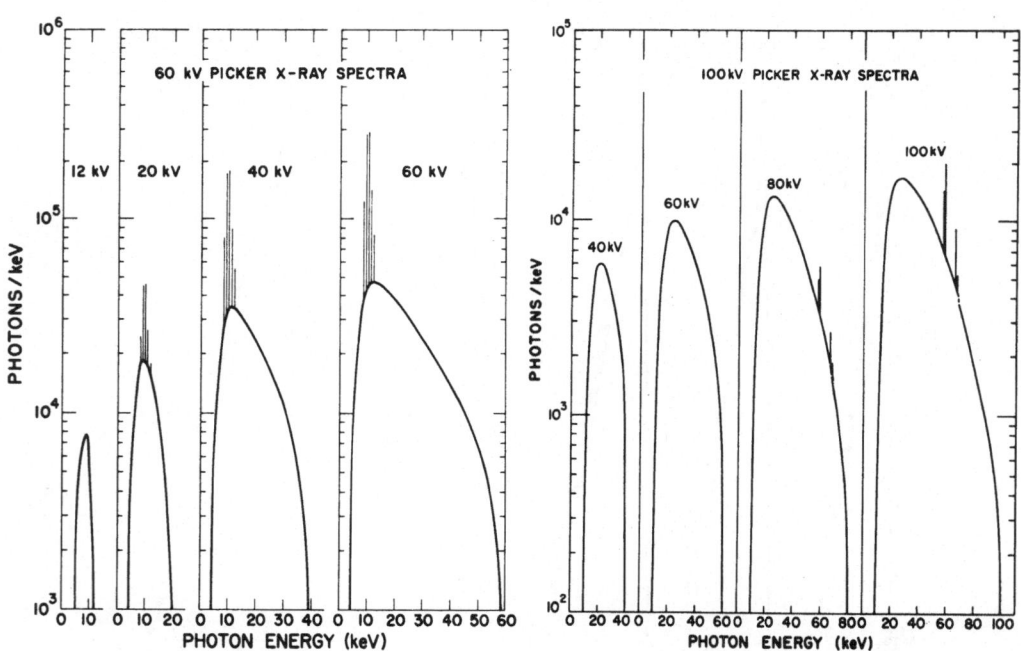

Fig. 9.   Undistorted spectra
from the 60-kV Picker unit.

Fig. 10.   Undistorted spectra
from the 100-kV Picker unit.

TABLE II. EXPERIMENTAL CONDITIONS

X-Ray Unit	Detector Area (cm^2)	Time (sec)	Energy Interval (keV)	Tube Current (mA)	X-Ray Target to Detector Distance (cm)	Intervening Material (g/cm^2)	Inherent Filtration (g/cm^2)
300-kV Norelco	0.00941	510	1	2	4920	0.046 Be 0.021 Mylar 0.052 Air	0.44 Be 1.4 Oil 0.71 Araldit plastic 0.65 Pyrex glass
275-kV Triplett-Barton	0.00941	510	1	1	4797	0.046 Be 0.021 Mylar 0.046 Air	0.425 Be
100-kV Picker	0.00941	510	1	2	4787	0.046 Be 0.021 Mylar 0.036 Air	0.28 Pyrex glass
60-kV Picker	0.0100	480	1	2	4655	0.046 Be 0.021 Mylar 0.037 Air	0.046 Be

The Target Spectra

The final estimates of the undistorted spectra incident on the detectors were corrected for the attenuation(10) of the intervening material, including inherent filtration; corrected for the solid angle in steradians subtended by the detector; and converted to units of photons/sec-mA-keV-sr. The information necessary to perform these corrections is summarized in Table II, and the results for the four x-ray units are given in Figs. 11 to 14.

The 100-kV Picker and Norelco spectra fall off sharply below 10 and 20 keV, respectively, because of their relatively large inherent filtration. Consequently, the L-line and continuum intensities below these energies were indeterminable. Although the inherent filtration of the 60-kV Picker and Triplett-Barton units consisted only of beryllium windows, there is a large uncertainty in the continuum intensity below 15 keV for these units also, because the L-line intensity must first be estimated and subtracted. Both the 60-kV Picker and Triplett-Barton units show a peak in the continuum below 10 keV which appeared after the L-line removal and filtration corrections. These peaks may represent tungsten L-edge structure in the continuum which was not clearly resolved because of the superimposed L lines.

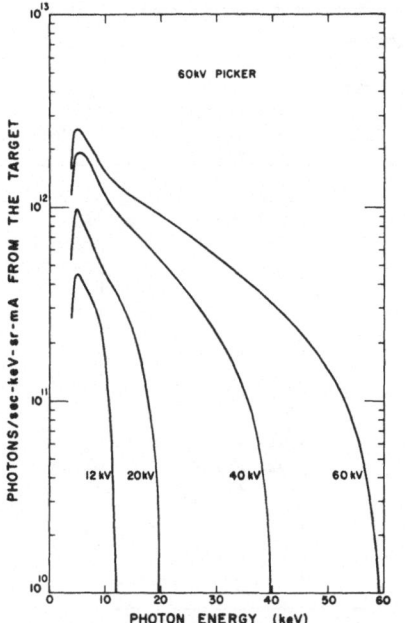

Fig. 11. Photon flux from the
60-kV Picker target as a function of photon energy. Characteristic lines have been omit-
ted from Figs. 11 through 14.

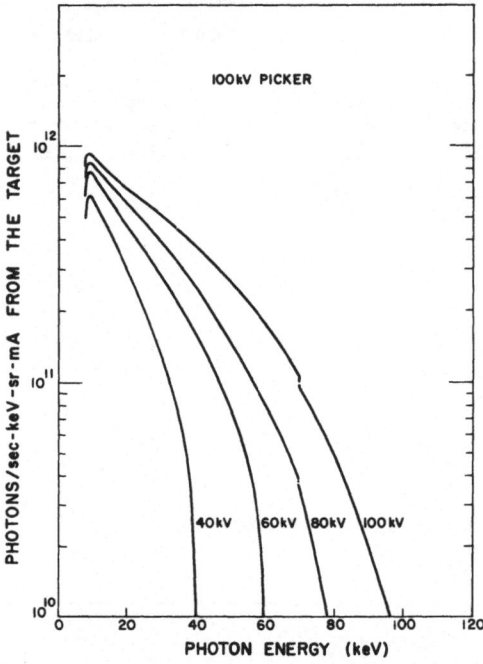

Fig. 12. Photon flux from the
100-kV Picker target as a function of photon energy.

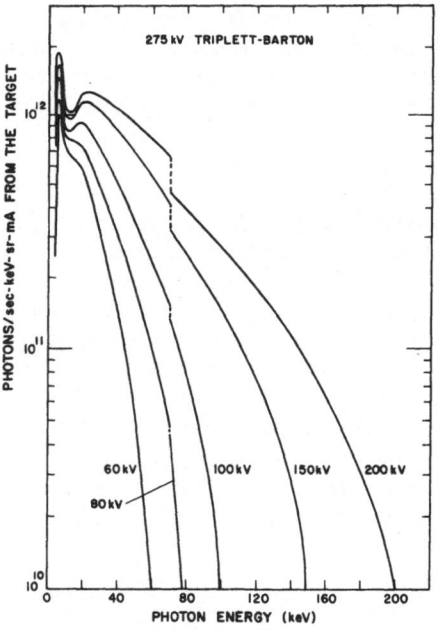

Fig. 13. Photon flux from the
275-kV Triplett-Barton target
as a function of photon energy.

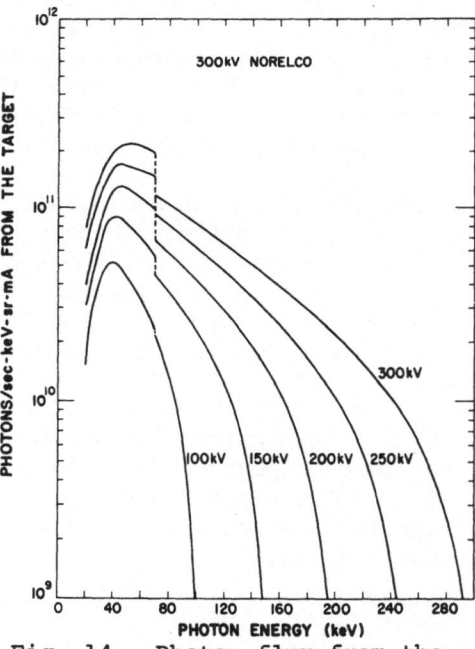

Fig. 14. Photon flux from the
300-kV Norelco target as a function of photon energy.

Integrated Photon and Energy Flux

The total number of continuum photons emitted from the target
was obtained by integrating over the photon energy. The total L-
and K-line intensities were calculated except for the 100-kV Picker
and Norelco L-line intensities which could not be determined because
of the large inherent filtration. The continuum and line totals in
terms of ergs/sec-mA-sr are plotted as functions of tube potential
in Figs. 15 and 16, respectively.

The 100-kV Picker, Triplett-Barton, and Norelco continuum and
K-line totals appear to be following a similar trend. However, the
60-kV Picker continuum and L-line totals lie a factor of 2 to 3
above the others. This may be due in part to the 45° target angle
of the 60-kV Picker unit, compared to the 20 to 22° target angles of
the other units. Photons produced in the tungsten target are attenu-
ated by a thickness $x/\tan \theta$ as they leave the target, where x is the
target thickness traversed by the electron before photon emission.
Consequently, the 60-kV Picker photons suffer less attenuation than
the photons from the other units. In addition, one would expect the
60-kV Picker and Norelco constant potential units to have greater
intensity than the two self-rectified units. These factors will be
discussed further in a separate paper in which the measurements will
be compared to previous measurements and to theory.

The ratio of the L- and K-line fluxes to the total fluxes may
be found in Ref. 9. The ratios increase rapidly above the edges,
reach a maximum at a tube potential two to three times the edge
energy, and then level off as the tube potential increases further.

X-Ray Production Efficiency

The efficiency for x-ray production, ε, was defined by Compton and
Allison(11) as

$$\varepsilon = \frac{\text{x-ray energy}}{\text{cathode-ray energy}} = kZV \quad , \tag{1}$$

where Z is the target atomic number, V is the tube potential, and k
is a constant of proportionality in units of reciprocal volts. Using
measurements made before 1935, Compton and Allison concluded that k
was approximately 1.1×10^{-6} kV^{-1}. After studying later measurements
and calculations, Evans(12) concluded that a better value for k was
0.7×10^{-6} kV^{-1}. From the line and continuum intensities, one can
obtain the total x-ray energy, and thus the efficiency and proportionality constant. The x-ray-production efficiency and k are given
in Table III for each x-ray unit and measured tube potential. The
value of k varies in the range $(0.23 \text{ to } 0.73) \times 10^{-6}$ kV^{-1}. The x-ray-
production efficiency increases from 0.026% at 12 kV to 1% at 300 kV.

TABLE III. X-RAY-PRODUCTION EFFICIENCY

X-Ray Unit	Potential (kV)	Efficiency	k (kV^{-1}) × 10^6
60-kV Picker	12	0.00026	0.30
	20	0.00088	0.59
	40	0.0022	0.73
	60	0.0032	0.71
100-kV Picker	40	0.00070	0.24
	60	0.0011	0.24
	80	0.0013	0.23
	100	0.0018	0.24
275-kV Triplett-Barton	60	0.0015	0.34
	80	0.0020	0.34
	100	0.0028	0.38
	150	0.0042	0.38
	200	0.0052	0.34
300-kV Norelco	100	0.0023	0.31
	150	0.0045	0.41
	200	0.0067	0.45
	250	0.0092	0.50
	300	0.012	0.54

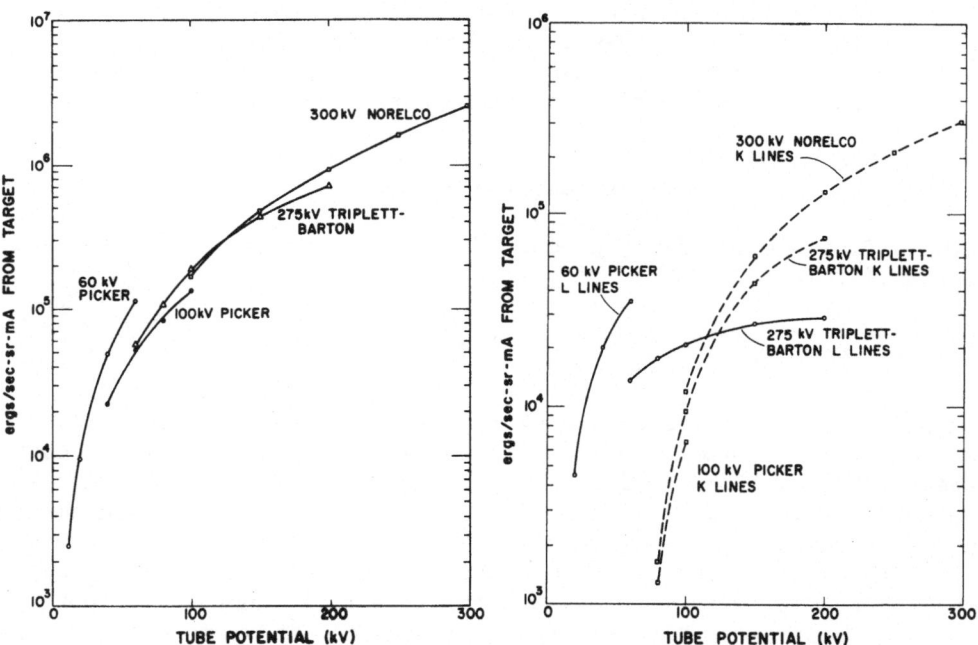

Fig. 15. Total continuum energy flux emitted from the target as a function of tube potential.

Fig. 16. Total L- and K-line energy flux emitted from the target as a function of tube potential.

REFERENCES

1. International Commission on Radiation Units and Measurements (ICRU) Report 10b, "Physical Aspects of Irradiation," 1962.

2. International Commission on Radiation Units and Measurements (ICRU) Report 17, "Radiation Dosimetry: X-Rays Generated at Potentials of 5 to 150 kV," 1970.

3. N. A. Dyson, "The Continuous X-Ray Spectrum from Electron-Opaque Targets," Proc. Phys. Soc. 73, 924 (1959).

4. R. C. Placious, "Dependence of 50- and 100-keV Bremsstrahlung on Target Thickness, Atomic Number, and Geometric Factors," J. Appl. Phys. 38, 2030 (1967).

5. M. J. Ehrlich, "Scintillation Spectrometry of Low-Energy Bremsstrahlung," J. Res. Natl. Bur. Std. 54, 107 (1955).

6. G. Hettinger and N. Starfelt, "Bremsstrahlung Spectra from Roentgen Tubes," Acta Radiol. 50, 381 (1958).

7. M. H. Unsworth and J. R. Greening, "Experimental Continuous and L-Characteristic X-Ray Spectra for Tungsten Target Tubes Operated at 15 to 30 kV," Phys. Med. Biol. 15, 631 (1970).

8. H. I. Israel, D. W. Lier, and E. Storm, "Comparison of Detectors Used in Measurement of 10- to 300-keV X-Ray Spectra," Nucl. Instr. Meth. 91, 141 (1971).

9. E. Storm, H. I. Israel, and D. W. Lier, "Bremsstrahlung Emission Measurements from Thick Tungsten Targets in the Energy Range 12 to 300 kV," Los Alamos Scientific Laboratory Report 4624 (1971).

10. E. Storm and H. I. Israel, "Photon Cross Sections from 1 keV to 100 MeV for Elements Z = 1 to Z = 100," Nuclear Data Tables A7, 565 (1970).

11. A. H. Compton and S. K. Allison, X-Rays in Theory and Experiment, D. Van Nostrand Co., New York, 1935, p. 89-90.

12. R. D. Evans, The Atomic Nucleus, McGraw-Hill, New York, 1955, p. 616.

*Work performed under the auspices of U.S. Atomic Energy Commission

ELEMENTAL X-RAY CROSS SECTIONS AT SELECTED WAVELENGTHS[*]

Bobby L. Bracewell and William J. Veigele[+]

Kaman Sciences Corporation

Colorado Springs, Colorado 80907

ABSTRACT

Revised tables of x-ray mass attenuation and absorption cross sections have been prepared for the elements with atomic numbers one to 94 at selected wavelengths of interest to spectroscopists, microprobe users,
diffraction workers, etc. The tables are derived from an
x-ray cross section compilation to be published during
the summer of 1971. The new compilation is a revision of
a previous work and provides photoelectric, coherent
scattering, incoherent scattering, absorption component of
incoherent scattering and total attenuation cross sections for 94 elements for the photon energy range 0.1 keV
to 1 MeV (0.0124 Å to 124 Å). The cross sections for
energies greater than 1 keV (12.4 Å) were determined from
experimental attenuation data and theoretical scattering
cross sections calculated using form factors and incoherent scattering functions based on a relativistic self-
consistent field method. For hydrogen, the photoelectric
absorption cross sections were calculated exactly. Least
squares procedures were used to interpolate and extrapolate for elements and photon energies where no experimental data were found. Cross sections for the energy
range 0.1 keV to 1 keV (12.4 Å to 124 Å) were calculated
using nonrelativistic, single electron, self-consistent
field theory with Herman-Skillman bound state wavefunctions. A brief description is given of the assumptions and methods used in preparing the revised compilation, and estimated uncertainties in the cross section
are reported. Examples illustrating agreement between

tabulated values and experimental data are given and a
comparison is made with values taken from earlier x-ray
cross section compilations.

INTRODUCTION

The tables of interest to this conference are derived from an extensive compilation of x-ray cross
sections (1) covering the photon energy range 0.1 keV to
1 MeV. Due to space limitations, the complete set of derived tables at selected wavelengths is not presented
here. Information is given below for those wishing to
obtain copies of the tables.

In order to establish the usefulness and limitations
of the derived tables, the parent compilation must be
described.

X-RAY CROSS SECTION COMPILATION

The parent compilation presents photoelectric, coherent scattering, incoherent scattering, absorption
component of incoherent scattering and total attenuation
cross sections for elements of atomic number one to 94
for photon energies between 0.1 keV and 1 MeV. All
cross sections are given in both atomic units (barns/
atom) and mass units (cm^2/gm). Due to the range of elements and photon energies covered, it is necessary to
utilize both experimental and theoretical cross sections
in establishing a consistent and most likely set of
values. The recommended values presented by the compilation are based on extensive analysis of both old and new
experimental data, well established theoretical calculations and many new calculations at very low photon
energies. The compilation is believed to constitute the
most accurate and consistent set of cross sections presently available for such a large number of elements and
wide energy range.

In order to reduce confusion, we have followed the
recommendations of the International Commission on
Radiation Units and Measurement (ICRU) in labeling the
various cross sections. Accordingly, "total attenuation
cross section" is used to denote the total photon interaction cross section and "absorption cross section" is
used to denote true absorption, i.e., photoelectric
absorption plus the absorption part of incoherent
scattering.

Collection and Evaluation of Data

A search for experimental data published between 1920 and 1970 provided 153 papers containing approximately 9000 measured values. The bulk of these data were total attenuation cross sections, which formed the basis for the compilation. The attenuation values and their errors were first adjusted, if necessary, to be consistent with recent and accepted values for physical constants (2), atomic weights (3), and wavelengths and energies (4). Data were discarded which were grossly inconsistent with the bulk of the available points or which were reported in a fashion making evaluation of their accuracy impossible. The final number of experimental values used in the compilation was approximately 8000.

Determination of Best Values

The attenuation data were assessed for reliability and weighting factors were assigned to them for use in a least squares adjustment. Since it was much more accurate (and convenient) to do the least squares adjustment using photoelectric rather than attenuation cross sections, the desired values were obtained as follows. First the total scattering cross section was calculated for all elements and energies for which experimental attenuation points were found. The scattering values were calculated using the recently computed incoherent scattering functions and form factors of Cromer (5), based on a relativistic self-consistent field method. This type of calculation is well established and documented and is very accurate for the photon energy region of interest. The calculated scattering cross sections were then subtracted from the experimental attenuation cross sections point by point to obtain the desired photoelectric data.

For photon energies greater than 1 keV, the photoelectric points were adjusted and extended by extrapolation and interpolation using a least squares computer program. Since experimental data for hydrogen are scarce and inconsistent, its photoelectric cross section was calculated exactly for photon energies greater than 1 keV. These calculated values were entered into the least squares adjustment program as if they were very reliable experimental points. The end result of the adjustment was a set of internally consistent photoelectric cross sections for all elements between 1 keV and 1 MeV. The "smoothing" was done with the cross sections expressed

in atomic units (barns/atom) assuming that the cross
sections varied smoothly, according to simple power laws,
with photon energy and atomic number (assumption of hydrogenic behavior). This last assumption was known to be
incorrect for low photon energies, especially for elements
with M and N shell electrons. An example of this situation is shown for the element krypton in Fig. 1. The dots
represent experimental data and the dashed line the least
squares adjusted attenuation cross section derived assuming a hydrogen-like model.

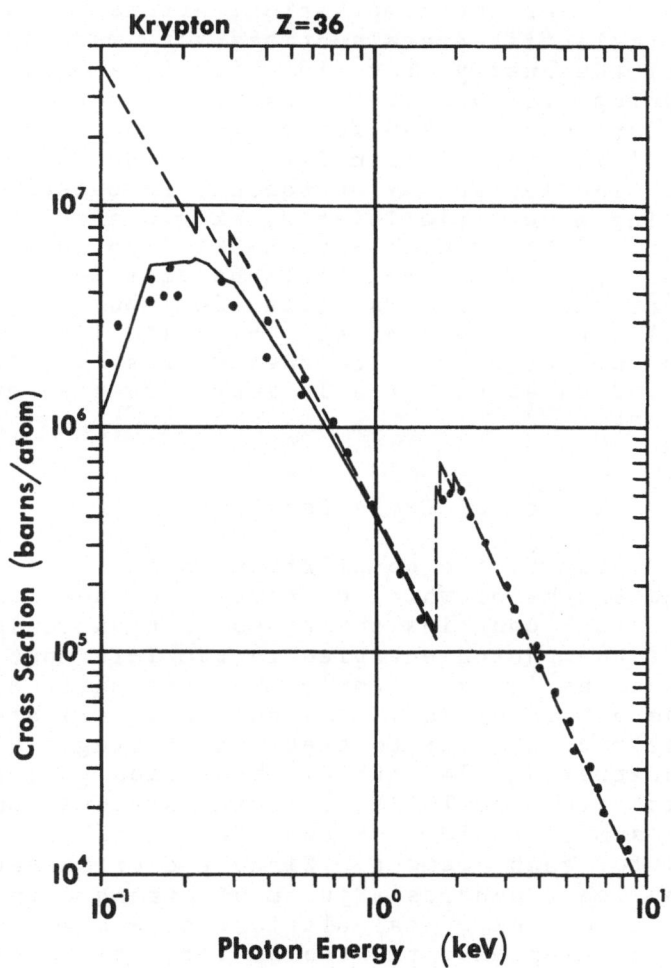

Figure 1. Cross Section vs. Energy Graph for krypton
           showing experimental data (·), least squares
           fit using hydrogenic approximation (dashed
           line), and low energy calculations (solid
           line).

Low Energy Cross Sections

The agreement shown in Fig. 1 between the data and
the dashed curve is excellent except in the low energy
region near the M absorption edges. In this low energy
region, scattering interactions contribute a negligible
amount to the total photon interaction cross section –
it is, for all practical purposes, all photoelectric
interaction. Unfortunately, for most elements, experimental cross sections in this low energy region are either
scarce or non-existent. It is, therefore, necessary to
use calculated photoelectric values to define the cross
section curves. For the compilation, extensive calculations of photoelectric cross sections were made for all
elements over the energy range in which they exhibited
this "non-hydrogenic" behavior. The low energy limit of
the calculations was 0.1 keV for all elements, while the
high energy limits varied from 1 to 8 keV depending on
the element. The low energy photoelectric calculations
were done using a non-relativistic, single electron, self-
consistent field theory with Herman-Skillman bound state
wavefunctions (6). These calculations provide a reliable
and consistent set of cross sections for many of the
elements at low photon energies. The solid line in Fig.
1 represents the calculated low energy cross section for
krypton and indicates the greatly improved agreement
with experiment.

Tables of Cross Sections

The main body of the compilation consists of a set
of tables and graphs of which examples are shown in Figs.
2 and 3. Figure 2 contains the cross section table for
krypton. For the photon energies in the left most column
the cross sections in both atomic and mass units are
listed in the other columns. The column labels are PHOTO
for photoelectric, COH for coherent scattering, INC for
incoherent scattering, INC AB for absorption part of incoherent scattering and TOTAL for total attenuation.
TOTAL is the sum of PHOTO, COH and INC. A table of this
type appears for each element. Since the cross sections
determined by least squares adjustment extend down to
1 keV, and the low energy calculations extend above this
energy for some elements, the tables for these elements
list both sets of cross sections in the overlapping
region to provide an idea of the extent of the non-hydrogenic behavior.

```
KRYPTON           Z = 36   AT WT =   83.80000    RHO = 3.4840E-03 G/C**3    K-JUMP =    6.920

ENERGY      PHOTO    COH      INC      INC AB    TOTAL     PHOTO    COH      INC       INC AB   TOTAL
KEV         B/A      B/A      B/A      B/A       B/A       C**2/G   C**2/G   C**2/G    C**2/G   C**2/G
  .1000     1.14+6   8.61+2   8.78-3   2.33-6    1.14+6    8.16+3   6.19+0   6.31-3    1.67-8   8.17+3
  .1500     5.34+6   8.61+2   1.80-2   7.26-6    5.34+6    3.84+4   6.19+0   1.29-4    5.22-8   3.84+4
  .2000     5.55+6   8.60+2   3.09-2   1.67-5    5.55+6    3.99+4   6.18+0   2.22-4    1.20-7   3.99+4
  .2161     5.21+6   8.60+2   3.59-2   2.10-5    5.21+6    3.74+4   6.18+0   2.58-4    1.51-7   3.74+4
  .2175     5.63+6   8.60+2   3.64-2   2.14-5    5.63+6    4.05+4   6.18+0   2.61-4    1.54-7   4.05+4
  .2452     5.24+6   8.59+2   4.59-2   3.05-5    5.24+6    3.77+4   6.17+0   3.30-4    2.19-7   3.77+4
  .2466     5.21+6   8.59+2   4.64-2   3.10-5    5.21+6    3.74+4   6.17+0   3.33-4    2.23-7   3.74+4
  .2893     4.33+6   8.58+2   6.29-2   4.93-5    4.33+6    3.11+4   6.17+0   4.52-4    3.54-7   3.11+4
  .2897     4.50+6   8.58+2   6.35-2   5.00-5    4.50+6    3.24+4   6.17+0   4.56-4    3.59-7   3.24+4
  .3000     4.32+6   8.58+2   6.80-2   5.54-5    4.32+6    3.10+4   6.17+0   4.88-4    3.99-7   3.10+4
  .3193     3.95+6   8.57+2   7.67-2   6.66-5    3.96+6    2.84+4   6.16+0   5.51-4    4.79-7   2.84+4
  .3207     3.93+6   8.57+2   7.74-2   6.75-5    3.93+6    2.82+4   6.16+0   5.56-4    4.85-7   2.83+4
  .4000     2.73+6   8.55+2   1.19-1   1.30-4    2.73+6    1.96+4   6.14+0   8.55-4    9.31-7   1.96+4
  .5000     1.79+6   8.51+2   1.84-1   2.50-4    1.79+6    1.29+4   6.11+0   1.32-3    1.80-6   1.29+4
  .6000     1.24+6   8.46+2   2.61-1   4.24-4    1.24+6    8.88+3   6.08+0   1.87-3    3.05-6   8.89+3
  .7000     8.89+5   8.40+2   3.49-1   6.63-4    8.90+5    6.39+3   6.04+0   2.51-3    4.76-6   6.40+3
  .8000     6.62+5   8.34+2   4.48-1   9.71-4    6.63+5    4.76+3   5.99+0   3.22-3    6.98-6   4.77+3
  .9000     5.08+5   8.27+2   5.54-1   1.34-3    5.09+5    3.65+3   5.94+0   3.98-3    9.65-6   3.65+3
 1.0000     3.98+5   8.20+2   6.68-1   1.80-3    3.99+5    2.86+3   5.89+0   4.80-3    1.29-5   2.87+3
 1.2000     2.59+5   8.03+2   9.13-1   2.92-3    2.60+5    1.86+3   5.77+0   6.56-3    2.10-5   1.87+3
 1.4000     1.79+5   7.84+2   1.17+0   4.35-3    1.80+5    1.29+3   5.63+0   8.42-3    3.12-5   1.29+3
 1.6000     1.29+5   7.65+2   1.45+0   6.09-3    1.30+5    9.28+2   5.50+0   1.04-2    4.37-5   9.33+2

 1.0000     4.67+5   8.44+2   7.07-1   1.91-3    4.68+5    3.36+3   6.06+0   5.08-3    1.37-5   3.36+3
 1.5000     1.49+5   7.75+2   1.31+0   5.19-3    1.50+5    1.07+3   5.57+0   9.42-3    3.73-5   1.08+3
 1.6750 L3  1.09+5   7.58+2   1.55+0   6.81-3    1.10+5    7.84+2   5.44+0   1.11-2    4.90-5   7.89+2
 1.6750     5.46+5   7.58+2   1.55+0   6.81-3    5.47+5    3.92+3   5.44+0   1.11-2    4.90-5   3.93+3
 1.7270 L2  5.03+5   7.52+2   1.62+0   7.34-3    5.03+5    3.61+3   5.41+0   1.17-2    5.27-5   3.62+3
 1.7270     7.04+5   7.52+2   1.62+0   7.34-3    7.04+5    5.06+3   5.41+0   1.17-2    5.27-5   5.06+3
 1.9210 L1  5.28+5   7.32+2   1.89+0   9.44-3    5.28+5    3.79+3   5.26+0   1.36-2    6.79-5   3.80+3
 1.9210     6.33+5   7.32+2   1.89+0   9.44-3    6.34+5    4.55+3   5.26+0   1.36-2    6.79-5   4.56+3
 2.0000     5.68+5   7.24+2   2.00+0   1.04-2    5.68+5    4.08+3   5.21+0   1.44-2    7.47-5   4.09+3
 3.0000     1.90+5   6.24+2   3.34+0   2.52-2    1.90+5    1.36+3   4.48+0   2.40-2    1.81-4   1.37+3
 4.0000     8.71+4   5.39+2   4.56+0   4.47-2    8.77+4    6.26+2   3.87+0   3.28-2    3.22-4   6.30+2
 5.0000     4.77+4   4.70+2   5.62+0   6.75-2    4.81+4    3.42+2   3.38+0   4.04-2    4.85-4   3.46+2
 6.0000     2.91+4   4.14+2   6.55+0   9.27-2    2.95+4    2.09+2   2.98+0   4.71-2    6.66-4   2.12+2
 8.0000     1.34+4   3.26+2   8.11+0   1.49-1    1.37+4    9.61+1   2.35+0   5.83-2    1.07-3   9.85+1
10.0000     7.31+3   2.61+2   9.42+0   2.12-1    7.58+3    5.26+1   1.87+0   6.77-2    1.52-3   5.45+1
12.6550 L3-K 3.87+3  1.97+2   1.08+1   3.02-1    4.08+3    2.78+1   1.42+0   7.79-2    2.17-3   2.93+1
14.3300 K   2.76+3   1.69+2   1.16+1   3.61-1    2.95+3    1.99+1   1.21+0   8.34-2    2.59-3   2.12+1
14.3300     1.91+4   1.69+2   1.16+1   3.61-1    1.93+4    1.37+2   1.21+0   8.34-2    2.59-3   1.39+2
15.0000     1.69+4   1.59+2   1.19+1   3.85-1    1.71+4    1.22+2   1.14+0   8.55-2    2.77-3   1.23+2
20.0000     7.75+3   1.08+2   1.36+1   5.65-1    7.88+3    5.57+1   7.78-1   9.74-2    4.06-3   5.66+1
30.0000     2.49+3   6.11+1   1.54+1   9.05-1    2.57+3    1.79+1   4.39-1   1.11-1    6.50-3   1.85+1
40.0000     1.09+3   3.92+1   1.63+1   1.21+0    1.15+3    7.86+0   2.82-1   1.17-1    8.69-3   8.26+0
50.0000     5.72+2   2.71+1   1.67+1   1.47+0    6.16+2    4.11+0   1.95-1   1.20-1    1.06-2   4.43+0
60.0000     3.35+2   2.00+1   1.69+1   1.70+0    3.72+2    2.41+0   1.44-1   1.21-1    1.22-2   2.67+0
80.0000     1.43+2   1.21+1   1.67+1   2.08+0    1.72+2    1.03+0   8.73-2   1.20-1    1.49-2   1.24+0
100.0000    7.40+1   8.21+0   1.63+1   2.37+0    9.85+1    5.32-1   5.90-2   1.17-1    1.70-2   7.08-1
150.0000    2.21+1   3.94+0   1.52+1   2.86+0    4.13+1    1.59-1   2.83-2   1.09-1    2.05-2   2.97-1
200.0000    9.41+0   2.30+0   1.42+1   3.14+0    2.59+1    6.76-2   1.65-2   1.02-1    2.26-2   1.86-1
300.0000    2.85+0   1.06+0   1.25+1   3.42+0    1.64+1    2.04-2   7.60-3   8.98-2    2.46-2   1.18-1
400.0000    1.23+0   6.06-1   1.13+1   3.52+0    1.31+1    8.84-3   4.36-3   8.09-2    2.53-2   9.41-2
500.0000    6.47-1   3.92-1   1.03+1   3.55+0    1.14+1    4.65-3   2.82-3   7.42-2    2.55-2   8.16-2
600.0000    3.86-1   2.74-1   9.56+0   3.54+0    1.02+1    2.77-3   1.97-3   6.87-2    2.54-2   7.35-2
800.0000    1.73-1   1.55-1   8.42+0   3.45+0    8.75+0    1.24-3   1.11-3   6.05-2    2.48-2   6.29-2
1000.0000   9.38-2   9.95-2   7.58+0   3.35+0    7.77+0    6.74-4   7.15-4   5.45-2    2.40-2   5.59-2
```

Figure 2.   Sample cross section table from parent
             compilation.

Figure 3 contains a graph of the cross sections for
krypton. A graph of this type appears for each element
and illustrates the attenuation, photoelectric, coherent
and incoherent scattering and low energy calculated cross
section curves. There is a supplement to the tables (7)
and graphs of the compilation which lists all pertinent
information for each element which was used in preparing
the compilation. This supplementary information includes
such things as all experimental data, their assigned

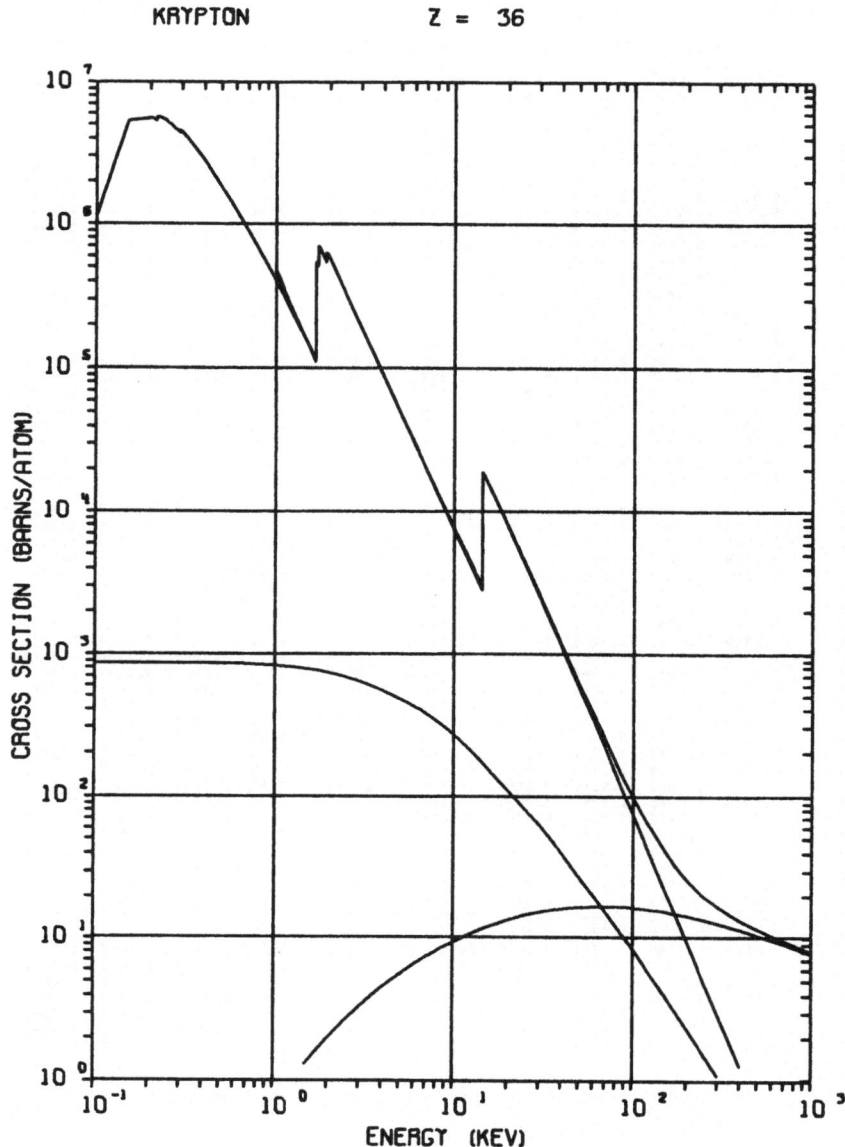

Figure 3. Sample cross section graph from parent compilation.

weights for the least squares adjustment, sources of each data point, coefficients for the photoelectric fits, fluorescence yield, incoherent scattering functions and form factors used in scattering calculations, etc.

Uncertainties in the Cross Sections

Based on (a) the quantity of data, (b) agreement of
the data among different experiments and, (c) the internal
consistency of the compiled values, estimates were made
of the uncertainties associated with the reported attenuation cross sections for energies above 1 keV. These
estimates were averaged over broad regions for each
element, that is, above the K absorption edge was one
region, between the K and L edges another region, and so
on. For each region one of the following designations
was chosen:

$$A = \pm (2-5)\%$$

$$B = \pm (5-10)\%$$

$$C = \pm (10-20)\%.$$

Within each group of edges, such as the region between
the L_I and L_{III} edges or between the M_I and M_V edges,
data are less reliable in general than in other regions.
Therefore, an uncertainty of $C = \pm (10-20)\%$ is suggested
for these regions for all elements. Figure 4 illustrates
the uncertainty table applying to the compilation.
These assignments may overestimate the uncertainties
because, even though experimental data are lacking, least
squares adjustments generally act to reduce uncertainties.
The uncertainty assignments C, however, focus attention
on those elements and energies for which experiments
would be valuable. In the low energy region, the uncertainties in calculated cross sections are not established because of the few experimental data, but are
probably accurate to approximately a factor of two.

CROSS SECTIONS AT SELECTED WAVELENGTHS

From the compilation just described, tables of mass
attenuation and mass absorption coefficients for the 94
elements of the compilation have been interpolated at
281 emission wavelengths. These wavelengths include the
following lines:

(a)   46 weighted $K\alpha_1\alpha_2$ lines for emitters boron
      through tin,
(b)   31 $K\beta_1$ lines for emitters neon through
      zirconium,

TABLE IV

UNCERTAINTIES ON ATTENUATION CROSS SECTIONS FROM 1 keV TO 1 MeV

ELEMENT	>K	K—L	L—M	ELEMENT	>K	K—L	L—M	M—N
H (1)	A			Cd (48)	A	A	B	
He (2)	A			In (49)	A	A	B	
Li (3)	A			Sn (50)	A	A	B	
Be (4)	A			Sb (51)	A	A	B	
B (5)	A			Te (52)	A	A	B	
C (6)	A			I (53)	A	A	B	
N (7)	A			Xe (54)	A	A	B	
O (8)	A			*Cs (55)	A	A	B	
F (9)	A			Ba (56)	A	A	B	
Ne (10)	A			La (57)	A	A	B	
Na (11)	A	A		Ce (58)	A	A	B	
Mg (12)	A	A		Pr (59)	A	A	B	
Al (13)	A	A		Nd (60)	A	A	B	
Si (14)	A	A		*Pm (61)	A	A	B	C
P (15)	A	A		Sm (62)	A	A	B	C
S (16)	A	A		*Eu (63)	A	A	B	C
Cl (17)	A	A		Gd (64)	A	A	B	C
Ar (18)	A	A		*Tb (65)	B	B	B	C
K (19)	A	A		*Dy (66)	B	B	B	C
Ca (20)	A	A		Ho (67)	B	B	B	C
Sc (21)	A	A		Er (68)	B	B	B	C
Ti (22)	A	A		*Tm (69)	B	B	B	C
V (23)	A	A		Yb (70)	B	B	B	C
Cr (24)	A	A		Lu (71)	B	B	B	C
Mn (25)	A	A		Hf (72)	B	B	B	C
Fe (26)	A	A		Ta (73)	A	A	B	C
Co (27)	A	A		W (74)	A	A	B	C
Ni (28)	A	A		*Re (75)	A	A	B	C
Cu (29)	A	A		*Os (76)	A	A	B	C
Zn (30)	A	A	C	Ir (77)	A	A	B	B
*Ga (31)	A	A	C	Pt (78)	A	A	A	B
Ge (32)	A	A	C	Au (79)	A	A	A	B
*As (33)	A	A	C	Hg (80)	A	A	A	B
Se (34)	A	A	C	Tl (81)	A	A	A	B
Br (35)	A	A	C	Pb (82)	A	A	A	B
Kr (36)	A	A	B	Bi (83)	B	A	A	B
*Rb (37)	A	A	B	*Po (84)	B	B	C	C
*Sr (38)	A	A	B	*At (85)	B	B	C	C
*Y (39)	A	A	B	*Rn (86)	B	B	C	C
Zr (40)	A	A	B	*Fr (87)	B	B	C	C
Nb (41)	A	A	B	*Ra (88)	B	B	C	C
Mo (42)	A	A	B	*Ac (89)	B	B	C	C
*Tc (43)	A	A	B	Th (90)	B	B	B	C
*Ru (44)	A	A	B	*Pa (91)	B	B	B	C
Rh (45)	A	A	B	U (92)	B	B	B	C
Pd (46)	A	A	B	*Np (93)	B	B	C	C
Ag (47)	A	A	B	Pu (94)	B	B	C	C

*NO EXPERIMENTAL DATA FOUND FROM 1 keV TO 1 MeV

Figure 4.   Table of estimated uncertainties from parent compilation.

(c)   73 Lα_1 lines for emitters calcium through
      uranium,
(d)   73 Lβ_1 lines for emitters calcium through
      uranium,
(e)   Mα lines for 29 emitters between lanthanum
      and uranium,
(f)   Mβ lines for 29 emitters between lanthanum
      and uranium.

Wavelengths for these lines were taken from Bearden (4).
An example of the tables for the K$_\alpha$ lines is given in
Fig. 5. For the emitters listed across the top, the mass
attenuation or absorption coefficients (in cm^2/gm) are
given for the absorbers listed on the left. Note that
the attenuation coefficients are what many spectroscopists
have usually called "mass absorption coefficients". To
agree with recommended terminology of the ICRU we have
used the title "attenuation cross sections". The
separate tables of absorption coefficients are for true
absorption, i.e., photoelectric absorption plus the
absorption part of incoherent scattering. Each entry in
the tables gives the cross section to three significant
figures followed by a power of ten so that, for example,
1.23+4 means 1.23×10^4 cm^2/gm.

Due to space limitations, it is not possible to
present here the sets of tables for all 281 wavelengths.
The complete tables may be obtained by request from the
authors.

COMPARISON WITH OLDER TABLES

Due primarily to the new low energy calculations
included in the compilation, the values of the cross
sections for low energy lines are generally different
from those provided by older compilations and tables,
especially for absorbers of high atomic number. Figure
6 shows a comparison between cross sections derived from
a compilation (8) based on hydrogenic approximations
(dashed line) and cross sections from the present tables
(solid line). The cross sections are for the K$_\alpha$ emission
energy of carbon at 0.277 keV (44.76 Å). The difference
is quite dramatic for atomic numbers greater than 18.
These large differences between old and new values
disappear for lines with energies above 8 to 10 keV.

MASS ATTENUATION COEFFICIENTS FOR K ALPHA LINES

EMITTER	K	CA	SC	TI	V	CR	MN	FE	CO	NI	CU	ZN	GA	GE
WAVELENGTH	3.74+0	3.36+0	3.03+0	2.75+0	2.50+0	2.29+0	2.10+0	1.94+0	1.79+0	1.66+0	1.54+0	1.44+0	1.34+0	1.25+0
ENERGY (KEV)	3.31+0	3.69+0	4.09+0	4.51+0	4.95+0	5.41+0	5.90+0	6.40+0	6.93+0	7.47+0	8.04+0	8.63+0	9.24+0	9.88+0

ABSORBER

	K	CA	SC	TI	V	CR	MN	FE	CO	NI	CU	ZN	GA	GE
1	5.17-1	4.77-1	4.51-1	4.36-1	4.21-1	4.13-1	4.06-1	4.01-1	3.97-1	3.94-1	3.91-1	3.89-1	3.87-1	3.86-1
2	1.57+0	1.17+0	9.06-1	7.55-1	5.97-1	5.15-1	4.40-1	3.92-1	3.50-1	3.22-1	2.97-1	2.81-1	2.66-1	2.54-1
3	5.52+0	3.92+0	2.86+0	2.26+0	1.63+0	1.31+0	1.02+0	8.39-1	6.82-1	5.83-1	4.93-1	4.35-1	3.84-1	3.45-1
4	1.60+1	1.13+1	9.21+0	6.42+0	4.54+0	3.57+0	2.69+0	2.15+0	1.67+0	1.38+0	1.11+0	1.08+0	9.35-1	6.69-1
5	3.47+1	2.46+1	1.79+1	1.40+1	9.87+0	7.74+0	5.80+0	4.58+0	3.53+0	2.87+0	2.26+0	1.88+0	1.54+0	1.28+0
6	6.75+1	4.82+1	3.51+1	2.75+1	1.95+1	1.57+1	1.15+1	9.05+0	6.96+0	5.62+0	4.41+0	3.63+0	2.95+0	2.43+0
7	1.11+2	7.98+1	5.84+1	4.58+1	3.26+1	2.57+1	1.93+1	1.52+1	1.17+1	9.44+0	7.39+0	6.07+0	4.91+0	4.02+0
8	1.68+2	1.21+2	8.69+1	7.00+1	5.00+1	3.95+1	2.97+1	2.35+1	1.81+1	1.46+1	1.15+1	9.41+0	7.60+0	6.21+0
9	2.23+2	1.61+2	1.18+2	1.05+2	6.73+1	5.32+1	4.02+1	3.19+1	2.46+1	1.99+1	1.56+1	1.28+1	1.04+1	8.48+0
10	3.14+2	2.29+2	1.70+2	1.34+2	9.68+1	7.67+1	5.82+1	4.62+1	3.58+1	2.90+1	2.28+1	1.86+1	1.52+1	1.24+1
11	3.96+2	2.89+2	2.15+2	1.70+2	1.23+2	9.78+1	7.43+1	5.91+1	4.59+1	3.72+1	2.93+1	2.41+1	1.96+1	1.60+1
12	5.18+2	3.79+2	2.83+2	2.25+2	1.63+2	1.30+2	9.90+1	7.89+1	6.14+1	4.99+1	3.93+1	3.24+1	2.63+1	2.15+1
13	6.33+2	4.67+2	3.50+2	2.79+2	2.04+2	1.63+2	1.24+2	9.94+1	7.75+1	6.30+1	4.98+1	4.11+1	3.34+1	2.74+1
14	7.94+2	5.84+2	4.38+2	3.49+2	2.55+2	2.03+2	1.55+2	1.24+2	9.67+1	7.87+1	6.22+1	5.14+1	4.18+1	3.43+1
15	9.20+2	6.81+2	5.12+2	4.09+2	3.00+2	2.40+2	1.84+2	1.47+2	1.15+2	9.41+1	7.45+1	6.17+1	5.02+1	4.13+1
16	1.12+3	8.28+2	6.28+2	4.99+2	3.67+2	2.94+2	2.26+2	1.81+2	1.42+2	1.16+2	9.22+1	7.64+1	6.22+1	5.12+1
17	1.25+3	9.28+2	7.02+2	5.62+2	4.14+2	3.33+2	2.56+2	2.06+2	1.62+2	1.32+2	1.05+2	8.72+1	7.12+1	5.86+1
18	1.37+3	1.04+3	7.85+2	6.28+2	4.63+2	3.72+2	2.87+2	2.30+2	1.81+2	1.48+2	1.18+2	9.77+1	7.97+1	6.57+1
19	1.79+2	1.23+3	9.43+2	7.56+2	5.61+2	4.51+2	3.49+2	2.81+2	2.22+2	1.82+2	1.45+2	1.20+2	9.85+1	8.13+1
20	2.34+2	1.60+2	1.08+3	8.74+2	6.54+2	5.28+2	4.10+2	3.31+2	2.61+2	2.14+2	1.71+2	1.43+2	1.17+2	9.66+1
21	2.34+2	1.75+2	1.32+2	8.74+2	6.85+2	5.52+2	4.34+2	3.51+2	2.78+2	2.29+2	1.83+2	1.53+2	1.25+2	1.03+2
22	2.65+2	1.97+2	1.50+2	1.20+2	8.90+1	6.06+2	4.75+2	3.86+2	3.07+2	2.54+2	2.04+2	1.71+2	1.41+2	1.17+2
23	2.92+2	2.17+2	1.93+2	1.55+2	9.82+1	7.67+1	5.20+2	4.27+2	3.39+2	2.80+2	2.25+2	1.86+2	1.55+2	1.28+2
24	3.42+2	2.54+2	2.15+2	1.73+2	1.28+2	9.36+1	7.29+1	4.93+2	3.92+2	3.24+2	2.61+2	2.18+2	1.80+2	1.49+2
25	3.81+2	2.83+2	2.48+2	1.99+2	1.48+2	1.19+2	9.29+1	7.55+1	4.21+2	3.89+2	3.16+2	2.35+2	1.94+2	1.61+2
26	4.40+2	3.27+2	2.73+2	2.19+2	1.63+2	1.32+2	1.03+2	8.33+1	6.01+1	3.89+2	3.53+2	2.65+2	2.19+2	1.83+2
27	4.04+2	3.60+2	3.14+2	2.52+2	1.87+2	1.51+2	1.17+2	9.53+1	6.63+1	5.43+1	5.06+1	2.96+2	2.44+2	2.03+2
28	5.58+2	4.14+2	3.14+2	2.70+2	2.01+2	1.62+2	1.26+2	1.03+2	7.58+1	6.20+1	5.46+1	3.15+2	2.64+2	2.21+2
29	5.90+2	4.44+2	3.36+2	2.99+2	2.23+2	1.80+2	1.41+2	1.14+2	9.12+1	6.75+1	5.71+1	4.58+1	2.78+2	2.33+2
30	6.53+2	5.52+2	3.71+2	3.16+2	2.36+2	1.93+2	1.50+2	1.22+2	9.69+1	7.56+1	6.12+1	5.15+1	2.60+2	2.50+2
31	7.29+2	5.52+2	3.91+2	3.78+2	2.57+2	2.08+2	1.62+2	1.32+2	1.04+2	8.02+1	6.48+1	5.44+1	4.51+1	3.78+1
32	7.96+2	6.58+2	4.67+2	4.05+2	2.84+2	2.30+2	1.79+2	1.45+2	1.15+2	8.61+1	6.94+1	5.83+1	5.03+1	4.04+1
33	8.68+2	7.05+2	5.01+2	4.48+2	3.04+2	2.46+2	1.92+2	1.55+2	1.24+2	9.54+1	7.70+1	6.47+1	5.36+1	4.48+1
34	9.29+2	7.79+2	5.56+2	4.85+2	3.37+2	2.73+2	2.12+2	1.72+2	1.36+2	1.02+2	8.25+1	6.93+1	5.74+1	4.80+1
35	1.03+3	7.99+2	6.05+2	5.34+2	3.60+2	2.91+2	2.26+2	1.84+2	1.46+2	1.13+2	9.09+1	7.63+1	6.32+1	5.28+1
36	1.08+3	8.83+2	6.05+2	5.34+2	3.95+2	3.18+2	2.47+2	2.00+2	1.58+2	1.21+2	9.74+1	8.18+1	6.78+1	5.67+1
37	1.19+3	9.59+2	7.24+2	5.80+2	4.28+2	3.45+2	2.68+2	2.17+2	1.72+2	1.31+2	1.05+2	8.81+1	7.28+1	6.08+1
38	1.30+3	1.05+3	7.92+2	6.35+2	4.69+2	3.78+2	2.93+2	2.37+2	1.88+2	1.42+2	1.14+2	9.55+1	7.90+1	6.59+1
39	1.42+3	1.14+3	8.60+2	6.88+2	5.07+2	4.08+2	3.16+2	2.55+2	2.00+2	1.52+2	1.25+2	1.04+2	8.63+1	7.20+1
40	1.55+3	1.25+3	9.41+2	7.53+2	5.55+2	4.46+2	3.45+2	2.79+2	2.21+2	1.66+2	1.34+2	1.12+2	9.24+1	7.70+1
41	1.69+3	1.33+3	1.00+3	8.02+2	5.90+2	4.74+2	3.67+2	2.96+2	2.34+2	1.82+2	1.46+2	1.22+2	1.01+2	8.41+1
42	1.81+3	1.40+3	1.06+3	8.47+2	6.25+2	5.03+2	3.89+2	3.15+2	2.49+2	1.92+2	1.54+2	1.29+2	1.06+2	8.86+1
43	1.89+3	1.51+3	1.14+3	9.10+2	6.71+2	5.40+2	4.18+2	3.38+2	2.67+2	2.05+2	1.65+2	1.38+2	1.14+2	9.49+1
44	2.00+3	1.60+3	1.21+3	9.66+2	7.14+2	5.75+2	4.46+2	3.61+2	2.86+2	2.26+2	1.77+2	1.48+2	1.22+2	1.02+2
45	1.77+3	1.72+3	1.31+3	1.05+3	7.70+2	6.18+2	4.78+2	3.86+2	3.05+2	2.36+2	1.90+2	1.59+2	1.31+2	1.09+2
46	1.37+3	1.72+3	1.31+3	1.05+3	7.70+2	6.18+2	4.78+2	3.86+2	3.05+2	2.51+2	2.01+2	1.68+2	1.39+2	1.15+2
47	4.14+2	1.52+3	1.39+3	1.11+3	8.22+2	6.61+2	5.11+2	4.13+2	3.27+2	2.69+2	2.16+2	1.81+2	1.49+2	1.24+2

Figure 5.   Sample from tables at selected wavelengths.

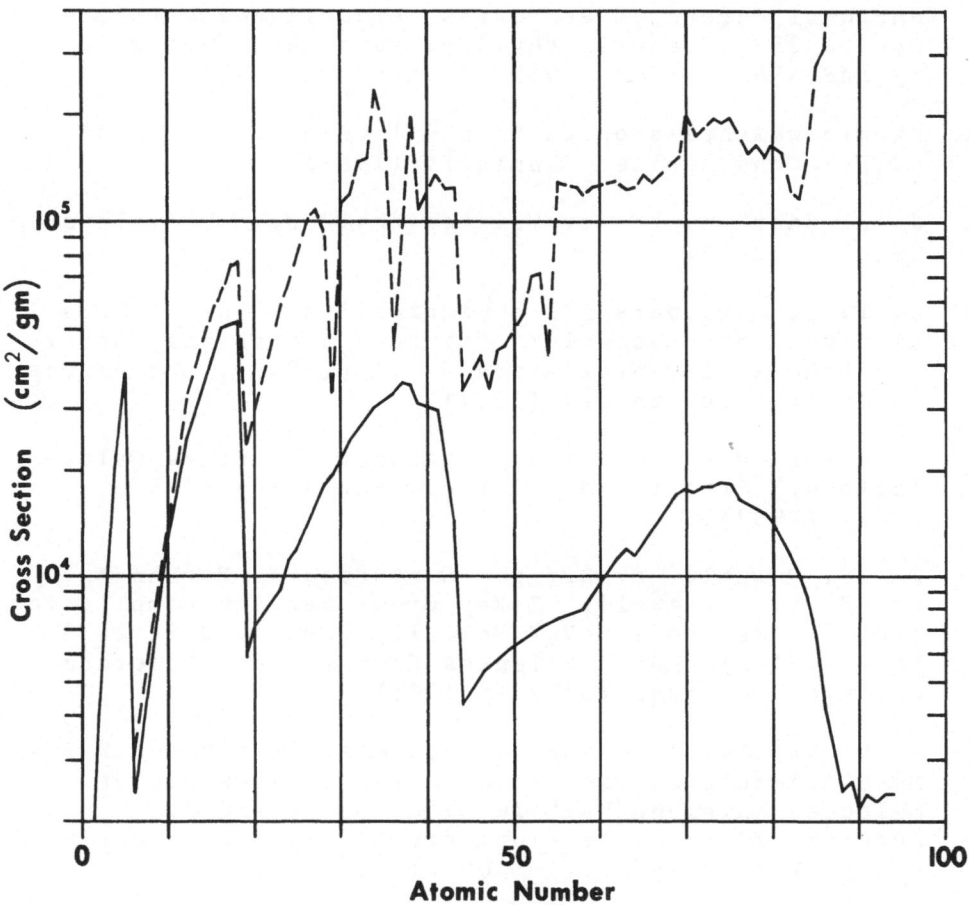

Figure 6.   Comparison of old (dashed line) and revised
            (solid line) cross sections for carbon K$_\alpha$ line.

     In general, the reliability of the values in the
tables must be assessed from the table given in Fig. 4,
which lists the suggested uncertainties for the parent
compilation, or in the manner suggested in the section
"Uncertainties in the Cross Sections" above.

                        REFERENCES

1.   Wm. J. Veigele, E. Briggs, L. Bates, E. M. Henry
     and B. L. Bracewell, "X-Ray Cross Section Compilation
     from 0.1 keV to 1 MeV," Vol. I, Revision 1, KN-71-
     431(R). Kaman Sciences Corporation, Colorado Springs,
     Colorado. (31 July 1971).

2.   National Bureau of Standards, Miscellaneous Publi-
     cation 253, "General Physical Constants Recommended
     by NAS-NRS." (Nov. 1965, Revised May 1967).

3.   Atomic weights adopted by the International Union
     of Pure and Applied Chemistry (1961).

4.   J. A. Bearden, "X-Ray Wavelengths," Rev. Mod. Phys.
     39, 78 (1967).

5.   D. T. Cromer, personal communication. The methods
     used were discussed in: "Compton Scattering Factors
     for Spherically Free Atoms." LA-DC-8819, Los Alamos
     Scientific Laboratory (1967).

6.   F. Herman and S. Skillman, "Atomic Structure Calcu-
     lations," Prentice-Hall, Inc., Englewood Cliffs,
     N. J. (1963).

7.   Wm. J. Veigele, E. Briggs, L. Bates, E. M. Henry
     and B. L. Bracewell, "X-Ray Cross Section Compilation
     from 0.1 keV to 1 MeV," Vol. II, Rev. 1, DASA 2433,
     KN-71-431(R), Kaman Sciences Corporation, Colorado
     Springs, Colorado (31 July 1971).

8.   B. L. Bracewell and Wm. J. Veigele, "Tables of X-Ray
     Mass Attenuation Coefficients for 87 Elements at
     Selected Wavelengths," in E. L. Grove and A. J.
     Perkins, Editors, Developments in Applied Spectro-
     scopy, Vol. 9, p. 357 - 400 Plenum Press (1971).

FOOTNOTES

* Work supported by the Defense Atomic Support Agency.

+ Also at University of Colorado, Colorado Springs,
  Colorado.

A COMPUTERIZED TECHNIQUE OF PLOTTING A COMPLETE POLE FIGURE BY AN X-RAY REFLECTION METHOD

J. J. Klappholz, S. Waxman, and C. Feng

Picatinny Arsenal

Dover, New Jersey   07801

ABSTRACT

The technique of plotting a complete pole figure composed of data points in both longitude and latitude from 0 to 180 degrees by a computer program is described. X-ray data were obtained by a reflection method from a specimen cut into three sections mutually perpendicular to one another. The computer program calculates each position in the pole figure based on the time rate of change of the tilt angle Φ and the spin angle α which are transformed into rectangular coordinates.

The advantage of the present technique is to minimize the x-ray intensity loss due to geometric defocusing, since each section of a given specimen is required to tilt not more than 55 degrees. Due to the fact that a complete pole figure is plotted, one is allowed to examine the symmetry or lack of symmetry in a given specimen with respect to a set of references axes.

INTRODUCTION

Pole figures may be constructed from data obtained by either transmission or reflection x-ray methods. On numerous occasions, complete pole figures were plotted with a combination of the two, i.e. the central portions were plotted from reflection data whereas the outer portions were plotted from transmission data. To determine accurately the x-ray intensity for the various locations in a pole figure, provisions must be made to correct for losses of intensity due to mass absorption and geometric defocusing. For transmission specimens, the loss of intensity is primarily due to

365

absorption whereas for reflection specimens, the loss of intensity is primarily due to defocusing (1-7). Difficulties will arise, however, in establishing a uniform system to represent the x-ray intensity of the two regions if the data are to be obtained by two different techniques and corrected according to two different principles. It is, therefore, desirable to employ one type of specimen. In this regard, successful methods have been devised to construct a complete pole figure using the reflection techniques (8-10).

Holland, Engler and Powers (11) developed a computerized program of plotting a quadrant of the pole figure from the reflection data of three orthogonally cut specimens. The data are not suitable to plot a complete pole figure without assuming symmetry with respect to both longitude and latitude, which may not be the case, particularly for materials with a more complex texture. In this paper the method of Holland, Engler and Powers (11) is modified to plot a complete pole figure by computer, showing data points from 0 to 180 degrees along longitude and latitude lines with corrections for background and intensity loss due to geometric defocusing.

EXPERIMENTAL PROCEDURE

To plot a complete pole figure by the reflection method, a minimum of three specimens are required. These three specimens are cut from one sample being investigated. For the present study, they were cut mutually perpendicular to one another, with reference axes N, L and T which correspond to the Cartesian coordinates (x, y, z) as shown in Figure 1.

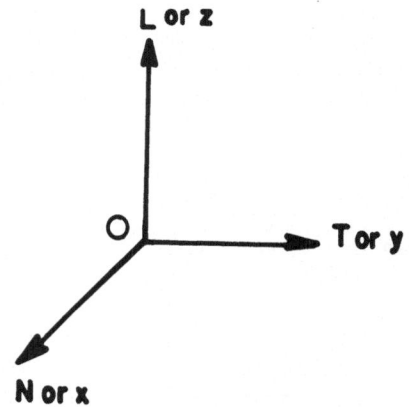

The "zero" or starting position, i.e. $\alpha = 0$; $\Phi = 0$, of these specimens are shown in Figures 2a, 2b and 2c. During the experiment, each specimen was allowed to oscillate and to rotate clockwise about its surface normal. The angle of rotation was α. The specimen was also simultaneously tilted counterclockwise through

an angle Φ, which was measured between the surface normal and the
plane of x-rays. The maximum tilted angle for each specimen was
55 degrees. This provided sufficient data for a complete pole
figure.

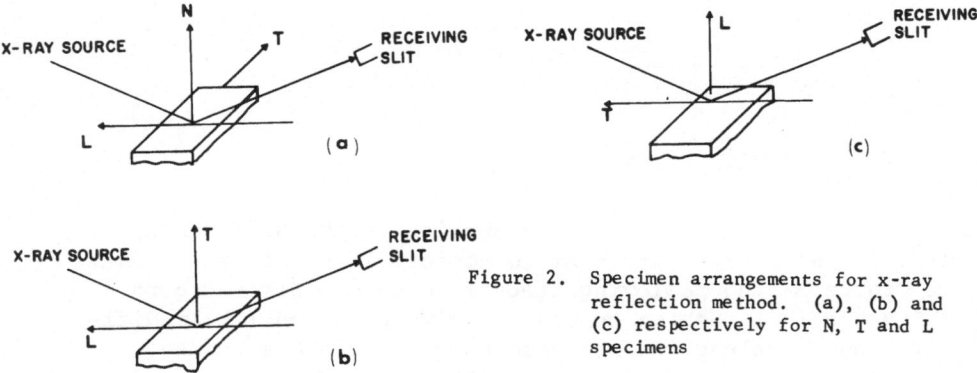

Figure 2.   Specimen arrangements for x-ray
reflection method. (a), (b) and
(c) respectively for N, T and L
specimens

COMPUTER PLOTTING TECHNIQUE

The computer was programmed to analyze the x-ray data collected on magnetic tape at a constant time interval. The program
accepted data in the following sequence: the normal (N) specimen,
followed by the transverse (T) specimen, and finally the longitudinal (L) specimen. From the input which consisted of the
constant time rates of change for α and Φ, the intensity data, and
the time interval of sampling, the program was able to calculate
an α and a Φ value associated with a respective intensity. From
these data, the values α and Φ can be transformed into rectangular
coordinates using the transformation equations (11).

$$\left. \begin{array}{l} Y = (\sin \Phi \cos \alpha)/(\cos \Phi + 1) \\ Z = (\sin \Phi \sin \alpha)/(\cos \Phi + 1) \end{array} \right\} \qquad (1)$$

$$Y = \frac{\sin \alpha \sin \Phi}{\sin \Phi \cos \alpha + 1}$$

$$\left.\vphantom{\frac{a}{b}}\right\} \qquad (2)$$

$$Z = \frac{-\cos \Phi}{\sin \Phi \cos \alpha + 1}$$

$$Y = \frac{\sin \alpha \sin \Phi}{\sin \Phi \cos \alpha + 1}$$

$$\left.\vphantom{\frac{a}{b}}\right\} \qquad (3)$$

$$Z = \frac{\cos \Phi}{\sin \Phi \cos \alpha + 1}$$

Equations (1), (2) and (3) were applied to the N, T and L specimens, respectively. In order to conform with the master chart shown in Figure 3, the following steps were used. Since Figure 3 was constructed in a fashion consistent with equation (1), modification of α and Φ values was not necessary for the N specimen. For the T specimen, however, equation (2) was used where Y = -Z with an operation on Φ depending on the associated α. If α was less than or equal to 90 degrees, and also greater than 270 but less than or equal to 360 degrees, the corresponding Φ was unchanged. On the other hand, if α was greater than 90 degrees but less than or equal to 270 degrees, the corresponding Φ was changed by subtracting 180 degrees. For the L specimen, equation (3) was used. Φ was changed depending upon the corresponding α in the same manner as the T specimen. Now the program has transformed the values of α and Φ associated with respective intensities to rectangular coordinates. The next task for the program was to correct the intensity values for background and geometric defocusing which is known to be a primary function of Φ (3, 4, 7).

In the present study after correction for background, the method described in reference (4) for correcting the loss of intensity due to geometric defocusing was used, where

$$I/I1 = 1 - [W \cos \theta \tan \Phi / D\Delta(2\theta)] \qquad (4)$$

$$I/Io = D\Delta(2\theta)/4W \tan \Phi \cos \theta \qquad (5)$$

Equation (4) applies to the condition when $I/Io < 0.5$, whereas equation (5) applies to the condition when $I/Io \geq 0.5$. Io is intensity of a random sample at $\Phi = 0$, I is intensity of a random sample at some other angles where $\Phi \neq 0$. W is the effective width of x-ray beam which was determined to be 0.12 cm; θ is the Bragg angle; $\Delta 2\theta$ is the line breadth, and D is the focal distance which was 17 cm for the present study. The values of W, $\Delta 2\theta$, D, θ, the background correction, and a value of Φ at $I/Io = 0.5$ were used as input data for the program. The program then used equations (4)

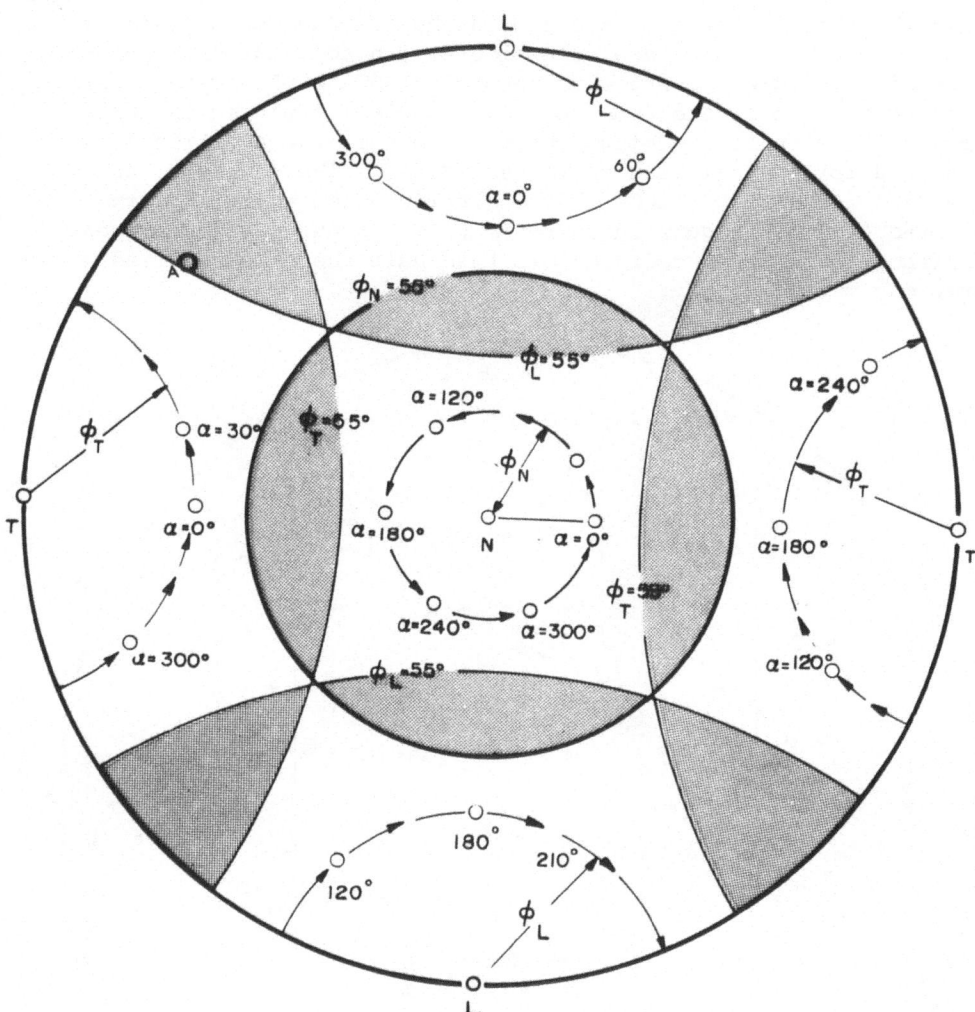

Figure 3. Schematic diagram showing the
positions of α and Φ with respect
to N, T, and L specimens.

and (5) to calculate I/Io. Next the background was subtracted
from the intensity which was divided by I/Io to get a corrected
intensity for both background and geometric defocusing. Finally,
the program determined the maximum corrected intensity, divided
this into each of the other intensities, multiplied by 100 to get
values in percentage form. At this point the computer had enough
information to plot the pole figure.

The overall computer operation required two steps. First, a CDC 6500 computer performed the calculations as described previously from the input x-ray data and stored the information on magnetic tape. To plot an accurate pole figure with adjacent data points spreading 5 degrees apart with respect to Φ and 15 degrees apart with respect to α, the computer will complete the task in approximately 29 seconds. Doubling the data points, the additional time required for the calculation is negligible. The magnetic tape was then fed into a CALCOMP 770 digital plotter which in turn plotted the pole figure in approximately 5 minutes. The present program, therefore, required less than half the time required by previous methods (13).

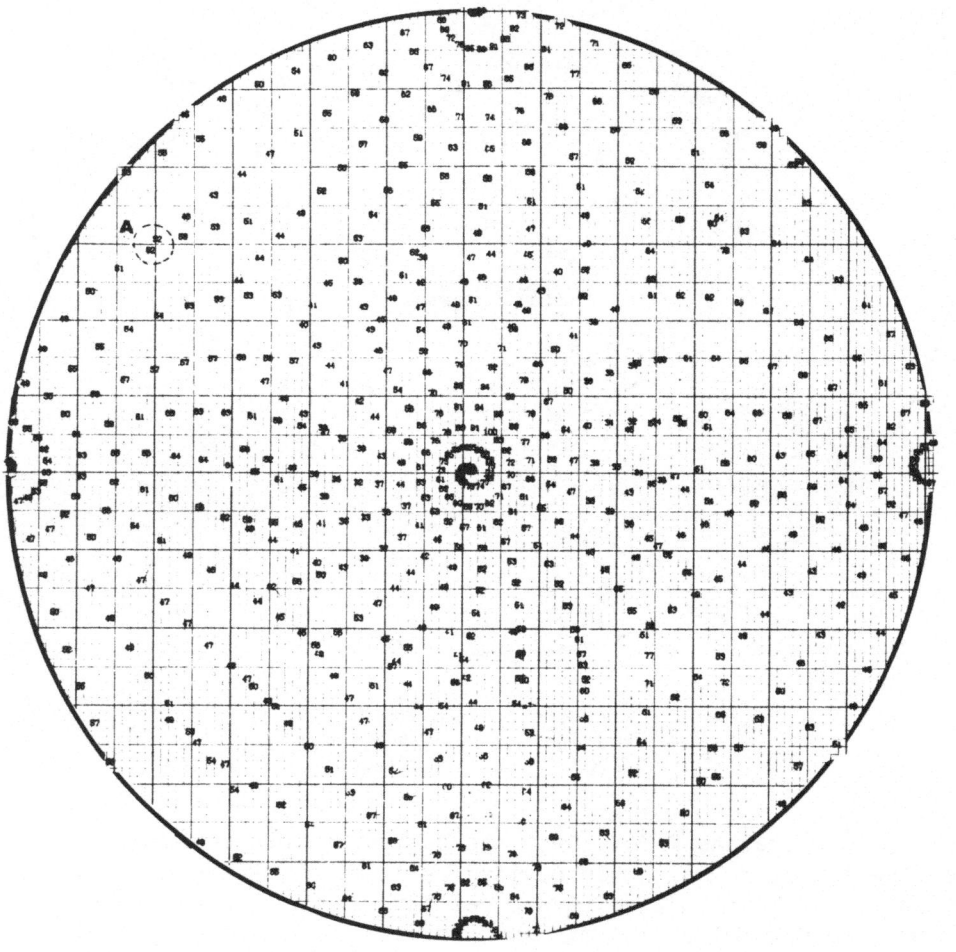

Figure 4. The computer plotted (111) pole figure of an annealed copper cone.

RESULTS AND DISCUSSION

The results of the computer output for a copper cone shaped structure after annealing at 875°F is shown in Figure 4. Several important facts may be found by superimposing Figures 3 and 4. The surface normal, N, at a given point of the cone is located at the center of the two superimposed figures. Following the arrangements shown in Figures 2a, 2b, and 2c, the corresponding positions of L, a direction in the cone surface and the projection of which is the cone axis, and T, a direction tangent to the cone surface and parallel to the base, are located at the top and the left end of the equator in the superimposed figure, respectively.

Identification of texture elements may be made by following the standard methods of locating zones of equal intensity from Figure 4. For the annealed cone in this study, the major texture element was identified as $\{321\}$ $<1\bar{2}1>$, meaning that the $\{321\}$ planes are parallel to the tangent of the cone surface and that the L direction is coincident with a $<211>$ direction. This was found to be in general agreement with those reported in the literature for annealing copper (12).

It is interesting to note that the shaded area in Figure 3 may be used to check how well the intensity has been corrected, since the x-ray data from more than one specimen are located at the same region in the pole figure. For example, the area circled in A of Figure 3 represented a reading of T specimen when $\Phi \sim 35$ deg., $\alpha \sim 80$ deg., also a reading of L specimen when $\Phi \sim 55$ deg., $\alpha \sim 130$ deg. This same region, A, is also circled in Figure 4. For the T specimen, the correction was calculated on the basis of $\Phi \sim 35$ degrees, and for the L specimen, the correction calculation was based on $\Phi \sim 55$ degrees. The two intensities after correction were both 52. This indicates that the method of intensity correction is satisfactory.

ACKNOWLEDGMENTS

This study was partially supported by the Quality Assurance Directorate, Picatinny Arsenal. The authors are grateful to D. L. Bagnoli, who originated this project.

REFERENCES

1.   B. F. Decker, E. T. Asp and D. Harker, "Preferred Orientation Determination Using a Geiger Counter X-ray Goniometer," J. Appl. Phys. <u>19</u>, 388-392, 1948.

2.   L. G. Schulz, "Determination of Preferred Orientation in Flat
      Transmission Samples Using a Geiger Counter X-ray Spectrometer,"
      J. Appl. Phys. 20, 1033-1036, 1949.

3.   W. P. Chernock and P. A. Beck, "Analysis of Certain Errors in
      the X-ray Reflection Method for the Quantitative Determination
      of Preferred Orientations," J. Appl. Phys. 23, 341-345, 1952.

4.   C. Feng, "Determination of Relative Intensity in X-ray Reflec-
      tion Study," J. Appl. Phys. 36, 3432-3455, 1965.

5.   K. Aoki, S. Hayami and M. Matsuo, "Improvement of Accuracy in
      Presentation of Conventional Pole Figures," J. B. Newkirk and
      G. R. Mallett, Editors, Advances in X-ray Analysis, Vol. 10,
      342-353, 1966.

6.   R. H. Bragg and C. M. Packer, "Quantitative Determination of
      Preferred Orientation," J. Appl. Phys. 35, 1322-1328, 1964.

7.   E. Tenckhoff, "Defocusing for the Schulz Technique of Determin-
      ing Preferred Orientation," J. Appl. Phys. 41, 3944-3948, 1970.

8.   J. T. Norton, "A Technique for Quantitative Determination of
      Texture of Sheet Metals," J. Appl. Phys. 19, 1176-1178, 1948.

9.   M. Field and M. E. Merchant, "Reflection Method of Determining
      Preferred Orientation on the Geiger Counter Spectrometer,"
      J. Appl. Phys. 20, 741-744, 1949.

10.   L. G. Schulz, "A Direct Method of Determining Preferred Orien-
      tation of a Flat Reflection Sample Using a Geiger Counter X-ray
      Spectrometer," J. Appl. Phys. 20, 1030-1033, 1949.

11.   J. R. Holland, N. Engler and W. Powers, "The Use of Computer
      Techniques to Plot Pole Figures," W. M. Mueller, Editor,
      Advances in X-ray Analysis, Vol. 4, 74-84, 1960.

12.   P. A. Beck and H. Hu, "The Origin of Recrystallization Texture,"
      ASM Seminar on Recrystallization, Grain Growth and Textures,
      October 1965, Cleveland, Ohio, 393-433.

13.   Hung-Chi Chao, "Direct Printout X-ray Pole Figures from Digital
      Computers," C. S. Barrett, J. B. Newkirk and G. R. Mallett,
      Editors, Advances in X-ray Analysis, Vol. 12, 391-403, 1968.

PROTON-INDUCED X-RAY EMISSION SPECTROSCOPY IN ELEMENTAL TRACE ANALYSIS

T.B. Johansson, R. Akselsson and S.A.E. Johansson

Department of Nuclear Physics

Lund Institute of Technology, Lund, Sweden

ABSTRACT

Using protons in the MeV range as excitation source and a high resolution Si(Li) detector, X-ray emission spectroscopy is shown to be capable of analysing many elements with $Z > 15$ simultaneously at the 10^{-12} g level. This work discusses a theoretical lower limit of detection at moderate proton energies and gives examples of possible applications: analysis of the elemental composition of air-borne particles as a function of particle size, oil slick identification, and analysis of water and blood serum.

INTRODUCTION

X-ray emission analysis has long been applied to trace element analysis. Photons and electrons have hitherto mostly been used as excitation sources and the characteristic X-rays have been analyzed by crystals or low resolution proportional and scintillation counters.

The introduction of high resolution semiconductor detectors has, however, opened the way for fast and efficient detection of the X-rays. This has made multi-elemental analysis much more attractive since the time required is greatly reduced. In recent years, heavier particles have also been used for excitation of characteristic X-rays from a sample.

Important factors determining lower limits of detection are X-ray production cross sections, the available flux of exciting particles or photons and background conditions.

The cross sections for photon excitation are rather large. When photons are used, the limitations come from the difficulty of producing strong radioactive sources and from the background of scattered photons.

Electrons are used in micro-analysis in connection with electron-microscopes and extremely small amounts ($\sim 10^{-14}$ g) have been reported as the lower limit of detection. Very small samples are then analyzed, however, so the concentrations detectable tend to remain in the 0.01 % region. A serious drawback with electrons is the background arising from bremsstrahlung.

In a report to this conference last year, Needham and Sartwell (1) discussed proton excitation in the energy region of 100 keV. In a recent report we have discussed the use of MeV protons in trace analysis (2). Watson et al have investigated the analytical capabilities of high energy alpha particles as exciting particles (3). Heavy particles produce considerably less intense bremsstrahlung than electrons thereby reducing the background. The ionization cross sections are also higher, thus making these sources of excitation attractable. Contrary to photons, intense beams of charged particles can easily be produced.

The cross sections for X-ray production by heavy ion irradiation are two to three orders of magnitude larger than the cross sections for protons in the same energy range (4). Extremely small quantities should in principle be detectable using heavy ions although the amount of material that can be analyzed is limited by the small penetration depth.

In the present paper we give a further discussion of proton excitation and give a few examples of possible applications.

THEORETICAL

A theoretical lower limit of detection can be calculated in the following way. The total number of counts in a peak corresponding to A atoms of an element in a thin sample is

$$N = A\; n\; \sigma\; \omega\; k(\Omega/4\pi)\varepsilon \tag{1}$$

where n is the total number of protons per cm^2, σ the cross section for ionization of the shell in question, ω the fluorescence yield, k the relative transition probability for the particular X-ray peak used in the measurement, Ω the solid angle subtended by the detector and ε the efficiency of the detector.

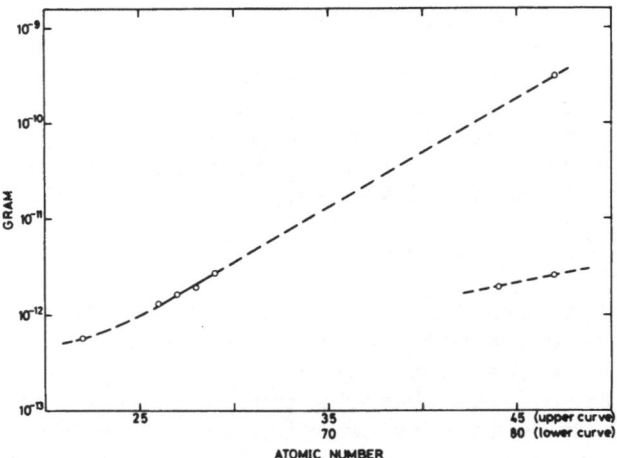

Figure 1. Lower limit of detection for different elements.

In figure 1 we present lower limits of detection for a part of the periodic table. The calculations are made for protons of 2.5 MeV, the energy given by our accelerator. For the beam current and analyzing time, we have chosen values of 1 µA and 30 minutes giving a total charge of 1.8 mC through the 4 mm diameter collimator. The detector is also assumed to be placed 50 mm from the target.

The number of counts, N, needed to give a well defined peak is assumed to be 100. This is a conservative figure at X-ray energies above 10 keV but at lower energies the background contains several peaks from contamination in the sample backing which in some cases will present problems.

The calculations are made for the most intense peak from each element using relative X-ray transition probabilities from Hansen et al (5). The X-ray production cross sections used are taken from recent measurements in this laboratory (6). They are in agreement with measurements by Bissinger et al (7).

For heavy elements, the L X-rays are used since the L X-ray production cross sections are much larger than the corresponding K X-ray cross sections and the L X-ray energy is more suitable for analyzing with Si(Li) detectors.

As demonstrated by Bissinger et al the ionization cross sections increase with the proton energy and have a maximum which for medium-heavy elements e.g. nickel occurs at $\sim$ 15 MeV, where the cross section is 1500 barns compared to our value of 200 barns at 2.5 MeV. Thus, by increasing the proton energy it is possible to reduce the lower limits of detection further.

EXPERIMENTAL

For quantitative analysis it is necessary to know the
distribution of target atoms over the irradiated surface and
the distribution of particles in the beam. If one of these
quantities has a rectangular distribution the calculations are
simplified. From formula (1) we have

$$N = \text{const} \int n(s) \, A(s) \, dS \qquad\qquad (2)$$

where $n(s)$ and $A(s)$ are functions describing the distribution
of protons and target atoms over the irradiated area, S. As we
usually do not know the distribution of the target atoms, we
have chosen to arrange so that the protons have a rectangular
distribution. This is made by sweeping the well-collimated beam
in two perpendicular directions using deflecting voltages in
the form of symmetrical triangles with frequencies of 116 Hz
in one direction and 1250 Hz in the other. This arrangement
gives the desired proton distribution (8). Equation (2) then
simplifies to

$$N = \text{const} \, n \int A(s) \, dS = \text{const} \, n \, A_{tot} \qquad\qquad (3)$$

which simply requires that all the atoms to be analyzed are in
the irradiated area. This condition is fulfilled in the measurements.

The set of collimators just before the target includes a
quartz glass with a hole slightly larger than the collimators.
By turning a protecting tantalum collimator away, the intensity
distribution in the beam can be observed and the sweeping voltage adjusted to a proper value. In selecting this, it is important to bear in mind that the momentary counting rate in
detector and electronics is determined by the momentary beam
current.

The charge passing through the last collimator is collected
in an isolated cup (see figure 2) earthed through a commercial
current digitizer. Thus the current passing through the target
can be measured accurately. The current digitizer has been
checked to be better than 1 %. Behind the last collimator a
ring of pure aluminum is placed. We keep this plate at a potential of -150 V to suppress secondary electrons from the target,
which would otherwise interfere with the current measurement.
This interference is in the order of 10 %.

The samples to be analyzed are placed on thin self-support-
ing carbon foils $(20 \, \mu g/cm^2)$, which cover a hole in a pure
aluminum frame. Six frames are placed on a ladder, which is
inserted in the irradiation chamber through a vacuum lock.

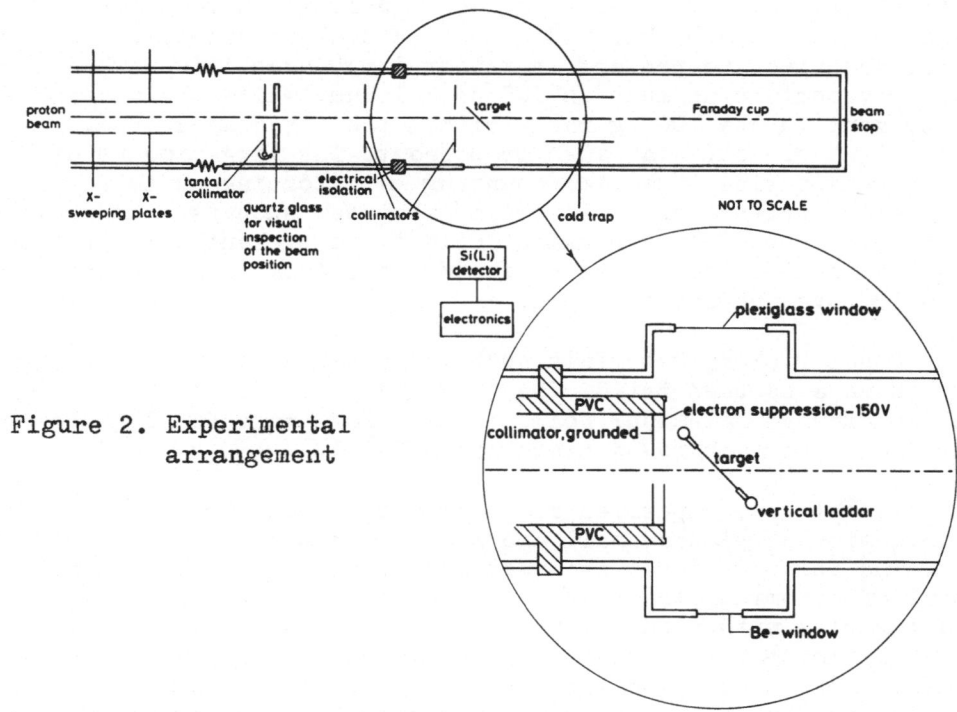

Figure 2. Experimental
          arrangement

The samples are thin to protons and X-rays. This is implicity assumed in formula (1), where we should otherwise have to integrate over the proton.

The detector is a Si(Li) detector with a cooled FET-transistor in the first preamplifier stage. The pulses are amplified in conventional amplifiers and analyzed with a multi-channel analyzer. The detector area is 26 mm^2. The system resolution is 0.26 keV FWHM. The spectra are punched on paper tapes for further processing in a computer. The analyzing program is a version of SAMPO (9). Some confusion in analyzing the spectra may arise from interferences between K_α and K_β of adjacent elements or between K and L radiation from different elements. For example $L_{\alpha 1}$ of lead and $K_{\alpha 1}$ of arsenic are separated by only 8 eV. When such ambiguity arises, it is necessary to make use of the relative X-ray transition probabilities to assist in unravelling the spectrum.

Background

A low background is necessary if small quantities are to be measured. The background from our carbon foils is 300 counts per keV and mC at energies over 10 keV with a target-detector

distance of 5 cm. At lower energies impurities in the carbon
foils increase the background. Elements such as calcium, iron,
copper and zinc are present in rather large quantities. The
foils may contain as much as 100 ng calcium, while the others
are present in the 100 pg range within the irradiated area.
Other light elements may also be present. The large amount of
calcium gives rise to a high counting rate forcing us to use
lower beam currents or to insert an absorber such as aluminum.
The preparation of pure backing foils is at present a major factor limiting the usefulness of the method. Work is now going
on to improve the carbon foils.

Other backing materials such as hostaphan or thin aluminum
foils have also been tried. The background from hostaphan
(350 $\mu g/cm^2$) and aluminum foils (1 μm thick) were found to be
four times the background from carbon foils.

To get some experimental information about the lower limit
of detection we placed small amounts of chromium on a carbon
foil. The foil was first analyzed and found to contain a small
amount of chromium. The background in this energy region is
unfavourable, with a chromium peak situated on the side of a
huge calcium peak. This latter peak was reduced by the use of
aluminum absorbers but was nevertheless about 1000 times as
high as the chromium peak. Various amounts of chromium were
then placed on a carbon foil using a diluted water solution
of $Cr(NO_3)_3$, molybdenum being added to the solution as a
carrier. Figure 3 shows the peak imposed on the side of the
calcium peak in a run after the addition of 10 pg chromium.
The peak was obtained in a two hours run with a total charge
of 3.5 mC passing through the target. The collimator diameter
was 5.5 mm and the target-detector distance 50 mm with a
30 μm aluminum-absorber inserted. The peak in figure 3 corresponds to about 30 pg chromium. Hence most of it is background.

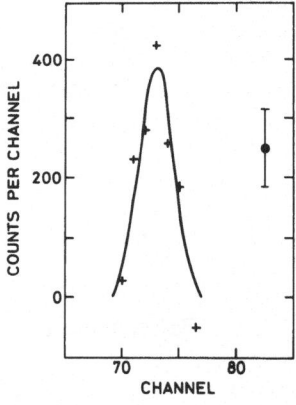

Figure 3. A chromium-peak after background subtraction.
The peak corresponds to 30 pg of chromium.

This demonstrates that the detection limit is a few times 10^{-12}g but that it is difficult to make use of this low value because of background problems.

APPLICATIONS

Work is in progress on a number of different problems to investigate various possible applications of the technique. Some preliminary results will be reported here.

Analysis of Air-borne Particles

In a previous work we showed a spectrum from a carbon foil which had been placed outdoors for one day (2). In all, 13 elements could be detected. This method collects sedimenting

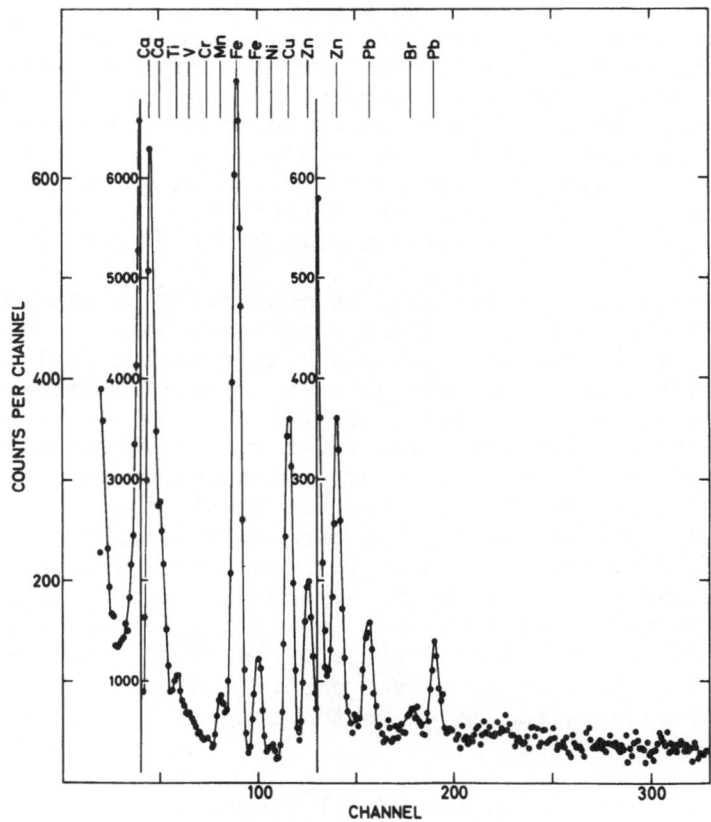

Figure 4. A pulseheight spectrum from analysis of air-borne particles with a diameter of 5-9 μm. The background has not been subtracted. Mn, Zn, Br and Pb are present in the particles. The other elements are impurities in the carbon foil.

Table 1
The mass of some elements in air-borne particles of different
sizes from one cubicmeter of air. The sample is collected a
rainy day in June 1971 in the outskirts of Lund.

Aerodynamic particle size interval (μm) (theoretical)	Mass per particle size interval x 10^{-10}g per cubicmeter air			
	Pb	Br	Mn	Zn
1 - 2	25	6	0.3	7
2 - 5	3	2	0.1	5
5 - 9	2	0.5	1.5	13
> 9	-	-	0.6	1

particles and integrates over all sizes of sedimenting particles. Filters of various kinds are generally used in the
study of air-borne particles. These sampling methods give the
total mass but do not give information about the size distribution. The aerodynamic size distribution, for example, determines the amount of particles that is deposited in different
parts of the air-passages (10). A method to collect particles
differentiated in size classes is to use cascade impactors
(11,12). Generally only the weight distribution has been
studied, but recently chemical analysis has been reported (13).

For a preliminary experiment we constructed a cascade
impactor. The air is drawn through successive holes of decreasing diameter and a glass slide covered by a carbon foil is
placed perpendicular to the air jet. The cut-off diameters have
not been measured but calculated (11). After collecting particles the carbon foils are transferred to the aluminum frames
and irradiated. Figure 4 shows a typical spectrum obtained in
this type of analysis. The spectrum was obtained in 45 min and
1 mC of 2.5 MeV protons passed the target. The target-detector
distance was 50 mm and a 30 pm aluminum absorber was inserted.
The spectrum corresponds to an aerodynamic particle diameter
of 5-9 μm. The peaks not mentioned in table 1 belong to the
background. Table 1 gives an example of the kind of information
obtained by this type of air-borne particle studies. In the
table a difference is indicated in the composition of particles of different sizes.

Identification of Oil Samples

In an extensive study Bryan et al (14) have shown the
possibility of identifying oil slicks by using their elemental
content as "finger prints". These workers used neutron activation analysis to investigate the elemental content of 40 oil

Table 3
The concentrations of some elements in a water sample.

Element	Measured "concentration" not corrected for background	Concentration present in the sample (after subtraction of background) from the carbon foil
Ca	3 ppm	1 ppm
Ti	17 ppb	-
Mn	4 "	4 ppb
Fe	9 "	-
Ni	14 "	14 "
Cu	31 "	-
Zn	20 "	-
As	190 "	190 "
Br	21 "	21 "
Sr	47 "	47 "
Mo	11 "	10 "

target-detector distance was 50 mm and a 30μm aluminum-absorber was used. Table 3 presents the total concentrations as well as the concentrations after subtraction of the background.

Analysis of Blood Serum

Determination of elements in blood serum is another problem we are investigating (15).

In a recent report Niedermeier et al (16) describe an emission spectrometric method of analyzing blood serum and other biological fluids. After wet ashing the sample was excited in a 10 a dc arc. For example a threshold value for iron as low as 1.6 μg/100 ml was reported.

Evaporation of blood serum on a carbon foil as in preparation of water samples does not work since the contration of the serum tears the foil. Hostaphan foils (350 μg/cm^2), are, however, strong enough to withstand these forces and can be used. Another problem arises when irradiating the samples. There is a tendency that the sample will become detached from the backing. Therefore we had to cover the serum sample, after evaporation of the water, with another hostaphan foil. This arrangement worked quite well. An advantage of this preparation is that no chemistry has to be made on the sample. In figure 6 we show spectra with and without serum for iron. The spectra were obtained in 60 minutes runs with a beam current as low as 50 nA, due to the risk of destroying the hostaphan foils. The amount of serum was 15 mg. The iron content in the serum sample was

assumptions made earlier in this paper assuming an analyzed
amount of 1 mg of oil. These limits indicate that about the
same information that is obtainable with neutron activation
analysis can also be gathered in a single run with our method.

Water Analysis

Water samples are easily prepared. A drop of water is
placed on a carbon foil and after evaporation of the water
the sample is ready to be irradiated. In order to develop a
method for geological prospecting we have made some work on
ground water samples. These samples were first pre-concentrated
seven times by evaporation. A drop was then analyzed. In figure
5 we show a typical spectrum obtained in a run with 2.5 MeV
protons in 130 min. A charge of 8 mC passed the target. The

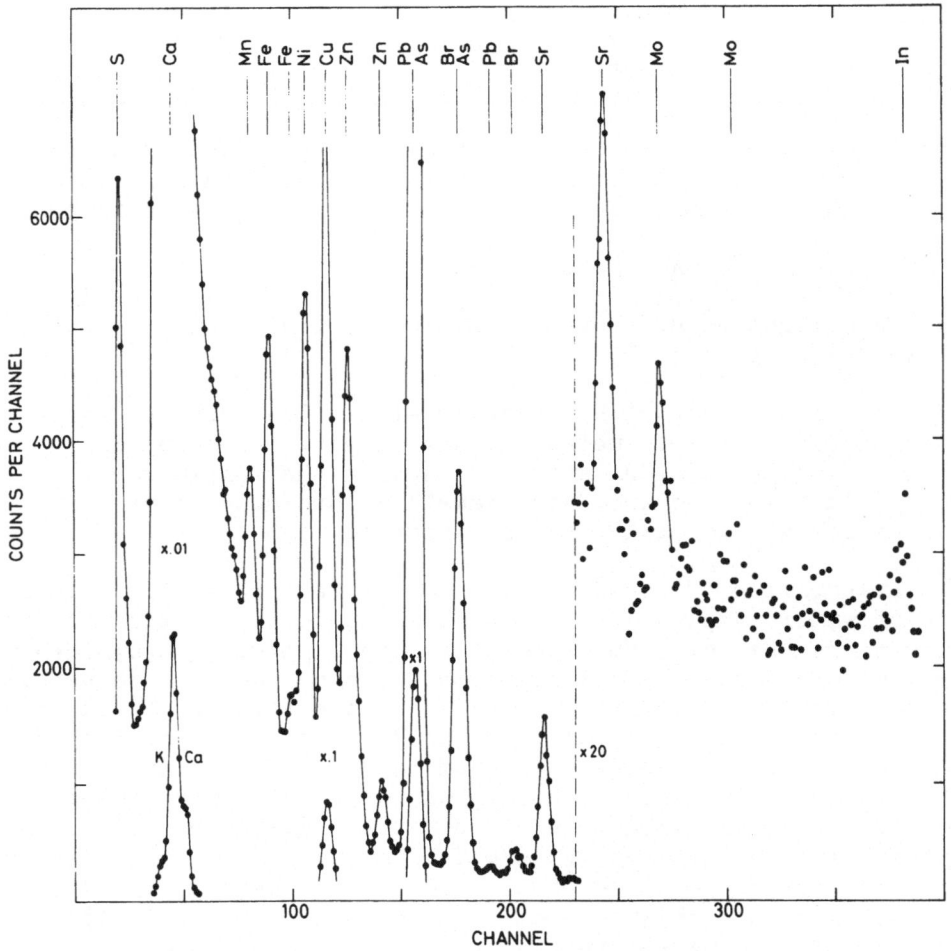

Figure 5. A pulseheight spectrum from analysis of a water
          sample. The background has not been subtracted.

Table 3
The concentrations of some elements in a water sample.

Element	Measured "concentration" not corrected for background	Concentration present in the sample (after subtraction of background from the carbon foil)
Ca	3 ppm	1 ppm
Ti	17 ppb	-
Mn	4 "	4 ppb
Fe	9 "	-
Ni	14 "	14 "
Cu	31 "	-
Zn	20 "	-
As	190 "	190 "
Br	21 "	21 "
Sr	47 "	47 "
Mo	11 "	10 "

target-detector distance was 50 mm and a 30 μm aluminum-absorber was used. Table 3 presents the total concentrations as well as the concentrations after subtraction of the background.

Analysis of Blood Serum

Determination of element in blood serum is another problem we are investigating (15).

In a recent report Niedermeier et al (16) describe an emission spectrometric method of analyzing blood serum and other biological fluids. After wet ashing   the sample was excited in a 10 A dc arc. For example a threshold value for iron as low as 1.6μ g/100 ml was reported.

Evaporation of blood serum on a carbon foil as in preparation of water samples does not work since the contraction of the serum tears the foil. Hostaphan foils ($2.5 \mu g/cm^2$), are, however, strong enough to withstand these forces and can be used. Another problem arises when irradiating the samples. There is a tendency that the sample will become detached from the backing. Therefore we had to cover the serum sample, after evaporation of the water, with another hostaphan foil. This arrangement worked quite well. An advantage of this preparation is that no chemistry has to be made on the sample. In figure 6 we show spectra with and without serum for iron. The spectra were obtained in 60 minutes runs with a beam current as low as 50 nA, due to the risk of destroying the hostaphan foils. The iron content in the serum sample was

calculated to be 53 μ g/100 ml compared with the figure of
61 μ g/100 ml obtained with standard atomic absorbtion methods.

From the spectra in figure 6 a lower limit of detection
about 1 μg/100 ml of serum can be estimated, assuming as the
criterion of detection that the iron-peak is to be increased
to four times the standard deviation by addition of the serum.
It should be possible to reduce the background further and thus
decrease the lower limit of detection.

In our spectra we have also observed elements such as
copper, bromine and zinc. As the sensitivity of existing
methods generally is sufficient for routine analysis, the
advantage in using the present method is that most elements
will be determined simultaneously and that elements occurring
in very low concentrations can also be detected.

DISCUSSION

The use of proton-excited characteristic X-rays in trace
analysis has shown to be applicable down to 10^{-11} g when
2.5 MeV protons are used. A limiting factor is the purity of
the backing material which, at the present moment, prevents
us, especially for medium-heavy elements, making full use of
the high sensitivity of the method. As shown in the above
theoretical considerations a lower limit of detection at the
10^{-12} g level is readily achieved when backing material
without larger impurities is available. The ionization cross
sections can be increased by a factor of at least 5 by choosing

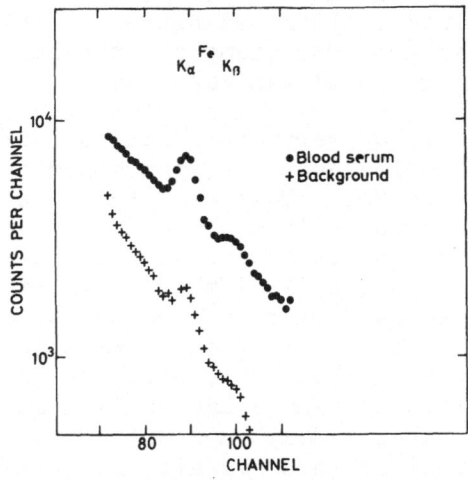

Figure 6. A pulseheight spectrum from analysis of iron in
blood serum.

higher proton energy. This, and the possibility of using higher beam currents in some applications, as well as increasing the analyzing times may reduce the lower limit of detection further. The very high ionization cross sections obtainable with heavy ions might also permit a further reduction in the lower limit of detection.

Methods for trace analysis generally use known samples to establish a calibration curve. The parameters determining the response from the proton beam are all known or possible to determine for a run. Thus it is possible to determine the present amounts from a simple calculation. Especially when multi-element analyses are made, this simplifies the procedure.

Compared with most methods of trace analysis, one great advantage of our technique is its capability to determine most of the elements present in a single run. This is already demonstrated in the spectra shown above, where, in spite of a rather poor resolution compared with that of the most recent detectors, many elements have been determined simultaneously. Since the interest in determining all the elements present in a single run is great, this reduces considerably the time needed for analysis. The improved resolution of the detectors also gives narrower peaks, thereby increasing the signal-to-background ratio.

A further advantage of this method compared with many other methods of trace analysis is the rather small variation in sensitivity between different elements.

Together with the lower limit of detection expressed in weight terms, the same quantity expressed in concentration terms is of interest. For example, the very low amounts of an element detectable with the electron microprobe must be present in very small samples, thus the concentration of the elements must be rather high to be detected.

The stopping power of material is in the order of 100 keV/mg/cm^2 for 2.5 MeV protons, which allows us to use rather thick samples without destroying the thin target assumption. This makes it possible not only to detect small amounts but also small concentrations.

ACKNOWLEDGEMENTS

Mr K. Håkansson and Mr C. Nilsson provided the proton beam and Mr K. Sjöberg built our apparatus. Their skilful assistance is greatly acknowledged. The work has been sponsored by the Swedish Board for Technical Development.

REFERENCES

1.   P.B. Needham, Jr. and B.D. Sartwell, "X-ray production efficiencies for K-, L-, M-, and N-shell excitation by ion impact", in C.S. Barrett, J.B. Newkirk and C.O. Ruud, Editors, <u>Advances in X-ray Analysis</u>, Vol. 14, p. 184-213 (1970)

2.   T.B. Johansson, R. Akselsson and S.A.E. Johansson, "X-ray analysis: Elemental trace analysis at the 10^{-12} g level", Nucl. Instr. Meth. <u>84</u>, 141-143 (1970)

3.   R.L. Watson, J.R. Sjurseth and R. W. Howard, "An investigation of the analytical capabilities of X-ray emission induced by high energy alpha particles", Nucl. Instr. Meth. <u>93</u>, 69-76 (1971)

4.   F.W. Saris, "Characteristic X-ray production by heavy ion-atom collisions", Thesis Leiden 1971

5.   J.S. Hansen, H.U. Freund and R.W. Fink, "Relative X-ray transition probabilities to the K-shell", Nucl. Phys. <u>A 142</u>, 604-608 (1970)

6.   R. Akselsson and T.B. Johansson, to be published

7.   G.A. Bissinger, J.M. Joyce, E.J. Ludwig, W.S. McEver, and S.M. Shafroth, "Study of the production of K X-rays in Ca, Ti, and Ni by 2-28 MeV protons", Phys. Rev. <u>A 1</u>, 841-847 (1970)

8.   R. Akselsson and T.B. Johansson, "A beam mapping method", Nucl. Instr. Meth. <u>91</u>, 663-664 (1971)

9.   J.T. Routti and S.G. Prussin, "Photopeak method for the computer analysis of gamma-ray spectra from semiconducter detectors", Nucl. Instr. Meth. <u>72</u>, 125-142 (1969)

10.   Task group on lung dynamics, "Deposition and Retention Models for Internal Dosimetry of the Human Respiratory Tract", Health Physics <u>12</u>, 173-207 (1966)

11.   R.D. Cadle, P.L. Magill, A.A. Nichol, H.C. Ehrmantraut, and G.W. Newell, "Sampling procedures" in P.L. Magill, F.R. Holden and C. Ackley, Editors, <u>Air Pollution Handbook</u>, p. 10-26 - 10-30 (1956)

12.   R.E. Lee and J.P. Flesch,"A gravimetric method for determining the size distribution of particulates suspended in air", APCA No. 69-125, June 1969

13. R.E. Lee and R.K. Pattersson, "Size determination of
     atmospheric phosphate, nitrate, chloride,and ammonium
     particulate in several urban areas," Atmos. Envir. 3,
     249-255 (1969)

14. D.E. Bryan, V.P. Guinn, R.P. Hackleman,and H.R. Lukens,
     "Development of nuclear analytical techniques for oil
     slick identification", GA-9889 Gulf General Atomic, 134 p.

15. R. Akselsson, D. Brune and T.B. Johansson, to be published

16. W. Niedermeier, J.H. Griggs and R.S. Johnson, "Emission
     spectrometric determination of trace elements in biolo-
     gical fluids", Appl.Spectr. 25, 53-56 (1971)

USE OF A SOLID-STATE DETECTOR FOR THE ANALYSIS OF X-RAYS EXCITED IN SILICATE ROCKS BY ALPHA-PARTICLE BOMBARDMENT*

Ernest J. Franzgrote

Jet Propulsion Laboratory

California Institute of Technology, Pasadena, California

ABSTRACT

The analysis of alpha-excited X-rays has been studied as a possible addition to the alpha-scattering technique used on the Surveyor spacecraft for the first in situ chemical analyses of the lunar surface.

Targets of pure elements, simple compounds, and silicate rocks have been exposed to alpha particles and other radiation from a curium-244 source and the resulting X-ray spectra measured by means of a cooled lithium-drifted silicon detector and pulse-height analysis.

Alpha-particle bombardment is a simple and efficient means of X-ray excitation for light elements. Useful spectra of silicate rocks may be obtained in a few minutes with a source activity of 50 millicuries, a detector area of 0.1 cm^2 and a sample distance of 3 cm. An advantage over electron excitation is the higher characteristic response relative to the bremsstrahlung continuum. Peak-to-background ratios of greater than 100 to 1 have been obtained for elemental targets. Relative efficiencies of X-ray excitation by alpha particles and by X-rays from the curium source have been determined.

---

*This paper presents the results of one phase of research carried out at the Jet Propulsion Laboratory, California Institute of Technology, under Contract No. NAS 7-100, sponsored by the National Aeronautics and Space Administration.

Resolution of the detector system used is approximately 150
eV for the lighter elements. This is sufficient to resolve the $K\alpha$
X-rays of the geochemically important elements, Na, Mg, Al, and Si
in silicate rocks. Although these and lighter elements are analyzed as well or better by the alpha-scattering and alpha-proton
technique, the X-ray mode enables results to be obtained more
quickly.

The study shows that the addition of an X-ray mode to the
alpha-scattering analysis technique would result in a significant
improvement in analytical capability for the heavier elements. In
particular, important indicators of geochemical differentiation
such as K and Ca (which are only marginally separated in an alpha-
scattering and alpha-proton analysis) may be determined quantitatively by measuring the alpha-excited X-rays. An X-ray detector is
under consideration as an addition to an alpha-scattering instrument now under development for possible use on a Mars-lander mission.

INTRODUCTION

Energy-dispersive analysis of X-rays excited by an alpha
source is being studied as a possible addition to the alpha-
scattering technique developed by A. Turkevich (1,2) and co-
workers for the first in situ chemical analyses of the lunar surface. The alpha-scattering instruments, as operated on three
Surveyor missions, used solid-state detectors for the energy analysis of scattered alpha particles and of protons produced in alpha-
proton nuclear reactions to provide elemental abundances of lunar
material at two mare locations and one highland site (3,4,5). The
precision of the method is better than 1% by atom for the major
rock-forming elements (6). Individual light elements can be resolved by the method but elements of mass greater than approximately 40 (calcium) must be grouped with neighboring elements into
mass-ranges. Because of this limitation, the addition of an X-ray
mode to the technique is under consideration for future missions.
The objective of the initial phase of these studies has been to
determine whether analysis of alpha-excited X-rays using a cooled
silicon detector can give a significant improvement in the analysis
of the heavier elements in rocks.

EXPERIMENTAL

The apparatus used in this study is a Kevex spectrometer system, Series 3000/4000, and a Nuclear Data Model 130 512-channel
pulse-height analyzer. The spectrometer system includes a lithium-
drifted silicon detector with an area of 0.1 cm^2 and a depletion
depth of 0.3 cm, an optical-feedback preamplifier, and a cryogenic

system for cooling the detector to liquid nitrogen temperatures.
The detector is operated in vacuum behind a 25-micron thick beryllium window.

The source-sample-detector geometry is shown in Fig. 1.   The
alpha source, provided by Dr. James H. Patterson of Argonne
National Laboratory, consists of approximately 50 mc of curium-244
($t_{1/2}$, 18 years; E_α, 5.8 MeV) deposited on a platinum disk and
mounted in a stainless steel capsule. An opening of 0.8 cm diameter in the end of the capsule that is directed toward the sample
serves to collimate the alpha particles. Two aluminum oxide films,
each 1200 Å thick, are stretched across the opening to prevent
radioactive contamination of the sample by recoil. The original
energy of the curium-244 alpha particles is reduced by self-
absorption and absorption in the aluminum oxide films so that the
average energy of emergent alpha particles is approximately 5.4
MeV. To prevent further loss of energy, the source and sample were
located in a vacuum chamber and operated under vacuum conditions or
at fairly low gas pressures. The source was located 2.8 cm along
its centerline from the surface of the sample at an angle 23 deg
from a line normal to the sample.

The sample holder shown in Fig. 1 is made of carbon with a
circular depression 2.9 cm in diameter by 0.15 cm deep to contain
the sample. Ten of these sample holders were mounted on a turntable

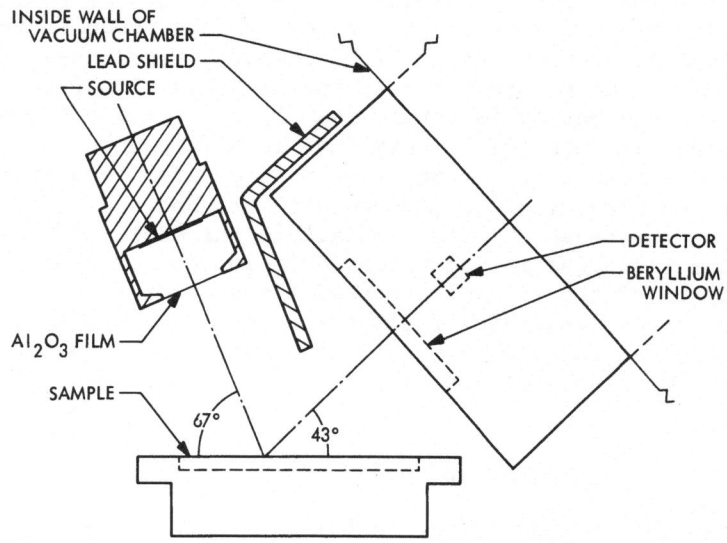

Fig. 1.   Schematic diagram showing the geometrical
          arrangement of the source, sample, and
          detector used in this study.

that could be rotated to bring each sample to rest beneath the
source. Reproducibility of sample positioning was approximately
0.05 cm in the vertical direction and 0.1 cm horizontally. Most of
the samples used in this study were powdered oxides and standard
rocks. They were prepared simply by loosely filling the circular
depressions in the sample holders with the powdered material and
leveling with a spatula.

The detector was mounted through a feed-through in the wall of
the vacuum chamber at a distance of 2.9 cm along its centerline
from the sample, at an angle 47 deg from a line normal to the
sample. Radiation from the sample reached the detector at an
average angle of 110 deg from the original direction of the incident
radiation. The centerlines of source and detector intersected the
sample at nearly the same point, approximately 0.5 cm from the
center of the sample. A sheet of lead, 0.15 cm thick, was used to
shield the detector from X-rays produced in the aluminum oxide
films and stainless steel source collimator and to reduce higher
energy X-rays from the source itself.

The overall resolution of the detector and electronic system
for the 5.90-KeV $K\alpha$ X-ray of manganese from an iron-55 source was
170 eV. The resolution at 1 KeV was 145 eV. The detector and preamplifier were shielded from vibration and acoustical noise during
operation to reduce electronic noise from microphonic pickup. The
stability of the system as measured by changes in the energy scale
was better than 0.1% per day. The electronic threshold was set at
a level equivalent to an energy of 700 eV.

A spectrum of X-rays excited with the curium source in a sample
of ferric oxide and measured with the silicon detector is shown as
the solid curve in Fig. 2 to illustrate some of the basic features
of the experimental data. The intensity of the X-rays in counts
per channel per 10 min is plotted on a logarithmic scale as a function of energy in KeV. The energy scale in this spectrum and other
sample spectra of this report is 20.4 eV/channel.

As the characteristic iron X-rays from the sample strike the
silicon detector, most of them give rise to photoelectric absorption
in which the energy of an incident X-ray is distributed between the
kinetic energy of a photoelectron and the potential energy of an
excited silicon atom. This energy is usually completely absorbed
within the silicon, giving an electronic pulse from the detector
which is proportional to the original energy of the incident X-ray.
Thus, the main peaks in the resulting pulse-height spectrum of
Fig. 2 are due to the $K\alpha$ (6.40 KeV) and $K\beta$ (7.06 KeV) X-rays of
iron. Occasionally an excited silicon atom of the detector will
emit an X-ray that escapes from the detector before reabsorption.
The resulting pulse from the detector will thus be lower than that
for an incident X-ray that is completely absorbed by an amount

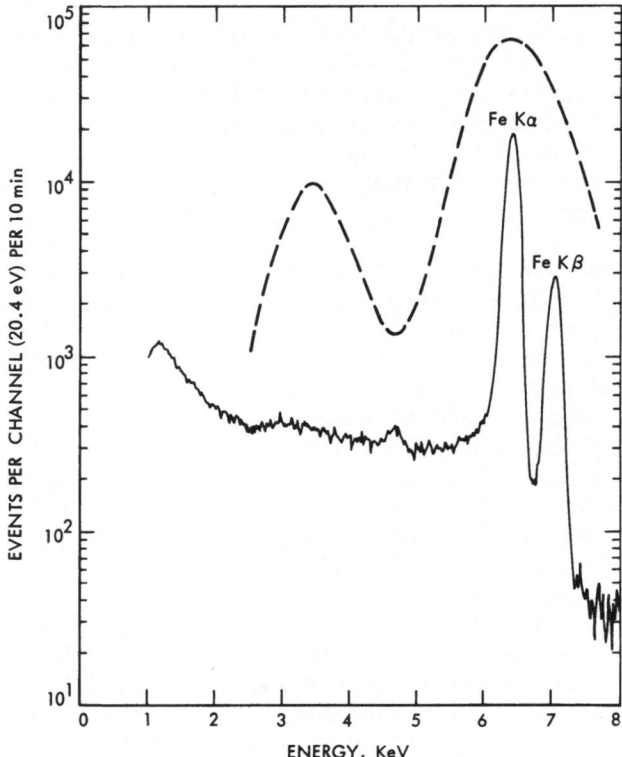

Fig. 2.   Spectra of the K-series X-rays of iron.
The solid curve is the spectrum obtained
by excitation by the curium-244 source
using the cooled silicon detector. The
broken curve is a spectrum obtained with
a gas-filled proportional counter.

proportional to the energy of the escaping X-ray. This phenomenon
is responsible for the small "silicon escape peak" visible in the
spectrum at 4.66 KeV (the energy of the iron Kα X-rays, 6.40 KeV,
minus the energy of the silicon Kα X-rays, 1.74 KeV).

The relatively low-intensity continuum at energies less than
6 KeV is associated with the detection of the iron X-rays, probably
due to partial charge collection within the detector. The iron
peaks and associated continuum are superimposed on another continuum, seen in Fig. 2 only at energies above that of the iron Kβ
peak, due to partial charge collection of X-rays from the curium
source. X-rays excited in the oxygen of the Fe_2O_3 are of too low
an energy to penetrate the 25-micron beryllium window.

The resolution of the silicon detector system is significantly
better than that obtained in most of the earlier energy-dispersive

work using gas-filled proportional counters. The broken curve of
Fig. 2 is a spectrum of iron X-rays obtained with a counter filled
with argon and methane, shown at the appropriate energy but at an
arbitrary intensity. The $K\alpha$ and $K\beta$ X-rays are not resolved. The
relatively prominent feature between 3 and 4 KeV is the "argon
escape peak" of the iron K X-rays.

The generation of characteristic X-rays by alpha particles has
been known since the early days of work with radioactive sources
but only in recent years have the results of quantitative studies
been reported (7,8,9). Other work related to the use of alpha-
excited X-rays for chemical analysis has been reported by Robert
(10), Imamura et al. (11), and Trombka et al. (12,13). One major
difference relative to excitation by electrons is the nearly complete absence of a bremsstrahlung continuum (7). Also, the probability of alpha-particle interactions with materials leading to the
production of characteristic X-rays increases rapidly with decreasing atomic number, Z.

Characteristic X-rays are excited in samples by two types of
radiation from curium-244. Alpha-particles from the source are
efficient for exciting X-rays in the lighter elements whereas the
L-series X-rays of curium are more efficient for the heavier elements. When a polyethylene film of 0.025 cm thickness is inter-
posed between the curium source and the sample, alpha particles
from the source are stopped without appreciable effect on the
X-rays from the source. In this way, relative efficiencies of the
two types of excitation for characteristic X-rays from several elements have been determined. For elements lighter than silicon,
nearly 100% of the excitation is due to the alpha particles from
the source. Approximately 80-90% of the X-rays from potassium and
calcium are generated by the alpha particles. For elements of Z
greater than 22 (titanium), most of the characteristic radiation is
produced by the X-rays from the source. For elements from Z = 11
(sodium) to Z = 28 (nickel) the overall effect is to give X-ray
yields, as a function of weight percent of element, that do not
vary among different elements by more than a factor of three. The
observed yields of K X-rays for alpha-particle excitation are close
to those calculated by Sellers and Ziegler (9) from the results of
Ogier et al. (14) for proton excitation.

The energy region of most interest for the measurement of the
K-series X-rays from the abundant rock-forming elements is 0-8
KeV. A typical background or continuum from the curium source in
this energy region is shown in Fig. 3 for a sample of sodium car-
bonate. The only structure in the spectrum in addition to the sodium
peak at 1.0 KeV is a small peak at 3.3 KeV from potassium contamination in the sample. Figure 4 shows the background at higher
energies in a spectrum obtained from a sample of iron oxide. The
energy scale of this spectrum is 82.4 eV/channel. The peak at

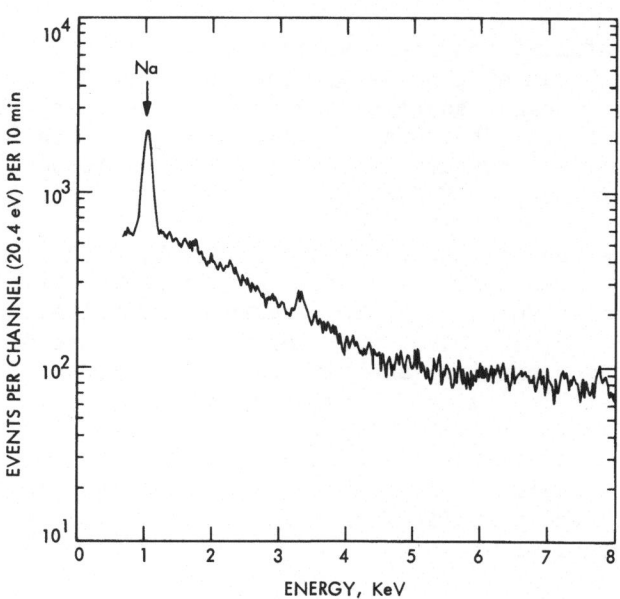

Fig. 3.   "Background" spectrum, obtained with
          a sample of sodium carbonate, showing
          the continuum in the region 1-8 KeV
          associated with the detection of
          scattered L-series X-rays (14-22 KeV)
          from the curium source.

8.6 KeV is due to $K\alpha$ X-rays from zinc contamination in the sample,
the features at 9.4 and 11.1 KeV are platinum $L\alpha$ and $L\beta$ X-rays excited in the source substrate and scattered from the sample, and
the peaks at energies between 14 and 22 KeV are the L-series X-rays
from curium and its daughter product, plutonium, scattered from the
sample.

    Characteristic X-ray spectra excited by the curium-244 source
were obtained from a series of elements and simple compounds and
from seventeen rock standards of known composition. Seven of the
rock samples were U.S. Geological Survey silicate rock standards:
(1) W-1, a basalt; (2) G-2, a granite; (3) GSP-1, a granodiorite;
(4) AGV-1, an andesite; (5) PCC-1, a peridotite; (6) DTS-1, a dunite;
and (7) BCR-1, a basalt. Three of the samples were obtained from
the National Bureau of Standards; (8) argillaceous limestone, 1a;
(9) burnt refractory, 76; and (10) chrome refractory, 103a. Five
of the standards were from the Centre de Recherches Petrographiques
et Geochimiques: (11) granite BR; (12) granite GH; (13) granite
GA; (14) basalte BR; and (15) biotite, MICA-Fe. Two of the samples
were from the Canadian Association for Applied Spectroscopy:

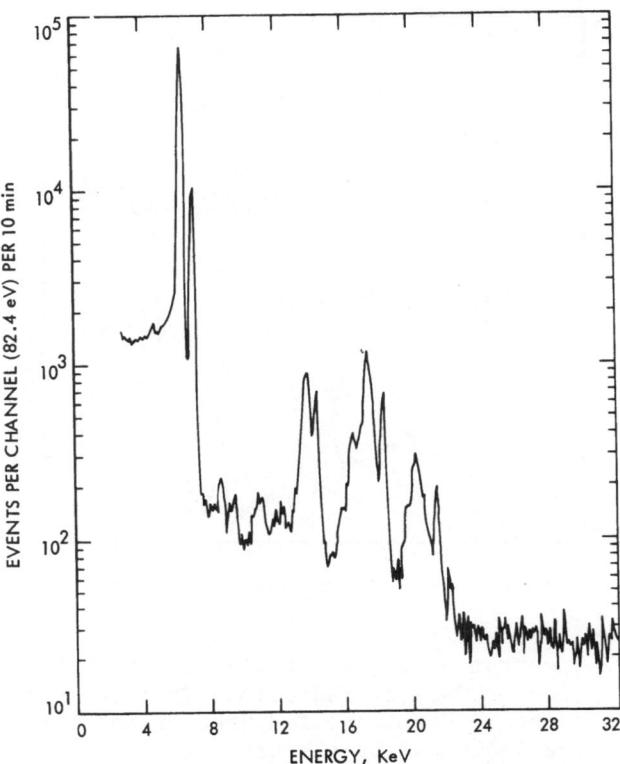

Fig. 4.   Spectrum from a sample of iron oxide
in the energy region 0-32 KeV, showing scattered X-rays from the curium
source at energies above 7 KeV.

(16) a syenite rock; (17) a sulphide ore. The composition of these
standards were taken from several publications (15-18).

A typical spectrum of X-rays from these rocks is shown in Fig.
5 for sample 5, PCC-1, a peridotite. Some of the main features of
the spectrum are identified by element symbols. The intensity of
the continuum is low enough so that X-rays from several of the
minor elements are readily seen. These elements include chromium
(0.29 wt %), manganese (0.09 wt %), and nickel (0.24 wt %). In
general, the rock spectra in the energy region 1-8 KeV are dominated by the K-series X-rays of elements sodium through nickel; the
L-series X-rays of copper through hafnium also occur in this region,
but, because of the generally lower abundances of the heavier elements in rocks, their intensities are usually negligible.

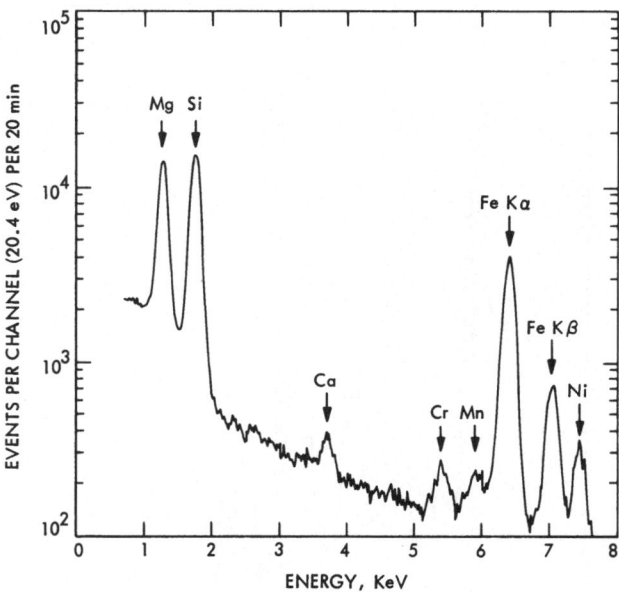

Fig. 5. Spectrum of X-rays excited in an ultrabasic
         rock (PCC-1) by radiations from the curium
         source. Prominent features of the spectrum
         are indicated by element symbols.

DATA ANALYSIS

To be able to determine quantitatively the abundances of the
elements in the rocks from the X-ray spectra, the intensities of
the characteristic peaks must be proportional to the corresponding
abundances, or the deviations from proportionality must be known.
Figure 6 shows the energy regions of the K X-rays from potassium
and calcium, important indicators of geochemical differentiation,
as measured in four U.S. Geological Survey standard rocks. Visual
inspection of the spectra indicates that, to the first approximation, proportionality holds for these elements.

One of the ways of determining the composition of the rocks is
to find the contributions to the whole spectrum by each of the elements present, correcting for deviations from proportionality.
This can be done by a least squares fit of the elemental spectra to
the complex spectra as was done for the Surveyor alpha-scattering
analyses (3,4,5,6), and by Trombka et al. (12, 13) for their analyses of alpha-excited X-ray spectra obtained from rocks with a proportional counter.

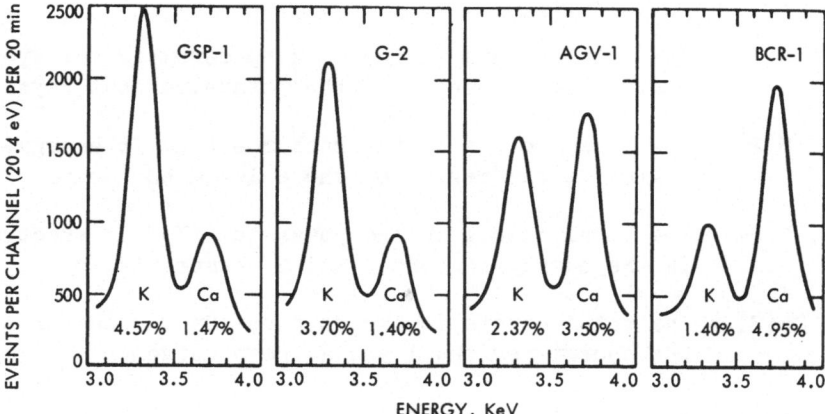

Fig. 6.   Spectra of the K X-rays of potassium and calcium
          excited in four U.S. Geological Survey standard
          rocks by the curium source. The values listed
          below the element symbols are the abundances in
          weight % as determined by conventional chemical
          analyses.

For the results reported here, a simplified data analysis was
used. First, a close approximation of the intensity of the characteristic X-rays contributed by each of nine elements was obtained
and then a correction for absorption of these X-rays in the different rocks was applied, if appropriate. The corrected intensities
for each element were then plotted versus the known compositions to
determine the average scatter of the analysis for the suite of
rocks.

The intensities of the X-rays from potassium and calcium were
calculated by establishing two energy regions centered about the
$K\alpha$ peaks of the two elements. The spectra of K_2CO_3 and $CaCO_3$ were
used to determine the contribution of potassium and calcium to each
of these two energy regions in the elemental spectra. For example,
although the major intensity from potassium is in its $K\alpha$ peak, the
potassium $K\beta$ X-rays contribute to the intensity in the region of
the calcium $K\alpha$ peak. On the basic assumption that the contribution
of heavier elements to these two energy regions could be estimated
from the intensity of the continuum at an energy just greater than
that of the Ca $K\beta$ peak, and then subtracted, the elemental spectra
were used to estimate the contributions of each of the two elements
to the rock spectra by means of the following equations:

$$S(1) = xK(1) + yCa(1) \tag{1}$$

$$S(2) = xK(2) + yCa(2) \tag{2}$$

where

S(1) = The total counts in the rock spectrum in energy-
region 1, the region of the potassium $K\alpha$ X-rays

S(2) = The total counts in the rock spectrum in energy-
region 2, the region of the calcium $K\alpha$ X-rays

K(1), K(2) = The total counts in the potassium (K_2CO_3) spectrum
in the energy-regions 1 and 2, respectively

Ca(1), Ca(2) = The total counts in the calcium ($CaCO_3$) spectrum in
the energy-regions 1 and 2, respectively

x, y = The fractions of the total counts in the two energy
regions of the potassium (K_2CO_3) and calcium
($CaCO_3$) spectra, respectively, that satisfy equations (1) and (2)

In the solution of the two equations for the unknowns x and y,
the contribution of potassium to the intensity in the region of the
potassium $K\alpha$ peak of the rock spectrum was calculated as xK(1) and
the intensity due calcium in the region of the calcium $K\alpha$ peak of
the rock spectrum was calculated as yCa(2).

Simple calculations of this type were used to determine the
contributions of sodium, magnesium, aluminum, silicon, potassium,
calcium, titanium, iron, and nickel to the X-ray intensities in the
appropriate energy regions of the rock spectra. The elements (K,
Ca) and (Fe, Ni) were determined in pairs, Ti as a single element,
and (Na, Mg, Al, Si) in a 4 x 4 matrix. For each element, or group
of elements, the continuum was estimated from that observed just
above the heaviest element in the group, using the shape of the
background spectrum of Fig. 3 to estimate its intensity at lower
energies.

RESULTS

Because of the two types of excitation by the radiations from
the curium source, some of the elements in the sample may be expected to behave as "thick targets" and some as "thin targets," with
some between the two. The characteristic K X-rays of the elements
titanium, iron, and nickel are excited primarily by the 14-22 KeV
X-rays from the source. Because the curium X-rays excite these
heavy elements at depths greater than the "critical thickness" at
which most of their characteristic X-rays can escape, they behave
as infinitely thick samples, and corrections for absorption are
necessary. On the assumption that the samples were homogeneous,
corrections were made for these heavy elements by calculating an

average mass absorption coefficient at each X-ray energy for each
of the rocks, using the known compositions, and normalizing the
intensities to that of the average mass absorption coefficient of
all the rocks. The resulting corrected intensities for these three
elements in each of the rocks are plotted versus their known abundances in weight percent in Fig. 7. The corrected intensities were
found to give better fits to a 45 deg straight line than the uncorrected values.

The K-series X-rays of potassium and calcium are produced
mainly by the alpha particles from the source, at depths somewhat
less than the "critical thicknesses" (tens of microns) for these
elements in the rocks. Thus, for K and Ca, the samples were assumed to be "infinitely thin" targets, and absorption corrections
were not made. The uncorrected intensities for K and Ca are
plotted versus their known abundances in the rocks in Figs. 8 and 9.

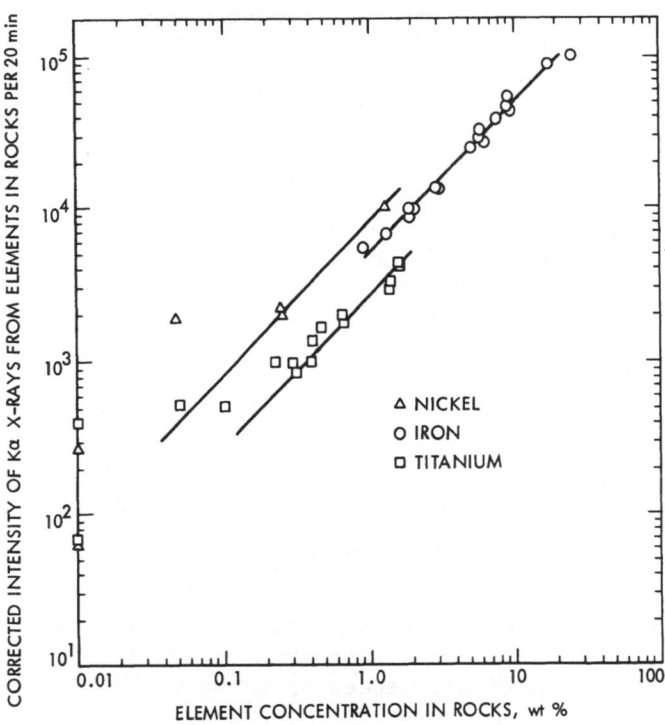

Fig. 7. Corrected intensities of Kα X-rays from
         titanium, iron, and nickel, excited in
         silicate rocks by the curium source,
         versus the known abundances of these
         elements in the rocks.

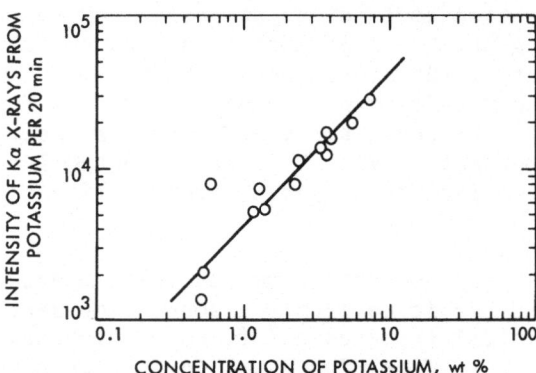

Fig. 8.   Intensity of Kα X-rays from potassium
          in the rock standards versus the
          known abundances in these rocks.

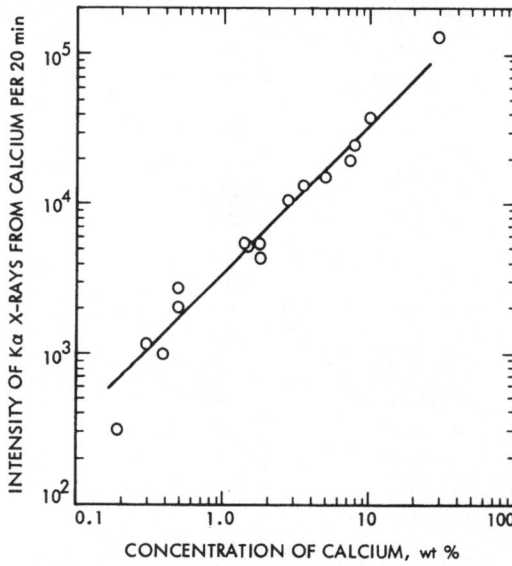

Fig. 9.   Intensity of Kα X-rays from calcium
          in the rock standards versus the
          known abundances in these rocks.

In Fig. 10, the ratio of the calculated intensities of potassium to calcium are plotted versus their known weight ratios in the
rocks. The fit to a 45 deg straight line is good over a fairly
large range of relative abundances.

For lighter elements, absorption of X-rays in the sample again
becomes important as the effective range of the X-rays approaches

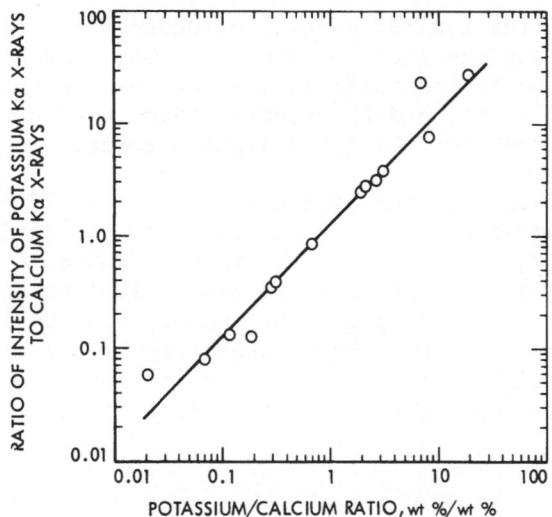

Fig. 10.   Intensity ratio of potassium to
calcium X-rays versus the abundance
ratio of potassium to calcium in
the rocks.

that of their production by the alpha particles. For example, the
"critical depth" of sodium X-rays is approximately 1 micron in
silicate rocks. For the light and intermediate elements, a rig-
orous solution of the problem would have to take into account the
variation in depth of production of the X-rays by the alpha particles as a function of alpha-particle range in the sample.   In
Fig. 11, as an example of the results for the light elements, the

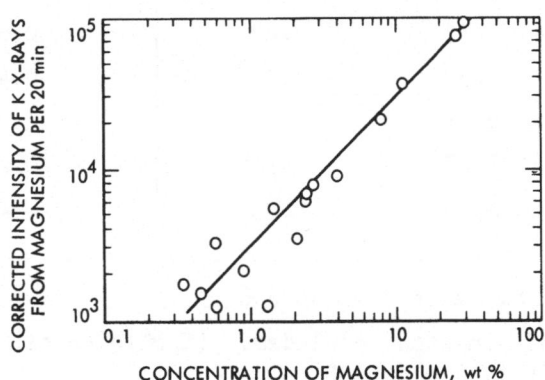

Fig. 11.   Corrected intensity of K X-rays
from magnesium in the rock
standards versus the known abundances in these rocks.

K X-ray intensities for magnesium, corrected only for absorption, are plotted versus the known abundances. The range of concentration of magnesium in the rocks is greater than that of sodium, aluminum, and silicon, and the scatter about a 45 deg line is somewhat less than that for the other light elements.

The 1σ scatter for the nine elements in the 17 rocks is given in Table 1. The scatter was calculated from the deviations of the data from straight-line fits of the type of those of Figs. 7, 8, 9, and 11. The errors in the first column, listed as wt %, have been used to estimate the errors on a percent-by-atom basis for column 2. In column 3 are values for the scatter for these elements for the alpha-scattering/alpha-proton technique (6), calculated by comparing the alpha-scattering results with those of conventional chemical analyses.

Table 1. Scatter of the data

Element	Alpha/X-ray analysis		Alpha-scattering
	Wt %	Atom %	Atom %
Na	0.80	0.72	0.12
Mg	0.75	0.63	0.96
Al	1.28	1.01	0.23
Si	3.40	2.43	0.75
K	0.53	0.28	0.68[a]
Ca	0.59[b]	0.30[b]	0.41[a]
Ti	0.11	0.05	0.25
Fe	0.27[c]	0.10[c]	0.25
Ni	0.08	0.03	----[d]

[a] Recalculated from Table 1 of Ref. 6

[b] Contribution of limestone excluded. (Inclusion of limestone gives values of 1.06 and 0.55)

[c] Contribution of sulphide ore excluded. (Inclusion of sulphide ore gives values of 0.57 and 0.22)

[d] Not determined

The data from the two techniques are comparable in the sense
that measurements on similar series of rock samples under the same
experimental conditions would be expected to give the same scatter,
even though the present results represent only a comparison of
calculated intensities with the known compositions, and not, as in
the case of the alpha-scattering analyses, a measure of the absolute
fractions of elemental spectra needed to fit the rock spectra.

DISCUSSION AND CONCLUSIONS

The simplified method of data analysis used in this initial
phase of the study does not take into account several sources of
error that undoubtedly contribute to the scatter of the results
shown in Table 1. The precision of the results depends on the
absolute intensities of X-rays in the spectra. These intensities
are a function of many factors, including a number of geometrical
and instrumental instabilities that can be reduced in later work.
For example, the present electronic system in the first few hours
after turn-on exhibits a change in the measured intensity of a constant source of X-rays of approximately 20%; for this reason the
equipment was kept operating continuously. Nevertheless, the 1σ
variation in count rate from magnesium oxide and iron oxide standards measured several times during the course of the rock analyses
exceeds 5%. The electronics also exhibit a count-rate instability
at rates near or exceeding those of the rock spectra ($\sim$500/sec).
Variations in the position of the samples should not have caused
changes in the count rates greater than approximately 3%. Except
for elements at concentrations less than approximately 0.1 wt %,
statistical variations in the data are small compared with these
instrumental errors.

The simplified data analysis does not take into account structure in the subtracted continuum from the silicon-escape peaks and
from the L-series X-rays of heavy elements. These effects can be
accounted for in a least-squares analysis using a complete library
of elemental spectra. Similarly, corrections for enhancement effects and more sophisticated absorption corrections can be incorporated in a least-squares computer program.

One source of error that cannot be corrected for is inherent
in the assumption of homogeneity of the samples. Corrections for
X-ray absorption based on the average mass absorption coefficient
of a rock do not account for variations in the concentration of
elements in the different mineral phases. The assumption of homogeneity is valid only for the heavy elements whose X-rays originate
at depths in the sample much greater than the average particle size.

Even though errors due to instrumental instabilities and incomplete analysis of the data are relatively large, the study

shows that the addition of an X-ray mode to the alpha-scattering analysis technique would result in a significant improvement in analytical capability for the heavier elements. In particular, important indicators of geochemical differentiation such as potassium and calcium (which are only marginally separated in an alpha-scattering and alpha-proton analysis) may be determined quantitatively by measuring the alpha-excited X-rays.

An X-ray detector is under consideration for use in a pre-prototype alpha-scattering instrument now under development by A. Turkevich at the University of Chicago for possible use on a Mars lander mission. Further studies with the laboratory X-ray system are planned to improve the precision of the results, to assess the value of adding alpha-excitation to a combined X-ray diffraction/fluorescence experiment, and to determine the operating characteristics of the detector at temperatures higher than 77°K.

ACKNOWLEDGMENTS

The author wishes to acknowledge the assistance of George O. Ladner, Jr., in obtaining the experimental data. Thanks are also due to James H. Patterson for providing the alpha source and to Allan S. Jacobson, Albert E. Metzger, and Anthony L. Turkevich for helpful comments and suggestions.

REFERENCES

1. A. L. Turkevich, K. Knolle, R. A. Emmert, W. A. Anderson, J. H. Patterson, and E. Franzgrote, "Instrument for Lunar Surface Chemical Analysis," Rev. Sci. Instrum., 37, 1681-1686 (1966).

2. A. L. Turkevich, K. Knolle, E. Franzgrote, and J. H. Patterson, "Chemical Analysis Experiment for the Surveyor Lunar Mission," J. Geophys. Res., 72, 831-839 (1967).

3. A. L. Turkevich, E. J. Franzgrote, and J. H. Patterson, "Chemical Composition of the Lunar Surface in Mare Tranquillitatis," Science, 165, 277-279 (1969).

4. E. J. Franzgrote, J. H. Patterson, A. L. Turkevich, T. E. Economou, and K. P. Sowinski, "Chemical Composition of the Lunar Surface in Sinus Medii," Science, 167, 376-379 (1970).

5. J. H. Patterson, A. L. Turkevich, E. J. Franzgrote, T. E. Economou, and K. P. Sowinski, "Chemical Composition of the Lunar Surface in a Terra Region Near the Crater Tycho," Science, 168, 825-828 (1970).

6. T. E. Economou, A. L. Turkevich, K. P. Sowinski, J. H. Patterson, and E. J. Franzgrote, "The Alpha-Scattering Technique of Chemical Analysis," J. Geophys. Res., 75, 6514-6523 (1970).

7. E. Merzbacher, H. W. Lewis, "X-Ray Production by Heavy Charged Particles," Handbuch der Physik, 34, 166-192, Springer Verlag, Berlin/Göttingen/Heidelberg (1958).

8. L. S. Birks, R. E. Seebold, A. P. Batt, and J. S. Grosso, "Excitation of Characteristic X-Rays by Protons, Electrons, and Primary X-rays," J. Appl. Physics, 35, 2578-2581 (1964).

9. B. Sellers and C. A. Ziegler, "Generation and Practical Use of Monoenergetic X-Rays from Alpha-Emitting Isotopes," 353-373, Proceedings of the Symposium on Low-Energy X- and Gamma Sources and Applications, Oct. 1964, Chicago, Illinois, published by the Atomic Energy Commission (1965).

10. A. Robert, "Contributions to the Analysis of Light Elements Using X-ray Fluorescence Excited by Radioelements," Commissariat a L'Energie Atomique, Rapport CEA-R 2539, 1964.

11. H. Imamura, K. Vehida, H. Tominaya, "Fluorescent X-Ray Analyzer with Radioactive Sources for Mixing Control of Cement Raw Materials," Radioisotopes, 11, 4 (1965).

12. J. I. Trombka, I. Adlér, R. Schmadebeck, and R. Lamothe, "Non-Dispersive X-Ray Emission Analysis for Lunar Surface Geochemical Exploration," Report No. X-641-66-344, Goddard Space Flight Center, Greenbelt, Maryland (1966).

13. I. Adler and J. I. Trombka, "Geochemical Exploration of the Moon and Planets," 67-74, 175-210, Springer-Verlag, Berlin/ Heidelberg (1970).

14. W. T. Ogier, G. J. Lucas, J. S. Murray, and T. E. Holzer, "Soft X-Ray Production by 1.5 MeV Protons," Phys. Rev., 134, A1070-1072 (1964).

15. F. J. Flanagan, "U.S. Geological Survey Silicate Rock Standards," Geochim. et Cosmochim. Acta, 31, 289-308 (1967).

16. National Bureau of Standards Certificate of Analyses of Standard Samples 1A, 76, and 103a, Department of Commerce, Washington, D.C., (1931, 1927, 1962).

17. M. Roubault, H. de la Roche and K. Govindaraju, "Rapport sur quatre roches etalons geochimiques: Granites GR, GA, GH et Basalte BR," Sciences de la Terre, Vol. XI, 105-121, Nancy (1966).

18. "Report of Nonmetallic Standards Committee, Canadian Association for Applied Spectroscopy," Applied Spectroscopy, 15, 159-160 (1961).

STUDIES OF X RAYS INDUCED BY CHARGED PARTICLES[*]

Jerome L. Duggan and William L. Beck
Oak Ridge Associated Universities,[†] Oak Ridge, Tn.

Larry Albrecht and Lee Munz
ORTEC, Inc., Oak Ridge, Tn.

James D. Spaulding
Pacific Union University, Angwin, California

ABSTRACT

Characteristic x rays have been produced for a variety of samples by bombardment with protons in the energy range from 75 keV to 5 MeV. The experiments were performed with two accelerators. For the low-energy studies (less than 150 keV), a Cockcroft-Walton accelerator was used. The higher-energy studies were done with a 5 MV Van de Graaff. The x rays were measured with high-resolution Si(Li) and Ge(Li) detectors. Yields for the cross section of characteristic K- and L-shell ionizations were measured for titanium, vanadium, chromium, manganese, iron, cobalt, nickel, copper, silver, gold, bismuth, and uranium. The experimental cross-sections have been compared to the theoretical predictions of the Born approximation for an interaction of this type.

Trace element analysis by 4-MeV proton bombardment of samples in the 10^{-12} gm range has also been performed. Some comments with regard to analysis with these sensitivities will be made.

---

[*] Some of the measurements reported in this paper were made on the 5 MV Van de Graaff accelerator at the Oak Ridge National Laboratory which is operated by Union Carbide Corporation for the USAEC.

[†] ORAU is supported in part by the USAEC.

INTRODUCTION

Researchers have known for years that the characteristic x rays of an element could be produced by bombardment with charged particles. The first definitive work in this area was done by Chadwick in 1912 (1). In this early work, x rays were produced by alpha particle bombardment of several elements. The advent of the particle accelerator in the 1930's provided a copious source of positive ions and, hence, many different kinds of projectiles have been used since that time.

The early workers were somewhat restricted because their measurements had to be made with Geiger counters and ionization chambers (2-5). However, some of these measurements did give x-ray yields that compared quite well with the earlier calculations of Henneberg (6). Interest in the field increased somewhat with the development in the early 1950's of the sodium-iodide detector and scintillation spectroscopy (7-11). In 1958 Merzbacher and Lewis (12) wrote an excellent review which included most of the experimental measurements that had been made to that time and theoretical interpretations of the interactions.

Most of the work that has been reported in the last ten years was done with protons and, in fact, Khan and Jopson at the Lawrence Radiation Laboratory have published a large fraction of the available experimental results (13-22). Table I is a partial summary of the work that has been done with protons (13-30), the detector used, elements studied, and the energy range. Table II gives the same data for other projectiles (31-46).

In the United States there are perhaps 500 accelerators that could be used for studying charged-particle-induced x rays. In Europe and the rest of the world, there are probably 200 more accelerators that are suitable for this application. By far, the greatest number of these accelerators are of the Cockcroft-Walton or Van de Graaff type. Most of these machines have ion sources of radio-frequency type, coupled with magnetic analysis systems that for the most part limits the analyzed projectiles to protons, deuterons, or alpha particles. Other ions can be made in rf type sources but with reduced efficiency (47-48). Different types of ion sources are, however, now being developed for ion implantation and heavy-ion studies with accelerators. A summary of some of these applications can be found in the literature (49-50). With pumping modifications, etc., these ion sources can usually be installed on the Cockcroft-Walton type accelerators. Van de Graaff accelerators offer a different problem, since with them sufficient power and space are usually more difficult to obtain.

TABLE I
SUMMARY OF RECENT LITERATURE
(Protons Only)

Investigator	Date	Detector	Elements	Energy Range (MeV)
Jopson et al. (13)	1962	NaI(Tℓ)	26 elements from Z=22 to Z=92	0.10–0.50
Khan and Potter (14)	1964	P.C.	Mn, Al, Cu	0.060–0.50
Khan et al. (15)	1964	P.C.	Cu	0.10–0.50
Khan et al. (16)	1964	P.C.	Nd, Sm, Gd, Tb, Dy, Ho.	0.025–0.100
Khan et al. (17)	1964	P.C.	Y	0.030–0.100
Khan et al. (18)	1965	P.C.	Cu, Mg, Al, Nd, Sm, Gd, Tb, Dy, Ho	0.015–1.90
Sterk (19)	1965	P.C.	C, O, Al	0.060–0.110
Khan et al. (20)	1966	P.C.	Al	0.100
Khan et al. (21)	1966	P.C.	Cu	0.025–1.70
Khan et al. (22)	1966	P.C.	Al, Cu	0.070–0.100
Christensen et al (23)	1967	P.C.	C, Be, Al, Cu, Yb	0.050–0.100
Khan et al. (24)	1967	P.C.	Al, Cu, W	0.250–1.56
Sterk et al. (25)	1967	P.C.	C, Al, V, Cu, Mo, Sn, Fe, Ta, Pb, U	0.075
Hart et al. (26)	1968	P.C.	O_2	0.100
Hart et al. (27)	1969	P.C.	O, Al	0.020–0.100
Bissinger et al. (28)	1970	P.C.	Ca, Ti, Ni	2–28
Johansson et al. (29)	1970	Si(Li)	Ti, Cu	1.5
Spaulding (30)	1970	Si(Li)	Ti, V, Cr, Fe, Co, Ni, Cu	0.100–0.150

TABLE II

(Heavy Charged Particles)

Investigator	Date	Detector	Target	Projectile	Energy Range (MeV)
Sharma et al. (31)	1965	NaI	Sm, Te, Ce, Sm-144, Sm-152 Sm-154, Gd-160 W-186, Pb	Alphas	3 and 4
Brandt et al. (32)	1966	P.C.	Mg, Al, Cu	H^+, He-3 He-4	0.020-0.200
Marks et al. (33)	1967	P.C.	O, Al, Au	H+, Ar+	0.045-0.075
Sellers et al. (34)	1967	P.C. and Solid State	C, N, O, F, Na, Mg, Al, V, Cu	Alphas	2.75
Der et al. (35)	1968	P.C.	C	H+, He+, C+, N+, O+, Ne+, Ar+, Kr+, Xe+	0.020-0.080
Brandt et al. (36)	1969	P.C.	Al	H+, He-3, He-4	0.025-0.200
Fortner et al. (37)	1969	P.C.	C	C+	0.020-1.50
Richard et al. (38)	1969	Si(Li)	Cu, Ni	H+, 0-4+	6 and 15

TABLE II (Contd.)

Investigator	Date	Detector	Target	Projectile	Energy Range (MeV)
Brandt et al. (39)	1970	P.C.	Al, Ne	N, O, Ar, C, Ne, Al	0.100–3.2
Burch et al. (40)	1970	Si(Li)	Ca, V	H+, 0–4+	6–15
Cunningham et al. (41)	1970	P.C.	Ar	Ar+	0.050–0.330
Richard et al. (42)	1970	Si(Li)	Cu	H+, 0–4+	6–19
Stein et al. (43)	1970	P.C.	Yb, Te	I+	30
Watson et al. (44)	1970	Si(Li)	Fe, Ni, Cu, Zr, Nb, Rh, Pd, Ag, Sn, Sm, Tm, Ta, Au	Alphas	30–80
Lewis et al. (45)	1971	Si(Li)	Ti, Cu, Au	α, H–2+	6.25 & 20 MeV/amu
Mokler (46)	1971	Si(Li)	Mo, Yb, Au	I+	15–60

In summarizing work done with charged-particle production of
x rays, it might be said that we have just scratched the surface.
It will be shown later in this paper that the cross sections are so
large in magnitude for many reactions that very-low-beam currents
are required to make the indicated measurements. These high cross-
sections [for example 45 MeV oxygen-16 ions on copper (K - α 1) has
a cross section of the order of 10^5 barns] also mean trace analysis
and surface analysis studies can easily be made with this method.
In Figure 1 we compare the sensitivity of trace analysis using x-ray
fluorescence induced by protons (from a 5-MeV Van de Graaff accelerator) with other analytical techniques. With heavy ions and higher energy, sensitivity would increase two orders of magnitude.
Therefore, this method of trace analysis is potentially one of the
most sensitive analytical techniques.

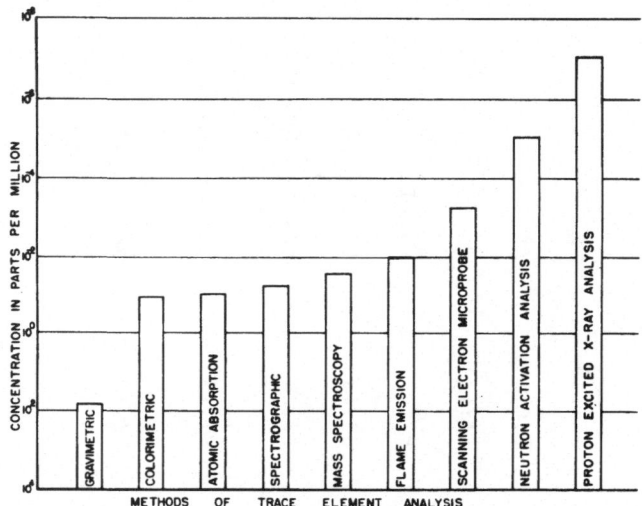

Figure 1. Proton-Excited X-Ray Fluorescence Compared to Other
Analytical Techniques

In what follows we will describe some experiments that were
made at energies of less than 150 keV with a Cockcroft-Walton accelerator and at energies up to 5 MeV with the Van de Graaff at the
Oak Ridge National Laboratory.

EXPERIMENTAL PROCEDURE

Experimentally, studies of x rays produced by charged particles
are easy to perform. Figure 2 shows the basic experimental arrangement used for most of the measurements. The beam spot from the accelerator was defined by a circular aperture of 2-mm diameter. The
beam then entered the Faraday cup and impinged onto the target.
The thin targets were either self-supported foils of approximately

50 μgm/cm^2 thickness or targets of that thickness evaporated onto
20 μgm/cm^2 carbon foils. The procedures for preparing thin foils
for accelerator research have been well documented in the literature
(51-52). The x rays produced at the target passed through a 5 x 10^{-3}
inch Be window and into either a Si(Li) or Ge(Li) detector. For
some experiments the Si(Li) detector was coupled directly to the
beam tubing with an O-ring connection.* Under these conditions the
x rays produced at the target had to pass through only 5 x 10^{-4}
inches of Be before striking the silicon surface of the detector.
In order to avoid unwanted background in some of the trace analysis
studies, the target support was machined of high-purity, reactor-
grade graphite. The amplified output from the detector was fed into
a 4096-channel analyzer, which was multiplexed to a Control Data
Model 3200 Computer for plotting and data analysis. The number of
incident charged particles was determined with a precision current-
integrator.

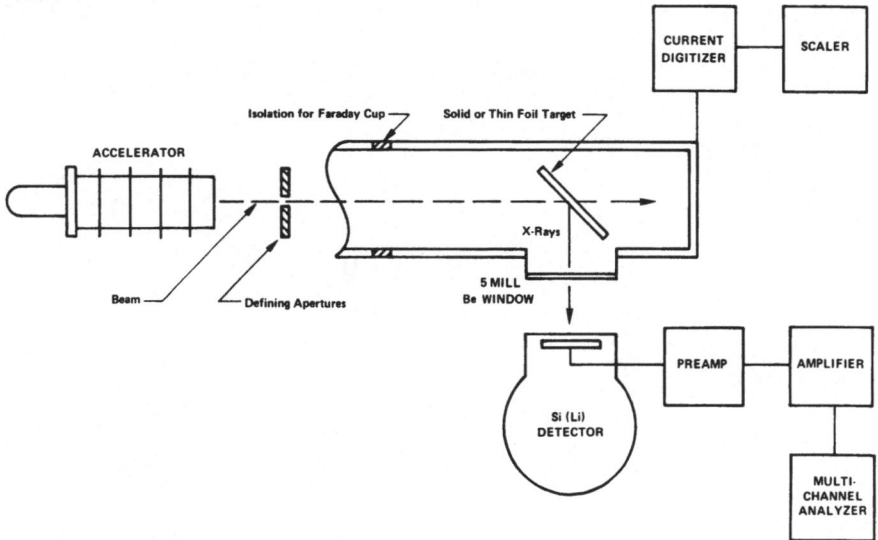

Figure 2.   Experimental Arrangement for Charged-Particle X-Ray
            Fluorescence

     The absolute efficiency of the detector system and its energy
calibration were determined by placing calibrated radioactive sources
at the position where the beam struck the target. The sources had
approximately the same finite size as the beam spot. The absolute
efficiency of the detector was determined to 7% for all measurements
in this paper. A detailed description of this calibration process

---

*The authors would be glad to furnish shop drawings of this arrange-
ment upon request.

for a Si(Li) detector used for a similar application at the Texas
A & M Cyclotron has been reported (53-56).

EXPERIMENTAL RESULTS AND DISCUSSION

<u>The Cockcroft-Walton and Low-Voltage Van de Graaff Data</u>. The
targets used for these measurements were thick targets and, hence,
thick target yields will be quoted. For an uncooled target, it was
necessary to keep the beam current to less than 1 microampere or
risk target evaporation. However, water-cooled targets have been
used for similar applications with beam currents as high as 10 mil-
liamperes (49-50).

Figure 3 shows the characteristic spectra that were measured
for several elements at an incident proton energy of 350 keV. These
measurements were made with a 400-keV Van de Graaff (teaching type)
accelerator.

Figure 3. X-Ray
Spectra from Thick
Targets Bombarded
with 350-keV Protons

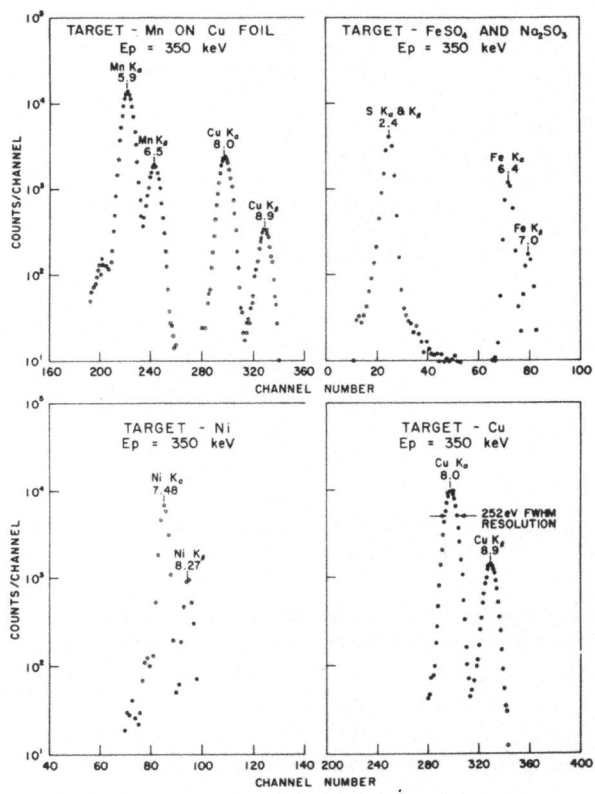

Table III shows the results of the ionization cross-sections
that were measured at low energy. Shown also in this table are the
theoretical values that were calculated from the Born approximation.
As expected, the theoretical values are about a factor of ten larger

TABLE III

X-Ray YIELDS AND K SHELL IONIZATION CROSS SECTIONS FOR COCKCROFT–WALTON MEASUREMENTS

Element	E_p (keV)	$I\mu \times 10^{11}$ (X rays/ proton)	$dI\mu/dE_p \times 10^{12}$ (X rays/ proton–keV)	σ_I (measured)$\times 10^{28}$ (cm²)	σ_I(theor)$\times 10^{26}$ (cm²)	$\dfrac{\sigma_I(\text{theor})}{\sigma_I(\text{measd.})}$
Titanium	100	348.00	244.00	410.0	37.0	9.0
	125	1540.00	783.00	1200.0	86.0	7.2
	150	4730.00	1880.00	2900.0	160.0	5.5
Vanadium	100	205.00	146.00	220.0	20.0	9.1
	125	934.00	487.00	690.0	49.0	7.1
	150	2970.00	1220.00	1700.0	95.0	5.6
Chromium	100	63.80	79.60	100.0	11.0	11.0
	125	532.00	302.00	370.0	28.0	7.6
	150	1560.00	815.00	960.0	56.0	5.8
Manganese	100	50.00	35.00	37.0	6.4	17.0
	125	241.00	137.00	140.0	16.0	11.0
	150	883.00	421.00	420.0	33.0	7.9
Iron	100	31.30	35.90	34.0	3.6	11.0
	125	249.00	154.00	140.0	9.7	6.9
	150	891.00	378.00	350.0	20.0	5.7
Cobalt	100	13.80	14.70	13.0	2.0	15.0
	125	110.00	72.50	61.0	5.8	9.5
	150	445.00	213.00	180.0	12.0	6.7
Nickel	100	7.81	9.15	6.0	1.1	18.0
	125	75.20	53.70	35.0	3.4	9.7
	150	340.00	175.00	110.0	7.6	6.9
Copper	125	45.90	35.00	21.0	2.0	9.5
	150	222.00	117.00	69.0	4.7	6.8

than the experimental values, and the best agreement is with the
lowest Z and highest energy. The absolute errors on our work are
±30%.

In regions where our results overlap with the previous work of
Messelt (10) and Khan and Potter (14), the agreement is ±10% which
gives some credentials to the measurements. One possible reason for
the failure of the Born approximation to give results which are in
agreement with the measurements is that it does not consider the
deflection of the incident particle by the Coulomb field of the target atom. For low energies this is quite important. Bang and
Hansteen's semiclassical description, which assumes a classical hy-
perbolic trajectory for the incident particle, gives better results
at low energy (14). It has been suggested (57) that calculations
using Hartree wave functions for atomic electrons and deflection
trajectory for the incident particles would yield more accurate results at low energies, but to date these calculations have not been
made.

With low-energy measurements there is an excellent potential
for work with Cockcroft-Walton type machines in the area of surface
analysis with heavy ions. Since heavy ions penetrate at most only
a few hundred Angstroms into solids, the phenomenon can be used to
examine surface deposits. An ideal application of current interest
to the electronic industry is the measurement of the quantity and
depth of implanted ions in semiconductors. Several informative
papers have been written describing ion probe measurements (58-60).

The High-Energy Data (1- to 5-MeV Van de Graaff Measurements).
Figure 4 shows the cross-section measurements that have been made
for gold, silver, bismuth, and uranium in the energy range from 1-
to 5-MeV. The solid lines on the figure are the theoretical predictions of the Born approximation. The normalization constant K'
is the average value that the experimental data were multiplied by
in order to fit the theoretical predictions. Since the uranium
target thicknesses were not known, the values obtained for it are
only relative ionization cross-sections that have been normalized
to the Born approximation at 3.4 MeV. As in the case of the Cockcroft-Walton data, it is expected that the theoretical description
would be improved with Hartree wave functions for the atomic electrons and some considerations of the deflection trajectory for the
incident charged particles.

Trace Analysis. Johansson, et al. (29) have described trace
analysis by x-ray fluorescence using 1.5-MeV protons. Our only
improvement over these earlier measurements is the factor of 20
gained from cross-section considerations in going from 1.5 to 4.5
MeV (see Figure 4). For our measurements 10 nanograms of zinc was
uniformly evaporated onto a 50 $\mu gm/cm^2$ carbon foil, over an area of

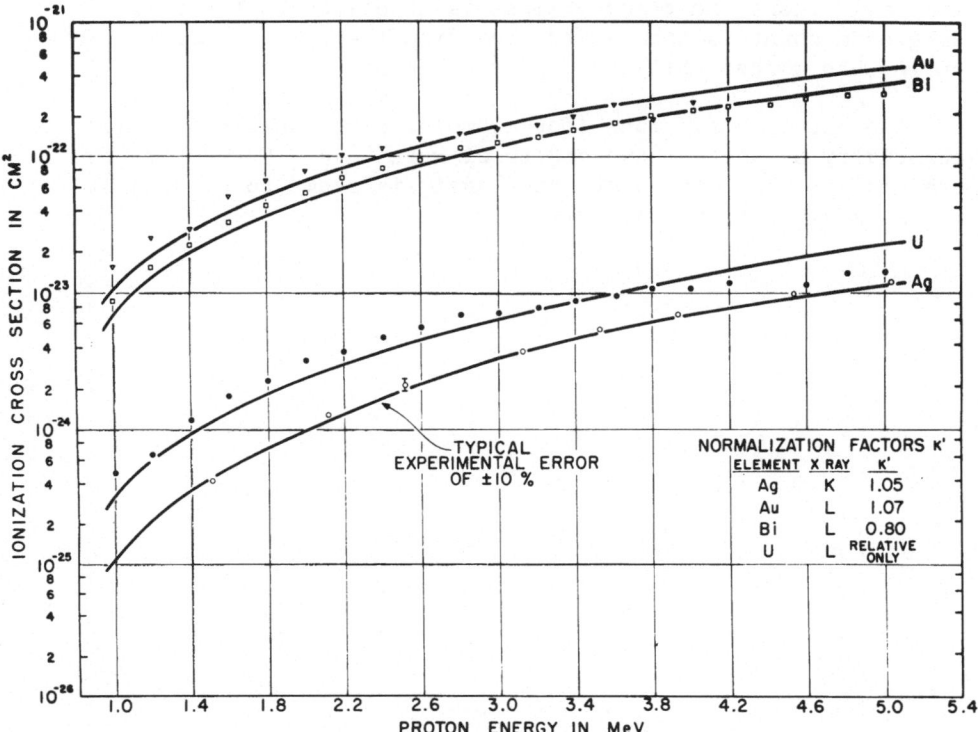

Figure 4.   Ionization Cross-Sections for the K_α Line from Silver
and the L_α Lines from Gold, Bismuth, and Uranium

100 mm^2. The beam was collimated so that it impinged onto only 1
mm^2 of this area, or 1 x 10^{-10} gm of zinc. An accumulation time of
30 minutes gave ~ 20,000 counts under the Zn – $K_{\alpha 1}$ peak. Hence, a
bombarded sample of 1 x 10^{-12} gm would give 200 counts in 30 minutes
which is clearly enough for identification and reasonable quantitative measurements. Since spectra of charged-particle x rays are
usually quite free from background interference (see Figure 3),
better statistics can be obtained simply by increasing the running
time.

For our measurements the background was considerably reduced
by fabricating the target-holding apparatus from pure, reactor-
grade graphite. The Coulomb scattered protons from the defining
apertures impinged onto only graphite. For trace analysis studies,
effects such as scattering, contaminant build-up on the target, etc.,
are quite important. However, once these experimental problems are
solved, one should be able to analyze samples on a routine basis.

The best material to use as a backing for trace targets is
self-supported carbon. The contamination problem associated with
carbon foils can be solved by determining the background count on

each foil before the trace element is deposited on the foil. The background count is then subtracted from the trace spectra. This routine has worked well for our measurements.

The use of heavy ions (for example, neon) should increase the sensitivity by another two orders of magnitude. Currently, the potential of this method of trace analysis seems to be unequaled by most analytical techniques.

REFERENCES

1. J. Chadwick, "The α Rays Excited by the β Rays of Radium," Phil. Mag. $\underline{24}$, 594 (1912)

2. W. Bothe and H. Franz, "Products of Atomic Disintegration, Reflected α-Particles and X-Rays Excited by α-Particles", Z. Physik $\underline{49}$, 1-26 (1928)

3. H. A. Barton, "Comparison of Protons and Electrons in Excitation of X-Rays by Impact", J. Franklin Inst. $\underline{209}$, 1-19 (1930)

4. C. Gerthsen and W. Reusse, "Anregumg von Charakteristischer Rontgenstrahlung durch Kanalstrahlenstob" or "Stimulation of X-Radiation by Canal-Ray Impacts," Phys. Z. $\underline{34}$, 478-482 (1933)

5. M. S. Livingston, F. Genevese, and E. J. Konopinski, "The Excitation of Characteristic X-Rays by Protons," Phys. Rev. $\underline{51}$, 835 (1937).

6. W. Henneberg, "Electron Scattering by Heavy Atoms," Z. Physik $\underline{83}$, 555-580 (1933)

7. H. H. Lewis, B. E. Simmons, and E. Merzbacher, "Production of Characteristic X-Rays by Protons of 1.7 - 3 MeV Energy," Phys. Rev. $\underline{91}$, 943 (1953).

8. E. M. Bernstein, and H. W. Lewis, "L Shell Ionization by Protons of 1.5 - 4.25 MeV Energy," Phys. Rev. $\underline{95}$, 83, (1954).

9. J. M. Hanstein, and S. Messelt, "Characteristic X-Ray Produced by Proton of 0.2 to 1.6 MeV Energy," Nucl. Phys. $\underline{2}$, 526 (1956/57)

10. S. Messelt, "K-Shell Ionization by Protons," Nucl. Phys. $\underline{5}$, 435, (1958).

11. B. Singh, "Excitation of Characteristic K X-Rays by Protons, Deuterons, and Alpha Particles," Phys. Rev. $\underline{107}$, 711 (1957).

12. E Merzbacher and H. W. Lewis, "X-Ray Production by Heavy Charged Particles," Encyclopedia of Physics, S. Flugge, Editor, Vol. 34, Berlin: Springer-Verlog, (1958).

13. R. C. Jopson, H. Mark, and C. D. Swift, "Production of Characteristic X-Rays by Low Energy Protons," The Physical Review, $\underline{127}$, 1612 (1962).

14. J. M. Khan and D. L. Potter, "Characteristic K-Shell X-Ray Production in Magnesium, Aluminum, and Copper by 60- and 500-keV Protons," Phys. Rev. $\underline{133}$, A890 (1964).

15. J. M. Khan, D. L. Potter, and R. D. Worley, "Characteristic X-Ray Production in the L_{III} Shell of Copper by Low Energy (100 to 500 KeV) Protons," Phys. Rev. 134, A316 (1964).

16. J. M. Khan, D. L. Potter, and R. D. Worley, "Characteristic X-Ray Production in the M Shell of Nd, Sm, Gd, Tb, Dy, and Ho by 25 - 100 keV Protons," Phys. Rev. 135, A511 (1964).

17. J. M. Khan, D. L. Potter, and R. D. Worley, "Characteristic X-Ray Production in the M_V Shell in Ytterbium by 30-100 keV Protons," The Physical Review. 136, A108, (1964).

18. J. M. Khan, D. L. Potter, and R. D. Worley, "Studies in X-Ray Production by Proton Bombardment of C, Mg, Al, Nd, Sm, Gd, Tb, Dy, and Ho," The Physical Review. 139, A1735 (1965).

19. A. A. Sterk, "X-Ray Production in the L Shell of Copper by 25-1700 eV Protons," The Physical Review. 145, 23 (1966).

20. J. M. Khan, D. L. Potter, R. D. Worley, and H. P. Smith, Jr., "Characteristic X-Ray Production in Single Crystals (Al, Cu) by Proton Bombardment I. Protons of 70-100 keV." Phys. Rev. 148, 413 (1966).

21. A. A. Sterk, C. L. Marks, and W. P. Saylor, "Production Efficiencies of X-Ray Emission Spectra by Proton Bombardment," in J. B. Newkirk and G. R. Mallett, Editors, Advances in X-Ray Analysis, Vol. 10, New York: Plenum Press, (1966).

22. J. M. Khan, D. L. Potter, R. D. Worley, and H. P. Smith, Jr., "Characteristic X-Ray Production in Single Crystals (Al, Cu, W) by Proton Bombardment, II, Protons of 250 to 1560 keV," Phys. Rev.163, 81 (1967).

23. L. J. Christensen, J. M. Khan, and W. F. Brunner, "Measurement of Microgram Surface Densities by Observation of Proton Produced S-Rays" Rev. Scien. Inst. 38, 20 (1967).

24. J. M. Khan, D. L. Potter, R. D. Worley and H. P. Smith, Jr. "Characteristic X-Ray Production in Single Crystals (Al, Cu, W) by Proton Bombardment II. Protons of 250 to 1560 keV." Phys. Rev., 163, 81 (1967).

25. A. A. Sterk, C. L. Marks, and W. D. Saylor, "Production Efficiencies of X-Ray Emission Spectra by Proton Bombardment" Advances in X-Ray Emission Spectra by Proton Bombardment" Advances in X-Ray Analysis, Vol. 10, 399 (1967).

26. R. R. Hart, N. T. Olson, H. P. Smith, Jr., J. M. Khan, "Oxygen Surface Density Measurements Based on Characteristic X-Ray Production by 100 keV Protons" J. of Appl. Phy. 39 5538 (1968).

27.   R. R. Hart, F. W. Reuter, III, H. P. Smith, Jr., and J. M. Khan, "Oxygen K Shell X-Ray Production in Thin Films of Aluminum Oxide by 20 to 100-keV Protons", Phys. Rev. $\underline{179}$ (1969).

28.   G. A. Bissinger, J. M. Joyce, E. J. Ludwig, W. S. McEver, and S. M. Shafroth, "Study of the Production of K X-Rays in Ca, Ti, and Ni by 2-28 MeV Protons," Phys. Rev. $\underline{A1}$, 841 (1970).

29.   T. B. Johansson, R. Akselsson, and S. A. E. Johansson, "X-Ray Analysis: Elemental Trace Analysis at the 10^{-12}g Level," Nucl. Inst. and Meth. $\underline{84}$, 141 (1970).

30.   J. D. Spaulding, "Charged Particle Excited X-Rays," Proceedings of the Second Oak Ridge Conference on the Use of Small Accelerators for Teaching and Research, Jerome L. Duggan, Editor, p. 113 (1970).

31.   R. P. Sharma, B. V. Thosar, and K. G. Prasad, "X-Ray Yields From K Shell Ionization by α Particles," Phys. Rev. $\underline{140}$, A1084 (1965).

32.   W. Brandt, R. Laubert, and Ivan Sellin "Characteristic X-Ray Production in Magnesium, Aluminum, and Copper by Low-Energy Hydrogen and Helium Ions", Phys. Rev. $\underline{151}$, 56 (1966).

33.   C. L. Marks, W. P. Saylor, and A. A. Sterk, "X-Ray Analysis by Proton Bombardment" Proceedings of 2nd Symposium on Low-Energy X- and Gamma-Sources and Applications. Published as ORNL-IIC-10, p. 587 (1967).

34.   B. Sellers, H. H. Wilson, and J. Papadopoulos, "Heavy Particle Excitation of Low Energy Characteristic X-Radiation-Analytical and Experimental Correlations", Proceedings of 2nd Symposium on Low-energy x- and Gamma Sources and Applications, ORNL-IIC-10, p. 576 (1967).

35.   R. G. Der, T. M. Kavanagh, J. M. Khan, B. P. Curry, and R. J. Fortner, "Production of Carbon Characteristic X-Rays by Heavy-Ion Bombardment", Phys. Rev. Letters, $\underline{21}$, 1731 (1968).

36.   W. Brandt, and R. Laubert, "Ionization of the Aluminum K Shell by Low Energy Hydrogen and Helium Ions", Phys. Rev. $\underline{178}$, 225 (1969).

37.   R. J. Fortner, B. P. Curry, R. C. Der, T. M. Kavanagh, and J. M. Khan, "X-Ray Production in C^{+}-C Collisions in The Energy Range 20 keV to 1.5 MeV", Phys. Rev. $\underline{185}$, 164 (1969).

38.   P. Richard, I. L. Morgan, T. Furuta, and D. Burch, "Observed K_β Energy Shift in Cu and Ni" Phys. Rev. Lets. $\underline{23}$, 1009 (1969).

39. W. Brandt and R. Laubert, "Pauli Excitation of Atoms in Colli-
     sion", Phys. Rev. Letters 24, 1037 (1970).

40. D. Burch, and P. Richard "X-Ray Spectra From Oxygen-Ion Bombard-
     ments on Ca and V at 15 MeV." Phys. Rev. Letters 25, 983 (1970).

41. M. E. Cunningham, R. C. Der, R. J. Fortner, T. M. Kavanagh, J.
     M. Khan, C. B. Layne, E. J. Zanaris, and J. D. Garcia, "X-Ray
     Spectra from Argon-Argon Collisions", Phys. Rev. Letters 24,
     931 (1970).

42. P. Richard, T. I. Bonner, T. Furuta, I. L. Morgan and J. R.
     Rhodes, "Cu Kα/Kβ X-Ray Production from Proton and Oxygen Bom-
     bardment," Phys. Rev. 1A, 1044 (1970).

43. H. J. Stein, H. O. Lute, P. H. Mokler, K. Sistemich, and P.
     Armbruster, "Impact-Parameter Dependence of Inner-Shell Vacancy
     Production by Heavy-Ion Bombardment", Phys. Rev. Letters 24,
     701 (1970).

44. R. L. Watson, C. W. Lewis, and J. B. Natowitz, "X-Ray Emission
     Induced by 30 to 80 MeV Alpha Particles," Nucl. Phys. A154, 561
     (1970).

45. C. W. Lewis, J. B. Natowitz, and R. L. Watson, "Precise Test of
     the Z^2 Dependence of X-Ray Emission Induced by α Particles and
     Deuterons," Phys. Rev. Letters 26, 481 (1971).

46. P. H. Mokler, "Energy Shift of Characteristic X-Rays Induced in
     Collisions Between 15 to 60 MeV I Ions and Mo, Yb, and Au
     Targets," Phys. Rev. Letters 26, 811 (1971).

47. G. L. Lockwood, "Production of Ions of Alkali Metals, Other
     Metals, and Halogens in an rf Source," Rev. Sci. Instruments
     Vol. 37, No. 2, p. 226.

48. R. S. Hall, D. H. Poole, M. S. Stagg, "Production of Metallic
     and Other Ions in rf Ion Source," Review of Sci. Instr., Vol.
     37, No. 7, p. 956.

49. Jerome L. Duggan, Editor, "Proceedings of The Second Oak Ridge
     Conference on The Use of Small Accelerators for Teaching and
     Research", Available from the Clearinghouse for Federal and Te
     Technical Information, National Bureau of Standards, Spring-
     field, Virginia 22151. Conf. No. 700322 (1970).

50. Jerome L. Duggan, Editor, "Proceedings of the Use of Small Ac-
     celerators for Teaching and Research," Oak Ridge, Tenn., Avail-
     able from Clearinghouse for Federal and Tech. Information,
     National Bureau of Standards, Springfield, Virginia 22151,
     Conf. No. 680411 (1968).

51. G. Dearnaley, "Preparation of Thin Self Supporting Carbon Foils"
     Rev. Sci. Instr. Vol. 31, No. 2, p. 197 (1960).

52. E. H. Kobisk, "Targets for Nuclear Research," Proceedings of
     the Conference on the Use of Small Accelerators for Teaching and
     Research, Jerome L. Duggan, Editor. Available from the Clearing-
     house for Federal Scientific and Technical Information, Nation-
     al Bureau of Standards, Springfield, Virginia, Conf. No. 680411,
     p. 426, (1968).

53. R. L. Watson, C. W. Lewis, and J. B. Natowitz, Nuclear Physics
     A 154, p. 561-575 (1970).

54. International Atomic Energy Agency, Vienna Calibrated Gamma
     Sources.

55. G. W. Grodstein, "X-Ray Attenuation Coefficients for 10 keV to
     100 meV," National Bureau of Standards Circular 583, Washington
     D.C. (1957).

56. H. A. Liebhafsky, H. G. Pfeiffer, E. H. Winslow, and P. D.
     Zemany, "Values of Mass Absorption Coefficients of Elements in
     the Region 0.1 Å, "X-Ray Absorption and Emmissions in Analytical
     Chemistry. New York: John Wiley and Sons, Inc. (1960).

57. G. S. Khandelwal, B. H. Choi, and E. Merzbacher, "Tables for
     Born Approximation Calculations of the K and L-Shell Ionization
     by Proton and Other Charged Particles," Atomic Data 1, 103 (1969)

58. J. A. Cairus, D. F. Holloway, and R. S. Nelson, "Heavy Ion-
     Induced X-Ray Generation as an Analytical Probe" Available from
     Solid State Division, Atomic Energy Research Establishment,
     Harwell Berkshire, England as Report No. AERE-R-6490.

59. D. M. Poole and J. L. Show, "Microanalysis with a Proton Probe,"
     in G. Mollensteadt and K. H. Gaukler, Editors, 5th International
     Congress on X-Ray Optics and Microanalysis, p. 319-324, Spring-
     er-Verlay, Berlin (1968).

60. J. A. Cairns, D. F. Holloway and R. S. Nelson, "Characteristic
     X-Ray Generation by Heavy Particle Irradiation of Copper," in
     D. W. Palmer, M. W. Thompson, and P. D. Townsend, Editors,
     Atomic Collision Phenomina in Solids, p. 541-552, North Holland
     Publishing Co. (1970).

EVALUATION OF X-RAY IMAGE INTENSIFIERS AS DETECTORS FOR X-RAY ASTRONOMY

W. S. Andrus, L. P. VanSpeybroeck, E. M. Kellogg and
H. Gursky

American Science & Engineering

Cambridge, Massachusetts   02142

ABSTRACT

Proposed experiments in X-ray astronomy require a detector capable of forming a high resolution image of a weak source. An X-ray image intensifier incorporating a microchannel plate (MCP) has been studied as a candidate for this application. Experiments have shown that the device is sensitive to single X-ray photons and has adequate quantum efficiency. Consideration has been given to the possibility of improving the quantum efficiency by deposition on the input surface of the MCP of a material with high photoelectric yield in the wavelength region of interest. Photographs of the output light from the image intensifier show than an X-ray photon detected at the MCP results in a spot on the film of about 60μM diameter. Since the position of the centroid of the spot could be determined to better accuracy, the device has sufficient resolution for the contemplated experiment. Detector noise is found to be so low that it would no limitation on the experiment's sensitivity. The image can be retrieved and transmitted to the ground by focussing a television camera on the face of the fiber optic output of the X-ray image intensifier; the data show that a Secondary Electron Conduction vidicon is sensitive enough to see the output resulting from a single X-ray photon. An alternative technique would sense charge pulses from the MCP on a fine wire grid, avoiding the need for a phosphor and providing superior time resolution and simpler data compression.

424

Since the detector is to be used in a satellite-borne experiment, its space qualification must be established. Two launches on sounding rockets have failed to impair the usefulness of one unit that has been studied over a period of more than a year. A long term test of continuous performance in high vacuum has so far shown encouraging results. A count life test has shown that an MCP can last through detection of orders of magnitude more counts than could be expected in the course of a mission. X-ray image intensifiers appear entirely suitable for high resolution X-ray astronomy.

INTRODUCTION

The study of soft X rays from celestial sources is capable of yielding astronomical information which is not otherwise obtainable. Some of the most interesting results in the next few years will come from the use of X-ray telescopes, which employ grazing incidence reflection to form a focussed image of the source (1). Since the weakness of the sources requires exposure times much longer than the jitter period of the space vehicle containing the telescope, visible light aspect data must be employed to determine the source position of each detected X-ray quantum. Thus a single quantum detector with high spatial resolution is essential to the program.

The most attractive candidate for this application is the microchannel plate (MCP), a wafer-like array of parallel glass tubes with diameters as small as 0.002 inches or less (2). Photoelectrons produced by an incident X-ray photon are accelerated down a tube by a potential of the order of one kilovolt. The photoelectrons strike the tube walls to produce secondary electrons, and an avalanche develops. The burst of electrons emerging from the tube can be detected, for example, by accelerating it through several kilovolts to a phosphor. We have studied the applicability of MCP's to X-ray astronomy and found the results very encouraging.

EXPERIMENTAL RESULTS

The experimental arrangement is illustrated in Figure 1. With an X-ray beam normally incident on the full MCP face, measurements can be made in two ways. The charge pulses at the phosphor are sensed by a charge-sensitive preamplifier and ultimately recorded in a pulse height analyzer. At the same time, a visible light image is formed on the output of a fiber optic element coupled to the phosphor. If an intense X-ray beam illuminates the MCP through a resolution mask, an image of the mask is clearly visible at the fiber optic output. For most of our experiments, the flux was lower and no mask was used; in this case individual flashes, each representing detection of a single X-ray photon, could be seen. By making a time-exposed photograph and later counting spots on the developed film, we obtained an independent measurement of the performance of the MCP.

The crucial parameter is the quantum efficiency, operationally defined as the probability that an incident X ray will result in a count in the data. We have measured the quantum efficiency at three wavelengths in the X-ray facility shown in figure 2. The diffusion-pumped source chamber contains tungsten filaments and an anode of aluminum or graphite. With accelerating voltage a little above the K-shell ionization potential, the resulting X-ray beam is dominated by the characteristic line. This beam entered the detector chamber, which housed the image intensifier, through a 3μM Mylar window. Since diffusion pump oil affects the properties of an MCP, the detector chamber is held at clean high vacuum by an ion pump-titanium sublimation well combination with sorption roughing. The beam intensity is monitored in the source chamber by a methane-filled flow proportional counter with a 3μM Mylar window; or by a sealed beryllium windowed counter. In either case the same thicknesses of the same

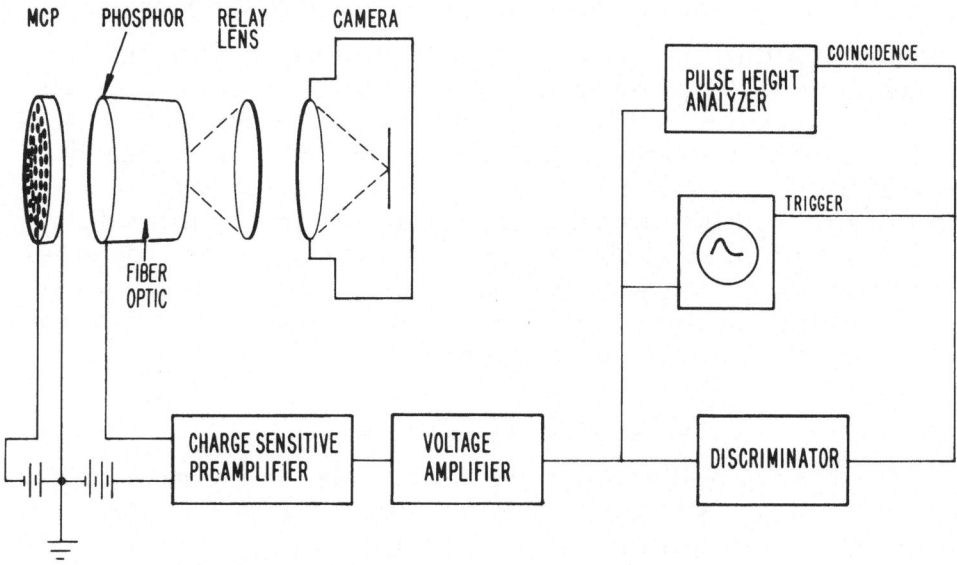

Figure 1. The electrical arrangement for image intensifier studies. The electrons emerging from the MCP excite a phosphor, and are simultaneously sensed by a charge sensitive preamplifier. Spectral information is stored in the pulse height analyzer while the visible light image is recorded on film.

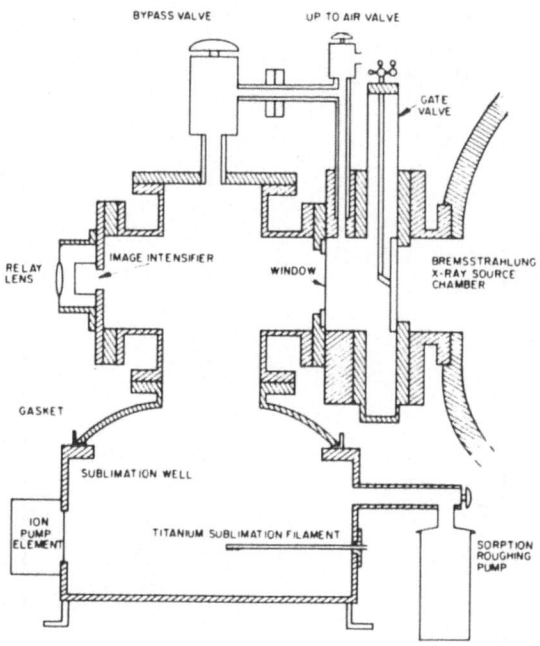

Figure 2.   The image intensifier test facility.   Clean high vacuum is maintained in the detector chamber at left, which is isolated from the diffusion-pumped source chamber by a thin organic window.

materials are traversed in reaching the MCP as in reaching the interior of the proportional counter, so that window transmission cancels out of the quantum efficiency measurements.   The counter is moved out of the beam during irradiation of the MCP.   The quantum efficiencies at three wavelengths are shown in figure 3.   The point at 12 Å was obtained with an Fe^{55} source.   The values obtained by counting spots in photographs of the fiber optic output agree well with those obtained from pulse height spectra.   In each case a threshold is present.   The pulse height spectrum from the MCP is exponential with no peak, because the gain is not high enough for saturation to occur.   In integrating the spectrum to obtain the total number of counts, a lower limit set by electronic noise must be observed.   The threshold for the photographic data is that gain for which a spot can be noticed in scanning the developed film.   The fact that the photographic and electronic measurements agree indicates that the thresholds are about the same.   A change in the potential across the MCP changes its gain and thus changes the observed quantum efficiency. A threshold-independent upper limit on the quantum efficiency can be obtained by extrapolating the spectrum to zero and to infinity and integrating.   The values we report are about 60% of this upper limit.

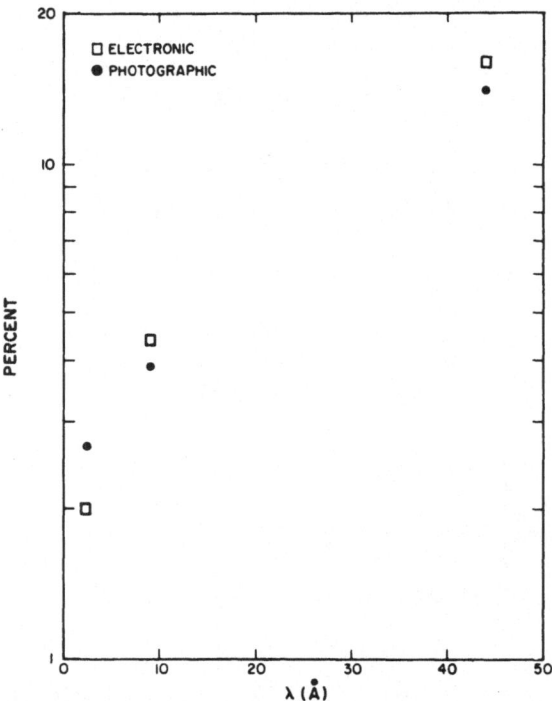

Figure 3.   The quantum efficiency of an MCP at three X-ray wavelengths.   The similarity of the electronic and photographic values
indicates that the effective threshold in each case is the same.

The photographic data provide good indications of the resolution
and noise of the device.   Celestial X-ray sources are generally so
weak that the probability of two overlapping counts in a single exposure is negligible.   The resolution therefore depends, not in the
size of a spot, which is about $60\mu M$, but on the accuracy with which
its centroid can be located; the uncertainty in the measurement is not
much greater than the channel diameter, so the MCP will not degrade
the resolution of an experiment using any X-ray telescope so far proposed.   The noise can be determined by counting the number of spots
in a photograph taken with no X rays incident on the MCP; it never
exceeds 2 counts per second over the full 1 inch diameter of the
MCP.   Since the image of a celestial source would normally be very
much smaller than this, MCP noise will be no problem.

READ-OUT TECHNIQUES

Although for sounding rocket work a film record of the image is
quite convenient, satellite-borne experiments will require other techniques.   One approach would be to take television pictures of the
fiber optic output, and return the video signal to the ground via

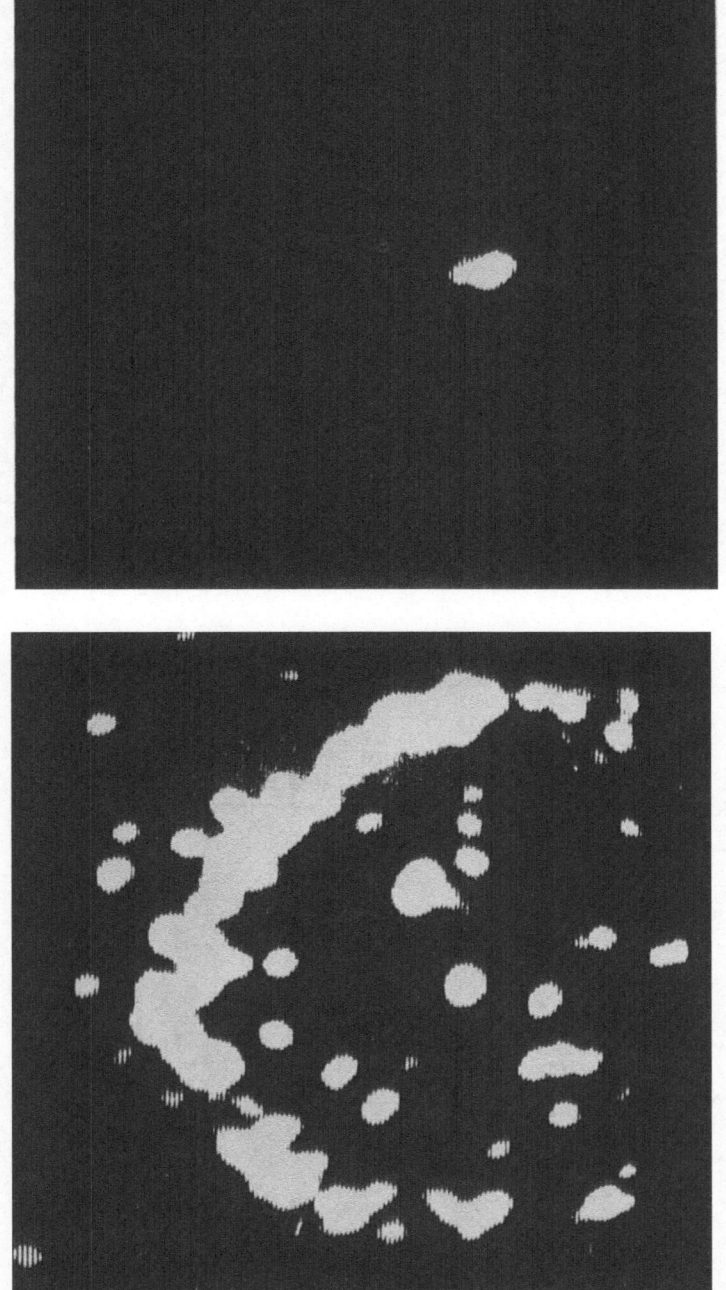

Figure 4. Photographs of a video monitor, showing the signal from a vidicon focussed on the fiber optic output of an image intensifier. The left hand picture shows the pattern with the MCP 0.25 inches from the focus of an X-ray telescope. The focussed image is shown in the right hand picture.

telemetry.   To test the practicality of this technique, we assembled
a system comprising a grazing incidence X-ray telescope, image intensifier, and Secondary Electron Conduction (SEC) vidicon.   The
telescope focussed X rays from a small source 220 feet distant onto
the MCP.   The vidicon viewed the fiber optic output through a relay
lens.   Figure 4a is a photograph of the monitor when the MCP was
deliberately placed 0.25 inch out of focus.   The ring-shaped pattern
is characteristic of an annular telescope.   It is clear that single
photon events do register in the video signal.   Figure 4b shows the
picture after moving the MCP into focus.

We are considering an alternative scheme which would have
certain advantages over use of a vidicon.   This approach would replace the phosphor by an array of closely spaced capacitively coupled
wires.   The wires would be at positive potential, so the electron
burst from a channel would be attracted to the array.   The charge ratio
between the end wires of the array is directly related to the position
at which the electrons were detected.   Two crossed arrays can be
used to obtain two-dimensional position information.   The electronic configuration is depicted in figure 5.

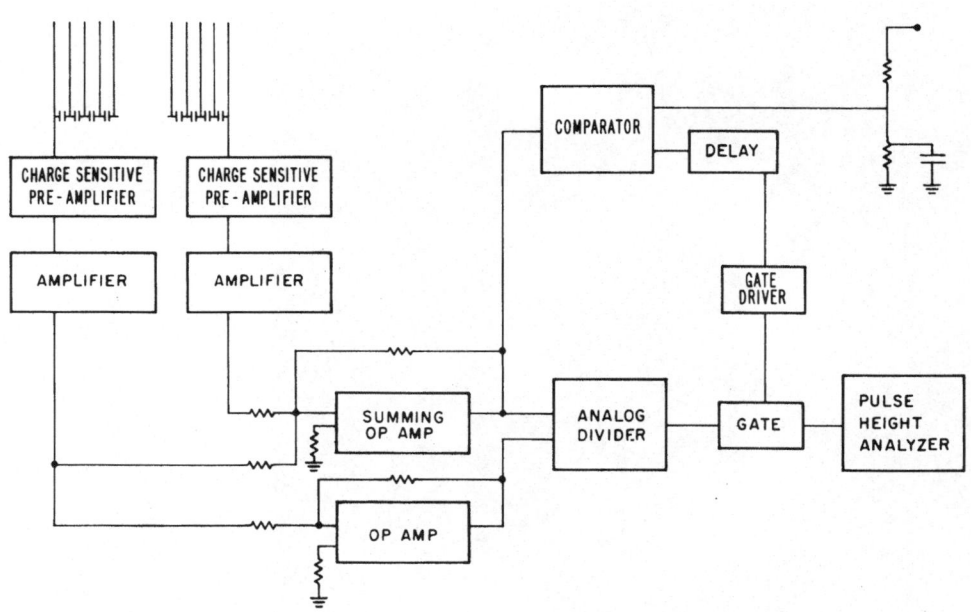

Figure 5.   An alternate read-out technique avoiding the need for a
vidicon.   The detected charge ratio between end wires of the array is
directly related to the position of the electron burst from the MCP.

SPACE QUALIFICATION

The detector in a satellite experiment must go through launch without damage and operate stably for up to two years without degradation due to time, vacuum, or count life. Although an MCP is easily broken by forces applied normal to its face, such forces are not great during launch because the mass of the MCP is very small. We have subjected an MCP to two Aerobee rocket launches and at least twenty vibration tests, as well as numerous vacuum cycles, considerable handling, and at least one incident of breakdown due to exposure to high voltage while in the corona region of pressure. The performance of this unit is still good, as is that of another which was stored in a room environment for four months. We consider the ruggedness of an MCP established.

A measurement of the count life of a demountable MCP has been performed. The MCP was mounted in the fixture shown in figure 6. Both contacts were polished stainless steel. Experience has shown

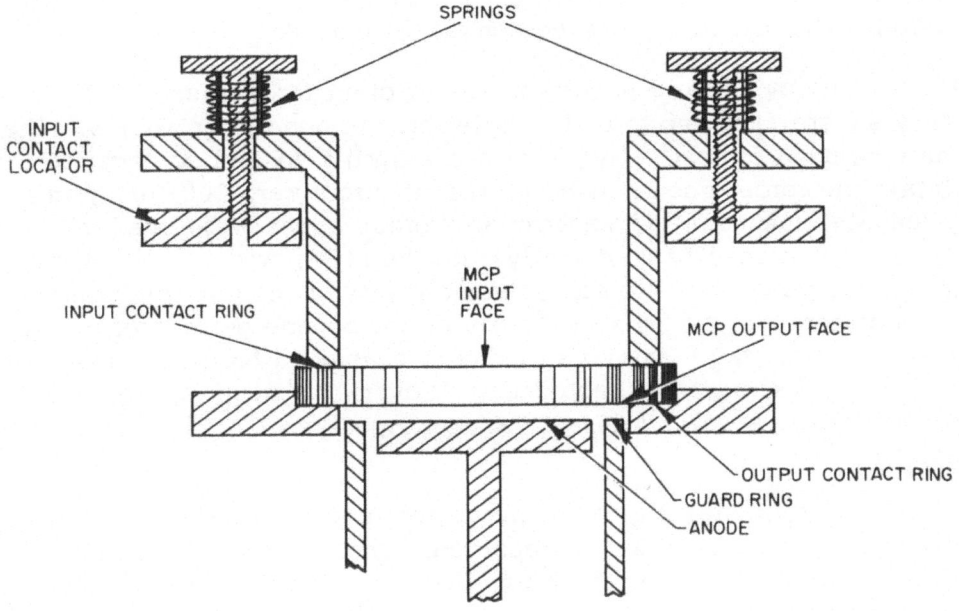

Figure 6. A fixture for testing demountable MCP's. The compression of the springs can be varied to provide the correct contact force. The guard ring eliminates edge effects at the anode.

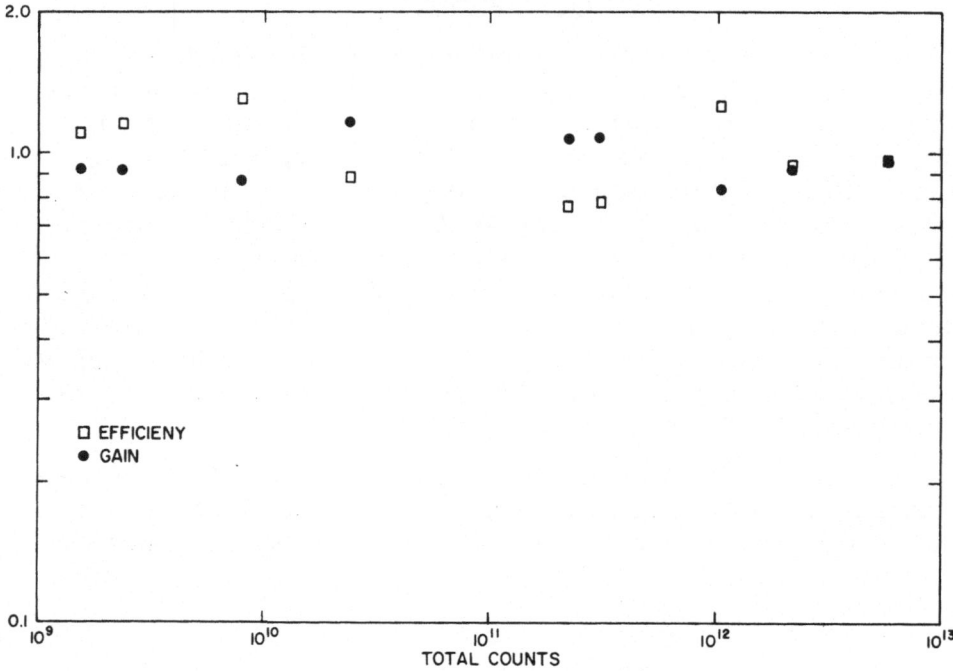

Figure 7. The relative gain and quantum efficiency of an MCP exposed to about 1000 times the counts expected in an X-ray astronomy mission. The measurements were made with an Fe^{55} source.

that the contact force is important; insufficient force caused sufficiently severe sparking to drill a hole through one of our MCP's. The fixture we used for collecting data had a spring-loaded contact to maintain the force recommended by the manufacturer, 200-300 grams. The tungsten filament and water cooled brass anode were used to shine a very intense beam directly onto the MCP, with a $6\mu M$ aluminum foil blocking ultra violet. Measurements of the gain and quantum efficiency for X rays from an Fe^{55} source were made periodically. As figure 7 shows, there was no significant change in the performance of the MCP even after it had recorded well over 10^{12} counts. The number of counts in a celestial X-ray astronomy mission could hardly exceed 10^{10}

The long term stability of an image intensifier operated continuously in high vacuum is being measured. The device is held in a stainless steel chamber with an ion pump-titanium sublimation well combination to keep the pressure below 10^{-8} Torr. An Fe^{55} source shines X rays onto the MCP at a rate similar to that expected in a mission. The test has been running for four months and is continuing.

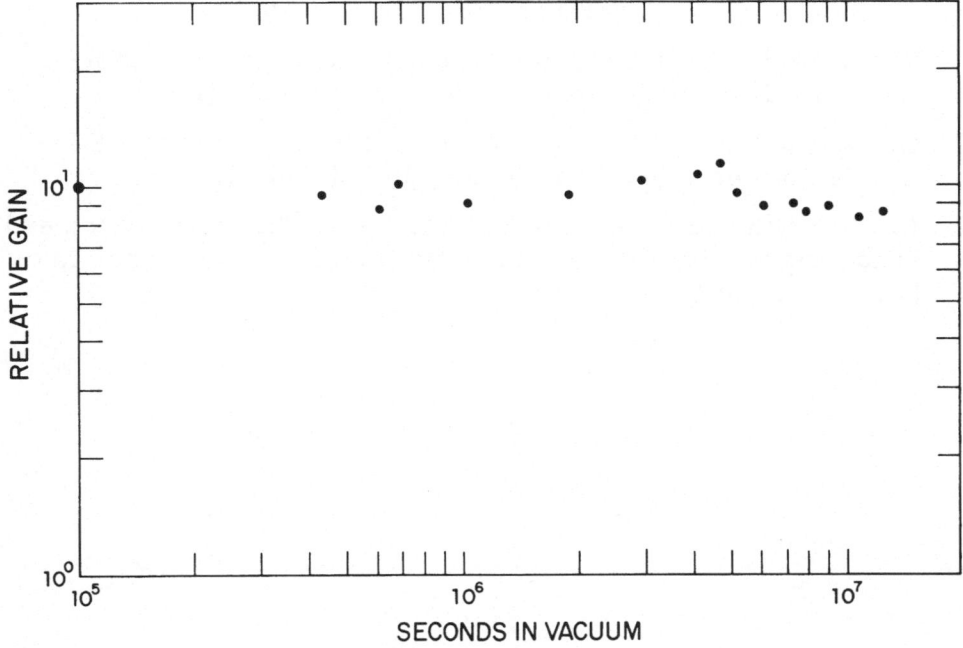

Figure 8.   The gain stability of an MCP in continuous operation.

No significant change has been noted in the performance of the MCP.
Figure 8 summarizes the data so far.

CONCLUSIONS

No factor has emerged in the course of our testing program to
suggest that an MCP will not work well in satellite-borne soft X-ray
astronomy.   The device has good spatial resolution, and enough gain
to be read out conveniently by a vidicon, wire grid, or other technique.   The wire grid could make excellent use of the rapid response
of the MCP in pulsar measurements, without sacrificing resolution or
efficiency.   The quantum efficiency, while less than ideal, is quite
adequate, particularly at the longer wavelengths, which are important for extra-galactic sources.   It is likely that the quantum efficiency can be improved by depositions on the MCP input surface of
efficient photoelectric materials (3).   Studies of space qualification
demonstrate the ruggedness, lifetime, and stability are such that
other elements in the payload will set the limits.

REFERENCES

1.   W. P. Reidi, "Soft X-ray Instrumentation for Space Environments", Advances in X-ray Analysis 13, 313-329 (1969)

2.   G. W. Goodrich and W. C. Wiley, "Continuous Channel Electron Multiplier", Rev. Sci. Instr. 33, 761-762 (1962)

3.   D. G. Smith and K. A. Pounds, Soft X-ray Photon Detection and Image Dissection Using Channel Multipliers", IEEE Trans. Nuc. Sci. NS 15, 541-550 (1968)

AN ELECTRO-OPTICAL X-RAY DIFFRACTION SYSTEM FOR GRAIN

BOUNDARY MIGRATION MEASUREMENTS AT TEMPERATURE

Robert E. Green, Jr.

The Johns Hopkins University

Baltimore, Maryland 21218

ABSTRACT

Considerable work has been undertaken in order to gain an understanding of the mechanisms responsible for the generation of recrystallization textures developed upon annealing of cold-worked metals. Most direct measurements have consisted of measuring the increase in average diameter of the largest grain growing into a polycrystalline aggregate. Experimental measurements of individual boundaries migrating into deformed single crystals, though of a more fundamental nature, have been made by far fewer investigators. This is probably due to the increased experimental difficulties associated with careful control of such experiments. Most previous investigators have made grain boundary migration measurements by the heat-cool-etch method, despite the fact that it has several marked disadvantages. Other investigators have constructed an X-ray goniometer furnace and used it to measure grain boundary migration rates while the test specimen was maintained at temperature. Since there have been no published reports of the use of such a system in the past thirteen years, it must be concluded that the technique was unsuccessful in general.

The system described in the present work is relatively simple in design and extremely simple to use. Not only does it permit absolute measurement of grain boundary position at temperature but it also permits boundary migration measurements to be made of extremely fast moving boundaries. The basic components of the system are as follows. A continuous spectrum X-ray beam is converted by a slit collimating system into a beam which is incident along the entire length of the test specimen. This beam is interrupted by a wire grid just prior to impingement on the test

specimen. The test specimen is supported vertically in a furnace
maintained at the temperature required for grain boundary migration. The various diffracted X-ray beams pass out of the furnace
through a highly reflecting insulating baffle made from very thin
aluminum foil and impinge on a fluorescent screen. This screen
converts the X-ray image into a visible one which is amplified and
recorded using the electro-optical system.

INTRODUCTION

Considerable work has been undertaken in order to gain an
understanding of the mechanisms responsible for the generation of
recrystallization textures developed upon annealing of cold-worked
metals. Two theories were proposed historically to account for the
experimentally observed textures, that of preferred growth(1,2) and
that of preferred nucleation(3). Numerous experiments have been
performed in order to test the validity of these theories. Studies
involved with detection and orientation determination of nuclei
have been limited due to the experimental difficulties present in
such investigations. The development of electron microscopic
techniques has opened new possibilities in this regard, but defi-
nite proof as to preferred nucleation accounting for the observed
recrystallization textures is still lacking.

On the other hand, experiments concerned with measurement of
grain boundary migration rates have been more prevalent. The
large majority of such experiments have consisted of measuring the
increase in average diameter of the largest grain growing into a
polycrystalline aggregate. Since selection of the largest grain
for observation in such experiments already biases the results,
the conclusions reached from these experiments may not give an
accurate picture of the origin of recrystallization textures.
Experimental measurements of individual boundaries migrating into
deformed single crystals, though of a more fundamental nature,
have been made by far fewer investigators. This is probably due
to the increased experimental difficulties associated with careful
control of such experiments. Review articles or books summarizing
the general state of knowledge in the area of recrystallization
have appeared in 1959(4), 1961(5-10), 1962(11), 1963(12), 1965(13)
and in 1966(14).

PREVIOUS MEASUREMENT SYSTEMS

Most previous investigators, with the exceptions of Graham and
Cahn(15,16) and Leighly, et al.(17), have made grain boundary migration rate measurements by one of two methods. The method used most
often is termed the heat-cool-etch technique. In this technique the
initial position of the grain boundary is recorded, the test

specimen is placed in a furnace maintained at the desired temperature, and grain boundary migration is allowed to take place for a
given interval of time. Then the specimen is removed from the
furnace, allowed to cool, and etched to reveal the new position of
the grain boundary. The test specimen is returned to the furnace
for a second given interval of time, removed, allowed to cool, and
etched. This process is repeated until sufficient data have been
obtained to determine the grain boundary migration rate. Although
this technique has been used most often it has several disadvantages. The time period during which the grain boundary actually
moves at temperature is uncertain since it is impossible to determine exactly at what temperature and time the grain boundary
begins to move and also it is impossible to determine exactly at
what temperature and time the grain boundary stops moving upon
removal from the furnace. Additional disadvantages are that the
repeated thermal cycles may alter the driving force for boundary
migration and that the etching, by its very nature of delineation
of the boundary position, can alter the surface tension by grooving
at the grain boundary - air interface and thus impede migration of
the boundary. Since boundary migration rates are generally made
using relatively thin test specimens, the influence of etching may
be a large one.

The second method used for grain boundary migration studies
serves to eliminate some of the disadvantages of the heat-cool-etch
method although other complications are introduced. This method
consists of using a number of identical bi-crystal specimens with
identical grain boundaries. It should be noted that this very
requirement imposes severe restrictions upon the applicability of
the method. The initial positions of the grain boundaries are
recorded and all of the test specimens are placed in the furnace
simultaneously. After a given interval of time one test specimen
is removed from the furnace, cooled, etched, and the new position
of the grain boundary measured. After a second given interval of
time another test specimen is removed from the furnace, cooled,
etched, and the new position of the grain boundary measured. This
process is repeated after successive intervals of time until all
of the specimens have been removed from the furnace. The grain
boundary position is then plotted as a function of time to determine the rate, with each test specimen contributing one data point
to the overall curve. This second method eliminates thermal
cycling and repeated etching, but introduces problems of its own
with regard to the ability of producing sufficient specimens
identical in all essential respects.

Graham and Cahn(15,16) constructed a goniometer furnace and
used it to measure grain boundary migration rates by means of
X-rays while the test specimen was maintained at temperature.
Essentially the same system was used several years later by

Leighly, et al.(17). The test specimen containing the grain
boundary is placed in the goniometer furnace and is oriented so
that one of the two grains on either side of the boundary is
suitably positioned with respect to the incident X-ray beam which
passes through a slit in the furnace. Next, the specimen is
rotated into proper Bragg reflection geometry so that a strong
diffracted X-ray beam passes out of the furnace and is detected
by a Geiger counter. When the specimen is moved in the goniometer
furnace parallel to its own length and the incident X-ray beam
strikes the second grain the counting rate will drop since in
general there will no longer be a strong X-ray beam diffracted
into the counter. The position of the grain boundary is determined
by the position of the test specimen at which the change in counting rate occurs. The general procedure used in operating the X-ray
goniometer furnace was a relatively complicated one and even then
absolute determination of boundary position was not possible. The
accuracy in rate of boundary migration was estimated to be about
10 percent. Since there have been no published reports of the use
of such a system in the past thirteen years, it must be concluded
that the technique was unsuccessful in general.

PRESENT MEASUREMENT SYSTEM

The system to be described in the present work is similar to
that constructed by Graham and Cahn and Leighly, et al. in that an
X-ray beam is used to determine the position of the grain boundary.
However, the present system is much simpler to use and permits
rapid interchange of test specimens. Not only does the present
system permit absolute measurement of grain boundary position with
an accuracy essentially limited to that of the resolving power of
the X-ray fluorescent screen, but migration measurements can be
made of extremely fast moving boundaries.

The basic components of the present system, depicted
schematically in Fig. 1, are as follows. A continuous spectrum
X-ray beam is converted by a slit collimating system into a beam
which is incident along most of the length of the test specimen.
This beam is interrupted by a wire grid just prior to impingement
on the test specimen. The test specimen is supported vertically
in a furnace maintained at the temperature required for grain
boundary migration. The various diffracted X-ray beams pass out
of the furnace through a highly reflecting insulating window made
from several very thin aluminum foil sheets and impinge on a fluorescent screen. Cemented across the vertical center of the
fluorescent screen on the side facing the incident X-ray beam is
a strip of copper of sufficient thickness to attenuate the direct
transmitted beam to an intensity level comparable with that of the
more intense diffracted beams. The fluorescent screen converts the

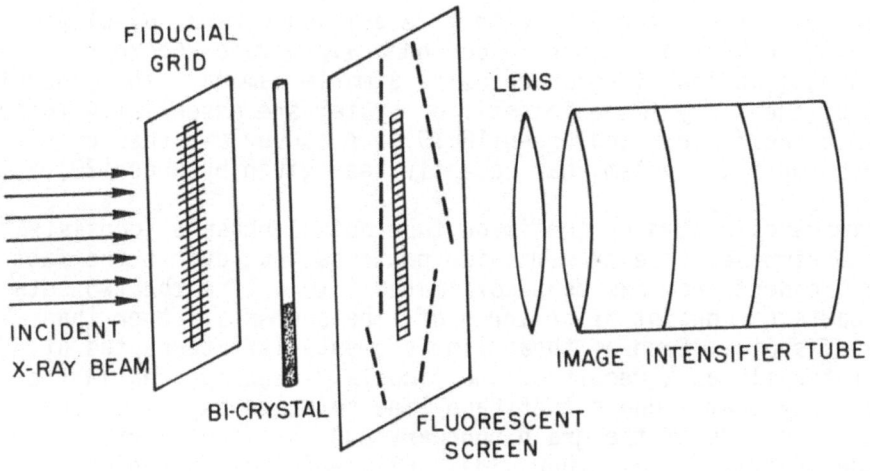

Figure 1.   Schematic of Electro-Optical X-ray
            Diffraction System

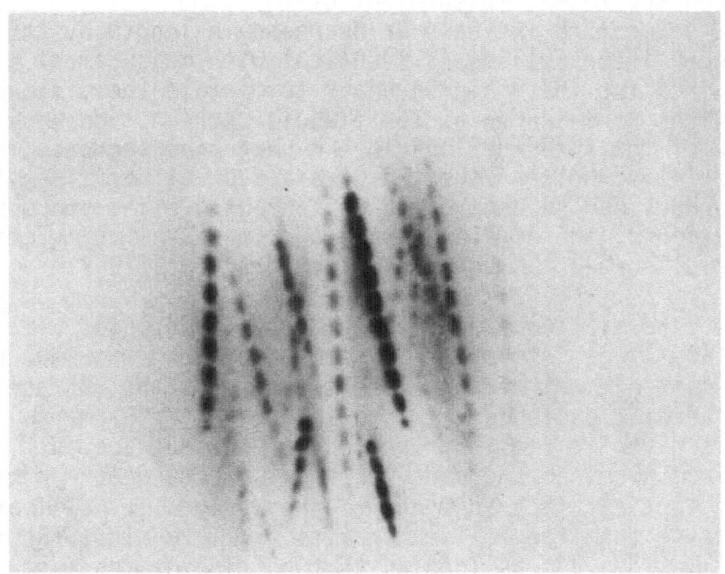

Figure 2.   Typical Bi-Crystal X-ray Diffraction Pattern
            (35 mm Tri-X Film, f/2.8, 1/30 sec)

X-ray image into a visible one, which is projected by means of a
very fast demagnifying lens on to the input photocathode of an
image intensifier tube. The intensified output image from the
image tube is either picked up with a television camera, displayed
on a television monitor, and recorded on a video tape recorder or
else is photographically recorded using a movie camera. The general
features of the X-ray image intensifier system are essentially those
described by Reifsnider and Green(18,19). A survey of other possible electro-optical systems has recently been given by Green(20).

The image displayed on the image tube output phosphor consists
of two superimposed Laue transmission patterns, but due to the fact
that the incident beam has been collimated into a line the two sets
of Laue patterns consist of an array of line segments rather than
an array of spots. Each of these line segments is interrupted at
regular intervals as a result of the fiducial screen placed in the
incident X-ray beam. One set of Laue line segments is due to the
crystal on one side of the grain boundary and the second set is
due to the crystal on the other side. Figure 2 shows a typical
diffraction pattern obtained from a bi-crystal specimen as recorded
photographically from the output phosphor of the image intensifier
tube. The photograph was taken on 35 mm Tri-X Film at f/2.8 with
an exposure time of 1/30 second. When the grain boundary moves,
one set of line segments increases in length and the other set
becomes shorter. Since the actual distances between wires on the
fiducial screen are known, measurement of the time required for
any Laue line segment to increase or decrease in length by the
distance between interruptions is identical with measurement of
the time required for the grain boundary to migrate the distance
between two wires as measured at the fiducial screen. Moreover,
by making use of the interruptions in the Laue line segments, the
orientation relation between the two crystals on either side of
the grain boundary can be determined continuously. The procedure
for determination of the crystallographic orientation using this
type of fiducial screen has been described previously(21).

Since the essential components and methods associated with the
X-ray system and image intensifier system have been described previously, the main element of the present system is the furnace
itself. A schematic exploded view showing the major elements of
the furnace is given in Fig. 3. The body of the furnace is
constructed from aluminum and marnite, and is rectangular in cross
section. The vertical face adjacent to the X-ray slit collimator
is countersunk so that the collimator fits firmly against the face.
A slit slightly wider and as long as that portion of the test
specimen not covered by the end guides is cut through the furnace
wall to permit the incident X-ray beam to strike the specimen.
This beam is interrupted by a wire grid just prior to impingement
on the test specimen. On either side of the test specimen, placed

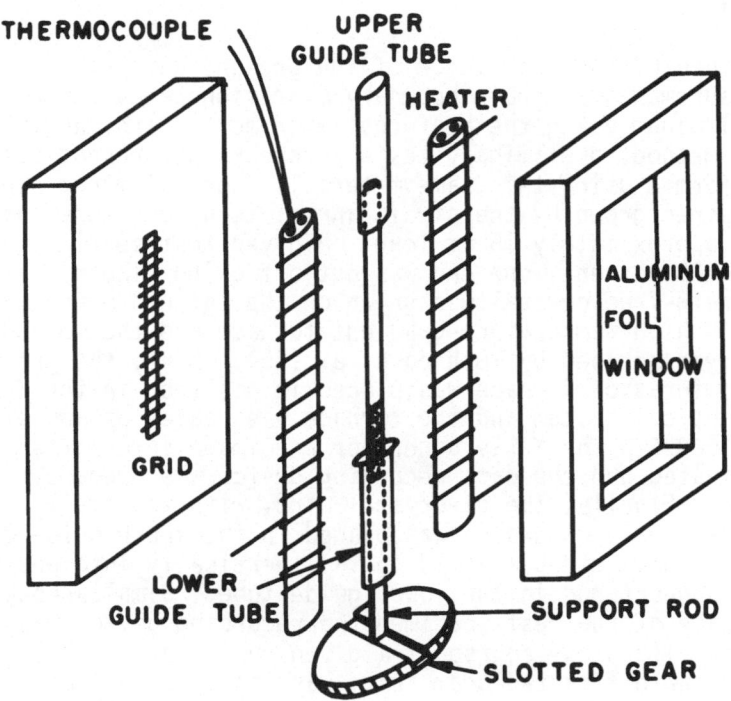

Figure 3. Schematic Exploded View of Main
Elements of Migration Furnace

parallel to it, are nichrome wire heating elements cemented on
alundum thermocouple tubes. The thermocouple which is used to
control the furnace temperature is inserted into one of the holes
in one of these thermocouple tubes. The heating elements are
placed as close as possible to the test specimen, with the pre-
caution that they do not interfere with the X-ray beams. The
specimen itself is supported vertically in an upper and lower
guide tube on top of a stainless steel support rod. The upper
end of the upper guide tube protrudes through the top of the
furnace. The lower end of the specimen support rod extends below
the bottom of the furnace and fits into a slot in a gear. This
gear can be rotated by means of a motor driven worm, which in turn
rotates the support rod and the bi-crystal test specimen. The
vertical face of the furnace adjacent to the image intensifier tube

has a window cut in it through which the diffracted X-ray beams
exit from the furnace. This window is covered by two separated
sheets of very thin aluminum foil which serve to reflect the heat
back into the furnace with very little absorption of the diffracted X-rays.

Because of the experience of the present author(22,23,24)
with measurements of grain boundary migration rates in technical
purity aluminum using the heat-cool-etch method and the multiple
specimen method, preliminary tests to check the present system
were performed using this same material. Long aluminum single
crystal wires grown by the strain-anneal technique were cut into
sections approximately 15 cm long. A given test section was
elongated 20 percent, one end was cut off using diagonal pliers,
and a strain-free crystal was grown on the cut end using a special
water-bath high-temperature-gradient furnace and the method
previously described by Yoshida et al.(25). Next, the grain
boundary migration furnace was placed in position in the X-ray
electro-optical system and the furnace was heated up and stabilized at 600°C. The X-ray generator and image intensifier tube
were activated and the data recording device was prepared for
recording. Finally, the bi-crystal wire, with the small
strain-free grain downward, was placed in the upper guide tube
atop the furnace and permitted to fall vertically into position
atop the support rod in the lower guide tube. Simultaneously
with release of the test specimen the recording device was
activated. If in the course of a given run a particularly
intense clear diffracted beam trace was situated so that it
crossed the direct beam attenuator or was in some other unfavorable position, then the motor which rotated the specimen was
activated and the specimen shifted into a more favorable
orientation.

Upon completion of the grain-boundary migration run, the data
recording device, the image intensifier tube, and the X-ray generator were cut off. The slotted gear was removed from beneath the
furnace and the support rod and test specimen permitted to fall out
of the furnace. In preparation for the next run, the gear was
placed back in position and the support rod was returned to its
proper position by dropping it in the upper guide tube. In this
manner tests were run one after another without the necessity of
cooling and reheating the furnace. The experimental results
obtained from one such preliminary run are shown in Fig. 4. This
figure shows the measured values for the distance the grain
boundary has moved as a function of time. It can be seen that the
boundary migration rate, as determined from the slope of the
experimental line, was constant at 2.5 mm/min except for two
positions where the boundary did not move for one minute. Also
plotted on the right vertical scale is the measured temperature
distribution in the furnace. Although this distribution occasioned

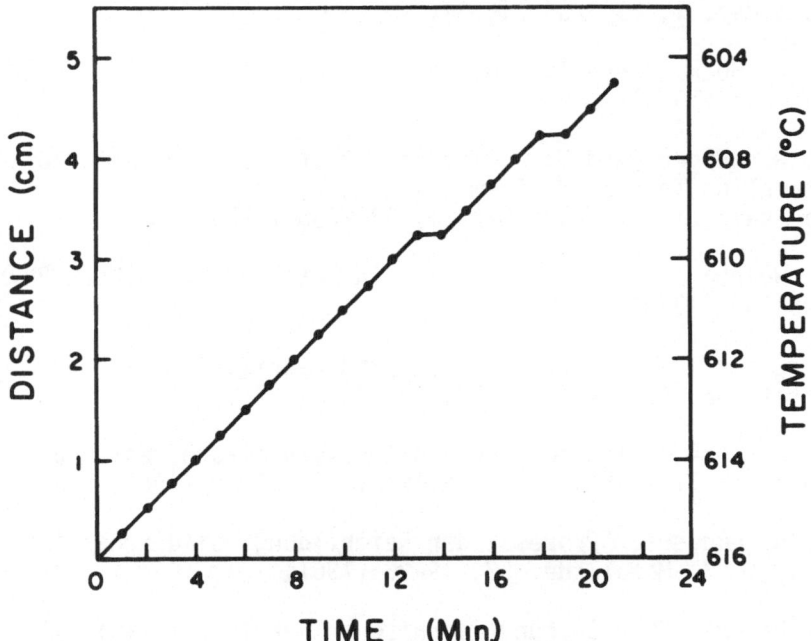

Figure 4. Experimental Results for Boundary Migration
           in Technical Purity Aluminum

a temperature gradient of 2.4°C/CM, this gradient apparently did
not influence the measured migration rate, since the rate was
essentially constant. Results of more extensive experiments and
direct comparison with measurements made by the heat-cool-etch
method will be presented elsewhere.

ACKNOWLEDGEMENTS

    The author would like to thank Russell Vane, III, for
technical assistance with all aspects of the experimental work.
He would like to thank Bernard Baker for his skill and infinite
patience in construction of the migration furnace. Finally, he
would like to thank Corinne Harness for typing the manuscript.

REFERENCES

1. P. A. Beck, "Notes on the Theory of Annealing Textures,"
    Acta Met. $\underline{1}$, 231-234 (1953).

2. P. A. Beck, "Annealing of Cold Worked Metals," Adv. in Phys. $\underline{3}$,
    245-324 (1954).

3. W. G. Burgers and T. J. Tiedema, "Notes on the Theory of
    Annealing Textures: Comments on a paper by P. A. Beck with
    the same title," Acta Met. $\underline{1}$, 234-238 (1953).

4. F. Weinberg, "Grain Boundaries in Metals," Prog. Met. Phys. $\underline{8}$.
    105-146 (1959).

5. K. Lücke, "Korngrenzenstruktur und Rekristallisation," Z.
    Metallkde. $\underline{52}$, 1-12 (1961).

6. P. A. Beck, "Die Bewegung von Grosswinkelkorngrenzen bei
    Rekristallisation," Z. Metallkde. $\underline{52}$, 13-19 (1961).

7. W. G. Burgers, "Prozesse der Keimbildung bei der Rekristallisa-
    tion," Z. Metallkde. $\underline{52}$, 19-26 (1961).

8. K. Detert, "Die Deutung Technischer Rekristallisation,"
    Z. Metallkde. $\underline{52}$, 27-34 (1961).

9. H. P. Stuwe, "Texturbildung bei der Primarrekristallisation,"
    Z. Metallkde. $\underline{52}$, 34-44 (1961).

10. U. Dehlinger, "Zur Theorie der Rekristallisation," Z.
    Metallkde. $\underline{52}$, 44-47 (1961).

11. K. T. Aust and J. W. Rutter, "Some Annealing Phenomena in
    High-Purity Metals," in Ultra-High-Purity Metals, Metals Park,
    Ohio (1962).

12. L. Himmel, Recovery and Recrystallization of Metals,
    Interscience Publishers (1963).

13. J. G. Byrne, Recovery, Recrystallization, and Grain Growth,
    The Macmillan Co. (1965).

14. Recrystallization, Grain Growth and Textures, American Society
    for Metals, Metals Park, Ohio (1966).

15. C. D. Graham, Jr., and R. W. Cahn, "Measurement of Grain
    Growth Rates in Recrystallization," Trans. Met. Soc. AIME $\underline{206}$,
    504-508 (1956).

16. C. D. Graham, Jr., and R. W. Cahn, "Grain Growth Rates and Orientation Relationships in the Recrystallization of Aluminum Single Crystals," Trans. Met. Soc. AIME 206, 517-521 (1956).

17. H. P. Leighly, Jr., R. A. McCune, and F. C. Perkins, "Research on the Recrystallization of Aluminum Single Crystals," Denver Research Institute, University of Denver, WADC Technical Report 58-634, ASTIA No. AD 212562 (1958).

18. K. Reifsnider and R. E. Green, Jr., "An Image Intensifier System for Dynamic X-ray Diffraction Studies," Rev. Sci. Instr. 39, 1651-1655 (1968).

19. K. Reifsnider and R. E. Green, Jr., "Dynamic X-ray Diffraction Study of the Deformation of Aluminum Crystals," Trans. Met. Soc. AIME 245, 1615-1619 (1969).

20. R. E. Green, Jr., "Electro-Optical Systems for Dynamic Recording of X-ray Diffraction Images," in C. S. Barrett, J. B. Newkirk, and C. O. Ruud, Editors, Advances in X-ray Analysis, Vol. 14, p. 311-337, Plenum Press (1971).

21. K. Reifsnider and R. E. Green, Jr., " X-ray Diffraction Macroscopic Study of Deformed Aluminum Crystals," Trans. Met. Soc. AIME 233, 932-936 (1965).

22. R. E. Green, Jr., H. Yoshida, and B. G. Liebmann, "Investigations on Recrystallization and Recovery of Strained Aluminum Single Crystals," Metals Res. Lab. Brown Univ. Tech. Note TN 59-381 AFOSR ASTIA No. AD 214 008 (1959).

23. R. E. Green, Jr., B. G. Liebmann, and H. Yoshida, "Influence of Thermal History on Preferred Orientations in the Recrystallization of Technically Pure Aluminum," Trans. Met. Soc. AIME 215, 610-613 (1959).

24. R. E. Green, Jr., "Some Factors Influencing Grain Boundary Migration in Aluminum," Trans. Met. Soc. AIME 233, 1954-1960 (1965).

25. H. Yoshida, B. Liebmann, and K. Lücke, "Orientation of Recrystallized Grains in Strained Aluminum Single Crystals," Acta Met. 7, 51-56 (1959).

PROPOSED FLASH X-RAY SYSTEM FOR X-RAY DIFFRACTION WITH SUBMICROSECOND EXPOSURE TIME

Francis M. Charbonnier

Field Emission Corporation

McMinnville, Oregon 97128

ABSTRACT

X-ray diffraction studies have generally used low current, low voltage X-ray tubes (e. g. , 50 mA, 50 kV). Consequently, exposure times ranging from seconds to hours have been required, limiting X-ray diffraction techniques to the study of essentially static situations. Using such an X-ray source with a very high-speed detector, R. E. Green was able to reduce the exposure time for laue patterns of aluminum crystals to a minimum of 0. 003 sec. Nanosecond exposure and pulse timing are required for certain events, e. g. , X-ray diffraction study of material under shock compression. Q. Johnson achieved this by using a low voltage, very high current flash X-ray source (approximately 50 kV, 50 kA, 30 nsec). However, such a system is difficult to build and to synchronize, and is not commercially available.

Published X-ray yield data, particularly the recent work of J. W. Motz, indicate that both continuum and characteristic X-ray yields increase rapidly with electron energy, reaching a maximum for an electron energy ranging from 150 keV for titanium to 300 keV for copper to nearly 1 MeV for molybdenum.

To take advantage of this property, a 300 kVp, 30 nsec pulse, 5 nsec jitter flash X-ray system has been built which appears to meet the intensity and timing requirements for single pulse laue or Bragg diffraction studies. System design, calculated output and initial tests are presented.

INTRODUCTION

Photographic or radiographic observations are greatly facilitated when the subject is static. Contrast and resolution can then
be optimized since long exposure times can be used to increase the
total radiation intensity. This is particularly true of most X-ray
diffraction work where high resolution patterns can be obtained
but exposure times are usually large.

There has been considerable recent effort to reduce X-ray
diffraction exposure times, as summarized in R. E. Green's
paper at last year's meeting (1). Most of this work shares two
common features: a) the subject is still essentially static, i.e.,
the reduction in exposure time is motivated by convenience rather
than by the need to freeze rapid motion; b) the main effort has concentrated on the development of improved detectors, capable of
much higher speed without excessive loss in resolution.

Consequently exposure times remain in the range of several
milliseconds to several seconds. The fastest diffraction system
reported to date is that of Reifsnider and Green (2). They use a
large fluorescent screen coupled by a lens to a three-stage image
intensifier whose output is displayed on a vidicon. This detection
system, used with a conventional 50 kV 50 mA X-ray source, can
produce laue patterns of aluminum single crystals in a minimum
of 4 milliseconds; of additional interest are the large field of view
(160 mm) and the fairly good resulting resolution ($>$300 lines).

X-ray diffraction techniques would add valuable insight in the
study of several high-speed processes, e.g., crystal phase
changes accompanying sudden temperature changes in certain
materials, or lattice plane changes under transient high stress
conditions (e.g., behind the front of a shock wave) or unstable
crystal phases which cannot be preserved for static observation.
These events usually occur so rapidly that they require extremely
short X-ray diffraction exposure times, in the microsecond or
even nanosecond range.

Such exposure times cannot be achieved by improved detectors
alone but require a major increase in the intensity of the X-ray
source, i.e., require the use of a flash X-ray system instead of
a dc X-ray source.

PAST FLASH X-RAY DIFFRACTION

The earliest reported flash X-ray diffraction experiments were performed by Tsukermann in Russia (3) who obtained laue patterns of aluminum single crystals in 1942 using a pulsed X-ray generator and millisecond exposures. During the next decade Schall (4) and Schaaffs (5, 6, 7) used high intensity 5 microsecond flash X-ray systems to record X-ray diffraction patterns in mono and polycrystalline specimens. This work was only partly successful because of the limited intensity and excessive jitter of the X-ray source.

In 1966 Tsukermann's group (8) sensed the potential advantage of using high voltage flash X-ray tubes to increase the characteristic X-ray yield from the target, and obtained 1 microsecond X-ray diffraction patterns of single and polycrystalline specimens.

The most successful recent flash X-ray diffraction work has been performed by Quintin Johnson at the University of California Radiation Laboratory (9, 10, 11, 12). Johnson and co-workers spent several years improving the performance and reliability of a relatively low voltage, extremely high current flash X-ray system using a Blumlein pulsed voltage generator and a vacuum arc flash X-ray tube. The nominal output is a 50 kV, 50 kA, 30 nsec X-ray pulse. Johnson has obtained Bragg reflection from single crystals and powder patterns from polycrystalline samples, using small scintillators or screen-assisted Polaroid film as detector. In this manner he made the first diffraction observations of crystal structure and lattice compression just behind a rapidly moving shock front. As Johnson points out, a most difficult part of the experiment was the need to synchronize the X-ray pulse precisely with the arrival of the shock front at the surface since only a thin surface layer (e.g., $\cong 0.1$ mm) contributes effectively to the diffracted pattern. Though Johnson was successful in the laboratory, it appears that widespread use of flash X-ray diffraction techniques requires a commercial flash X-ray source more compact and easier to synchronize than his Blumlein system.

X-RAY EMISSION YIELDS VERSUS ELECTRON ENERGY

Flash X-ray diffraction requires a very small X-ray source. This severely limits the amount of electron beam energy which the X-ray target can absorb without damage during a submicrosecond pulse. It is therefore particularly important to optimize the characteristic X-ray emission yield. This yield is usually measured by the number of characteristic output photons per unit solid angle and per electron incident on the target. The yield of interest is the net yield, i.e., after reduction of the photon intensity by partial reabsorption in the target and in the X-ray tube window. In the past it has been common practice to operate X-ray diffraction sources at voltages of the order of 50 kV since it was felt that higher voltages would reduce the "beam quality" (i.e., the intensity of characteristic emission relative to the continuous bremsstrahlung emission), hence would decrease the useful contrast in the diffraction pattern.

Motz and co-workers (13) have recently published a brief summary of their extensive measurements of characteristic X-ray yields and purity for metal targets bombarded by monoenergetic electron beams. Their measurements cover electron energies ranging from a few keV up to 3 MeV, thin and thick targets made of materials ranging in atomic number from 4 (beryllium) to 79 (gold), and have also investigated the dependence of the X-ray yield on direction of emission relative to the target surface.

Figures 1 and 2 are based on Motz's results. Figure 1 shows that for thick targets the normal yield (i.e., the number of K photons per steradian per electron emitted in the direction normal to the target surface) reaches its maximum at an electron energy which increases rapidly with target atomic number and is already above 300 keV for copper.

Since the emission of characteristic X-rays is isotropic, one would expect oblique yields to be nearly the same as the normal yield at low electron energies (hence at low target reabsorption) but to become less than the normal yield at high electron energies and to reach their maximum for a lower electron energy. This is indeed the case as shown in Fig. 2. The normal and 30° yields are experimental and their relative values agree well with a simple model: isotropic production of characteristic X-rays distributed along the electron range, followed by reabsorption in the target. This model has been used to calculate the oblique yields at small angles (15° and 7.5°) or for target materials not covered by Motz's measurements.

These results clearly indicate that the optimum electron energy is of the order of 120 to 150 keV for flash X-ray diffraction with the K emission from a copper target slanted to minimize the

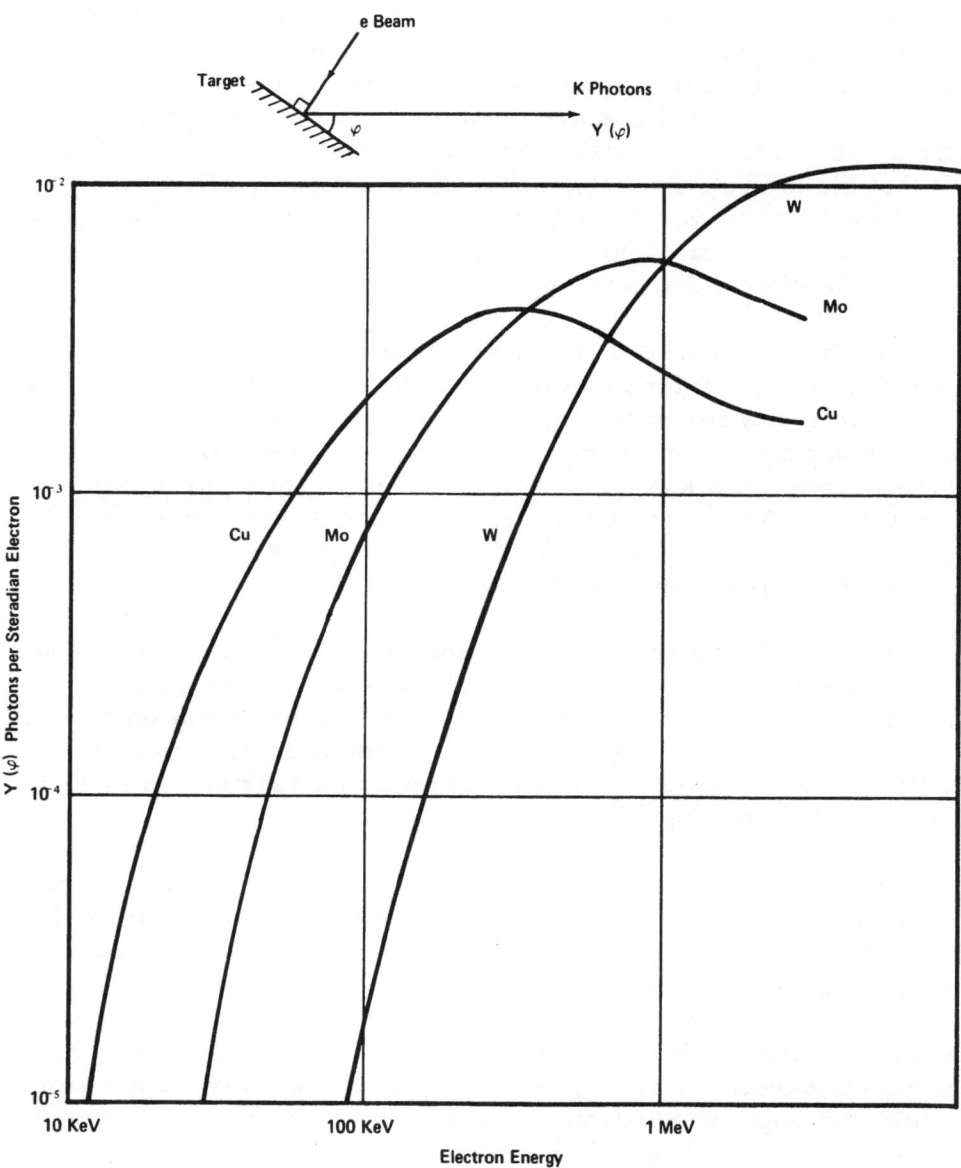

K X-ray Normal Yields ($\varphi = 90^O$)

Fig. 1.   Yield of characteristic X-ray emission in direction
normal to metal target versus electron beam energy.

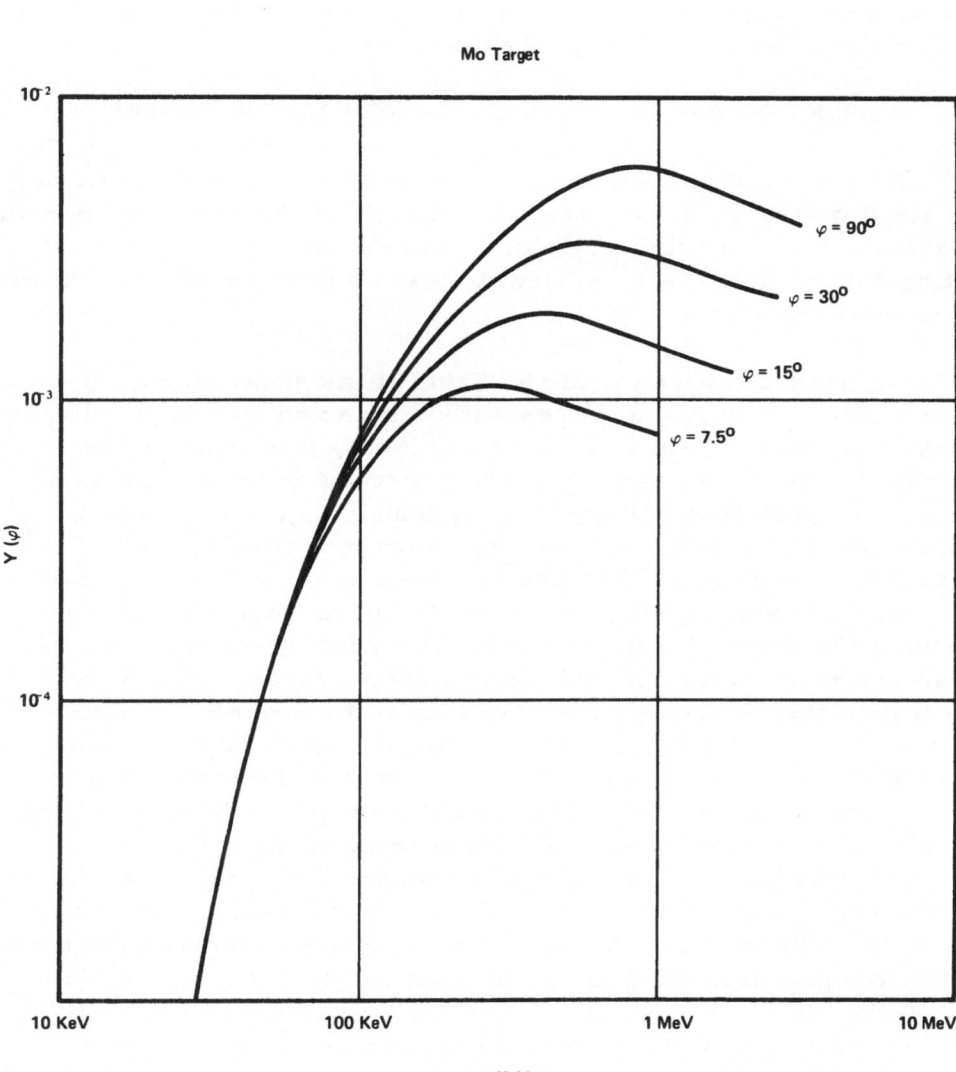

Fig. 2.   Characteristic X-ray yield as a function of angle of
emission for a molybdenum target bombarded by a
monoenergetic electron beam at normal incidence.

apparent X-ray source size in the direction of use. For molybdenum target an electron energy of approximately 300 keV maxi-
mizes the useful yield at an angle of 7.5°.

Though high voltages pose difficult insulation problems in dc
X-ray generators, they are easily achieved in the relatively compact short pulse flash X-ray tubes discussed below.

300 KV FLASH X-RAY SYSTEM FOR DIFFRACTION

Since the foregoing analysis indicated the desirability of using
an electron energy of the order of 300 keV to generate molybdenum
K radiation with maximum yield, it was decided to modify an existing 300 kV flash X-ray system to test its performance and X-ray
diffraction capability.

The standard system, Model 730-271, is shown in Fig. 3.
It consists of a Marx-surge generator connected by a high voltage
coaxial cable to a field emission flash X-ray tube in a remote
tubehead, with the trigger and timing circuits in the control console. The peak tube voltage is continuously adjustable up to a
maximum of nearly 400 kV corresponding to a peak current of
nearly 5000 amperes. The electron beam is converted to X-rays
in a cone-shaped tungsten target inside the vacuum tube. Figure
4 shows the output X-ray wave form recorded by means of a fast
response scintillator, photodiode and oscilloscope. This wave
form indicates an X-ray pulse duration of the order of 15 nsec at
half maximum intensity and 50 nsec at the base. At full output
the total electron beam charge in the tube is 370 microcoulombs,
i.e., approximately 0.4 mAs and the electrons striking the target
have a broad energy spread with an average of approximately
165 keV. With a standard trigger amplifier the X-ray pulse timing
can be controlled with a jitter of the order of 50 nsec, but an optional low jitter trigger circuit can be used which reduces the time
jitter to a guaranteed maximum of 5 nsec.

The standard X-ray tube is a high vacuum sealed-off tube
with a multiple needle cathode, a cone-shaped tungsten anode with
a 5.5 mm focus and a Kovar X-ray window. For X-ray diffraction
use the Kovar window was replaced by a .025" beryllium window
which is 90% transparent to 10 keV X-rays and 50% transparent
to 5 keV X-rays. Tubes have been built so far with tungsten and
molybdenum targets. The molybdenum target is useful for applications requiring high peak intensities of relatively high energy

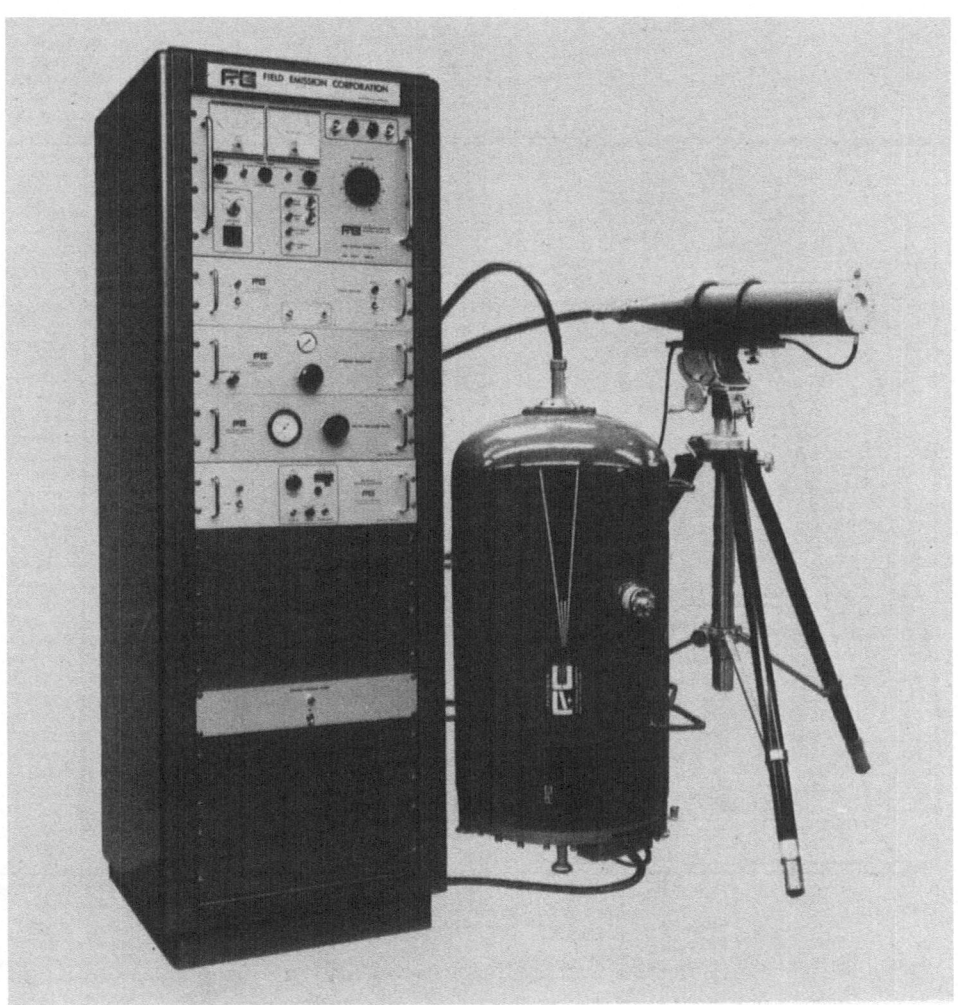

Fig. 3.   300 kV flash X-ray system Model 730-271.   Control
console, Marx-surge pulser and flash X-ray tube in
remote tubehead.

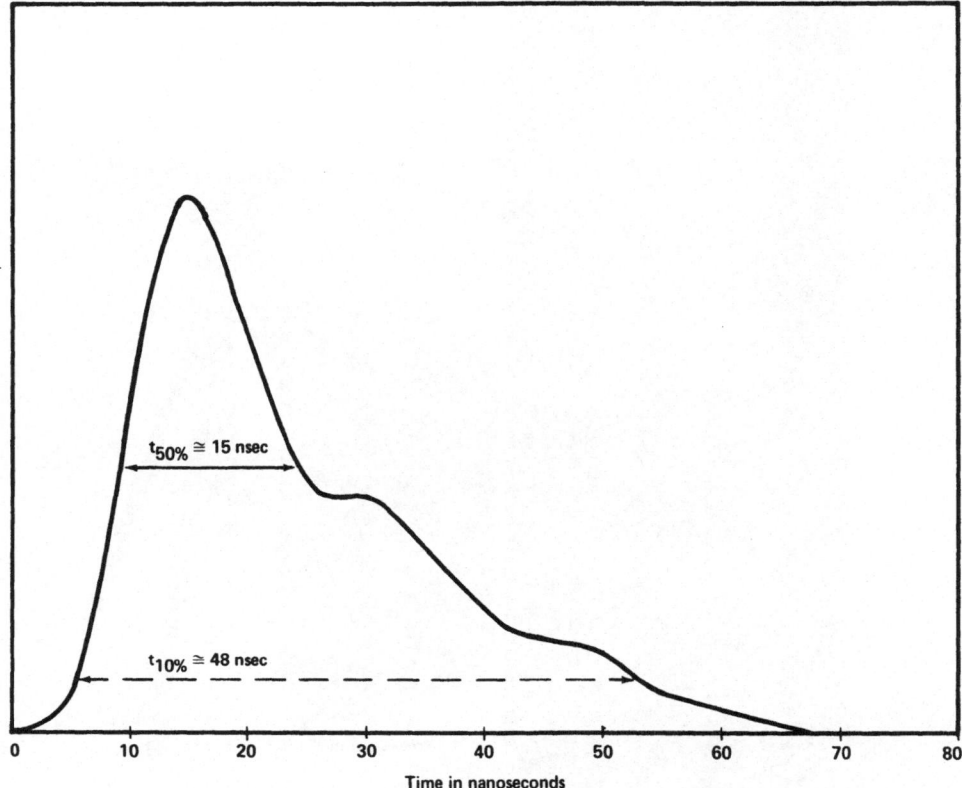

Fig. 4.   Output X-ray intensity wave form for Fexitron Model
           730-271.

characteristic radiation, whereas the tungsten target can be used either for its characteristic L emission at 8 keV or for its relatively intense soft bremsstrahlung output. Other target materials, e.g., copper, could of course be used in future tubes, e.g., to reduce bremsstrahlung background.

OUTPUT X-RAY SPECTRUM

The tube current and voltage wave forms have been measured by means of a shunt current viewing resistor and a capacitive voltage divider respectively. These wave forms were in turn used to calculate first the electron energy spectrum and then the resulting continuous bremsstrahlung spectrum and characteristic K X-ray intensity output of the tube, correcting for target and window absorption. Figure 5 shows the spectra calculated for molybdenum and tungsten targets. The spectrum discontinuities at 70 and 10 keV (W) and 20 keV (Mo) correspond to discontinuities in target absorption at the K and L edges. The tungsten bremsstrahlung output shows a broad peak near 40 keV and a sharp peak near 10 keV. The calculated tungsten K X-ray output is 40×10^{10} photons per steradian, i.e., 13% of the total photon output and 15 times the bremsstrahlung intensity per keV at 60 keV. By contrast the molybdenum target produces a more concentrated bremsstrahlung spectrum with a strong peak at 20 keV and also very intense K radiation: 160×10^{10} photons per steradian, i.e., 50% of the total photon output and 50 times the bremsstrahlung intensity per keV at 17.5 keV.

A direct experimental verification of the X-ray spectrum is difficult because of the high intensity and extremely short duration of the pulse. However two indirect tests were performed to provide a check on the calculated spectra. In the first test the total X-ray dose per pulse was calculated from the spectrum and compared to direct measurements with a low energy cut-off ionization chamber. The calculated total X-ray doses of 35 mR at 1 meter for the tungsten target and 50 mR for the molybdenum target were found to be in satisfactory agreement with the measured values. The second test compared the calculated and measured attenuation of the X-ray beams from the two targets in aluminum, and the results were also found to be in good agreement with predictions from the spectra. For the tungsten target tube the absorption rate was initially equivalent to that of 12 keV monoenergetic X-rays and increased rapidly to an equivalent 40 keV after 30 mils of aluminum. For the molybdenum target tubes the initial absorption

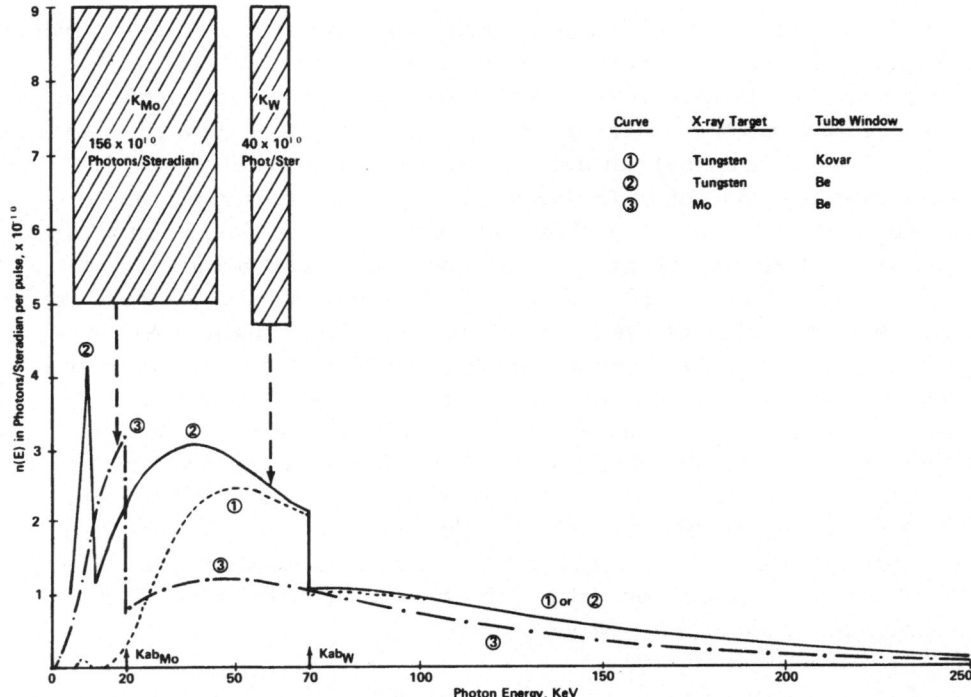

Fig. 5.   Calculated X-ray output energy spectrum n(E) for
           Fexitron Model 730-271 at maximum output voltage.
           The curves represent the continuous bremsstrahlung
           output in photons per steradian per keV.   The shaded
           areas are equal to the number of emitted K photons,
           in photons per steradian.

rate was equivalent to 16 keV and remained essentially constant at 18 keV for aluminum thicknesses ranging from 5 to 30 mils.

FLASH X-RAY DIFFRACTION TESTS

As discussed earlier, the fastest reported X-ray diffraction system is Reifsnider and Green's which requires only 4 ms at 50 mA, i.e., 0.2 mAs at 50 kV, to record laue patterns in aluminum single crystals. By comparison, the prototype diffraction source just discussed has a maximum output of 370 microcoulombs, i.e., 0.37 mAs at a peak electron energy of 380 keV and an average of 165 keV.

Hence, particularly considering the increase in photon yield per electron associated with higher electron energies, the flash X-ray source should have more than enough intensity for single pulse X-ray diffraction using Green's detector. An experimental program is planned at The Johns Hopkins University under the sponsorship of the Ballistics Research Laboratories, to evaluate the performance of the 300 kV flash X-ray source in combination with Professor Green's electro-optical detector for various types of X-ray diffraction, particularly laue and powder patterns. If, as expected, the system has surplus intensity it will be possible to collimate the X-ray beam more tightly and to increase the distances between tube, subject and detector. This will not only benefit resolution but also make it easier to achieve the tube and detector protection required in explosive events.

In the meantime a simple preliminary diffraction test was performed with the molybdenum target tube. Since powder patterns of polycrystalline samples require more X-ray intensity but are more generally useful than Bragg reflection studies from single crystals, and since film is a convenient and inexpensive detector, the test chosen was to record a single pulse exposure powder pattern from a polycrystalline lithium-fluoride pellet with a Polaroid XR-7 diffraction cassette. Figure 6 shows the experimental arrangement. The X-ray beam was collimated by two 1 x 4 mm slits, relatively short distances were used in view of the expected marginal X-ray intensity, and adequate lead shielding was used to shield the film from excessive exposure by the relatively hard bremsstrahlung primary beam.

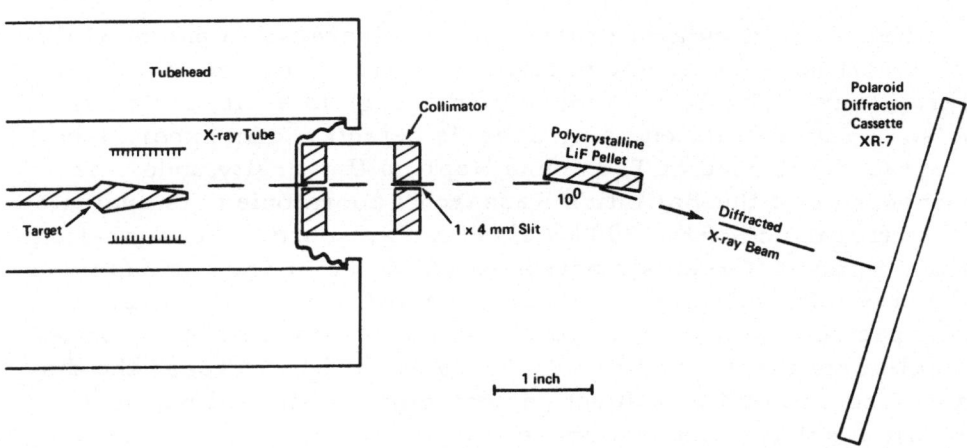

Fig. 6.   Experimental arrangement for flash X-ray powder
pattern of polycrystalline lithium-fluoride pellet.

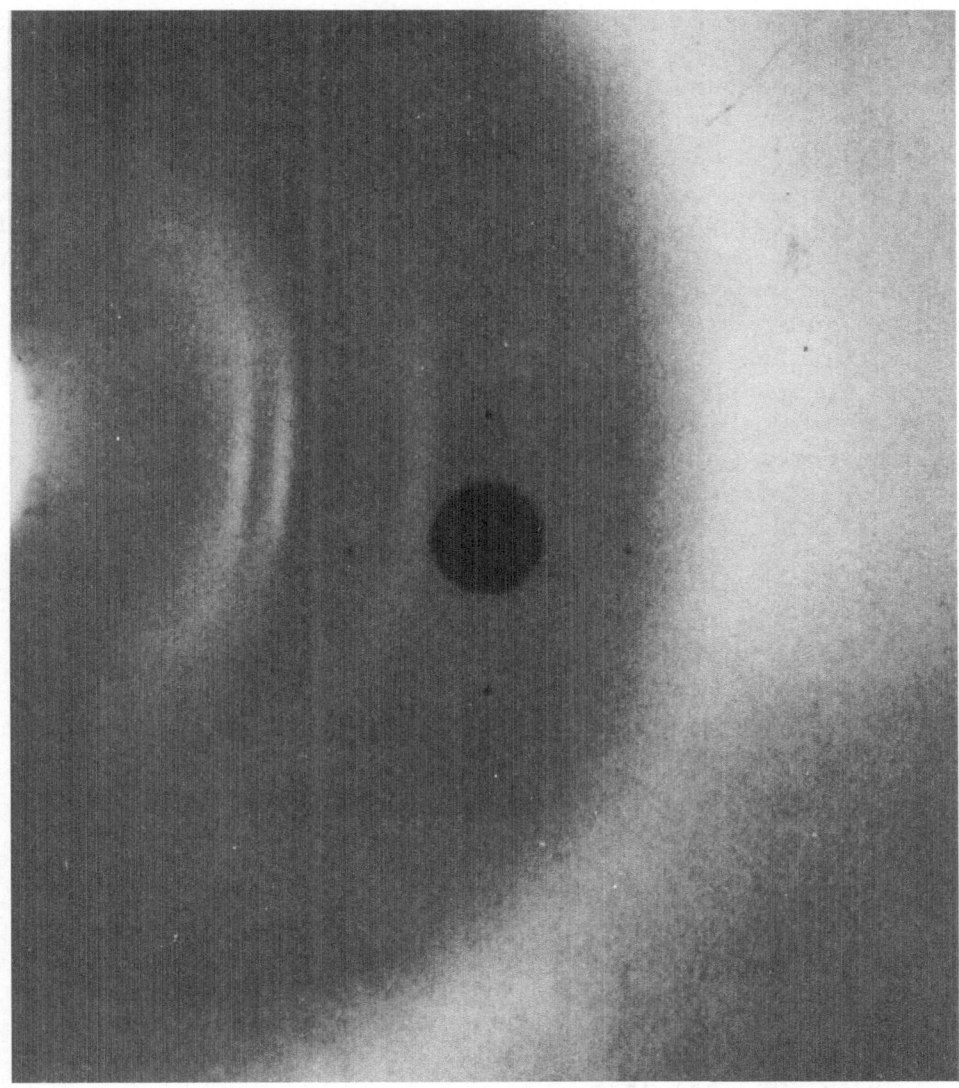

Fig. 7. Single pulse exposure (approximately 30 nsec) powder pattern using Polaroid XR-7 diffraction cassette.

Figure 7 shows the powder pattern observed. The flash X-ray system was operated at 85% of its maximum voltage since this was observed to reduce the background and improve contrast. Whereas the pattern has limited resolution and contrast, it still seems good enough to have diagnostic value.

CONCLUSIONS

Additional testing and improvement of the system are needed and planned. Nevertheless the initial results seem to substantiate the basic premise and calculations, namely 1) that the use of higher voltage is advantageous for submicrosecond flash X-ray diffraction because it increases the feasible X-ray intensity for a given target size and 2) that useful X-ray diffraction patterns can be obtained using a readily available, easily controlled flash X-ray source with exposure time short enough to allow time-resolved observations of very high-speed events.

ACKNOWLEDGMENTS

It is a pleasure to acknowledge the help of John Barbour and Frank Reece in performing the experimental flash X-ray diffraction test, stimulating discussions with Dr. Johnson and Professor Green, and the continued interest of Dr. Odell at the Ballistics Research Laboratories.

REFERENCES

1.   R. E. Green, Jr., "Electro-Optical Systems for Dynamic Display of X-Ray Diffraction Images," Advances in X-Ray Analysis, Vol. 14, Plenum Press (1971).

2.   K. Reifsnider and R. E. Green, Jr., "Image Intensifier System for Dynamic X-Ray Diffraction Studies," Rev. Sci. Instr. 39, 1651-1655 (1968).

3.   V. A. Tsukerman and A. I. Avdeenko, Zh. Tekhn. Fiz. 12, 185 (1942).

4.   R. Schall, Z. angew, Phys. 2, 83-88 (1950).

5.   W. Schaaffs, Z. Naturforsch. 5a, 631-632 (1950).

6.   W. Schaaffs, Erg. exakt. Naturwiss. 28, 1-46 (1954).

7.   W. Schaaffs, Z. angew. Phys. 8, 299-302 (1956).

8.   N. I. Zavada, M. A. Manakova, and V. A. Tsukerman, Prib. i Tekhn. Eksperim. No. 2, 434-438 (1966).

9.   Q. Johnson, R. N. Keeler, and J. W. Lyle, Nature 213, 1114-1115 (1967).

10. Q. Johnson, A. Mitchell, R. N. Keeler, and L. Evans, Trans. Amer. Cryst. Assoc. 4, 133-140 (1969).

11. Q. Johnson, A. Mitchell, R. Keeler, and L. Evans, Phys. Rev. Lets. 25, 1099-1101 (1970).

12. Q. Johnson, A. Mitchell, L. Evans, "X-Ray Diffraction Evidence for Crystalline Order and Isotropic Compression During the Shock-Wave Process," UCRL Report 73140, April 13, 1971.

13. J. W. Motz, C. E. Dick, A. C. Lucas, R. C. Placious, and J. H. Sparrow, "Production of High Intensity K X-Ray Beams", J. Appl. Phys. 42, No. 5, 2132-2133 (Apr 1971).

ANALYSIS OF SOLID SURFACES BY SOFT X-RAY APPEARANCE POTENTIAL SPECTROSCOPY

Robert L. Park and J. E. Houston

Sandia Laboratories

Albuquerque, New Mexico 87115

ABSTRACT

The total soft x-ray emission of an electron-bombarded surface exhibits abrupt changes at the threshold potentials for the excitation of core levels of surface atoms. These features represent the excitation probabilities of the core states superimposed on a smoothly increasing background and can be sensitively detected by differentiating the x-ray emission with respect to sample potential. The application of this technique to the study of surface composition, chemical shifts, and band structure is discussed.

INTRODUCTION

For at least half a century it has been known that if the potential applied to an electron-bombarded sample is increased to a value corresponding to a threshold for the core level excitation of surface atoms, the character of the soft x-ray emission changes. Unfortunately, the number of additional x rays per unit energy interval that make up this change when operating slightly above the threshold are generally quite small compared to the background, and attempts to use this effect to measure core electron binding energies of low-Z elements met with limited success.[1] Differentiating the x-ray emission with respect to the sample potential, however, has the effect of suppressing the slowly varying background relative to the comparitively sudden changes near the appearance potentials.

We have used this fact to design a nondispersive spectrometer
which employs an oscillating potential technique to obtain the derivative of a signal proportional to the total x-ray emission.(2,3)
This technique represents the simplest existing means of determining core level binding energies of surface atoms.

EXPERIMENTAL

A schematic of the spectrometer is shown in Fig. 1. Electrons
from a bare tungsten filament F impinge on the sample to be studied
S. A grid electrically separates the vacuum chamber into two parts.
X rays passing through the grid strike the walls of the chamber.
The resulting photoelectrons are collected on a straight wire electrode E. A resonant circuit consisting of a high-Q inductor L in
parallel with the distributed capacitance C of the circuit selects
that portion of the signal which varies at the frequency of the
oscillation superimposed on the target potential. This signal is
amplified and synchronously detected by a phase-lock amplifier.
The spectrum of a 304 stainless steel surface obtained by this
technique is shown in Fig. 2. The spectrum was taken in 40 min at
an emission current of 9 mA and a potential oscillation of 0.4
Vrms.

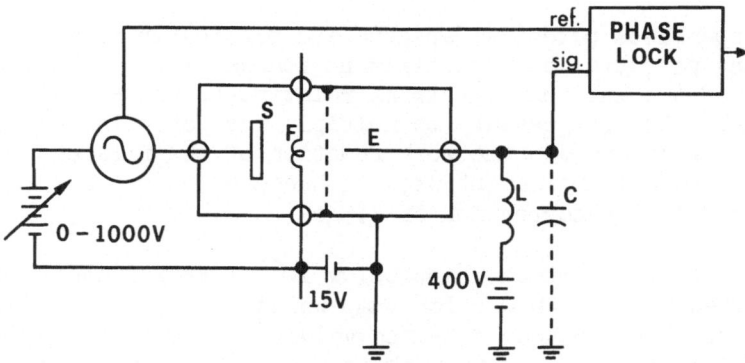

Fig. 1. Simplified schematic of the soft x-ray appearance
potential spectrometer.

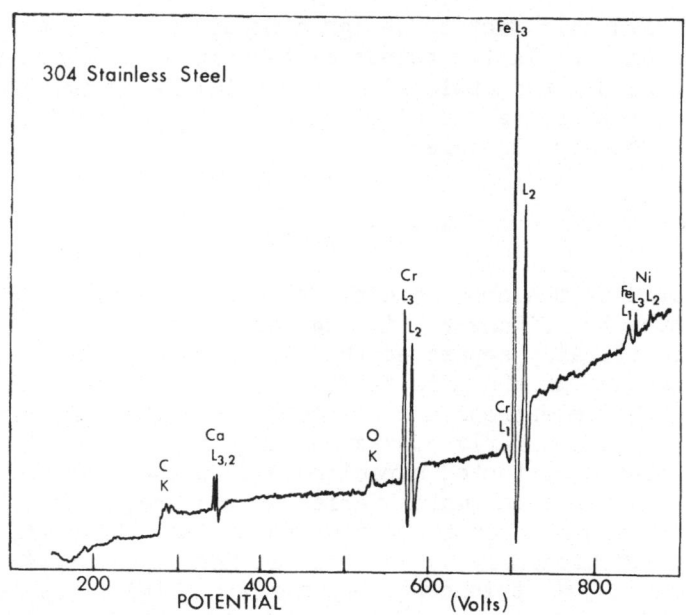

Fig. 2. Spectrum of a 304 stainless steel surface.

The ability of the apparatus to respond to very sudden changes in total fluorescence is ultimately limited by the energy spread of the incident electrons. This ultimate is, of course, achieved only when no significant broadening results from the potential oscillation. The total response width of the instrument, determined by the sharpest feature recorded thus far, is less than 1.5 eV. This is seen in Fig. 3 which expands the region around the calcium L_3, L_2 impurity peaks in Fig. 2. The peaks are separated by 2.8 volts and are about 1.3 volts in width.

The appearance potential spectra are plotted in terms of the accelerating potential between filament and sample. To convert to the energy of the incident electrons relative to the Fermi energy of the sample, it is necessary to multiply the potential by the electronic charge and add the work function of the filament.[4] For the pure tungsten filament used in these experiments we have used a work function correction of 4.5 eV.

The system pressure was kept below 10^{-9} Torr for the measurements reported here by an Orb Ion pump which could be isolated from the system by a bakeable 4-inch valve. The simplicity of the apparatus lends itself to ultrahigh vacuum techniques. High-purity metal samples in the form of polycrystalline discs 1/8 inch thick by 1/2 inch diameter were obtained from Materials Research Inc.

Clean surfaces were prepared by several cycles of sputtering and annealing and also, in some cases, by evaporating fresh films on the surface from directly heated high purity wires. Surface reactions were studied by admitting high purity gases through bakeable leak valves. During the admission of gases, the system was pumped by a separately valved liquid-nitrogen-trapped mercury diffusion pump to prevent contamination by pump reactions.

BAND MODEL

In the simple one-electron picture, assuming constant oscillator strengths, the excitation probability is proportional to the product of the density of initial and final electronic states. The initial states are the filled core levels which are to be emptied. The densities of final states, on the other hand, must take into account all possible combinations of positions for the two electrons (incident and excited core electron) which are allowed by conservation of energy. Thus, the transition probability will just be the product of the density of filled core states and the self-convolution of the density of empty states above the Fermi level, i.e.

$$T(E_B + E) \propto N_c(E_B) \int_0^E N(W)N(E - W)dW \qquad (1)$$

where $T(E_B + E)$ indicates the transition probability for incident electron energies $E_B + E$, $N_c(E_B)$ is the density of filled states for the core level at E_B, $N(E)$ is the density of unfilled states above the Fermi level (the zero of energy is taken to be the Fermi energy).

This is essentially the function one would expect for the x-ray excitation curves of a core level.(5) Excitation curves are usually obtained by setting an x-ray analyzer to accept a given characteristic line and varying the sample potential across the threshold for the excitation of that line. Excitation curves obtained in this way for nickel and cobalt have been reported by Burr(6) and by Dev and Brinkman(5) who also studied copper. If the total x-ray emission is plotted as a function of sample potential, the excitation curves of all the levels are superimposed on a smoothly increasing background. By plotting the derivative of the total x-ray emission in appearance potential spectroscopy, we obtain for each level a curve which should correspond to the derivative of equation (1) on a nearly constant background.

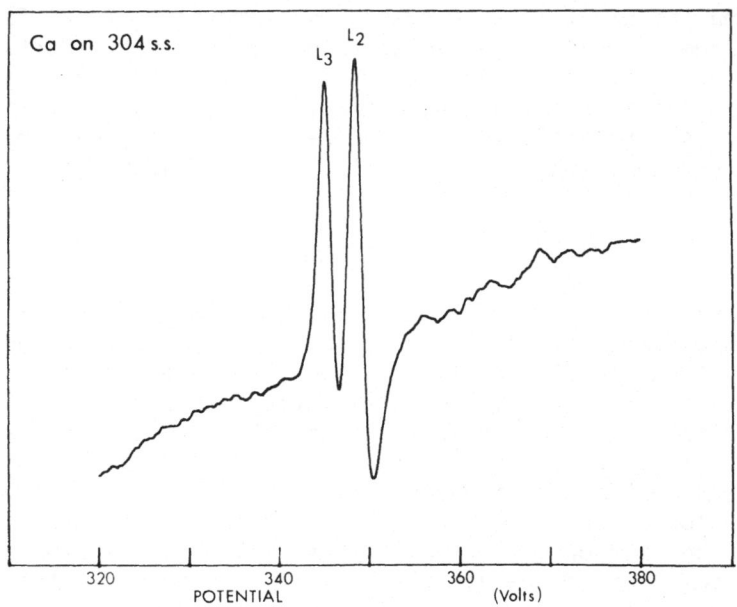

Fig. 3. Expanded view of the L_3, L_2 spectrum of a calcium impurity on 304 stainless steel.

RESULTS

For transition metals this highly simplified model predicts a sharp positive peak at each threshold, the width of which is essentially the width of the unfilled portion of the d band.(7) Indeed, near the threshold the shape of the curve should approximate the density of states. This is seen in the solid curve of Fig. 4 which shows the 2p spectrum (L_3 and L_2 levels) of a clean titanium surface. The apparent width of the L_3 threshold peak includes the instrument response function of about 1.5 eV. It is not possible therefore to make a quantitative comparison with theoretical calculations.

In the absence of the broad 4s band, the dips following the L_3 and L_2 threshold peak would be as great as the initial rise. The reduction of the negative dip thus measures the contribution of the s states. The L_3 threshold, after adding 4.5 eV to convert from sample potential to binding energy, occurs at 453.0 ± 0.5 eV as compared to the tabulated x-ray value of 455.5 ± 0.4 eV.(8) The x-ray values are, however, bulk values. Dev and Brinkman(5) point out that reduced coordination at the surface should shift the binding energies to lower values. They report early onsets for the L_3 levels of cobalt, nickel, and copper using the x-ray excitation technique. Burr(6) using the same technique also reports early onsets for the L_3 levels of cobalt and nickel as compared to those obtained by bulk sampling techniques.

The sensitivity of the technique to chemical effects is clear from a comparison of the clean L_3, L_2 spectrum with that from TiH_2 (dashed curve in Fig. 4). The hydride was obtained by in situ reaction of the clean surface with high-purity hydrogen. A chemical shift of 1.1 ± 0.1 eV in the L_3 level is observed. The width of the unoccupied portion of the 3d band, however, is nearly unchanged as indicated by the width of the initial peak. This is in agreement with the band structure calculations of Switendick.(9)

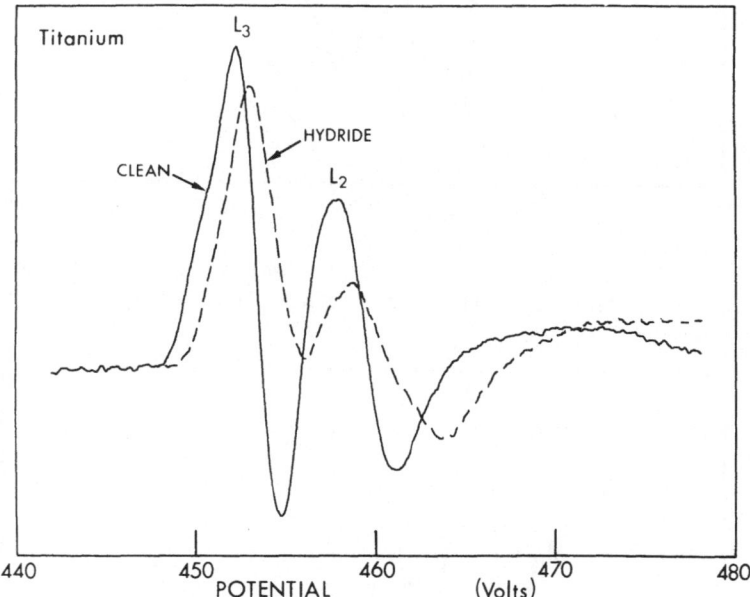

Fig. 4. The L_3, L_2 spectrum of a clean titanium surface (solid curve) contrasted with that of the same surface following reaction with hydrogen (dashed curve).

In situ oxidation of the clean titanium surface also increases the binding energy of the $2p_{3/2}$ electrons. The chemical shift in this case is 2.0 ± 0.1 eV as seen in Fig. 5. In addition, however, the width of the unoccupied portion of the 3d band increases about 0.5 eV as oxygen removes electrons from the metal. The shift in the Fermi level relative to the 3d band clearly exposes the low energy maxima which is present only as a weak shoulder in the clean surface spectrum. The double-peaked spectrum is also observed in the oxygen K level as shown in Fig. 6.

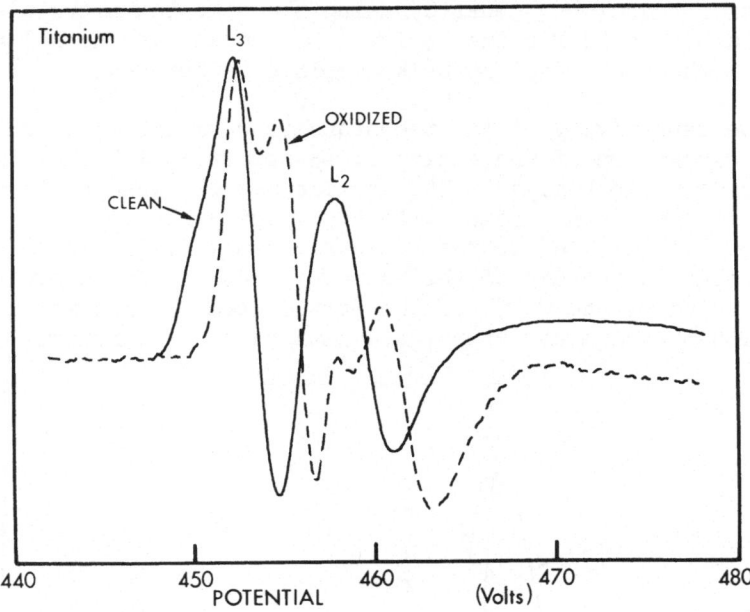

Fig. 5. The L_3, L_2 spectrum of clean titanium (solid curve)
contrasted with that of the same surface following oxidation
(dashed curve).

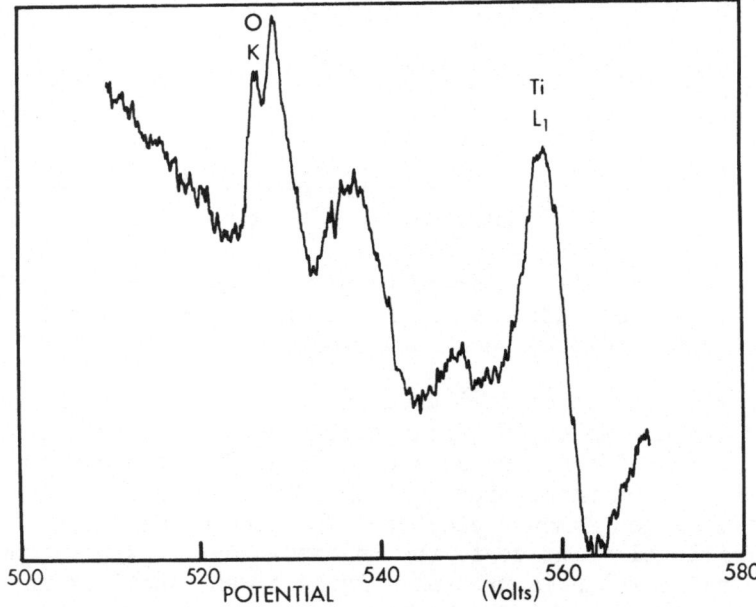

Fig. 6. The oxygen K spectrum and titanium L_1 spectrum of
an oxidized titanium surface.

SUMMARY

The results obtained thus far with soft x-ray appearance potential spectroscopy encourage further exploration of this simple approach to the study of chemical bonding at solid surfaces. Many questions remain to be answered concerning the dependence of sensitivity on Z and the role of final states interactions. It seems likely, however, that attempts to find answers will benefit the entire field of x-ray analysis.

REFERENCES

1. See for example: H. W. B. Skinner, "The Excitation Potentials of Light Elements. I. Lithium," Proc. Roy. Soc. A135, 84 (1932).

2. R. L. Park, J. E. Houston, and D. G. Schreiner, "A Soft X-Ray Appearance Potential Spectrometer for the Analysis of Solid Surfaces," Rev. Sci. Instr. 41, 1810 (1970).

3. R. L. Park and J. E. Houston, "Appearance Potential Spectroscopy on an Austere Budget," Sur. Sci. (in press).

4. H. Merz, "Isochromat Spectroscopic Determination of Work Functions," Phys. Stat. Sol. (a) 1, 707 (1970).

5. B. Dev and H. Brinkman, "L$_\alpha$ Radiation from Cobalt, Nickel, and Copper Crystal Surfaces," Nederlands Tijdscrift voor Vacuumtechniek 8, 176 (1970).

6. A. F. Burr, "The Application of X-Ray Data to the Determination of Atomic Energy Levels," in B. L. Henke, J. B. Newkirk, and G. R. Mallett, Editors, Advances in X-Ray Analysis, Vol. 13, Plenum Press (1970).

7. J. E. Houston and R. L. Park, "The Effect of Oxygen on the Soft X-Ray Appearance Potential Spectrum of Chromium," J. Chem. Phys. (in press).

8. J. A. Bearden and A. F. Burr, "Reevaluation of X-Ray Atomic Energy Levels," Rev. Mod. Phys. 39, 125 (1967).

9. A. C. Switendick, "Electronic Band Structures of Metal Hydrides," Solid State Comm. 8, 1463 (1970).

DETECTOR BACKGROUND AND SENSITIVITY OF SEMICONDUCTOR X-RAY

FLUORESCENCE SPECTROMETERS

F. S. Goulding, J. M. Jaklevic, B. V. Jarrett
and D. A. Landis
Lawrence Berkeley Laboratory University of California

Berkeley, California 94720

ABSTRACT

Degraded detector pulses are shown to contribute most of the background in spectra produced by semiconductor-detector X-ray spectrometers. A new guard-ring detector is used with appropriate circuits to reduce background by a large factor. We discuss the sensitivity of the new arrangement with various excitation sources.

INTRODUCTION

Semiconductor-detector X-ray fluorescence spectrometers are now commonly used for constituents present at levels over 10 ppm. Slightly lower limits of detection can be achieved in favorable circumstances, and if long counting times are employed. However, the sensitivity range below 1 ppm, which is of great interest in biological and environmental studies, is not easily accessible with existing X-ray spectrometers--the purpose of the work described here is to lower detection levels down to about 0.1 ppm.

In the past, work on semiconductor-detector spectrometers has been primarily aimed toward improving energy resolution. This was an essential step in order to achieve separation of peaks due to adjacent elements; furthermore, reduction of the width of peaks improves the peak-to-background ratio, thereby lowering the limit of detectability for trace elements. It is important to realize, however, that reducing background achieves the latter objective too, and this is the course pursued in the present work.

Figure 1 shows the X-ray spectrum produced by a conventional spectrometer using molybdenum X-ray excitation of a blood serum sample. The exciting radiation was derived from a molybdenum fluorescer excited by an ^{125}I ring source. In a system of this type, no low-energy radiation is produced by the exciting source, and the semiconductor detector should observe only fluorescent X-rays from the sample together with the exciting radiation back-scattered by the sample. In practice, well over 90% of the counts should occur in the backscatter peak at the high-energy end of the spectrum, but, if degraded in any way, they constitute a general background hiding the peaks characteristic of trace elements in the specimen. In the case shown in Fig. 1, about 20% of the back-scatter events are degraded to produce background.

We shall show that these degraded background pulses are the result of imperfect charge collection in detectors, and that they can be virtually eliminated by new detector techniques. For those unfamiliar with semiconductor detector methods and terminology, Ref. 6 will provide an introduction to the subject.

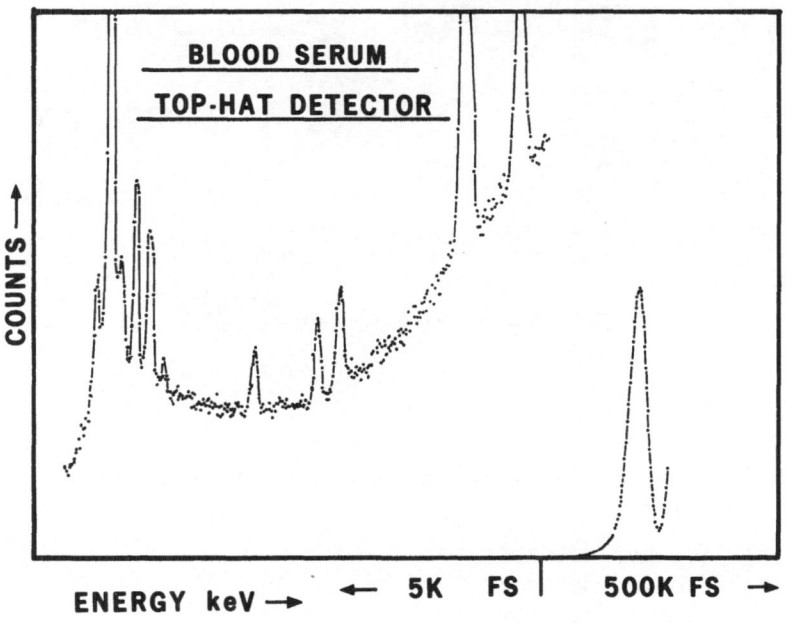

Figure 1. Fluorescence X-ray spectrum of a blood serum specimen using a conventional silicon detector spectrometer.

DETECTORS IN X-RAY SPECTROMETERS

Figure 2 shows three silicon detector configurations used in X-ray spectrometers. The first type, originally used by Miller (1) is referred to as the "top-hat" geometry, and is characterized by low-leakage current and excellent high-voltage behaviour. Both characteristics are desirable in high-resolution spectrometers, so this configuration is commonly used in these applications. The second type, the "grooved" detector, was used originally by E. Woo, and now employed in "Kevex" X-ray systems, possesses the same advantages as the "top-hat" geometry. The third geometry, generally referred to as "planar", exhibits higher leakage current and capacity than the other two types, and is therefore rarely used in X-ray spectrometers. However, its background properties probably deserve investigation. Llacer (2) analyzed the behaviour of these geometries with regard to their ability to sustain high-voltage operation, and to produce low leakage current. His results, and those obtained in our laboratory, indicate that an n-type surface channel normally exists on the surface of silicon detectors.

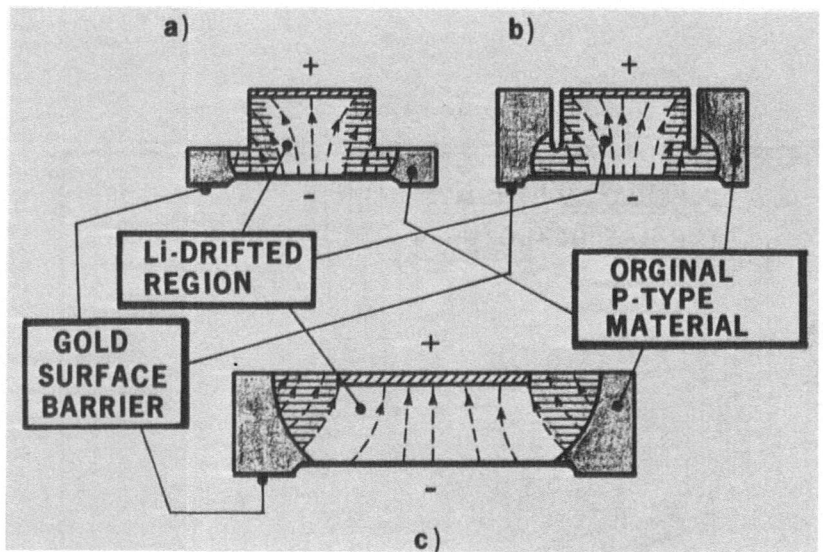

Figure 2. Types of detector configuration used in X-ray spectrometers:

    a)   Top-hat detector
    b)   Grooved detector
    c)   Planar detector

This channel acts as an extension of the n-type lithium-diffused region, and since it represents a poor junction to the bulk material, it contributes most of the leakage current, and sets the voltage limitation on detector operation. The fact that such channels can be "pinched-off" by internal electric fields normal to the surface explains the difference in behaviour between the structures of Fig. 2a and 2b and that of Fig. 2c.

These arguments fail to take into account the collection of charge produced by radiation in the bulk of the detector. The presence of n-type surface layers distorts the internal electric field pattern in the detector in such a way that collection of charge produced by X-rays interacting in some parts (shown by horizontal-line-shading in Fig. 2) is via the surface layers. This causes a loss of charge, so that signals that should appear in the back-scatter peaks appear in the general background in spectra. The tests presented in this paper show that this is the predominant source of background in existing spectrometers.

At first sight it may appear that collimation of X-rays to prevent their interaction in the poor field regions might reduce background, and, indeed, tests show that some improvement can be achieved by this method. It is also obvious that improvement results from increasing the detector area while collimating to a small central region, but the large consequent increase in detector capacity seriously degrades the system resolution--an intolerable price to pay. The degree of collimation that can be used is determined by the requirement for good sample-detector geometry; as shown in the typical geometry shown in Fig. 3, this implies a wide divergence of X-rays hitting the detector. Despite the possible auxiliary collimator shown in this figure, mounted on the detector face--an expedient rarely adopted as it is difficult to change this collimator to suit the energy range of interest--X-rays like that travelling from A to B still interact in regions of poor charge collection. On the other hand, the X-ray from C to D interacts in a region of good charge collection, and produces the correct signal.

As shown in Fig. 4, the background due to degraded pulses increases as the energy of the X-rays impinging on the detector increases. For cadmium X-rays, the total integrated background count approaches the total number of counts in the main X-ray peak. The proportion of counts in the background decreases considerably in the case of zirconium, and still further for lower energy X-rays.

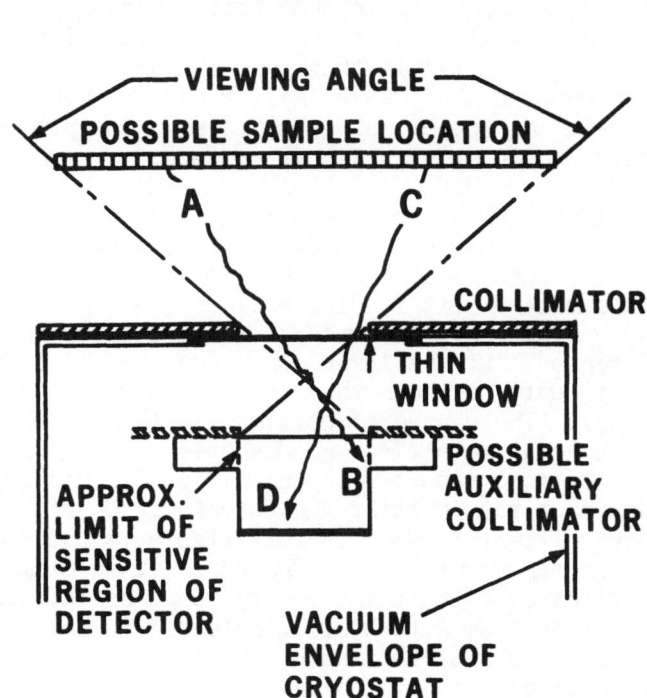

Figure 3.   Preferred collimation geometry for an X-ray spectrometer.

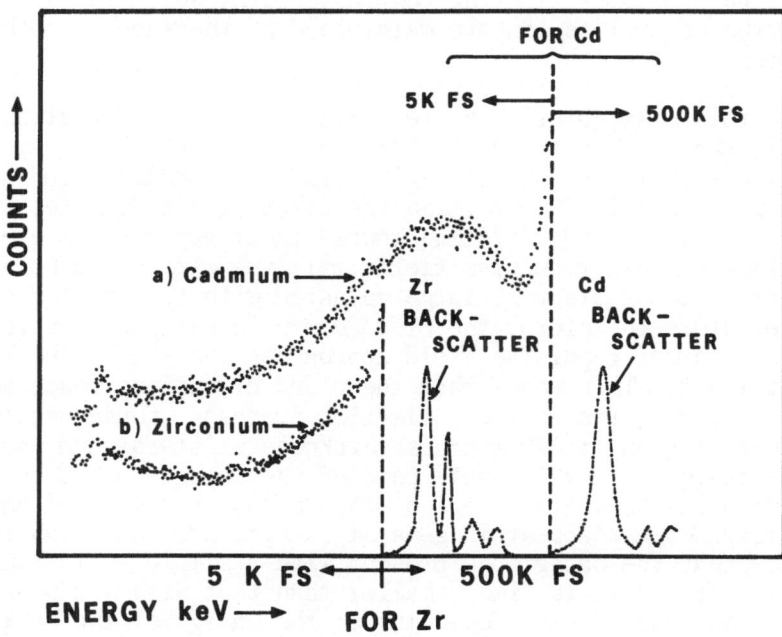

Figure 4. Showing the variation in background of a top-hat detector as a function of the energy of radiation striking the detector.

a) Cd X-rays scattered by lucite.
b) Zr X-rays scattered by lucite.

THE GUARD-RING DETECTOR METHOD

Guard rings have been employed for many years to overcome fringing field effects in standard capacitors, and have also found application (3) in semiconductor detectors as a device to reduce edge leakage. It therefore seems an obvious step to use a guard ring to define the boundary of the sensitive volume of a detector by internal electric field lines rather than by a physical surface with its unknown charge-trapping characteristics. This can also be considered as an electronic collimation technique. Figure 5a shows the simple implementation of the idea; note that the output signal is derived only from the central region, while the guard ring and the central region are maintained at the same dc potential (ground).

Even this configuration suffers from a signal degradation problem at the edge of the central region. The initial X-ray interaction in this peripheral region produces a dense cloud of charge $\simeq 5$ microns in diameter; in the electric field, holes and electrons are separated, drifting toward their appropriate electrode. The internal repulsive fields existing within the hole and electron clouds are very large compared with the drift field in the detector--therefore, the cloud dimensions rapidly increase until the internal repulsive field approaches the same value as the drift field. This means that the cloud dimensions reach about 100 microns during the charge collection process. Consequently, a peripheral region of 100 microns thickness exists around the sensitive region from which only part of the charge due to an event is collected in the central region--this means that many of the backscattered events appear in general background. Our measurements show that the background present with a simple guard-ring detector is from 2 to 10 times smaller than that with a top-hat detector, the exact factor depending on the energy of the back-scatter peak.

A further reduction in background is achieved by sensing coincident signals between the guard-ring and central regions, and rejecting the central-region signal when such a coincidence is registered. This "guard-ring reject" system effectively eliminates the partial collection from the peripheral region of the sensitive volume of the detector. With such an arrangement, we approach the background level expected due to electron escape from the detector surface. In our actual detector, shown in Fig. 5b, a double guard-ring is used, the outer ring serving to reduce edge leakage in the inner ring, and thereby improving its noise properties so that the inner ring signal discriminator can be set low to detect very small signals.

The improvement in background resulting from use of the guard-ring reject method is shown in Fig. 6. Using exactly the same geometry, cadmium X-rays scattered from lucite were used to irradiate a standard top-hat detector (a), and the guard-ring reject detector system (b). For the same number in the cadmium peak, the total counts recorded in the background is 40 times smaller for the guard-ring reject system than for the top-hat detector. Larger factors are obtained when the detectors are irradiated by radiation of higher energy than cadmium X-rays.

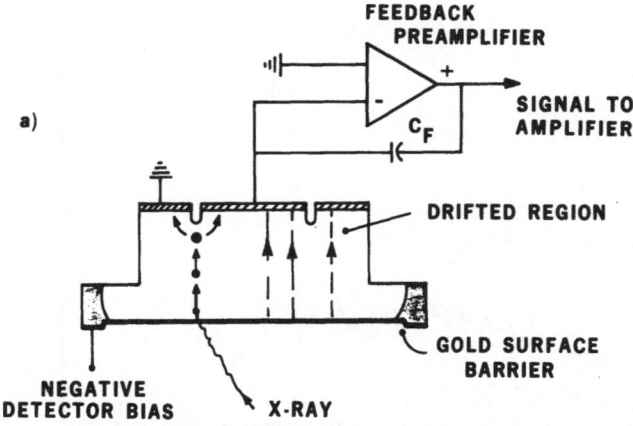

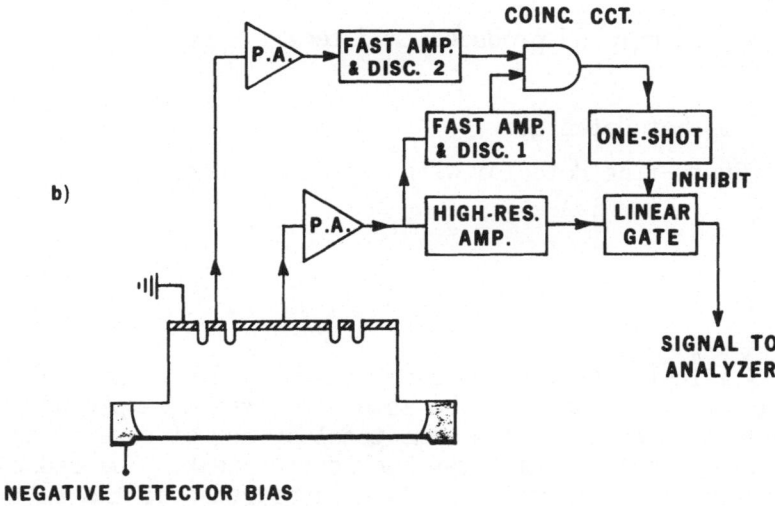

Figure 5.   Guard-ring detectors.

    a) Simple guard-ring approach showing the mechanism
       for degraded pulses.

    b) Double guard-ring detector with pulse-reject
       circuitry to remove degraded pulses.

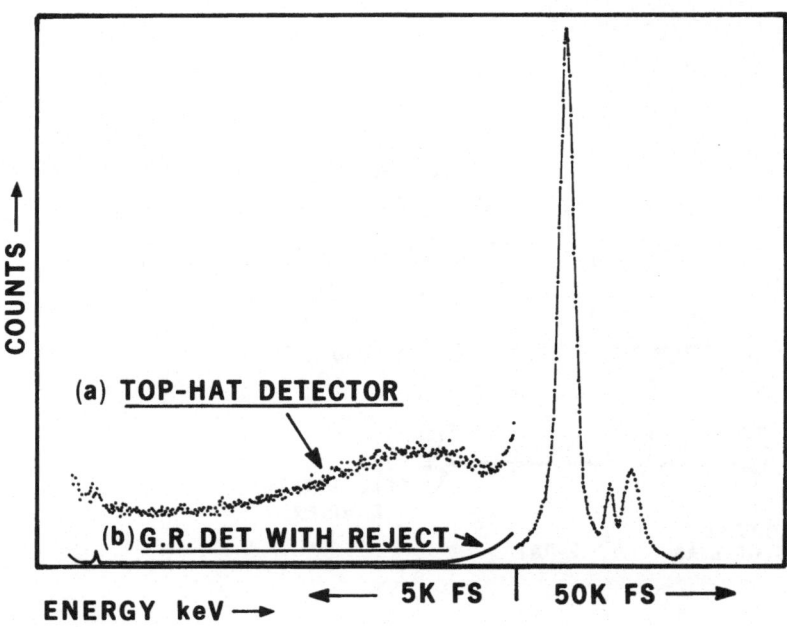

Figure 6.   Background produced in detector systems by Cd X-rays.

a)   Top-hat detector.

b)   Guard-ring detector with reject circuitry.

Figure 7 illustrates the improvement achieved by using this technique on a typical sample. The same blood serum sample used for Fig. 1 was examined, and the same total number of counts was accumulated in the molybdenum X-ray backscatter peak. The reduction in background seen in this spectrum, averaged over the full energy range, is about a factor of 15. Much larger factors, ranging up to about 60, have been observed for higher-energy excitation. Comparison of Fig. 7 with Fig. 1 shows the improvement in ability to see small traces of elements, such as nickel, present in the specimen at a level near 0.1 ppm. Better statistics realized by a longer count, or with more intense excitation, would further reduce the detection limit.

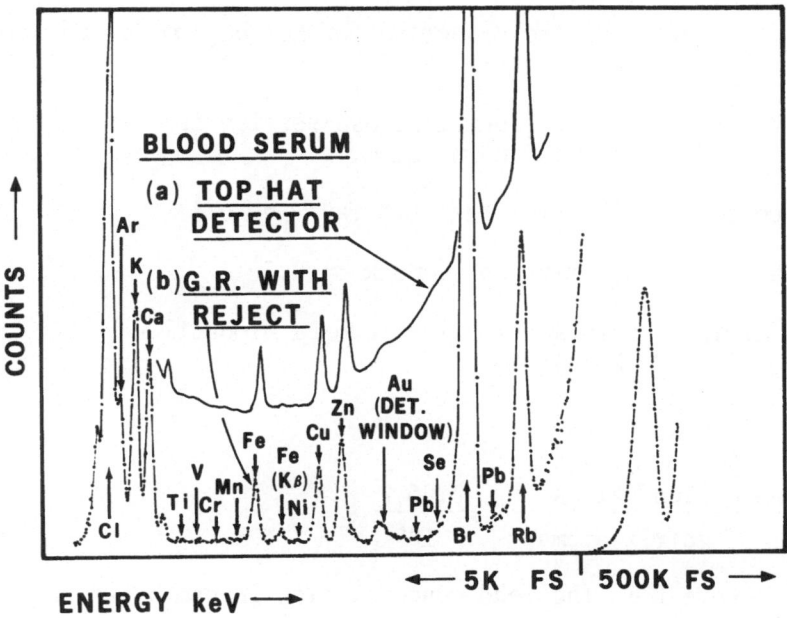

Figure 7. Fluorescence X-ray spectrum obtained on a blood
serum specimen using a spectrometer equipped with guard-
ring detector and reject circuitry (b). For comparison,
the spectrum obtained on the same sample with the same
geometry, and the same total counts in the scatter peak,
but with a simple top-hat detector, is also shown (a).
This is a smooth curve drawn through the points of Fig. 1.

CALCULATION OF SENSITIVITY LIMIT

It is interesting to calculate the detection limit for an
X-ray fluorescence spectrometer, and to relate this to detector
background. We will consider the case of traces of lead in an
organic matrix, using molybdenum Kα excitation, and will assume
that only the lead Lα peak is used for the analysis.

We have:

Total scattering cross-section for molybdenum Kα radiation in an organic matrix = 5 barns/atom.

Total L-shell photoelectric cross-section for lead = 3×10^4 barns/atom.

Fluorescent yield for lead L X-rays = 0.4

Fraction of L X-rays in Lα peak ≃ 0.5

∴Effective cross-section for lead Lα production ≃6×10^3 barns/atom.

Now

$$\frac{\text{No of lead atoms}}{\text{No of matrix atoms}} = \frac{0.06\ P}{10^6}$$

where P is the lead concentration in ppm by weight.

$$\therefore \frac{\text{No of Pb Lα's at detector}}{\text{No of scattered X-rays at detector}} = \frac{6 \times 10^3}{5} \cdot \frac{0.06\ P}{10^6}$$

$$= 7.2\ P \times 10^{-5} \qquad (1)$$

Now assume that N counts are accumulated in the backscatter, and that X% of these appear in a flat background from zero amplitude to the amplitude corresponding to the backscatter energy. Recognizing that the width of fluorescent lines is about 1% of the backscatter energy we find:

No of background counts beneath lead Lα peak $= \dfrac{N.X}{10^4}$

Experiment shows that a reasonable detection limit for the counts in a peak is twice the R.M.S. deviation in the background under the peak.

$$\therefore \text{Detectable limit} = 2 \times 10^{-2} \cdot \sqrt{N.X.}\ \ \text{counts} \qquad (2)$$

But Eq. 1 shows that the number of counts in the lead Lα peak is equal to $7.2\ P.N.\ 10^{-5}$.

∴Detection Limit, expressed in ppm by weight $= P_{LIM}$

$$= 300 \sqrt{\frac{X}{N}} \text{ ppm} \qquad (3)$$

Typically the value of N might be 10^6, and we might assume X = 20 for a top-hat detector, while X = 1 for a guard-ring reject system.

∴P_{LIM} for top-hat detector $\simeq 1.3$ ppm.

and P_{LIM} for G-R system $\simeq 0.3$ ppm.

Using a monochromatic X-ray tube (4), counts can be accumulated at a rate of 2×10^4 per second, limited by the counting rate capability of the pulsed-light feedback electronics we use. This means that the sensitivities we have derived are achieved in only 50 second counts. If the counting time is increased to ten minutes, a limit of detection of 0.1 ppm is comfortably achieved.

CONCLUSION

The guard-ring detector technique described here reduces background in X-ray fluorescence spectrometers by a large factor, the improvement increasing with higher exciting energies. The sensitivities that can be achieved using this technique, together with a monochromatic X-ray tube, extend the use of X-ray fluorescence spectrometers into the range of biological trace elements (5). The resulting instrument provides the capability of rapid analysis for a broad range of trace elements with only very simple sample preparation.

ACKNOWLEDGMENTS

We thank J. Walton and H. Sommer for their making the detectors used in this investigation, and R. Giauque for advice on analytical aspects of the problem.

This work was carried out as part of the program of the Nuclear Chemistry Division of the Lawrence Berkeley Laboratory under the United States Atomic Energy Commission Contract No. W-7405-eng-48.

REFERENCES

1. G. L. Miller, private communication referenced in Llacer's paper (Ref. 2) p. 99.

2. J. Llacer, "Geometric Control of Surface Leakage Current and Noise in Lithium Drifted Silicon Radiation Detectors", IEEE Trans. Nucl. Sci. NS-13, No. 1, p. 93 (1966).

3. W. Hansen and F. S. Goulding, "Leakage, Noise, Guard Rings and Resolution in Detectors", Proceedings of Asheville Conference NAS-NRC Report No. 32, 202 (1961).

4. J. M. Jaklevic, R. D. Giauque, D. F. Malone and W. L. Searles, "Small X-ray Tubes for Energy Dispersive Analysis Using Semi conductor Spectrometers", Lawrence Berkeley Laboratory Report No. LBL-10. Presented at the 20th Annual Denver X-ray Conference Aug. 11-13, 1971.

5. F. S. Goulding and J. M. Jaklevic, "Trace Element Analysis by X-ray Fluorescence", Lawrence Radiation Laboratory Report No. UCRL-20625, Berkeley, California 94720.

6. F. S. Goulding and Y. Stone, "Semiconductor Radiation Detectors", Science Vol. 170, 280 (1970).

THE MEASUREMENT OF SURFACE-LAYER STRESSES IN A POLYCRYSTALLINE

GLASS BY MEANS OF X-RAY DIFFRACTION

E.W. Kammer and C.L. Vold

Naval Research Laboratory

Washington, D.C. 20390

ABSTRACT

Many glass materials can be made more resistant to failure
if the surface layers are in a state of permanent compressive stress.
For purposes of monitoring production quality at the factory, as
well as to verify compliance with specifications before incorporation
into special devices, it is desirable to measure nondestructively
this stress magnitude. For those glass compositions in which the
stress generating process creates a residual, finely divided, and
uniformly dispersed crystalline phase (e.g. "Pyroceram"), a stress
measuring technique based upon X-ray diffraction has been adapted
and successfully used. Lattice spacings for near-surface
crystallites are determined under the stressed conditions present
in the final glass product. These lattice spacings are compared
with values obtained from a powdered sample of the identical glass
composition. In the powdered state the induced stress effects are
minimal. Hence, the shift in lattice spacings is a measure of the
compressive surface stress level attained in the manufacturing
process.

INTRODUCTION

The failure of glasses and other amorphous solids is generally
attributed to the presence of surface cracks or other flaws that
grow under the influence of an applied stress and/or chemical
attack. These materials can be made more resistant to failure if
the surface layers are in a permanent state of compressive stress.
This state can be accomplished in several ways. One process,

pertinent to this investigation, utilizes metal atom substitution at high temperature. The over-crowded surface lattice, especially if it has acquired at the same time a lower temperature coefficient of expansion than that of the unmodified substructure, can attain an exceptionally high surface stress level when the sandwich complex is restored to room temperature. At points sufficiently distant from the free edges of the specimen, the stress pattern is essentially uniform in all directions tangent to the surface.

The specific material being discussed here is known as polycrystalline glass, similar to the commercial product available under the name "Pyroceram". In general, the silicon dioxide host contains, together with other compounds, the oxides of aluminum, magnesium and titanium. After a preliminary heat treatment the material is submerged in a bath containing molten lithium oxide at sufficiently high temperature (about $1000^{\circ}C$) to favor substitution of the magnesium atoms in the surface layers by lithium atoms from solution on a one for two basis, respectively, as governed by their valences. The thickness of this substitution layer, controlled by a diffusion process, is dependent upon the time held in the fused lithium oxide, and typically is 100 to 150μm.

The subject of this paper is to establish a nondestructive method of evaluating the magnitude of the stresses present in the thin surface layers. This information is highly desirable for purposes of monitoring production quality, as well as to verify that the materials are in compliance with specifications. An X-ray diffraction procedure has been developed to measure the magnitude of these stresses and is described in the following section.

EXPERIMENTAL

An X-ray diffraction pattern of the surface layer reveals a sequence of crystal lattice spacings conforming to the mineral "beta-spodumene". This compound has the chemical formula $LiAl(SiO_3)_2$ and displays tetragonal symmetry. The compound is found in natural sources also, but only in the so-called low temperature or "alpha-" form, which is monoclinic. The presence of this finely divided crystalline component, the X-ray diffraction spectrum of which is composed of sharp lines (uncluttered by responses from other compounds making up the glass) is indeed fortunate, because it furnishes the stress analyst with ideal "strain gages" strategically placed in the region of interest.

In order to evaluate the lattice spacings under conditions free of macroscopic constraint, a small quantity of material was filed from the surface of the specimens, cleaned, and then further reduced to a uniform powder. From this finely divided state,

essentially free of large scale surface stress effects, a
reference diffraction pattern was obtained using conventional X-ray
diffractometry techniques. Two typical lattice spacings obtained
by the preceding method are presented in Table I. The same lattice
spacings were also determined from a surface area of a highly
stressed plate-shaped specimen using the same technique. Here, the
region examined was sufficiently isolated from any free edge of the
plate so as to sample the highly stressed surface. These spacings
are also presented in Table I. The lattice spacings for the
powdered state are significantly smaller than those measured on the
highly stressed plate, consistent with the interpretation that the
crystal planes producing the measured diffraction lines (these planes
being parallel to the surface of the plate) respond to the release
of surface compression by contracting in a direction normal to the
specimen surface (Poisson effect).

TABLE I   Strongly diffracting crystal lattice
          spacings for surface material in
          stressed plate and powder form
          together with corresponding lattice
          spacings for stoichiometric "beta-spodumene".

Plane	d_n (Å) Stressed Plate	d_o (Å) Powder	$(d_n-d_o)/d_o$
Surface Layer			
102,200	3.912	3.890	0.00565
201	3.481	3.465	0.00461
		average = 0.00513	
Beta spodumene (Ref. 1)			
102,200	-----	3.920	-----
201	-----	3.487	-----

For comparison, the lattice spacings for stoichiometric beta-spodumene, as determined by Skinner and Evans (1) are also listed in Table I. The differences exhibited by these spacings with respect to those observed from the powder derived from the stressed plates may be attributed to two possible effects. First, the beta-spodumene crystallites in the powder samples may be subject to a hydrostatic compression by the glassy matrix in which they are embedded, providing that the powder particle size is much larger than the beta-spodumene crystallite size; a condition fulfilled here. Second, the beta-spodumene phase field is quite broad (1) and exhibits sizable lattice parameter changes with composition, which would give rise to attendent lattice-spacing shifts. It is quite likely that both effects are operating.

There then remains the task of converting the observed lattice strains into corresponding surface stresses utilizing the appropriate elastic constants. A rectangular coordinate system is positioned with the x and y axes lying in the plane of the specimen surface at any convenient orientation. The principal stresses σ_x, σ_y, and strains ϵ_x, ϵ_y, associated respectively with the x and y axes are presumed to be equal in the biaxial stress field. Normal to the specimen surface the stress σ_z vanishes, but the corresponding strain ϵ_z is not zero. This latter quantity is related to the stresses in the x-y plane by the equation

$$\epsilon_z = -\frac{\nu}{E} (\sigma_x + \sigma_y), \tag{1}$$

where ν is Poisson's ratio and E is the Young's modulus. The value of ϵ_z is measured by the X-ray diffraction procedure and is expressed in terms of the lattice spacings by the relation

$$\epsilon_z = \frac{d_n - d_o}{d_o} \tag{2}$$

in which d_n is the spacing of the atomic planes when the lattice is under stress, and d_o is the spacing of the same planes in the absence of stress. Thus, the sum of the principal stresses for this special configuration is obtained from the relation

$$(\sigma_x + \sigma_y) = -\frac{E}{\nu} \left(\frac{d_n - d_o}{d_o} \right). \tag{3}$$

Meaningful values for E and ν should be those obtained from the lithium substituted glass layer alone. However, sufficiently large homogeneous specimens of this material needed for accurate measurements of these constants are not yet available.

Estimates of E and ν were, however, computed from velocities of ultrasonic waves propagating in a direction normal to the surface layers of the composite, in which only a few tenths of a millimeter

TABLE II   Comparative values of ultrasonic wave velocities,
           Young's modulus and Poisson's ratio for substrate
           and lithium treated surface material.

	V_L(cm/sec) (Longitudinal)	V_S(cm/sec) (Shear)	E(dynes/cm^2)	ν
Substrate	6.6×10^5	4.1×10^5	12.4×10^{11}	0.24
Surface Layer	11.9×10^5	6.9×10^5	25.5×10^{11}	0.18

of the sound path constituted the lithium substituted lattice.
Even though the transformed outer layers were very thin, being
between three and four percent of the total thickness, their
presence increased the apparent average ultrasonic wave velocity
for the composite material by about three percent relative to the
uncoated substrate. From these data an estimate of the wave
velocity in the surface layers alone could be computed. Table II
displays comparative results for the two media. The experimental
conditions presented by the samples, while certainly not conducive
to great accuracy, nevertheless permit establishing semi-quantitative
estimates for the surface material parameters E and ν, a goal
sufficient for the purposes here intended.

    Substituting the values obtained for E and ν for the surface
layer (Table II) and that for $\epsilon_z = (d_n - d_o)/d_o$ (Table I) into
equation 3, yields a compressive principal stress sum of 7.3×10^{10}
dynes/cm^2 (1×10^6 lbs/in^2). The measured lattice strain ϵ_z, averaging
about 0.5%, certainly implies extremely large stress levels in the
material of the type being considered here.

                        DISCUSSION

    Perhaps the greatest single contribution to error in the
preceding analysis arises from inaccuracies in the determination of
the appropriate Young's modulus and Poisson's ratio of the glasses
surface layer. The unfavorable experimental conditions for
determining these constants have already been considered in the
previous section. Also, the Young's modulus and Poisson's ratio,
as determined above, represent "quasi-isotropic" values for these
constants, resulting from the fact that the beta-spodumene
crystallites are randomly oriented over the surface layer of the
glass. It is clear that these "quasi-isotropic" values of E and ν
are not strictly equal to the constants which we would desire to

utilize —— namely, a modulus that is averaged over all directions
in the plane of the diffracting plane (hkl), and a Poisson's ratio
defined in a direction normal to the plane (hkl).

The magnitude of the error involved in the present stress
determination would then depend ultimately upon the degree of
elastic anisotropy of beta-spodumene, a question which cannot be
resolved until the six single crystal elastic constants are
determined. Considering the fact that beta-spodumene can be
produced artificially only under conditions of high temperature and
pressure (1), it is unlikely that large single crystal specimens
of this material will be produced to permit these measurements.

Nevertheless, the present study has shown that very large
compressive stresses do exist in the surface layers of these
glasses. Their magnitude also demonstrates the feasibility of
using X-ray diffractometer data as a quality control over the
production of this high strength glass.

REFERENCES

1. B. J. Skinner and H. T. Evans, Jr., "Crystal Chemistry of
    β-Spodumene Solid Solutions on the Join $Li_2O.Al_2O_3-SiO_2$",
    Am. J. Science, Bradley Volume 258-A, 312-324 (1960).

THE DETERMINATION OF THE AXIS OF LATTICE ROTATION

WITH RESPECT TO A CHANGE IN TEXTURE

C. Feng

Picatinny Arsenal

Dover, New Jersey   07801

ABSTRACT

It is known that a change in texture may take place in various metals and alloys following cold working, fabrication and recrystallization processes. Such change may be determined by an analysis of the pole figures obtained from the x-ray diffraction data. In numerous occasions, it is analogous to a lattice rotation.

In the present paper, a method is derived to identify the axis of rotation with respect to the starting and final texture of a given specimen. Once this axis is identified, the amount and the direction of the lattice rotation can be accurately determined. The techniques in obtaining a solution by mathematical calculation and by stereographic presentation will be described. Discussion will be made also of the various parameters which may affect the validity and accuracy of the determination.

INTRODUCTION

The changes in texture as a result of recrystallization and cold working of metals and alloys has been explained by a mechanism of lattice rotation. For a given specimen with known lattice structure, the determination of the axis and the amount of rotation, as it will be shown in the latter part of this paper, involves only simple and straight forward mathematical calculations when two or more reference points can be identified before and after the rotation. The determination, however, was not carried out in a routine texture study in which the immediate objective lies with the identification of the family planes and directions of the texture elements. Without the knowledge of the individual

489

plane and direction with respect to a set of reference axes, it is
not possible to determine the axis of rotation because of the
symmetry factor inherent in a given lattice.

Recently, a number of published articles have dealt with this
problem and effectively eliminated the crystallographic uncertainty
by constructing the multi-axial inverse pole figures (1,2). The
specific crystallographic planes and directions of the texture
elements could be readily identified from these pole figures, thus
allowing the investigator to pursue further regarding the direc-
tional properties of the various material (3,4). Another natural
application of the biaxial pole figure is the determination of the
axis and the amount of lattice rotation. The mathematical calculations and the stereographic techniques involved in the task will
be described in this paper.

STEREOGRAPHIC METHOD

The rotation of a solid body about its own axis may be considered as if it were the globe rotating about the North-South pole.
The determination of this axis may be accomplished by simply trans-
forming the body into a rotating sphere with the rotational axis
being the imaginary North-South pole and the center of the sphere
being the origin. One may keep in mind only that, as depicted in
Figure 1, a given point A on the surface of the sphere moves along
a constant latitude circle as it rotates. For one revolution, A
will ascribe a complete cone with the center of the sphere being
the apex. If A stops after reaching B, the amount of rotation
may be measured by the angle $\angle$ ACB, where $\overline{AC}$ and $\overline{BC}$ are both
vectors perpendicular to the axis of rotation.

Stereographically, the axis and the amount of rotation can be
determined if the starting and the final position of two reference
points on the sphere can be identified. Let the vectors of the
starting positions of these two reference points be $[H_1K_1L_1]$* and
$[H_2K_2L_2]$ and their final positions after the rotation be $[h_1k_1\ell_1]$
and $[h_2k_2\ell_2]$, respectively. Since the angle between the axis of
rotation $[uvw]$ and $[h_1k_1\ell_1]$ is equal to that between $[uvw]$ and
$[H_1K_1L_1]$, one may draw a great circle such that the angle measured
from any point along the great circle to $[H_1K_1L_1]$ will be equal
to the angle measured from the same point to $[h_1k_1\ell_1]$. A second
great circle bearing the similar significance respect to $[H_2K_2L_2]$
and $[h_2k_2\ell_2]$ may also be drawn. The intersection of the two great
circles will pinpoint the exact location of $[uvw]$.

---

*Note: The conventional Miller indices with three orthogonal axes
will be adopted throughout this manuscript.

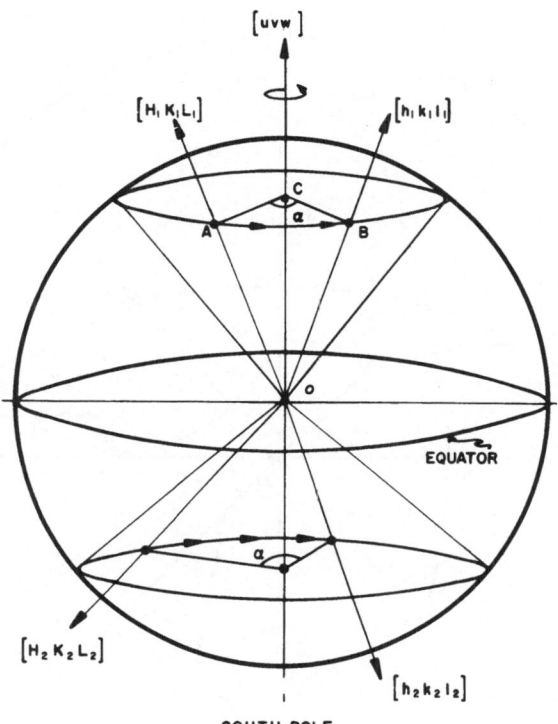

Figure 1.   Sketch depicting the axis of lattice rotation [uvw] to
be the North-South pole of an imaginary globe. The
angle of the rotation is α bound by AC and CB which
lie on the base of the cone.

An illustration is given in Figure 1 representing a case in
which textural change is involved in a medium carbon steel of bcc
structure under shock loading. Three major reference axes, namely,
the direction normal to the specimen surface, the longitudinal and
the transverse axes before and after shock, are listed in Table I.

TABLE I

Reference Axis	Before	After
Normal Direction	$[011]$	$[111]$
Longitudinal Direction	$[11\bar{1}]$	$[10\bar{1}]$
Transverse Direction	$[2\bar{1}1]$	$[1\bar{2}1]$

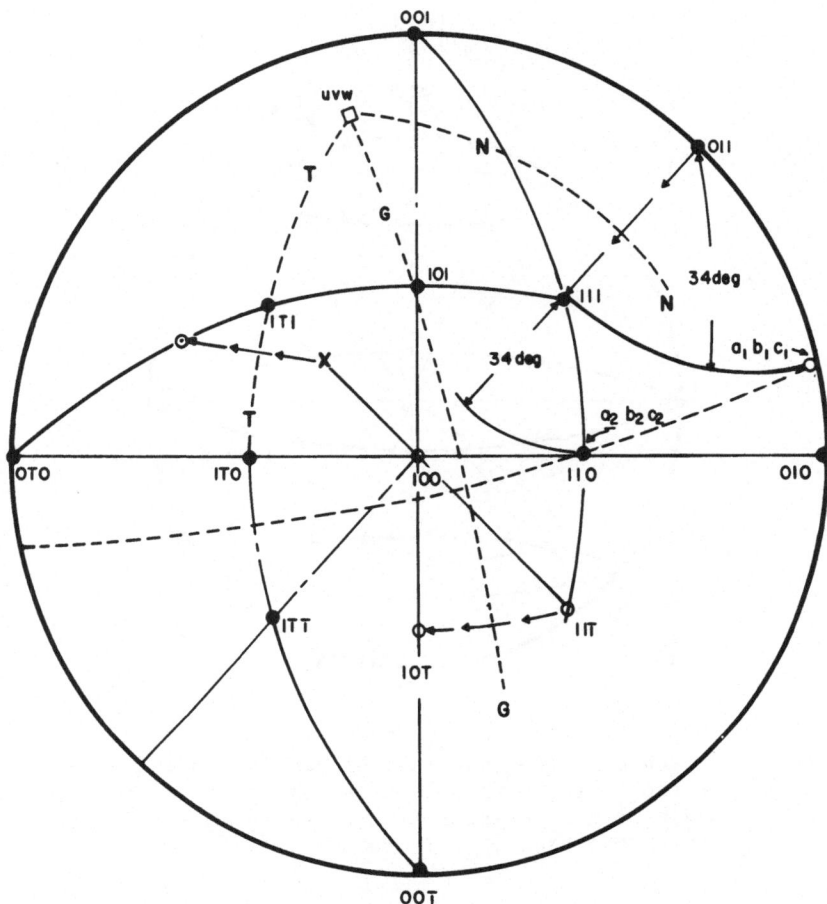

Figure 2.   The lattice rotation of a medium carbon steel shell as
            seen in a (100) standard projection. The surface
            normal shifted from [011] to [111] pole while the
            longitudinal axis shifted from [11Ī] to [10Ī] together
            with the transverse axis shifting from [2ĪĪ] to [1Ī1]
            pole. The axis of rotation is [uvw] which is the
            intersection of the three great circles NN, TT and GG.
            The locations of [a₁b₁c₁] and [a₂b₂c₂] are the inter-
            secting points of the 34 degree circles from [011] and
            [111] poles, respectively, with the 90 degrees great
            circle from [uvw]

The three mutually perpendicular axes are plotted in a standard
(100) projection, in which the surface normal before and after
rotation is placed within the first quadrant. This assumption
is valid for a cubic system. The longitudinal and transverse axes
may be located along a great circle 90 degrees from the surface
normal. Also shown in Figure 2 are the locations of the three
reference axes after shock. Portions of the three great circles,
NN, GG and TT, representing the loci making equal stereographic

angles with respect to the three reference axes before and after
shock loading. The intersection of the three great circles is
the axis of rotation [uvw], which is located in Figure 2 to be
approximately 15 deg. from the [001] pole lying between the [1$\bar{1}$5]
and [1$\bar{1}$6] directions. The amount of angular rotation of the three
references axes as a result of shock loading may be directly
measured by shifting [uvw] to the center of the stereographic
projection. This is shown in Figure 3. One may note that the
starting and the final positions of the three reference axes will
be shifted accordingly. The six positions after the shifting will
form three circles, each of which consists of the shifted positions
of one reference axis before and after shock loading. The angles
between each new pairs are the same and represent the actual amount
of angular rotation. This was measured to be approximately 44 deg.

MATHEMATICAL SOLUTION

Solutions to the axis and the amount of lattice rotation may
also be obtained by solving mathematical equations which are
compatible with the lattice system of the investigated material.
To be consistent with the example shown for the stereographic
analysis, we choose the cubic system for the mathematical method
of determination. Referring to Figure 1 and observing that a cone
is formed by the motion of the lattice point with respect to the
axis of rotation, one may write:

$$\frac{Hu + Kv + Lw}{hu + kv + \ell w} = \left(\frac{H^2 + K^2 + L^2}{h^2 + k^2 + \ell^2}\right)^{\frac{1}{2}} \tag{1}$$

where [uvw], [HKL] and [hkℓ] bear the same meanings as given
previously.

By substituting

$$A = \left(\frac{H^2 + K^2 + L^2}{h^2 + k^2 + \ell^2}\right)^{\frac{1}{2}} \tag{2}$$

equation (1) may be rewritten,

$$u(H-Ah) + v(K-Ak) + w(L-A\ell) = 0 \tag{3}$$

and further simplified into the following form,

$$mu + nv + pw = 0 \tag{4}$$

by assuming

$$\left.\begin{array}{l} m = H-Ah \\ n = K-Ak \\ p - L-A\ell \end{array}\right\} \tag{5}$$

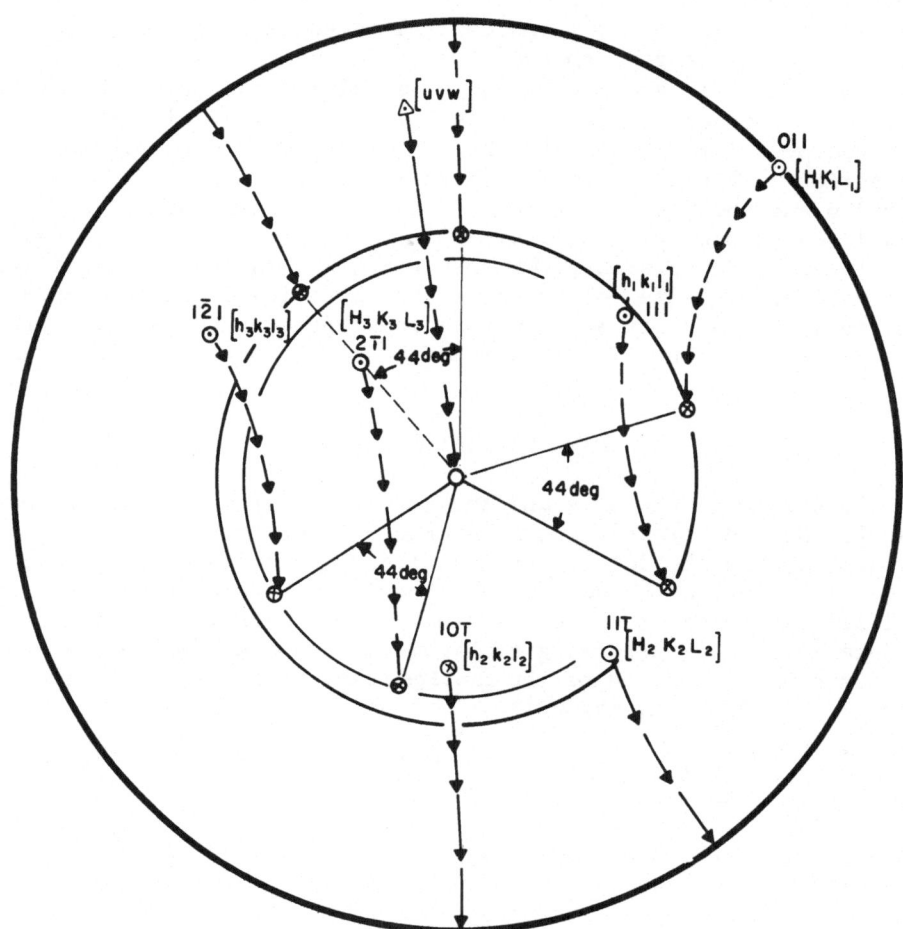

Figure 3. The angle of lattice rotation, α, is shown here by shifting
[uvw] to the center of the diagram. The starting and final positions of the
three reference axes shifted accordingly. α is measured approximately 44
degrees.

       Mathematically, equation (4) is determinate if a minimum of
two points and their subsequent positions after the rotation are
known. Let the starting position of the two points be $[H_1K_1L_1]$
and $(H_2K_2L_2)$ and their final position after the rotation be
$[h_1k_1\ell_1]$ and $[h_2k_2\ell_2]$, respectively. By following similar steps
as outlined previously, equations (4) and (5) may be expanded as:

$$m_1 = H_1 - A_1 h_1$$
$$n_1 = K_1 - A_1 k_1 \quad \Big\}$$
$$P_1 = L_1 - A_1 \ell_1 \qquad\qquad 4a$$

$$m_2 = H_2 - A_2 h_2$$
$$n_2 = K_2 - A_2 k_2 \quad \Big\}$$
$$P_2 = L_2 - A_2 \ell_2 \qquad\qquad 4b$$

and

$$m_1 u + n_1 v + p_1 w = 0 \quad \Big\}$$
$$m_2 u + n_2 v + p_2 w = 0 \quad \Big\} \qquad (6)$$

Solutions to equation (6) are given:

$$u = v \frac{\begin{vmatrix} n_1 & P_1 \\ n_2 & P_2 \end{vmatrix}}{\begin{vmatrix} m_2 & P_2 \\ m_1 & P_1 \end{vmatrix}}$$

$$v = w \frac{\begin{vmatrix} P_1 & m_1 \\ P_2 & m_2 \end{vmatrix}}{\begin{vmatrix} n_2 & m_2 \\ n_1 & m_1 \end{vmatrix}}$$

and

$$u = w \frac{\begin{vmatrix} P_1 & n_1 \\ P_2 & n_2 \end{vmatrix}}{\begin{vmatrix} m_2 & n_2 \\ m_1 & n_1 \end{vmatrix}} \qquad (7)$$

The axis of rotation [uvw] can be determined by solving
equation (7). The amount of rotation is equal to the angle α
bound by two vectors $[a_1 b_1 c_1]$ and $[a_2 b_2 c_2]$ which are projections
of [H K L] and [h k l], respectively, on a plane perpendicular to
[uvw], as illustrated in Figure 4. For a cubic system, α may be
determined from

$$\cos \alpha = \frac{(a_1 a_2 + b_1 b_2 + c_1 c_2)}{(a_1^2 + b_1^2 + c_1^2)^{\frac{1}{2}} (a_2^2 + b_2^2 + c_2^2)^{\frac{1}{2}}} \qquad (8)$$

and

$$b_1 c_2 - c_1 b_2 = u$$
$$c_1 a_2 - a_1 c_2 = v$$
$$a_1 b_2 - b_1 c_2 = w \qquad (9)$$

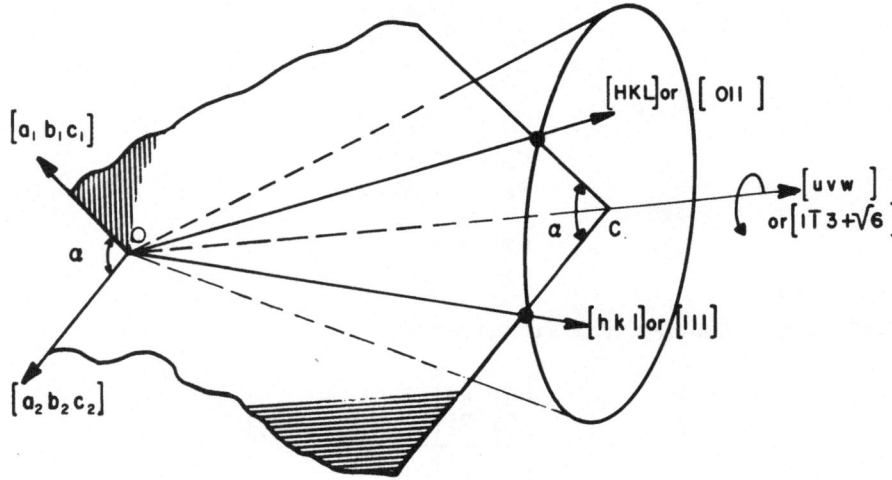

Figure 4. Sketch showing the angular relationships between
the five vectors: [uvw], [HKL], [a₁b₁c₁] and
[a₂b₂c₂]. Note that [uvw], [HKL] and [a₁b₁c₁]
and [uvw], [hkℓ] and [a₂b₂c₂] lie on two different
planes separated by angle α.

$$\left(\frac{a_1H + b_1K + c_1L}{a_2h + b_2k + c_2}\right)\left(\frac{a_2^2 + b_2^2 + c_2^2}{a_1^2 + b_1^2 + c_1^2}\right)^{\frac{1}{2}} = A \qquad (10)$$

If we choose the same example of the textural change in a shock loaded iron and make use of the data in Table I, we have in the following:

$A_1 = \sqrt{2/3}$	$A_2 = \sqrt{3/2}$
$-m_1 = \sqrt{2/3}$	$m_2 = 1 - \sqrt{3/2}$
$n_1 = 1 - \sqrt{2/3}$	$n_2 = 1$
$P_1 = 1 - \sqrt{2/3}$	$P_2 = -1 + \sqrt{3/2}$

Values of u, v, and w was obtained from equation (7) and of a_1, a_2, b_1, b_2, c_1 and c_2 from equation (9) and (10) are listed in Table II.

TABLE II. Mathematical Solutions for $[uvw]$, $[a_1b_1c_1]$ and $[a_2b_2c_2]$

Vector Notation	Values	Equation
$[u\ v\ w]$	$[1\ \bar{1}\ \ \ 3+\sqrt{6}]$	(7)
$[a_1b_1c_1]$	$\sim[1\ 1\ \ \ \ 0\ ]$	(9) and (10)
$[a_2b_2c_2]$	$\sim[0\ 5.5\ \ \ 1\ ]$	(9) and (10)

Solving equation (8) from values of $[a_1b_1c_1]$ and $[a_2b_2c_2]$, α may be found 45.9 deg. Similarly, the angle between $[0\bar{0}1]$ and $[1\ \bar{1}\ 3+\sqrt{6}]$, may be found 14.6 deg. These values agree well with the stereographic determination shown previously. The determination of $[a_1b_1c_1]$ and $[a_2b_2c_2]$ by stereographic method is shown in Figure 2. The stereographic angles between $[011]$ and $[a_1b_1c_1]$ and between $[111]$ and $[a_2b_2c_2]$ are 34 degrees which are the complementary angles between $[uvw]$ and $[110]$ and between $[uvw]$ and $[111]$, both of which are measured 56 degrees. The angular relationships between $[uvw]$, $[011]$ and $[a_1b_1c_1]$ may be referred to Figure 4. Also shown in Figure 2, $[a_1b_1c_1]$ and $[a_2b_2c_2]$ are stereographically located along the great circle 90 degrees from $[uvw]$. The locations of $[a_1b_1c_1]$ and $[a_2b_2c_2]$ are thus determined from the intersections between the 34 degree circles from $[011]$ and $[111]$, respectively, and the 90 degree great circle from $[uvw]$.

DISCUSSION

We have demonstrated that a satisfactory solution for the axis and the amount of lattice rotation may be obtained independently from either the stereographic analysis or mathematical calculation. An investigator may choose either method for the determination but in practice, the combined use of both methods is recommended. This will not only provide a check against possible errors but will also serve to simplify the process of determination, especially if one chooses to use mathematical calculation for the determination. In solving equations (7), (8) and (9) for $[uvw]$, $[a_1b_1c_1]$ and $[a_2b_2c_2]$, it is customary to assume one of the three Miller's indices for a given vector to be equal to unity and to determine the remaining two indices accordingly. Such assumption is valid so long as the particular Miller's index to be assumed as unity will not actually be zero. To check for this adverse possibility, the general location of the vector to be determined should be known and a stereographic plot should provide this information.

In the present study, textural change in association with lattice rotation was used as an illustration to determine the axis and the amount of rotation. Other physical phenomena in which atomic movements are involved, such as phase transformation and certain form of twinning operations, may also be related to lattice rotation. The possibility of using lattice rotation as a model to interpret these phenomena is attractive.

REFERENCES

1. R. O. Williams, "The Representation of the Textures of Rolled Copper and Aluminum by Biaxial Pole Figures", AIME Trans. 242, 105-115, 1968.

2. L. K. Jetter, C. J. McHargue and R. O. Williams, "Method of Representative Preferred Orientation Data", J. Appl. Phys. 27, 368-374, 1956.

3. N. L. Svensson, "Estimation of Yield Strength Anisotropy Due to Preferred Orientation", AIME Trans. 236, 1004-1009, 1966.

4. C. Feng and E. Krull "Anisotropy in Plastic Flow of a Ti-8Al-1Mo-1V Alloy" AIME Trans. 245, 1101-1109, 1969.

SIMULTANEOUS SPIRAL RECORDING OF POLE FIGURES ON POLAROID FILM FOR TEXTURE GONIOMETERS

H. Ebel and M. F. Ebel

Institut für Angewandte Physik

Technische Hochschule Wien, Vienna, Austria

ABSTRACT

The X-ray reflection method according to Schulz[1] is used for investigations of textures in rolled materials. The pole figures are measured either along spirals or along circles. Points of equal intensity are transposed from the record of X-ray intensity to a spiral diagram. Finally contour lines are delineated, pointing out regions of equal pole density. Three ways are known for simplification of the evaluation:

a) The results of the measurement are stored, evaluated by a computer and the pole figure is plotted. Points of equal pole density are represented by equal symbols [2,3].

b) The pole figure is recorded synchronously with the X-ray measurement along circles or spirals. Ranges of different pole density are characterized by different colors [4,5].

c) The pole figure is recorded on a photographic film along a spiral. The blackening depends on the measured countrate [6,7].

An outline on different instruments using photographic registration is given.

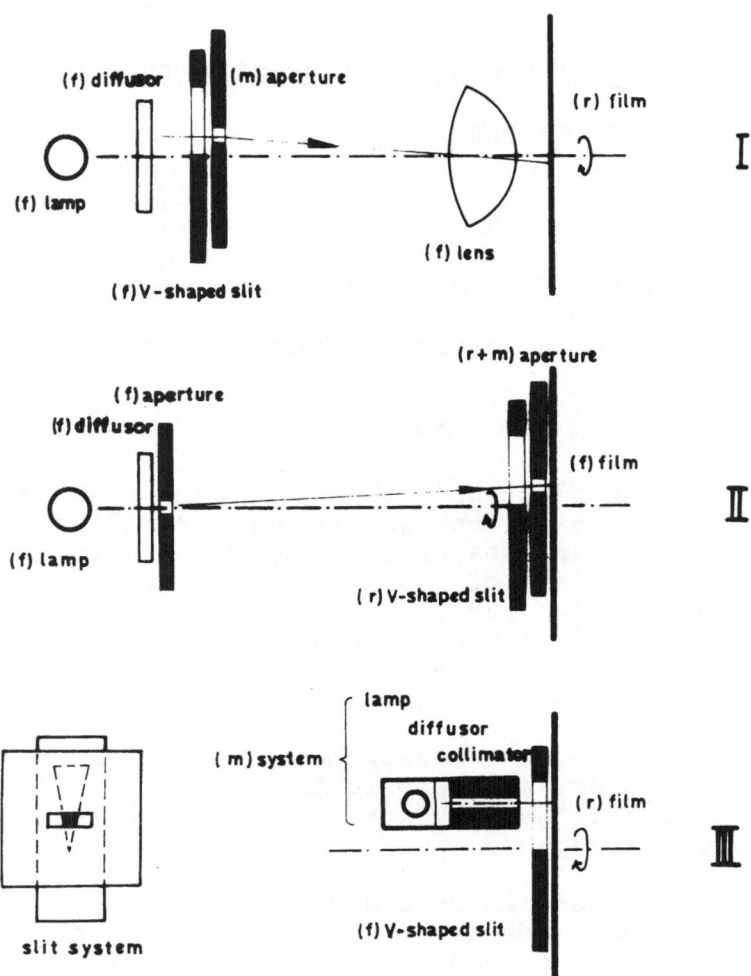

Figure 1. Three ways of photographic recording of pole figures in schematic representation (f...fixed, m... moving, r...rotating).

PRINCIPLE

A photographic film revolves with constant angular
velocity on an axis vertical at its surface. A light
beam which moves linearly across the film causes a
.spiral trace. Film and sample rotate with uniform angular
velocity. The intensity of the light beam is controlled
by the ratemeter output.

The three possibilities for photographic pole
figure recording are presented schematically in
figure 1.

   I) Lamp and diffusor are fixed.In front of the
      diffusor a V-shaped slit is fastened. A
      rectangular aperture moves with constant
      velocity parallel to the axis of the V-slit.
      Behind the fixed lens a film rotates on the
      optical axis.

  II) Lamp and diffusor are fixed.In order to avoid
      an optical system, the slit system is arranged
      near the fixed film. Only the V-shaped slit
      rotates and the rectangular aperture moves again
      parallel to the axis of the V-slit.

 III) Lamp, diffusor and collimator together perform
      a movement normal to the rotation axis of the
      film and parallel to the axis of the fixed
      V-slit.

For the arrangement I lenses or a camera are required. The slit-system of possibility II is - in
comparison to I and III - much more complicated. Therefore, the most simple and compact solution is realized
by arrangement III.

REMARKS

In all these cases the cusp of the V-shaped slit
is depicted in the center of the pole figure. When the
angular velocity is constant, the velocity of the light
beam on the film increases with increasing distance
from the center. The differences in exposure time are
compensated by the V-shaped slit. Thus, only by the
use of the V-slit the whole spiral shows a constant
blackening at constant light intensity. As mentioned
above, the light intensity is controlled by the rate-

meter output by means of a d.c.amplifier.

Disadvantages of This Method

1) Only a qualitative information is provided.

2) The spiral distance in the spiral diagram changes in dependence on the distance to the center. For our instrument an Archimedian spiral is used which causes a small distortion of the pole figure.

Advantage of This Method

It is a rapid and simple way to obtain information about preferred orientations.

EXAMPLE

Figures 2 and 3 show {111}-pole figures of cold rolled aluminium. Depending on the rolling conditions different types of textures and also of qualities of the rolled materials are found. So this technique seems to be a very good tool for a rapid production control.

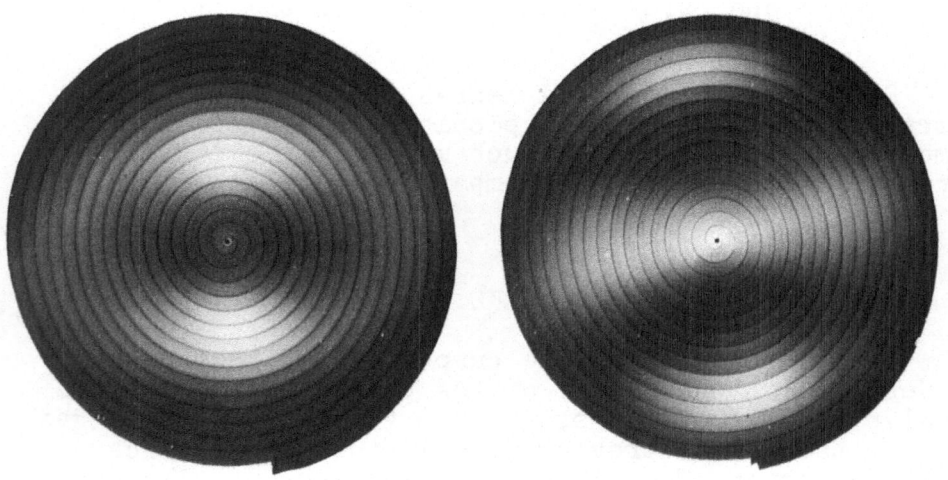

Figure 2.   Figure 3.
($\bar{3}$26) [$\bar{2}$6$\bar{3}$]   (001) [110]
{111}-pole figures of cold rolled aluminium.

ACKNOWLEDGMENTS

This work has been sponsored by "Forschungsförderungsfond der Gewerblichen Wirtschaft in Österreich".

The instrument is manufactured by Anton Paar K.G. Graz-Straßgang, Austria.

REFERENCES

1. L. G. Schulz, "A direct method of determining preferred orientation of a flat reflection sample using a Geiger counter X-ray spectrometer ", J. Appl. Phys. 20, 1030-1033 (1949)

2. H. Siemes, "Ein Rechenprogramm zur Auswertung von Röntgentexturaufnahmen ", Neues Jb. Mineralogie 2/3, 49-60 (1967).

3. A. Segmüller and J. Angilello, "Automatic pole figure evaluation ", IBM Research RC 2296, Nov.1968.

4. K. Lücke, Siemens Verkaufsprogramm Oktober 1968.

5. R. Baro and D. Ruer, "An Automatic pole figure recorder ", Sci. Instr. (J. Phys. E) 3, 541-543 (1970).

6. H. Ebel and B. Ortner, "Direct registration of pole diagrams for the determination of textures by X-rays ", J. Sci. Instr. 43, 959-960 (1966).

7. A. Wagendristel, H. Ebel, O. Marihart and W. Schneider, "Verfahren zur automatischen Darstellung von Polfiguren ", Materialprüfung 12, 337-340 (1970).

X-RAY DOUBLE CRYSTAL DIFFRACTOMETER INVESTIGATIONS OF IMPLANTED SILICON: D^+ AND N^+

E. H. teKaat* and G. H. Schwuttke**

International Business Machines Corporation

Route 52, Hopewell Junction, New York 12533

ABSTRACT

Double crystal diffractometer measurements on silicon bombarded to a fluence $>10^{16}$ ions/cm^2 with 1 MeV deuterium and 2 MeV nitrogen are reported. Such measurements provide insight into radiation damage in silicon through the observation of Bragg case pendelloesung fringes and double peak rocking curves. Bragg case pendelloesung fringes are used to determine non-destructively the projected range of ions in silicon. Double peak rocking curves are used to measure changes in lattice parameter with the ion dose. Finally, a model of radiation damage in silicon is presented.

INTRODUCTION

This paper reports on double crystal diffractometer investigations of ion implanted silicon crystals. Such measurements are shown to provide further insight into the crystal damage produced by heavy ions in silicon.

---

* On leave of absence from Physikalisches Institut der Universitaet Muenster, 44 Muenster, Schlossplatz 7, Germany.
**This work was supported in part under Air Force Contract No. F19(628)-68-C-0196, Air Force Cambridge Research Laboratories, Bedford, Massachusetts, U.S.A.

Previously, it was shown (1,2) that large dose
implantations of energetic ions, such as nitrogen, oxygen, carbon and others, implanted into silicon with an
energy of 1 MeV or larger, produce sub-surface layers
of heavily damaged silicon or even of amorphous silicon.
It was shown that these layers are embedded into a silicon matrix of good crystalline perfection. The heavily
bombarded crystal was compared to a bi-crystal, consisting of a perfect bulk and a perfect thin layer crystal
separated by an amorphous layer (3,4). It was also
shown that such bi-crystals produce characteristic fringe
systems in x-ray topographs. The topographs were analysed
by Bonse, et al experimentally (3) and theoretically (4)
as interference fringes formed by simultaneous diffraction of bulk and layer crystal. It is the purpose of
this investigation to study the perfection of the thin
"layer crystal" in more detail.

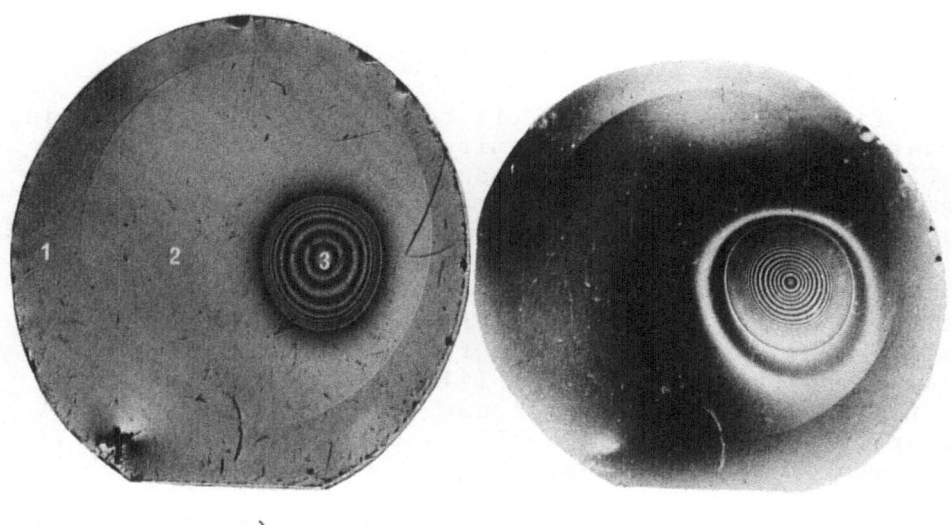

a)   b)

Figure 1a   Transmission Topograph of nitrogen
             bombarded (2 MeV, 5.5×10^{16} N^+/cm^2)
             silicon wafer, (111) reflection,
             Mo K_α radiation. Wafer size 25 mm.
             Shielded area marked 1, bombarded
             area marked: low dose 2 and high
             dose 3.

Figure 1b   Double Crystal Topograph of crystal
             shown in Figure 1a, (333) reflection,
             Cu K_α radiation.

EXPERIMENTAL

The silicon crystals used in these experiments
were of (111) orientation, zero dislocation density, and
of Lopex quality. They were bombarded with the AFCRL
van de Graaff generator to a fluence of about 10^{16} ions/
cm^2. The results discussed in this paper were obtained
after N^+ implantation (2 MeV, 5.5×10^{16} ions/cm^2) and
after D^+ implantation (1 MeV, 6×10^{16} ions/cm^2).

The measurements were made with a thermally stabilized double crystal diffractometer (5a). A non-dispersive
setting (+n;-n) and CuKα radiation was used. The monochromator crystal was cut for asymmetric reflection in
order to decrease the divergency of the exploring beam (5b).
The samples were aligned for symmetric reflection.
Double crystal rocking curves were recorded and in addition topographs were made on Ilford nuclear plates type
G5 and L4.

RESULTS

Figure 1 shows two (111) topographs of the nitrogen
bombarded sample. The transmission topograph is shown
in Figure 1a and the reflection topograph is shown in
Figure 1b. During the implantation, a circular mask
protected the wafer rim. The outline of the mask (area 1)
can be seen in the topographs. The inner area was exposed to a non-uniform ion beam having the intensity
maximum a little bit off center (area 3). The high dose
area 3 shows the interference fringes in the topographs.
In agreement with the theory (4), the fringe density in
the reflection topograph (Figure 1b) is higher by a
factor $(\cos\alpha)^{-1}=3$, where α∿71° is the angle between
both sets of {111} planes.

Figure 2 shows the (111) rocking curve of the same
crystal. The reflecting spot of approximately 0.7mm in
diameter for this curve was selected in the center of the
high dose area of Figure 1b. Note, that both wings of
the curve show subsidiary maxima. In the low absorption
wing (-θ) of the curve 13 maxima can be counted. Such
oscillations are attributed to the Bragg-case pendelloesung effect as predicted in von Laue's theory (6). Such
pendelloesung fringes were previously reported by Batter-
man and Hildebrandt (7), Renninger (8), and Uragami (9).
Bragg case pendelloesung fringes occur only in thin dynamical crystals. They can be obtained either for varying crystal thickness or for varying incident angle of the
radiation.

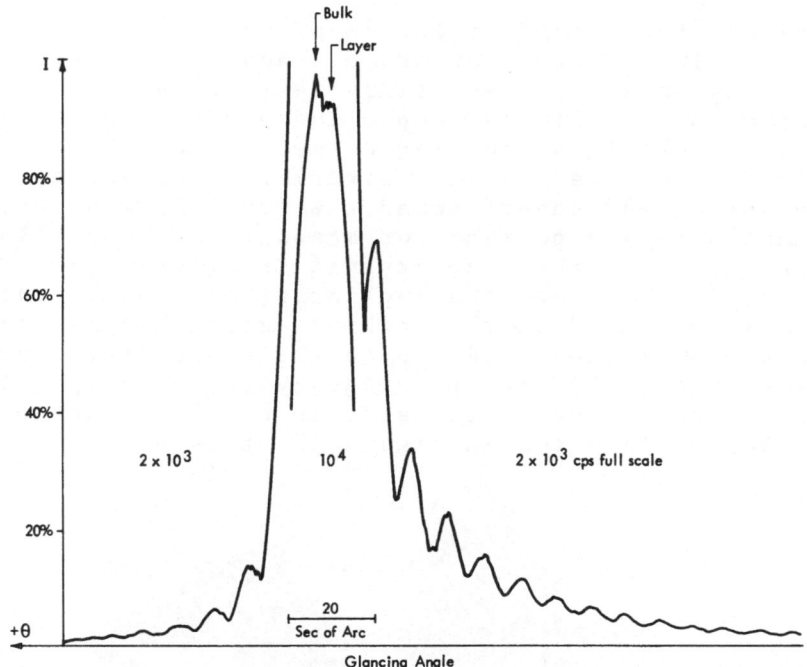

Figure 2   Double Crystal (111) Rocking Curve of
            crystal shown in Fig. 1. Reflecting
            spot in area 3 approximately 0.7 mm
            in diameter. Note Bragg pendelloesung
            fringes in wings of the curve. 13
            fringes on the low angle side (-θ) of
            the curve.

     Our experimental conditions are such that the inci-
dent angle varies. Therefore, the angular separation $\Delta\theta$
of adjacent minima or maxima - measured far out in the
wings of the rocking curve - is related to the layer
crystal thickness D as given by the simple expression
$D = \lambda/2(\Delta\theta) \cos\theta$, where λ = wavelength and θ = Bragg's
angle*. Using this expression for the N[+] implanted
sample, one calculates a layer thickness of 2.0 μm. This
is in good agreement with the value of 2.1 μm obtained
through direct optical measurements made on a similar
sample on a bevel (2).

─────────────

*This follows directly from the general formula given for
 a Bragg case reflection of a crystal of intermediate
 thickness by von Laue, if the term β' in his formula is
 assumed to be much larger than 1. See reference 6,
 page 373 formula 31.14.

Direct topographic evidence of such a pendelloesung
effect is given in the topographs shown in Figure 3.
These topographs represent (333) reflections of a
deuterium (D^+) implanted sample and were taken at the
low angle side of the rocking curve. Note, that the
topographs show the standard circular interference
fringe system and superimposed the pendelloesung fringes.
The simultaneous appearance of standard and pendelloesung
fringes indicates that the crystal is slightly bent as
a result of the large dose implantation. Consequently,
the "working point" on the rocking curve changes locally
across the bombarded area. Note the phase difference of
π between the interference fringe system displayed in the
two topographs. The center spot in the topograph of
Figure 3a is light and in Figure 3b it is dark.

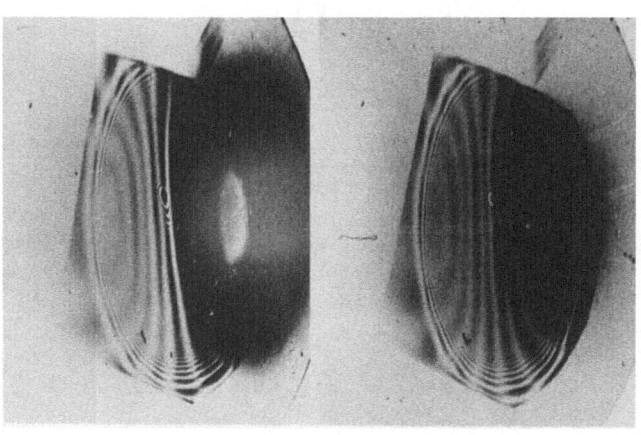

a)   b)

Figure 3   Double Crystal Topographs, (333) reflec-
             tion, CuK$_\alpha$ radiation, deuterium implanted
             (1 MeV, 6×10^{16} D^+/cm^2) showing standard
             interference pattern and superimposed
             pendelloesung fringes. Figures (a) and
             (b) both low angle side of rocking curve.

Another interesting result of the double crystal
diffractometer measurements is shown in Figure 4. The
rocking curve displayed in Figure 4 comes from the same
N^+ sample as shown in Figure 2, but this time a (333)
reflex was used for the recording. Instead of pendelloesung fringes, now one observes a single secondary
peak at the low angle side at an angular distance from
the main peak of about 23 sec of arc. This second peak
stems from the surface layer above the amorphous layer.

This is uniquely demonstrated in a series of topographs given in Figure 5. These topographs were taken at the top of the main peak (a) (Figure 5a), between both peaks (b) (Figure 5b), and at the top of the second peak (c) (Figure 5c).

Comparing the three topographs given in Figure 5 with the corresponding positions of the working point on the rocking curve, it follows that most of the intensity responsible for the topograph shown in Figure 5a comes from the bulk crystal while the topograph shown in Figure 5c is due to the intensity reflected from the

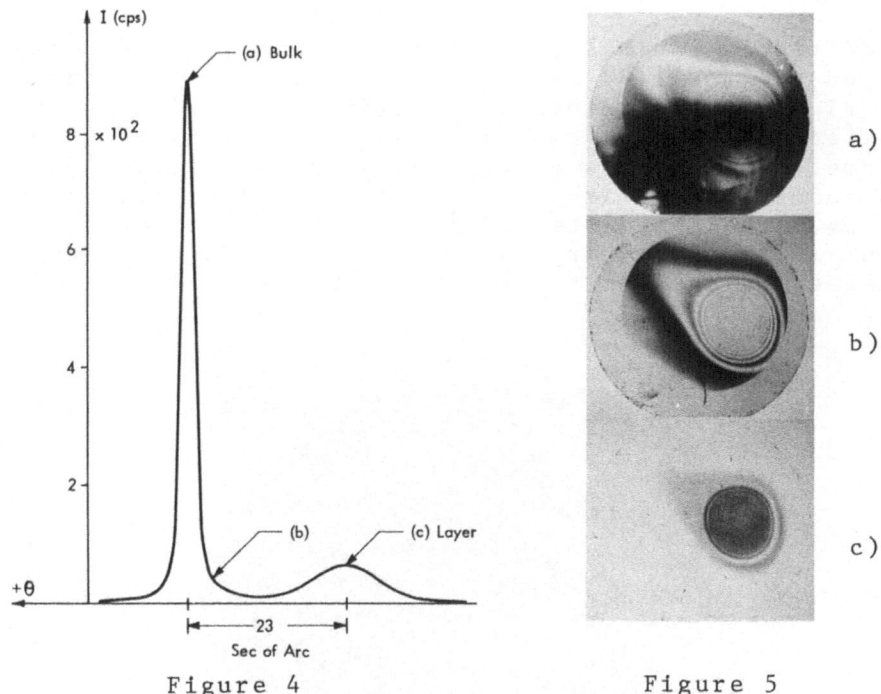

Figure 4   Figure 5

Figure 4   Double Crystal Rocking Curve of crystal
             shown in Figure 1, (333) reflection,
             CuK$_\alpha$. Reflecting spot in center of
             high dose area.

Figure 5   Double Crystal Topographs of crystal
             shown in Figure 1, (333) reflections,
             CuK$_\alpha$.

             a,b,c   Topographs recorded at positions
                     (a), (b), (c) respectively on
                     rocking curve.

layer crystal.   The topograph given in Figure 5b received
its intensity mainly from the low dose bombarded area
which is the crystal area showing no interference
fringes in the low order (111) topographs of Figure 1.

The (333) interference fringe system is best resolved
in Figure 5c. This is seen more clearly in Figure 6
which represents the topograph shown in Figure 5c at
higher magnification.   It can be seen that the fringe
density in the (333) topograph (Figure 6) is again higher
by a factor of three if compared to the (111) topograph
shown in Figure 1b.   In addition, fringes can be seen
even in the low dose area (compare Figure 5b).

Additional interesting information on lattice damage
after ion implantation can be obtained if the rocking
curve (Figure 4) is recorded for different positions of
the reflecting spot on the wafer surface.   This is shown
in Figure 7 which gives a graph of peak separation $\Delta\theta$
versus position superimposed upon the (333) topograph of
Figure 5b.   These results indicate a change of lattice
constant in the layer crystal with the ion dose.   Note
the plateau maximum in the graph where the ion beam was
most intense.

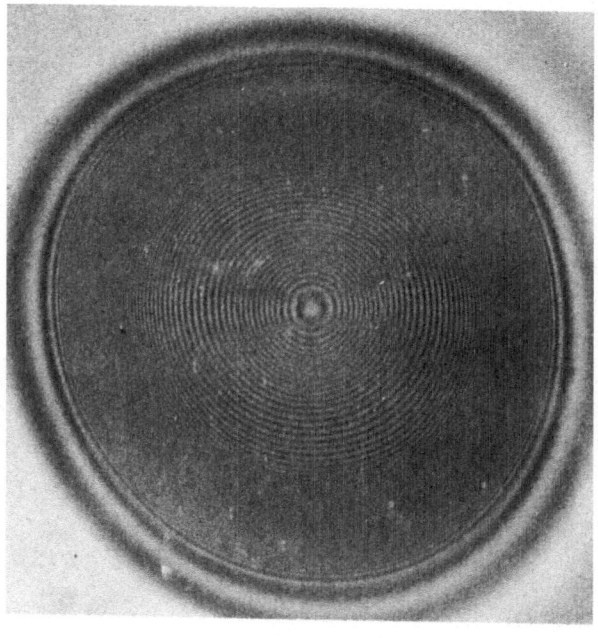

Figure 6   Magnified image of topograph
             shown in Figure 5c.

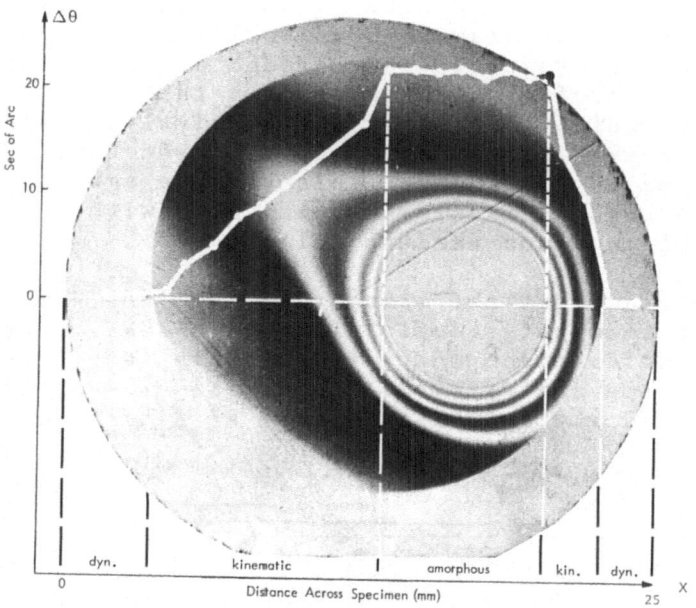

Figure 7   Double Crystal Topograph of
Figure 5b and superimposed Δθ
values recorded at different
positions across wafer

DISCUSSION

If the double crystal diffractometer measurements
are applied to the crystal damage model presented previously (3,4), the following refined model of radiation
damage in silicon after large dose implantation appears
to be reasonable. The main feature of interest in the
new model is a more pronounced stratification in the
damage structure. Qualitatively, we now differentiate
between five damage regions compared to three previously (3,4). These regions are characterized by their
ability to conduct x-ray wavefields which in turn is
determined by different degrees of radiation damage:

Starting with the layer crystal at the top, we sub-
divide the layer crystal into a dynamical and into a
kinematical zone. The dynamical zone is the upper part
and has suffered some radiation damage. The lower part
is the kinematical zone and contains various degrees of
heavy damage but is still single crystal silicon.

The bulk crystal is subdivided in a similar way into a dynamical and into a kinematical part. Consequently, the amorphous zone is surrounded by kinematical crystal layers. This model is drawn schematically in Figure 8a which represents a cross-section through an ion implanted wafer. A damage profile along an axis $\underline{z}$ through the amorphous layer which correlates qualitatively the number of displaced atoms N with the damage cross-section (Figure 8a) is given in Figure 8c. This profile agrees qualitatively with the LSS theory (10) and also with the calculations of Sigmund and Sanders (11).

It is obvious that the transition from one damage zone into the other must be gradual and not abrupt. But regarding x-ray diffraction there should exist two threshold values, separating the dynamical, kinematical and the amorphous crystal state on the N scale of Figure 8c.

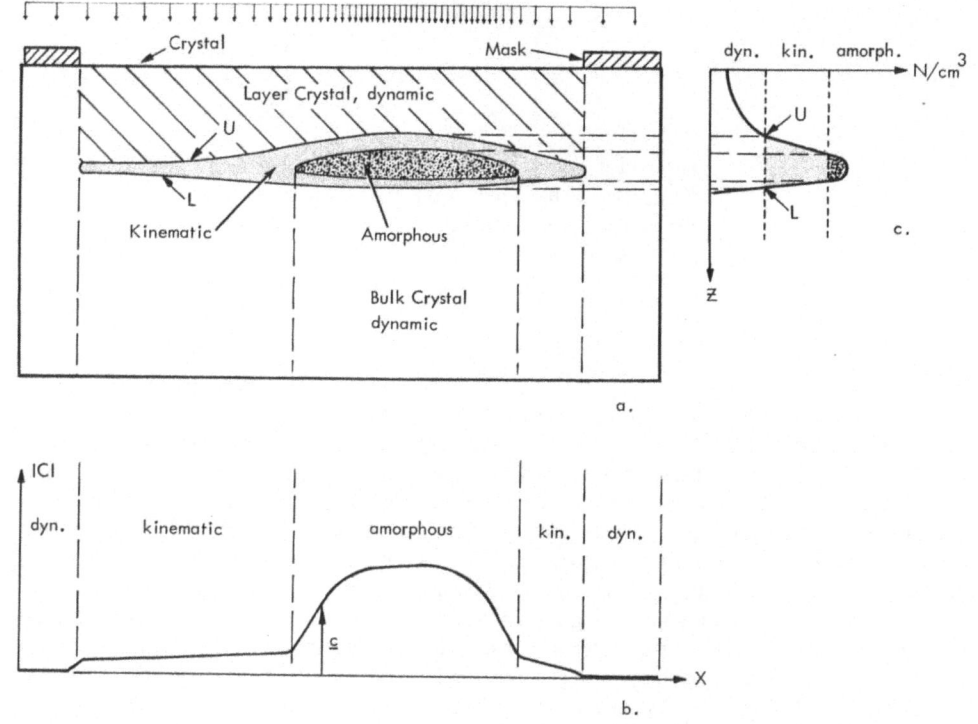

Figure 8   Model of radiation damage in silicon
            after ion implantation

            a. Cross-section through implanted wafer.
            b. Displacement profile across implanted
                wafer.
            c. Damage profile of implanted wafer.

In addition, we include a displacement profile in Figure 8b which pictures schematically the actual displacement c of atoms along the boundary U in Figure 8a relative to the unbombarded sample (3,4). Such a displacement must occur because the amorphous silicon needs approximately 10% more volume than the crystalline silicon (3,4). In the low dose area, the displacement is smaller because the kinematical layer expands less than the amorphous layer. Figure 7 relates also the topograph of Figure 5b to the defect structure as described in Figures 8a, b, and c.

The diffraction of x-rays in the bombarded crystal (Laue transmission) can now be described as follows: To simplify discussion, the wavefields are assumed to be generated on the bulk crystal side. Therefore, they travel first through the dynamical zone. We assume that close to the lower boundary L only the wavefield with lowest absorption is present due to the rather thick bulk crystal.

Approaching the kinematical zone, this wavefield will compensate for any long range lattice disturbance by a substantial redistribution of energy on each branch of its dispersion surface (tie-point migration). With increasing radiation damage resulting in more complex and extended defect structures such as amorphous clusters, the local damage gradients become so high that they are unmanageable for any wavefield propagation at all and the wavefield is forced to split into its two component waves. This is indicated by the lower boundary L.

Both waves emerging from the bulk crystal travel through the "gap" consisting either of only kinematical crystal material (region 2, Figure 1a) or of both kinematical and amorphous crystal (region 3, Figure 1a). Upon reaching the other side of the gap (approximately the boundary U in Figure 8a, the component waves meet about the same damage situation as in the lower boundary L and are able to rebuild wavefields for propagation in the layer crystal. Both wavefields in layer and bulk crystal are coherent (4); however, they are shifted in phase relative to each other due to the varying vector c (Figure 8b) as described in reference 3. Thus the transmitted x-ray intensity is modulated in accord with the expansion of the "gap" and gives rise to the interference fringes in the topographs. For other configurations of the incident beam, we refer to reference 3 and 4.

In the Bragg case, which has been used in our

experimental work, the discussion of wavefields is similar
but more complicated than in the Laue case; but in both
cases the detailed analysis of wavefield propagation impacts mainly the wave amplitude (tie point migration) and
the coherence calculations, which both determine the
visibility of the fringe system.

However, the shape of the fringe system is uniquely
defined by the actual distribution of the displacement
vector c in the boundary U and the operating diffraction
vector h and is therefore independent of wavefield considerations.

Support for the notion that a continuous amorphous
zone is not needed to produce a slightly expanded "gap"
in the crystal and, consequently, interference fringes
in x-ray topographs, follows directly from the (333)
topograph shown in Figure 5b. One notes widely spaced,
broad fringes in the low dose area (region 2 in Figure 1a)
of this cyrstal. This is in agreement with electron
transmission experiments on similar samples which showed
that a continuous amorphous layer does not exist in this
area (2). The start or end of the amorphous layer is
readily recognized in the topograph of Figure 5b due
to the sudden change in fringe density.

Interesting is also the plateau maximum in the curve
of Figure 7. The curve is superimposed on the (333) topograph of Figure 5. Evidently, the interference fringe
density continues to vary in the plateau area indicating
that the amorphous lens is changing in thickness while
the $\Delta\theta$ value has reached a maximum. The $\Delta\theta_{max}$ value indicates the maximum defect density the lattice can ac-
commodate in this case; above $\Delta\theta_{max}$ only the amorphous
phase is present. Saturation measurements of this kind
could provide more insight into the crystalline - amorphous
transformation through ion implantation. Direct quantitative measurements of defect production in silicon
through x-rays in relation to ion mass and energy appear
also feasible through this effect.

A final remark deals with the origin of the diffracted intensity responsible for the second peak in the
rocking curve of Figure 4. The $\Delta\theta$ - saturation (Figure 7)
suggests that most of this intensity comes from the kinematical zone in the layer crystal close to the amorphous
lens. Therefore, a nearly perfect dynamical crystal as
postulated (3,4) is compatible with the double crystal
measurements.

On the other hand, the best topograph of the fringe area (Figure 6) is obtained with the diffractometer aligned at the second peak of the rocking curve (Figure 4). This is evidently support for the necessity of wavefield adaptation in the long range strainfields caused by the kinematical and amorphous zones in the bombarded crystal.

REFERENCES

1.   G. H. Schwuttke, K. Brack, E. F. Gardner, and H. M. DeAngelis, in Radiation Effects in Semiconductors, ed. by F. L. Vook, Plenum Press, New York 1968, page 406.

2.   G. H. Schwuttke and K. Brack, Trans. Met. Soc. AIME, 245, 475, (1969).

3.   U. Bonse, M. Hart, and G. H. Schwuttke, Phys. Stat. Sol., 33, 361, (1969).

4.   U. Bonse and M. Hart, Phys. Stat. Sol., 33, 351, (1969).

5a.   U. Bonse and E. Kappler, Z. Naturforsch., 13a, 348, (1958).

5b.   M. Renninger, Z. Naturforsch., 16a, 1110, (1961); Advanc. X-Ray Analysis 10, 32, (1967).

6.   M. von Laue, Roentgenstrahlinterferenzen, Akademische Verlagsgesellschaft, Frankfurt (M) 1960.

7.   B. W. Battermann and G. Hildebrandt, Acta Cryst., A24, 150, (1968).

8.   M. Renninger, Acta Cryst., A24, 143, (1968).

9.   T. Uragami, J. Phys. Soc. Japan, 27, 147, (1969).

10.   J. Lindhard, M. Scharff, and H. E. Schiøtt, Kgl. Danske Videnskab Selskab. Mat. Fys. Medd., 33, No.14 (1963).

11.   P. Siegmund and J. B. Sanders, Proc. Int. Conf. Appl. Ion Beams Semiconductor Techn., ed. by P. Glotin, page 215, Editions Ophrys, Grenoble 1967.

X-RAY INVESTIGATIONS OF SPINEL SUBSTRATES

J. E. A. Maurits and A. M. Hawley

Union Carbide, Crystal Products

San Diego, California 92123

ABSTRACT

Recent advances in MOS integrated circuit technology have opened new, high-volume applications utilizing epitaxial silicon-on-insulating substrates. To provide good quality heteroepitaxial silicon films for the semiconductor industry, a development program has been established to improve both the crystalline quality and the fabrication techniques for the most promising current substrate, magnesium aluminum spinel.

The role of X-ray diffraction techniques in the investigation of crystal quality, substrate surface quality, and epitaxial film quality are discussed, and results interpreted as the program is traced from crystal growth to epitaxial film analysis.

INTRODUCTION

Several single crystal dielectric materials have been considered for heteroepitaxial films, but sapphire (a - Al_2O_3), and magnesium aluminum spinel ($MgAl_2O_4$) are the most promising. These are transparent, water-white, refractory, insulating dielectric single crystal oxides with a close lattice match to silicon. They are chemically inert and thermally stable at silicon deposition temperatures.

Early development work with silicon on insulating substrates used sapphire. The advantages of spinel as a substrate were first recognized in 1965 through the work of Manasevit and Forbes(1). A more recent and comprehensive review is given by Cullen(2).

516

A concentrated development program was started in 1969 within Crystal Products to refine the Czochralski growth techniques to obtain high quality stoichiometric spinel crystals, and to fabricate damage-free substrates suitable for epitaxy. X-ray diffraction analysis has proven to be an invaluable tool during the course of this program in monitoring crystal quality, fabrication damage, and film-substrate interactions.

CRYSTAL GROWTH

Magnesium aluminum spinel is a mixture of magnesia and alumina, which solidifies as a compound in single-crystal form from a congruent melt at 2150 deg. C. In the unit cell, the oxygen atoms form a close-packed cubic lattice structure with magnesium ions at the tetrahedral interstices and aluminum at the octohedral interstices. Spinel can exist as a single crystal over a wide composition range depending on the growth methods used; compositions ranging from alumina-rich (1 $MgO:3.3Al_2O_3$) to stoichiometric (1 $MgO:1Al_2O_3$) have been reported. Two growth techniques were considered for the program, the flame fusion process and the Czochralski or pulled-from-the-melt technique.

Figure 1.   Czochralski-Grown Spinel Crystals

Stoichiometric spinel is very difficult to obtain by the flame
fusion process, because the conditions required to control MgO
vaporization give problems with induced growth strain and cracking.
Near-stoichiometric crystals have been grown at RCA with reported
good yields and successful epitaxy(3). However, sizes have been
limited to 5/8" diameter because of the strain and cracking problem.

The Czochralski growth technique is the best suited for
commercial production of large size, high quality single crystals.
Crystals up to 2 1/4" diameter by 12" long have been achieved by
Crystal Products. The current production size is 1 3/8" diameter
by 12" long in the as-grown form; examples are shown in Figure 1.
The large size and uniform diameter control of the Czochralski
process is demonstrated.

Czochralski-grown spinel is also clearly superior in crystal
quality. Schulz-Wei patterns show misorientations in flame fusion
crystals on the order of 1/2 deg., but misorientations larger than
2 minutes are seldom seen in Czochralski crystals. Dislocation
counts revealed by chemical etching are about 1,000 per square
centimeter for Czochralski spinel, but one to two orders of magnitude
higher in flame fusion spinel. Comparison of the Laue back-
reflection patterns is very striking. The broken spots and
incomplete zone lines of the flame-fusion growth indicate poor
crystal quality and the presence of misoriented subgrains. The
Czochralski crystal Laue pattern has sharp, clear spots and complete zone lines indicating good crystal quality and the absence
of misoriented subgrains. Another striking comparison of quality
was shown by a two-circle goniometer X-ray diffraction analysis of
microtwins in the epitaxial film. Microtwinning, i.e., small areas
of twin lamallae imbedded in a single crystal matrix, are known to
occur in heteroepitaxial films on spinel(4). The integrated
intensity of the (311) diffraction peak is measured, then the (311)
twin across $(11\bar{2})$ is located, and its integrated intensity obtained.
The results are expressed as percent of microtwins per substrate
area. Values of 30 to 40% were obtained for films on flame-fusion
spinel, while films deposited on Czochralski material were 2 to 7%.

The Czochralski growth techniques having been refined to obtain
large, high-quality crystals, a quality assurance program was
established to detect any crystal defects and the subsequent effect
on epitaxial film quality. Each crystal is inspected for strain
birefringence under crossed polarizers, and the strain pattern
plotted. Highly strained areas are marked off and avoided for
fabrication of substrates. Surface strain in the substrate is
measured by diffractometer analysis of the lattice constant shift.
The (444) and (555) $MgAl_2O_4$ peak position are accurately charted,
and any change noted and related to epitaxial film quality. Figure
2 is a Berg-Barrett topograph of a highly-strained substrate

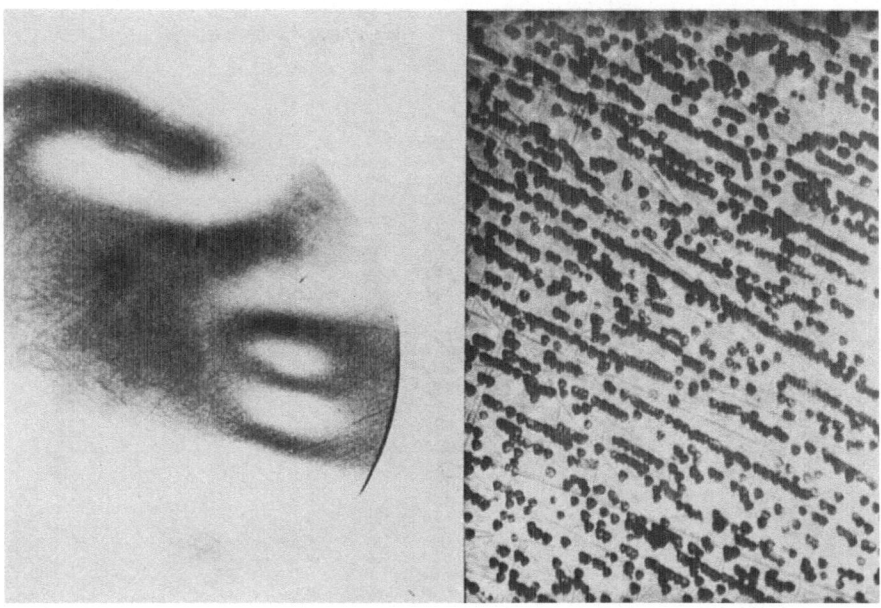

Figure 2. Topograph and Chemical Etch of Highly Strained Substrate

demonstrating "grown-in" strain. After a H_3PO_4 etch, dislocations aligned along slip planes are noted. This is also shown in Figure 2.

Macrobubbles are another growth defect that can occur. Figure 3 is a topograph of a substrate containing macrobubbles in the form of a triangular core. This substrate was cut from a portion of crystal showing this defect. The disturbed surface produced polycrystalline spots in the epitaxial films.

Figure 4 is a series of epitaxial films deposited on substrates cut sequentially through a selected section of crystal. A change in structure has occurred in this section, and polycrystalline nucleation follows the growth interface as the change occurs. A small polycrystalline spot appears in the center of the wafer, then grows to cover the entire surface as the conical-shaped growth interface is intersected.

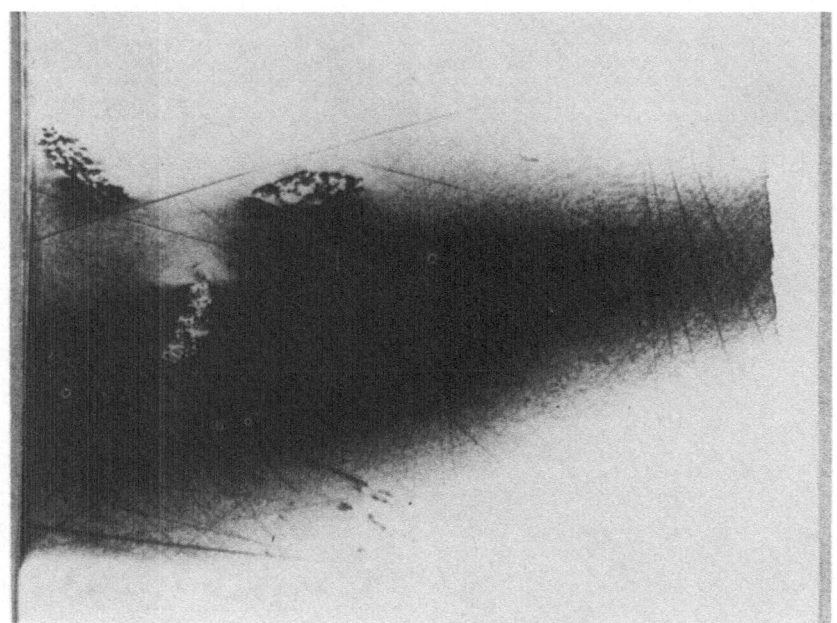

Figure 3.   Topograph of Macrobubble Defect

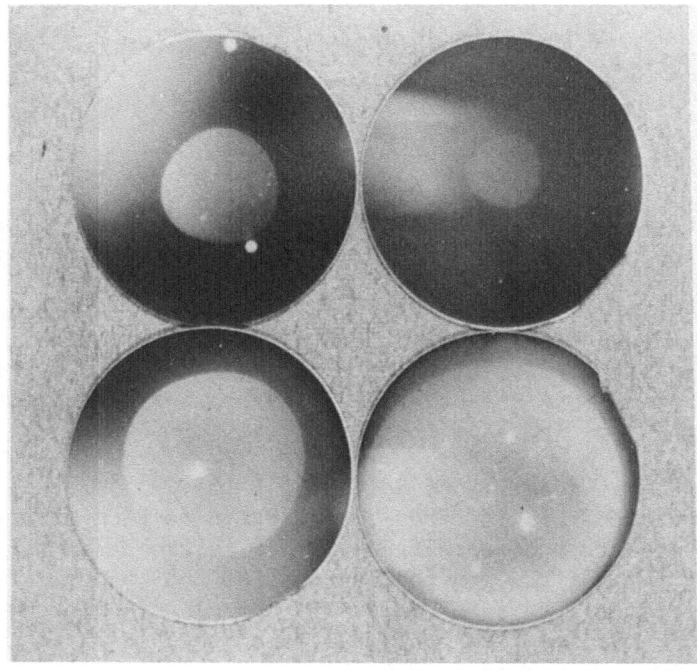

Figure 4.   Epitaxial Films on Substrates Intersecting
a Change in Crystalline Structure

SUBSTRATE FABRICATION

One of the requirements of a substrate surface for epitaxial deposition is that it be free of the damaged surface layer produced by the polishing. Chemical etches have been used to reveal crystal defects and fabrication damage process. These etch techniques are based on work in the literature (5-6).

The Berg-Barrett technique of X-ray topography has more recently been used to study the surfaces of the as-polished substrate and of the epitaxial wafer. With this method the substrates are mounted in a Lang-type camera and adjusted for diffraction from the (533). Slits and crystal are adjusted to allow diffraction of only the Ka_1 component of the copper X-radiation. The crystal and film are translated across the beam in a direction parallel to the plane of the substrate in order to expose as large an area as possible. Exposure times for a 1 1/4" substrate are about 12 hours. Assuming 50% of the integrated intensity of the diffracted beam is recorded, the rest being lost by scattering, the depth of the surface layer recorded is about 0.0019".

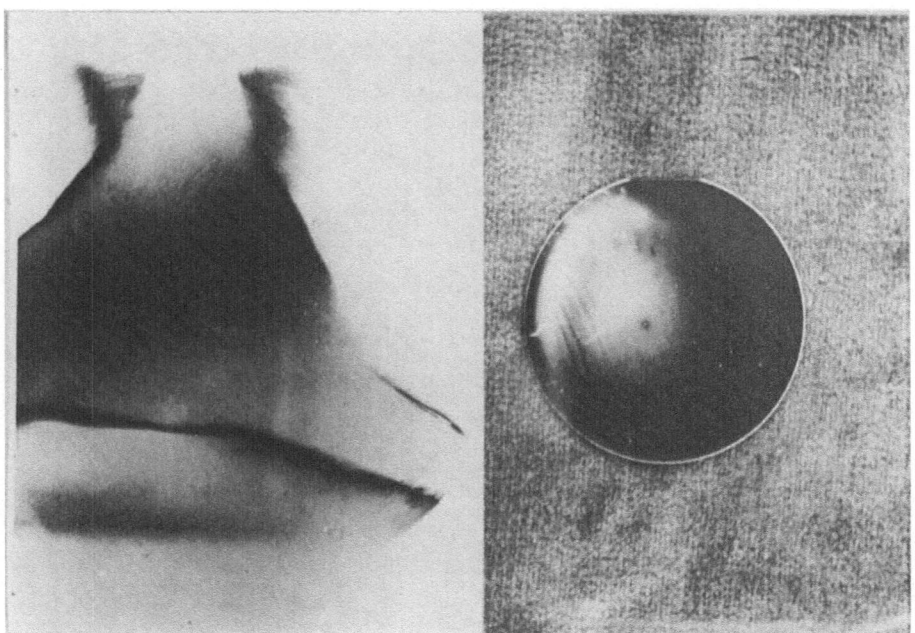

Figure 5.   Topograph of Saw Damage and Resultant Epitaxial Film

Slicing the crystal into substrate blanks is the first fabrication step. Figure 5 is a topograph showing severe strain induced by an excessively worn saw. The accompanying epitaxial wafer shows the gray polycrystalline areas resulting from saw damage.

After slicing, the next fabrication step is lapping. Sufficient stock is taken from the blank to assure removal of the saw-damaged material and prepare the blanks for the polishing operation. If insufficient stock is removed, attempts at epitaxy will result in a polycrystalline film.

During the polishing operation, damaged material is formed on the surface, as is commonly found when polishing oxide materials. Scratches become visible at 200X after a ten minute H_2SO_4 etch. As etch time increases, the scratches become larger, finally disappearing and being etched out after 45 minutes. This damaged layer is estimated at 1 to 2 microns deep. Once the work damage layer was detected, the various fabrication procedures could be evaluated. Figure 6 shows a chemically-etched substrate with a gross damage layer, and the corresponding topograph. Figure 7 shows the chemically etched improved finish and the corresponding topograph.

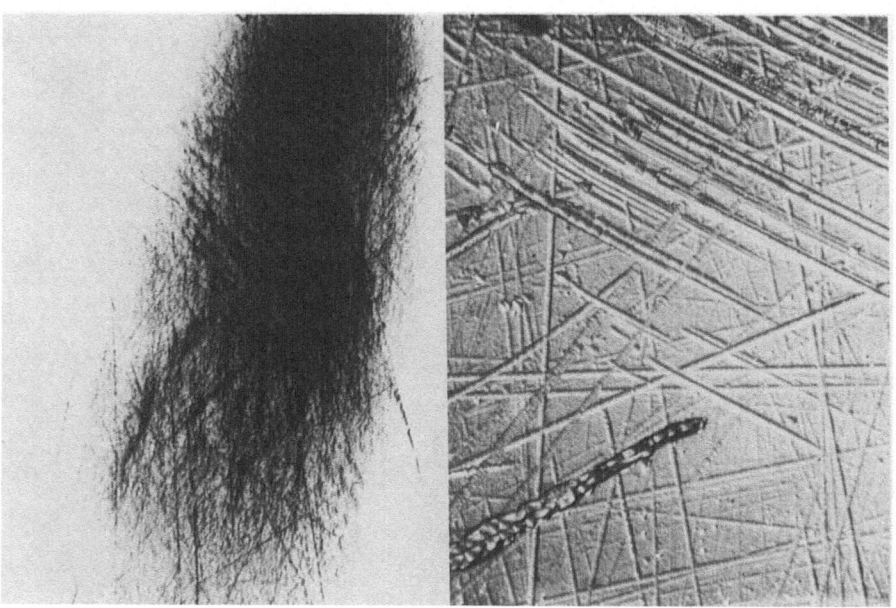

Figure 6.   Topograph and Chemical Etch of Gross Damage Layer

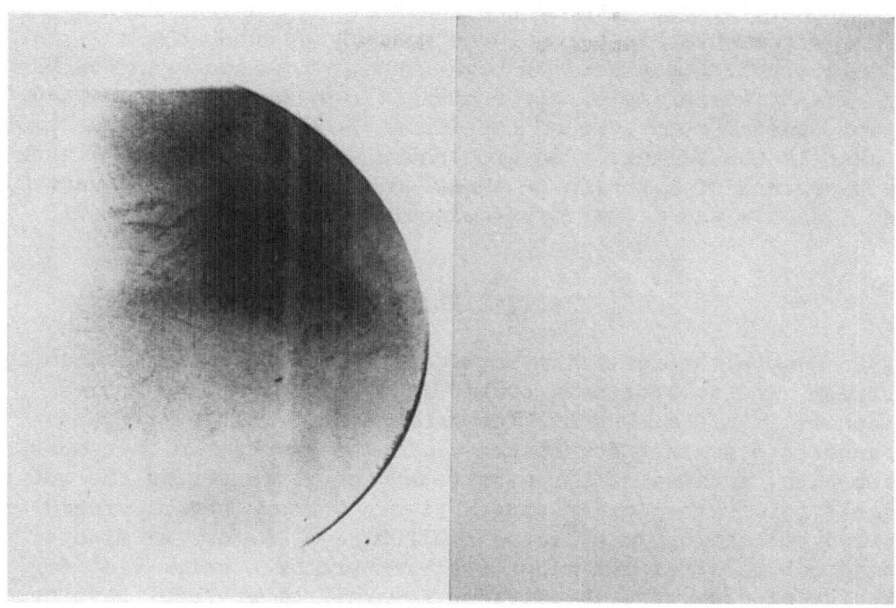

Figure 7.   Topograph and Chemical Etch of Minimized Damage

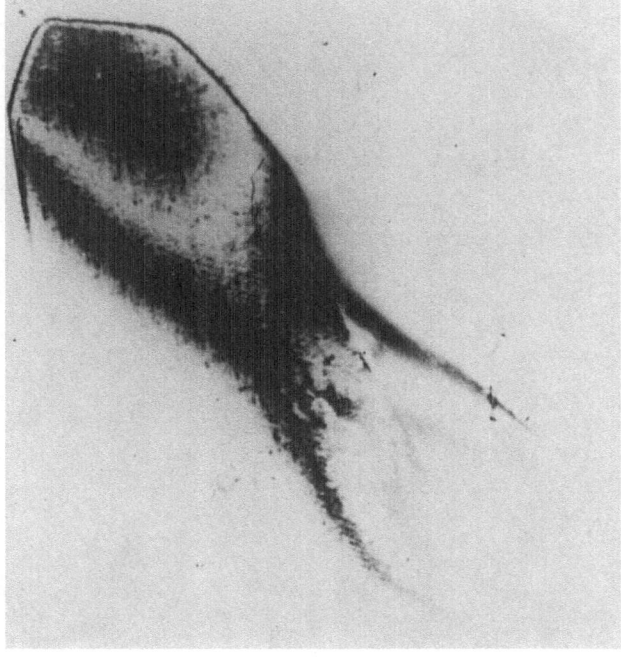

Figure 8.   Topograph of Substrate Annealed in Vacuum

Further improvements of the substrate finish by chemical and thermal treatments are being investigated. A good chemical-mechanical polish has not yet been found. Some improvements have been made by annealing of the substrate, which causes a rearrangement of damaged structure on the surface. However, great care must be taken in the selection on atmosphere and temperature. Figure 8 is a topograph of a substrate annealed at 1800 deg. C. in vacuum. Crystal damage has occurred by selective vaporization of MgO.

EPITAXIAL FILMS

To remove the remaining work damage from the final polishing operation, the substrate is etched in a hot, gaseous hydrogen atmosphere. This cleans off the damaged material and prepares the substrate surface for epitaxy. If the hydrogen etch temperature is too high, crazing of the substrate occurs, rendering the wafer unusable for further processing. Figure 9 shows severe crazing that occurred by etching in hydrogen at 1200 deg. C. for ten minutes. No crazing occurs after ten minutes at temperatures below 1185 deg. C. Extensive studies of etch rates at several temperatures have been done by Green(7).

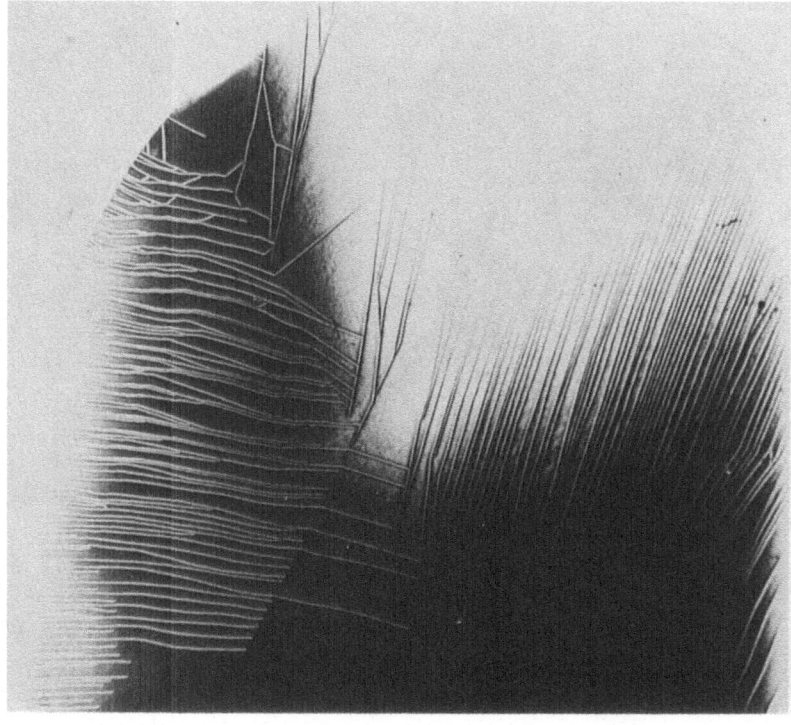

Figure 9.   Topograph of Crazed Substrate

The deposition of silicon is accomplished by decomposing silane, SiH_4, at 1000 deg. to 1100 deg. C. Silicon tetrachloride and the other chlorosilanes are not used because the spinel is attacked by chlorine at elevated temperatures. Figure 10 is a topograph of a (111) wafer where the silicon film was deposited from silicon tetrachloride. Cracking of the substrate occurs in the (111) cleavage plane directions as Y's, and polycrystalline nucleation of the film occurs at each junction of the Y's.

Epitaxial film quality is determined by diffractometer scans of the wafer. Orientations other than (111) are displayed as characteristic peaks on the scan. Correspondingly, these give a hazy appearance to the wafer, with the slightest amount of polycrystalline materials giving an easily visible haze, so that poor films may be readily rejected.

Figure 10. Topograph of Substrate Attacked by $SiCl_4$

ACKNOWLEDGMENTS

The authors wish to thank Mr. P. E. Otten, Dr. W. J. Alford, and Mr. H. D. Quandt of Union Carbide Crystal Products; Dr. S. H. Goodman of General Dynamics, Pomona, and Mr. R. Hays of Motorola for contributions to this program.

REFERENCES

1. H. M. Manasevit and D. H. Forbes, "Single Crystal Silicon on Spinel", Journal Appl. Physics 37 (2) 734-739 (1966).

2. G. W. Cullen, "The Preparation of Properties of Chemically Vapor Deposited Silicon on Sapphire and Spinel", Journal of Crystal Growth 9 107-125 (1971).

3. C. C. Wang, et al, "Single Crystal Spinel for Electronic Application", Technical Report AFML-TR-68-320 (1968).

4. H. M. Manasevit, et al, Transactions Metallurgical Society AIME 236 (1966).

5. W. J. Alford, D. L. Stephens, "Chemical Polishing and Etching Techniques for Al_2O_3 Single Crystals", Journal American Ceram. Soc. 46 (4) 193-194 (1963).

6. R. D. McBrayer, et al, "Chemical Etching of Defect Structures in Alumina-Rich Spinel Single Crystals", Journal American Ceram. Soc. 46 (10) 504-505 (1963).

7. J. M. Green, "The Etch Rate and Deterioration of Magnesium Aluminate Spinel in Hydrogen." Presented at the Electrochemical Society Meeting, Washington, D. C. 1971.

[Editors' note]:

The reader's attention is invited to Newkirk et al., J. Appl. Phys. 35, No. 4, 1362 (1964) and Jackson et al., J. Appl. Phys. 33, No. 7, 2301 (1962) which directly relate to the subject of this paper and which the authors have apparently overlooked.

CROSS LINKING OF COLLAGEN BY HYDROPHOBE BONDS

Edwin H. Shaw Jr.

The University of South Dakota

Vermillion, South Dakota 57069

ABSTRACT

Previous work on modification of the collagen lattice, presented at these meetings, has served to confirm
the Ramachandran proposal for the structure of this important protein, together with information on the lengths
and tensile strengths of these bonds, including hydrogen
bonds, hydrophobe bonds, and Vander Waals attractions.
Rat tail tendons were suspended in 2 M aqueous solutions
of dimethylsulfoxide, acetone, trimethylamineoxide and
1,4-cyclohexanedione, all of which showed evidence of
ligation by expansion of the collagen lattice and refinement of the 100 spot in the x-ray diffraction pattern run
by rotation on the collagen fiber axis. In the case of
trimethylamineoxide and cyclohexanedione, oriented crystallization also occurred, with the long direction of the
molecule in the plane perpendicular to the fiber axis of
collagen. With the monofunctional ligands, the -C=O
group forms a hydrogen bond to the N-H group in the Ramachandran standard revised model, leaving the methyl groups
to form the hydrophobe bond.

INTRODUCTION

The author has recently made use of the technique of
x-ray diffraction to study the modification of protein
structure that occurs when complexing agents are inserted
into the lattice. When these agents are solid, oriented
crystallization of the excess of complexing agent on the

hexagonal model of the collagen structure indicates the
position of the complexing agent with respect to the collagen fibers. In the case of amides, where the bond is
-N-H---O=C- (1), it has been shown the ligand parallels
   2.88A
the a-axis of collagen at intervals averaging 5.2 A along
the collagen c-axis with an expansion of the collagen a-
axis parameter from 13.5 A to an average of 15.0 A, corresponding to a core diameter of 7 A for the collagen
triple helix. Improvement of the collagen lattice is indicated by a sharpening of the collagen 100 spot to a
small circle as compared to an oval parallel to the c-
axis or fiber axis in the case of unmodified collagen.
When the ligand is -C=O---H-O-, (2), with the H-bond
length of 2.76 A, sharpening of the collagen lattice is
shown by refinement of the 100 spot with expansion of
the collagen a-parameter to 16.8 A, the diameter of the
core of the spirals remaining at 7.2 A (3). Recently the
author has expanded the collagen a-parameter to 18.3 A in
non-aqueous media by using liquid aliphatic diols, the
tensile strength being substantially increased in the
longer chains when curving increases the number of Vander
Waals contacts between the ligands (4).

EXPERIMENTAL

Weighted rat tail tendons were transferred from
0.9% Na Cl to 2 M aqueous solutions of the liquids dimethylsulfoxide and acetone contained in thin walled
1.2 mm pyrex capillary tubing, the ends sealed with bees-
wax, the x-ray diffraction runs being made without tension. In the case of the solids, trimethylamineoxide and
the covalent bonded 1,4-cyclohexanedione, tendons were
stretched in 2 M aqueous solutions of the reagents, dried
overnight, and x-ray diffractograms taken at 100 grams
tension. Tensile strengths were determined on tendons im-
mersed in 2 M solutions of the reagents and loaded with
50 gram increments, allowing 10 minutes for breaking between the next load increase, 3 determinations being
averaged. Results are reported in mg. per square micron,
equivalent to 2000 pounds per square inch. Diffractograms were made with Cu Kα radiation using General Electric XRD-1 equipment.

RESULTS

Experience with liquid ligands (4) led to trial of
dimethylsulfoxide which only has one grouping, the sulf-

oxide group, which is capable of hydrogen bonding to the
---H-N- group in proteins. On trial, the improvement and
expansion of the collagen 100 spot, (Figure 1), by dimethylsulfoxide and acetone indicated that the ligand in
each case was two ended, and the two methyl groups form
the hydrophobe bond

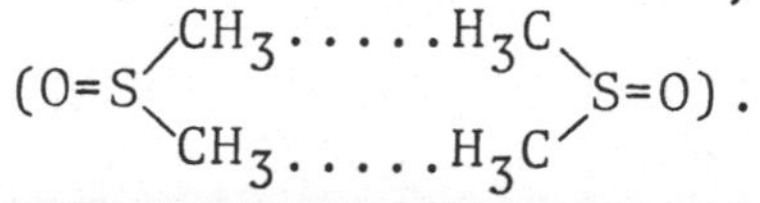

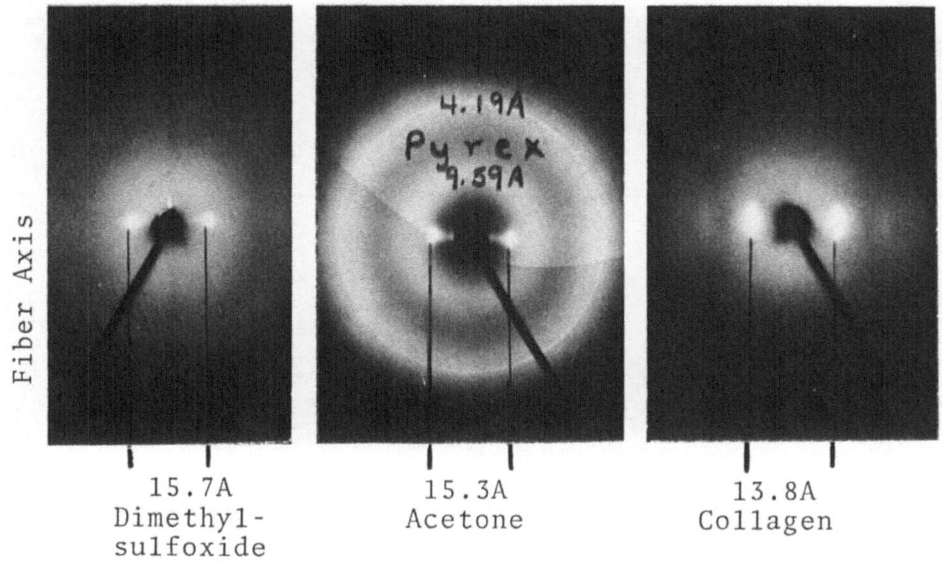

$$(O=S\underset{CH_3\ldots\ldots H_3C}{\overset{CH_3\ldots\ldots H_3C}{\diagdown\diagup}}S=O).$$

15.7A   15.3A   13.8A
Dimethyl-   Acetone   Collagen
sulfoxide

Figure 1.   Influence of ligands on the 100 spot of
            collagen.

In a similar manner, the two methyl groups on acetone
could attract each other, and the three methyl groups on
trimethylamineoxide would make an even stronger hydrophobe bonding.

        Diagrams of the fitting of these hydrophobe ligands
into the cross section of the Ramachandran "standard"
model of the cross-section of the collagen triple spiral
(5), are shown in Figure 2, dimethylsulfoxide in collagen,
Figure 3, acetone in collagen, Figure 4, trimethylamineoxide in collagen, Figure 5, 1,4-cyclohexanedione in collagen, which were drawn to scale before photographing.
Table I summarizes the data. The collagen equatorial
parameter $\underline{a}$ is the center-to-center distance between the
spirals. $\overline{\text{N}}$itrogenous collagen core diameter is measured
on the figures from the center of the spiral to the
N_3 spots and is 4.8 A. The length of the hydrophobe bond
is 3.5 $\pm$ 0.2 A as compared to 2.75 A for the 0-H---O bond,

2.89 A for the N-H---O bond, and the Vander Waals
approach of approximately 3.8 A.

Table I

EXPANSION OF THE COLLAGEN LATTICE
BY DIMETHYLSULFOXIDE AND ANALOGUES

	Collagen Equatorial Parameter $\underline{a}$ in A	Collagen Tensile Strength mg per sq. micron
2 M Dimethylsulf-oxide	15.7	1.2 $\pm$ 0.2
2 M Acetone	15.3	1.7 $\pm$ 0.3
2 M Trimethylamine-oxide	14.9	8.2 $\pm$ 1.1
2 M 1,4-Cyclohexane-dione	14.0	2.4 $\pm$ 0.5
Collagen in distilled water	13.2	

ACKNOWLEDGEMENT

    This work was supported in part by the Winner Com-
munity and VFW-Tice Cancer accounts. The gift of a
series of sulfoxides and sulfones from the Phillips Petroleum Company is gratefully acknowledged.

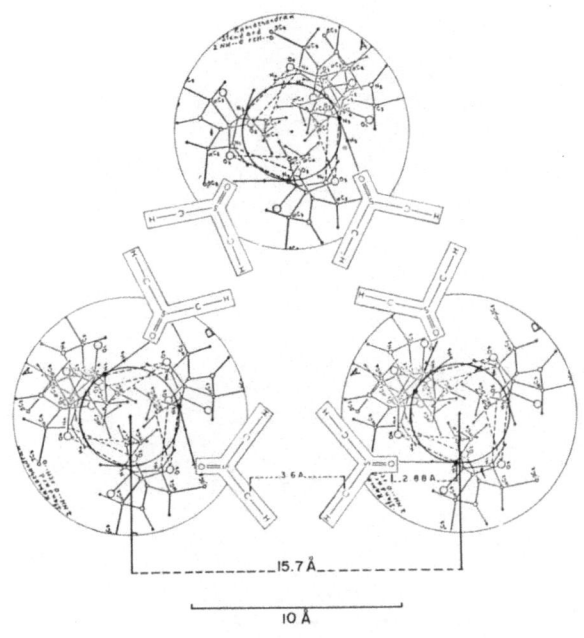

Figure 2.   Dimethylsulfoxide in Collagen

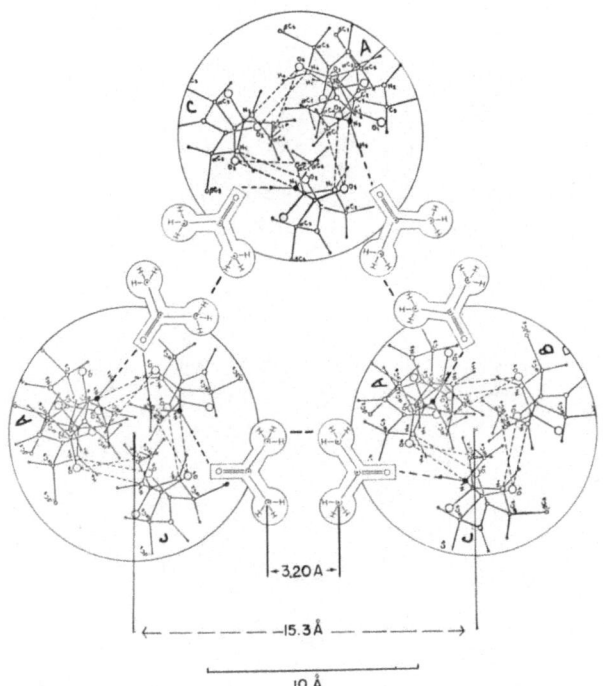

Figure 3.   Acetone in Collagen

E. H. Shaw, Jr.

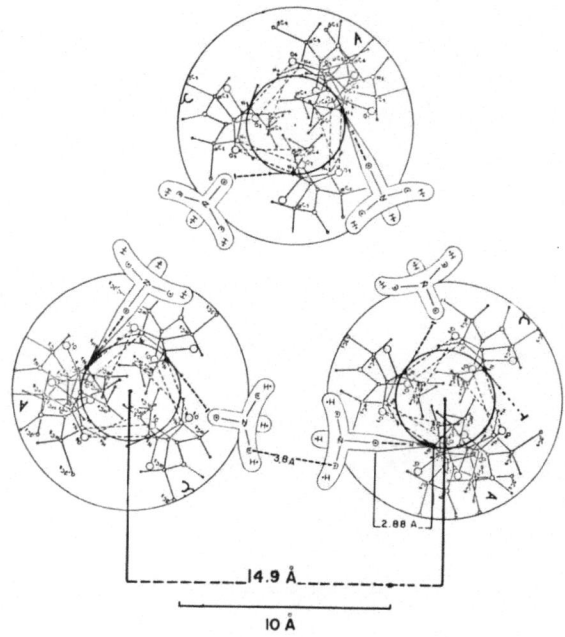

Figure 4.   Trimethylamineoxide in Collagen

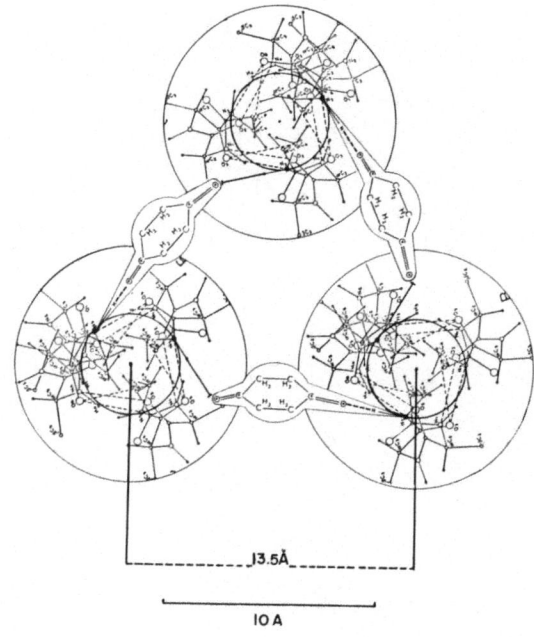

Figure 5.   1,4-Cyclohexanedione in Collagen

REFERENCES

1. E. H. Shaw, Jr., "Oriented Crystallization of Amides
    on Collagen with Modification of the Collagen Lat-
    tice", in W. M. Mueller, G. R. Mallett, and M. J.
    Fay, Editors, Advances in X-ray Analysis, Vol. 7,
    pp. 252-5, Plenum Press (1964).

2. E. H. Shaw, Jr. and Alton R. Christensen, "Modifica-
    tion of Collagen and Nylon Lattices by Resorcinol",
    in W. M. Mueller, G. R. Mallett, and M. J. Fay,
    Editors, Advances in X-ray Analysis, Vol. 8, pp. 175-
    179, Plenum Press (1965).

3. E. H. Shaw, Jr., as quoted by G. N. Ramachandran,
    "Treatise on Collagen, Vol. I, Chemistry of Collagen",
    Academic Press, N. Y., 1967, pp. 177-8.

4. E. H. Shaw, Jr., "Binding of Aliphatic Dihydroxy Com-
    pounds to Collagen with X-ray Diffraction Evidence for
    the Vander Waals Bond", in Charles S. Barrett, John
    B. Newkirk, and Clayton O. Ruud, Editors, Advances
    in X-ray Analysis, Vol. 14, pp. 268-274, Plenum Press
    (1971).

5. G. N. Ramachandran and V. Sasisekharan, "Refinement
    of the Structure of Collagen", Biochim. Biophys.
    Acta, 109, 314 (1965).

EFFECT OF ION EXCHANGE RESIN PARTICLE SIZE ON X-RAY FLUORESCENT
ANALYSIS.

A. L. Allen and V. C. Rose

University of Rhode Island

Kingston, Rhode Island 02881

ABSTRACT

The effect of resin particles on copper x-ray fluorescence
was studied. For any given resin size the relationship between
copper concentration and x-ray intensity was linear. As the
particle size decreased, the x-ray intensity increased for any
given copper concentration. The general shape of the curves are
similar to the ones predicted by Bernstein for a minor constituent
in a power sample. This study indicates that the variation in
intensity with particle size can be eliminated by using resins
with a mean particle diameter of 56 microns or less.

INTRODUCTION

Copper is introduced into streams by corrosion of cooling
equipment in the process and power industries, by effluents from
the metal finishing and chemical processing, etc.. Since small
amounts of copper have an adverse effect on living organisms,
including fish and shellfish, a quick accurate method of determining the amount of copper in water is necessary.

The usual technique for trace analysis of copper in water
consists of removing copper from the water with ion-exchange
resins. The copper is then eluted from the resin to give a concentrated solution and analyzed by routine chemical methods. In
recent work, ion-exchanger loaded filter papers, used to collect
metal ions, have been analyzed directly by x-ray fluorescence (1,2).
These papers have a limited capacity and require recycling

of the solution several times. In order to eliminate these problems, it was proposed to collect the ions on standard ion exchange
columns and to analyze a representative sample of the resin using
x-ray fluorescence. However, previous work (3,4,5) with other materials indicates that particle size affects the analytical results.
The work reported in this paper (6) indicates that proper selection
of resin size can minimize the effect of size on the analytical
results.

THEORY

In x-ray fluorescence analysis, it has been found that the
intensity of a constituent in a sample, within limits, is directly
proportional to its concentration in the sample. J.W. Meyer (7)
expressed this dependency with the following relationship:

$$I(e) = KC(e)/ \ C_i(\mu/\rho)_i$$

where $I(e)$ is the radiation intensity and $C(e)$ is the concentration
of the constituent in the sample, the sum of terms in the denominator is the mass attenuation coefficient for the entire sample and K
is a proportionality constant which includes machine characteristics and fluorescent yield of the element. If the absorption
coefficient for the entire sample is assumed to be constant this
relationship reduces to

$$I(e) = K'C(e)$$

Previous investigators have observed particle size effects on
x-ray intensity. They reached the general conclusion that the intensity of the fluorescence will increase with decreasing particle
size. Gunn (3) explained this dependence by the possible shielding
effects that a large particle could possess. Bernstein (5) thought
that the higher intensity was due to a decrease in the percent
voids as the particle size decreased. He found that the intensity
increased slowly as particle size decreased for particles much
greater than the effective depth of x-ray penetration (infinite
thickness). As the particle size approached infinite thickness
dimensions, the intensity increased rapidly with decreasing size.
The effect of particle size rapidly diminished as the particle size
becomes smaller than the infinite thickness.

EXPERIMENTAL PROCEDURES

The effect of five different particle size ranges on resultant
x-ray fluorescence was investigated for Dowex A-1, a weakly acidic,
chelating, cation exchange resin. In its sodium form, this resin

has a large preferential affinity for copper. Three size ranges
were used as received while the two smaller sizes were obtained by
grinding and air elutriation. The average particle size was determined by microscope observation of approximately 400 particles for
each range using procedures described by Herdan(8). The resin
samples for the four larger sizes were loaded with copper from
standard copper solutions using column techniques. Due to the
large resistance to flow through the column, batch techniques were
used for the smallest size resin. The resulting eluants were analyzed by atomic absorption spectrophotometry to be certain that
the copper was retained on the resin. The moist samples of resin
were analyzed with a General Electric Company XRD-S5 x-ray diffraction unit. The first order copper K-α radiation was detected using
a proportional flow counter with 90% methane, 10% argon gas mixture.

Data taken at different times were correlated by measuring
the sample's intensity relative to a standard sample. The relative
intensity is expressed by the following relationship:

$$R.I. = (I(r) - I(b))/(I(s) - I(b)).$$

where I(r), I(s) and I(b) are the x-ray intensity of a resin, a
standard sample and background respectively. Background was the
x-ray intensity for an unloaded resin sample.

RESULTS AND CONCLUSIONS

For each resin size and copper concentration three samples
were run. The average relative intensity for each set of runs is
reported in Table I. For a given resin size the relative intensity
varies linearly with the amount of copper present. A straight line
with an intercept of 0 represents the data fairly accurately. The
slopes of the lines increased with decreasing particle size.

As the resin particle size decreases, the x-ray intensity
emitted by the copper increases. As shown in Figure 1, the general
shape of these curves corresponds with that predicted by Bernstein
(4) for a minor constituent in a powder sample. For mean diameters
of 56 microns or less the effect of particle size on relative intensity appears to be negligible. This is in close agreement to
the value of the infinite thickness of the sample which was calculated as 30.9 microns. For mean diameters of 250 microns or
greater the relative intensity decreases very slowly with increased
diameter.

Thus, if resin mean particle diameters of 56 microns or smaller are used a maximum intensity may be obtained with no variation
due to particle size. If a lower intensity can be tolerated,

TABLE I

EFFECT OF PARTICLE SIZE ON RELATIVE INTENSITY

Amount Copper on Resin (mg)	Resin Weight Mean Diameter (microns)				
	478 (50-100 Mesh)	50 (100-200 Mesh)	112 (200-400 Mesh)	56	27
10	0.180	0.181	0.232	0.243	0.244
8	0.144	0.146	0.184	0.194	0.195
6	0.108	0.111	0.139	0.147	0.148
4	0.072	0.076	0.093	0.098	0.098
2	0.035	0.040	0.047	0.049	0.049

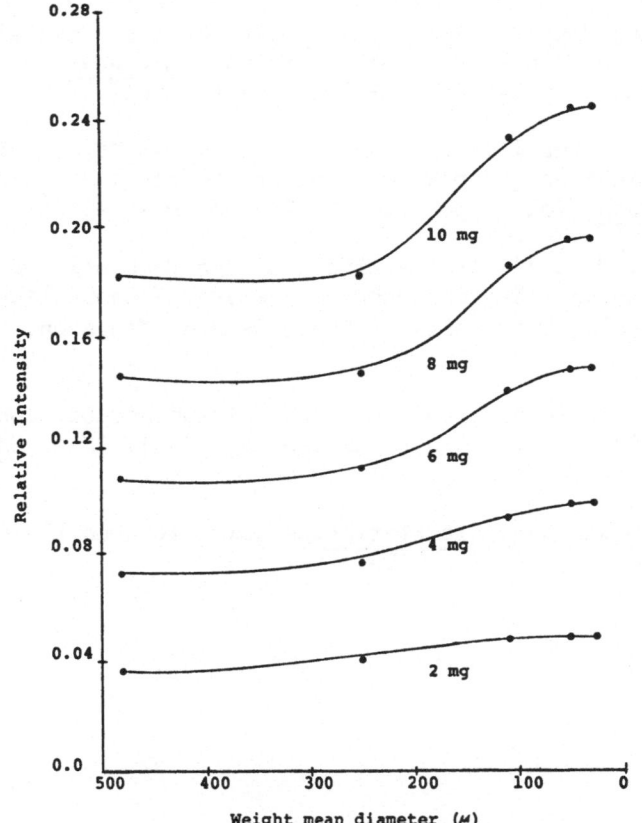

Figure 1 Effect of particle size on relative copper intensity.

100-200 mesh or larger commercial resin can be used to obtain intensities that are only slightly size dependent. Normal commercial resin sizes smaller than 100-200 mesh should never be used.

REFERENCES

1. W. J. Campbell, E. F. Spano and T. E. Green, "Micro and Trace Analysis by a Combination of Ion Exchange Resin-Loaded Papers and X-Ray Spectrography," Anal. Chem. 33, 787-996 (1966)

2. S. N. Ndam, "Trace Metal Analysis Using Ion Exchange Resin-Loaded Papers and X-Ray Fluorescence," Unpublished Master's Thesis, University of Rhode Island, Kingston, R. I., 1969.

3. E. L. Gunn, "The Effect of Particle and Surface Irregularities on the X-Ray Fluorescent Intensity of Selected Substances" in W. M. Mueller, Editor, Advances in X-Ray Analysis, Vol. 4, p. 382-400, Plenum Press (1960).

4. F. Bernstein, "Application of X-Ray Fluorescence Analysis to Process Control" in W. M. Mueller, Editor, Advances in X-Ray Analysis, Vol 5, p. 486-499, Plenum Press (1961).

5. F. Bernstein, "Particle Size and Mineralogical Effects in Mining Applications," in W. M. Mueller, Editor, Advances in X-Ray Analysis, Vol 6, p. 436-446, Plenum Press (1962).

6. A. L. Allen, "A Study of the Effect of Ion Exchange Resin Particle Size on X-Ray Fluorescent Analysis," Unpublished Master's Thesis, University of Rhode Island, Kingston, R. I., 1970.

7. J. W. Meyer, "Determination of Iron, Calcium and Silicon in Calcium Silicates by X-Ray Fluorescence," Anal. Chem. 33, 692 (1961).

8. G. Herdan, Small Particle Statistics, Elsevier Publishing Company, New York, N. Y., 1953.

ON-LINE PROCESS CONTROL COMPOSITIONAL ANALYSIS OF ALUMINUM FILMS CONTAINING A LOW PERCENTAGE OF COPPER

G. H. Glade, J. M. Matthews and F. R. Titcomb

IBM Components Division

Essex Junction, Vermont 05452

ABSTRACT

Aluminum film conducting stripes are widely used for semiconductor device interconnection networks. The addition of a low percentage of copper significantly increases their life. Composition must be controlled to maintain product quality.

The paper discusses various methods used to analyze the copper composition in the aluminum films, and adaptation of one of these methods for process control application. A portable instrument designed for field use was adapted for use as an on-line instrument.

INTRODUCTION

Multiple semiconductor devices are commonly manufactured on a single silicon chip. They are joined by an interconnection network which functions much in the manner of a printed circuit. Evaporated aluminum films are used for such interconnect networks. The addition of a low percentage of copper to the aluminum film stabilizes it and increases reliability (1).

During the manufacturing evaporation of aluminum-copper films, rapid analysis of copper content and film thickness are required for process control. Analytical

techniques using atomic absorption, x-ray spectroscopy,
and direct thickness measurement were developed to meet
the need.

ATOMIC ABSORPTION

Atomic absorption procedures were developed for
analysis of the aluminum-copper film. A standard laboratory atomic absorption spectro-photometer was used.
The conditions are given in Table I.

Aluminum-copper film samples were submitted on silicon wafers and on quartz and glass monitor substrates.
The films were dissolved from the wafers and substrates
with aqua regia and diluted with deionized water for
analysis. The specific dilution varied with the dimension of the sample.

Stock solutions of aluminum and copper were made
in the one gram per liter range, then diluted to the
part per million level for working standards. The
acidity of the standard was adjusted to match that of
the sample.

Table I.　Instrumental Conditions for the Atomic Absorption Analysis of Aluminum-Copper Film.

	Copper	Aluminum
Burner	10 cm, steel narrow slot	5.2 cm, nitrous oxide
Gasses	Air, Acetylene	Nitrous Oxide, Acetylene
Lamp	Copper Hollow Cathode	Aluminum Hollow Cathode
Source Current	15 mA	25 mA
Wavelength	3249 Å	3095 Å
Slit Width	0.3 mm, 2 Å	0.3 mm, 2 Å

A standard curve for each element was drawn by plotting the absorbance vs mg/1 of each standard on rectilinear graph paper. The absorbance of the sample was compared to the standard curve to determine its metal concentration (mg/1). Copper percentage was calculated from the copper concentration and the combined concentration of copper plus aluminum.

Precision and Accuracy

An aluminum-copper stock solution, which simulated a typical dissolved sample, was prepared from pure metals. Three dilutions of the stock solution to analytical concentration were made and analyzed daily for four days. The results of these analyses showed that the copper content of an aluminum-copper film could be determined with an observed standard deviation (67% confidence level) of 0.06 wt% copper.

The accuracy of the method was verified by:

(1) Materials balance experiments in which silicon wafers were weighed on a semi-micro balance before the deposition of aluminum-copper films, after deposition of the films, and after the films had been dissolved from the wafers. These weights were used to calculate the film weights. The film weights were compared to the weights obtained by atomic absorption analysis of the solution containing the dissolved films, and

(2) Analysis of duplicate samples by the atomic absorption method and by the dithizone extraction method.

In both cases the weights obtained agreed to within less than the one observed standard deviation determined above.

Application

The atomic absorption method is used for: 1) developmental analysis and special investigations, 2) making standards for x-ray analysis of the film, and 3) the reference method to which other methods are compared. However, it is too slow for routine process control.

X-RAY SPECTROSCOPY

X-ray spectroscopy provides the rapid analysis required for process control.

Aluminum characteristic x-radiation is highly absorbed by copper (2). Therefore, aluminum x-ray intensity measurements are sensitive to the amount and distribution of copper in the films. Copper characteristic x-radiation is only slightly absorbed by aluminum (2). Thus, copper x-ray intensity measurements and analysis are relatively insensitive to variations in film thickness (mass of aluminum) and copper distribution. For these reasons, only copper mass is determined x-ray spectrographically. Total film mass was originally determined gravimetrically and is now calculated from film thickness.

X-ray measurements are made with a laboratory, air path, x-ray spectrograph using the copper $K\alpha$ line. The conditions used for the analysis are given in Table II.

Copper is present in the spectrum of the primary x-ray tube. Therefore, an off-angle background correction, as is often used, would be erroneous. X-ray intensity measurements, made at the copper analytical angle from a pure aluminum film on a quartz substrate, are used for background corrections. However, for all routine analyses, gross copper intensities are used.

During the initial examination of the films on silicon wafers, significant variations were found in the x-ray intensities when the samples were rotated. This variation was caused by diffraction effects arising from the crystalline nature of the wafers. The problem was eliminated by the substitution of amorphous quartz disks, and later, pyrex glass for the wafers for all standards, routine experimental applications, and as process control monitors. When it is necessary to analyze films on silicon wafers, four measurements are made with the wafer rotated 90° between measurements and the values averaged.

Before standard making was undertaken, the uniformity and variation of films within deposition runs were investigated. Locations which produced nearly identical films were used for standardization. Locations in which there were a variation were used to produce a range of standards.

Table II. X-ray Spectrographic Analytical Conditions
           for Copper in Aluminum-Copper Films

Tube/Target	EA-75 Dual Target/Tungsten
KV/mA	60/50
Sample Mask	7/16 in. Square Lloyd Aperture
X-ray Path	Air
Crystal	LiF
Analytical Line/2Θ	$CuK\alpha I$/45.03°
Background	Blank Sample (see text)
Soller Slit	0.010 in.
Detector/Voltage	SPG-4 Scintillation/800
Amplifier Setting	16 x 0.90
E_L	1.5 V
E_U	23 V
Count Duration	40 sec

Films containing various amounts of copper were
evaporated to the standard thickness on quartz disks.
Multiple films at each copper level were prepared. Seven
x-ray runs were made to collect copper x-ray intensity
data. One film from each composition level was analyzed
by atomic absorption. The x-ray and atomic absorption
data for the analyzed films was used to set up working
curves for the standardization of the other films.

The films were standardized in terms of mass of
copper per unit area. This was done because variations
in film thickness cause variations in copper percentage
for any specific copper mass, and the x-ray analysis
gives no information about either aluminum mass or total
film mass.

Calculation Methods

Three methods were used to obtain total film mass. They were:

(1) The Additive (gravimetric) method. In it, preweighed numbered disks were weighed after deposition to obtain the film mass. The method was very time consuming, required record keeping, and was vunerable to chipping of the disks through handling.

(2) The Subtractive (gravimetric) method. In it, the monitors were weighed before and after the film was removed with aqua regia. This method was slower than the additive method because of the need for drying the disks before the second weighing.

(3) The Thickness formula method. This was the final method developed and is in current use.

As another process control parameter, film thickness is measured. Since total mass is proportional to film thickness and copper mass in the film, a formula was developed for calculating total film mass from these two measurements. Three assumptions were made in developing the formula:

(1) Aluminum and copper were treated as discrete films.

(2) Ninty-five percent density was assumed for aluminum.

(3) Bulk density was assumed for copper.

The formula is:

$$\frac{\left[\begin{array}{c}\text{Film Thick}\\ \text{Å}\end{array}\right] + \left[\left(\begin{array}{c}\text{Mass Cu}\\ \text{(gr)}\end{array}\right) \times k_2\right]}{k_1} = \left[\begin{array}{l}\text{Total Film}\\ \text{Mass (gr)}\end{array}\right]$$

Here, k_1 is the constant for converting an aluminum film of a measured thickness to its mass when its area is constant; k_2 is the constant for the perturbation introduced by a small amount of copper which is approximately three times denser

than the aluminum. The thickness formula calculates the total film mass. With total mass known,
copper percentage can be calculated.

Precision and Accuracy

A general survey of the precision and accuracy of
the routine analysis of the aluminum-copper film was
performed at the time that the thickness formula was
being introduced. Fifty-two routine, process control
samples were used. The samples were analyzed by the two
gravimetric methods. The thickness values of the samples were obtained and total masses and weight percent
copper were calculated using the thickness formula. The
solutions containing the stripped films from the subtractive method were analyzed by atomic absorption.

The values obtained by the various methods were
compared to those obtained by atomic absorption. A common x-ray mass of copper value was used for all calculations on each sample. Comparison was made on the basis
of weight percent copper. Difference is defined as
(Method -- A. A.), and s is the observed standard deviation of the differences.

The results of the comparison are given in Table
III. In all cases the average difference is less than
one observed deviation.

Table III.   Comparison of Various X-Ray Methods of
             Aluminum-Copper Analysis (Value in wt% Cu).

Instrument	Method	n	Avg. Diff.	s
Laboratory	Additive	38	+0.04	0.67
Laboratory	Subtractive	50	-0.21	0.70
Laboratory	Thickness	52	+0.05	0.47
On-Line	Thickness	66	-0.03	0.48

ON-LINE INSTRUMENT

Because of other commitments, process control analysis by the laboratory instrument was done on a batch basis. On occasion, operational errors were not discovered until several "out of spec" runs had been made. A dedicated instrument was indicated to overcome the problem. However, a laboratory instrument, like the one being used for analysis, was not economically justified.

A portable x-ray spectrograph, designed for field use, was purchased for the dedicated application. As a compact, low-cost instrument, its performance is limited as compared to the laboratory instrument. However, none of its limitations affected the aluminum-copper analysis. It, and the thickness formula, are being used to determine the copper content of the films in the same manner as the laboratory instrument.

A table was generated by computer giving copper percentage for all possible combinations of copper mass and film thickness. With the table, the operator performing the analysis can obtain the copper percentage directly without calculation. This significantly reduces the possiblity of error, speeds the analysis and reduces operator fatigue.

A precision evaluation, similar to that used to compare the various laboratory methods, was performed. The results are given in Table III. They are comparable to the results obtained with the laboratory instrument.

CONCLUSIONS

An atomic absorption method was developed for the analysis of the copper content of aluminum-copper films. It has excellent precision and accuracy and is used as the basic reference method. However, it is too slow for process control applications.

An x-ray spectrographic method to determine copper mass and a fomula for calculating total film mass from film thickness were developed. Together they provide a rapid means of compositional analysis of sufficient precision and accuracy for process control.

A portable x-ray spectrograph designed for field use, was installed as an on-line instrument. In this application, its precision and accuracy is comparable to the laboratory instrument.

REFERENCES

1. I. Ames, F. M. d'Heurle and R. E. Horstman, "Reduction of Electromigration in Aluminum Films by Copper Doping," IBM J. Res. and Develop., July 1970, pp. 461-463.

2. H. A. Liebhafsky, H. G. Pfeiffer, E. H. Winslow and P. D. Zemany, X-Ray Absorption and Emission in Analytical Chemistry, p. 313, John Wiley and Sons (1960).

FLAME TECHNIQUE FOR HIGH TEMPERATURE SINGLE CRYSTAL WEISSENBERG PHOTOGRAPHY (1000-3000°C)

M. A. Viswamitra and K. Jayalakshmi

Department of Physics, Indian Institute of Science

Bangalore 12, India

ABSTRACT

Oxy-acetylene flame heating has been employed with an ordinary Nonius Weissenberg camera for high-temperature single-crystal studies over the the range 1000-3000°C. The regulator blow pipe is mounted horizontally on the camera track and one end of the experimental crystal (1 cm long) is heated by the vertical flame issuing out of a 0.5 mm bore nozzle. A long slot cut at the top of the film-cassette allows the flame to escape out without touching the X-ray film. The film-cassette consists of two split-halves hinged on one side so that it can be opened and slipped over the screen tubes. This arrangement avoids the necessity for removing the flame each time a new film is mounted on the camera. The goniometer arcs can be positioned during high-temperature exposures, and there is no need to water-cool the cassette up to the highest temperature. The device has been used to take photographs of MgO single crystals at 2100°C. It permits a ready application to diffractometers based on Weissenberg geometry.

INTRODUCTION

Structure determinations, as are routinely done at room and low temperatures, have not been carried out at very high temperatures. Direct heating by a gas flame, which has a highly confined intense-heat zone, provides a simple and inexpensive method for reaching sample temperatures beyond 2000°C, compared to other methods employing radiation focussing, induction heating and electron beam bombardment. The only X-ray device to have made use

548

of this method, so far, has been that of Gubser, Hoffmann and
Nissen (1), for Buerger's precession camera. We have now devised
a gas flame device, suitable for routine single crystal Weissenberg photography, over the temperature range 1000-3000°C.   It has
been used with an ordinary Nonius Weissenberg camera for taking
pictures of MgO at 2100°C. The present device has been developed
as an extension of the authors' earlier devices, (2) and (3).

APPARATUS

Fig. 1 shows a schematic sketch of the experimental set up.
It consists of a fine jet of gas flame A issuing out of a metal
nozzle B with a 0.5 mm bore. B is connected to a standard oxy-
acetylene regulator C through a metal pipe D, held inside the
brass cylinder E. E fits exactly into the detachable screen
tube holder F.1. E has three through holes h1, h2 and h3 into
which the tube D can be inserted and securely clamped in any
desired position by screws G. The three holes allow three variable nozzle-to-crystal distances. The pin H projects into one of
the slots s1, s2 and s3 on the cylinder E. This restrains E from
rotating, but permits it to slide in and out inside the holder F.1.
E is clamped in any required position by screws I.

The experimental crystal J is fixed with high temperature
cement K in a quartz capillary L mounted on a standard goniometer
head M. The tip of the crystal is directly heated from the vertical flame issuing out of the nozzle. The temperature of the flame
is controlled by regulating the amount and the rate of flow of
oxygen and the gas, using the coarse-regulators N and the fine-
control needle-valves O.

Two main problems are encountered when the above arrangement
is adapted to ordinary room temperature Weissenberg cameras. The
first is the possibility of the flame coming into contact with the
film-cassette. This has been overcome by cutting a long slot
(14 cm x 2 cm) in the top of the cassette, as shown in Fig. 2.
The second problem is to eliminate the need for extinguishing the
flame and for the removal of the regulator-blow pipe every time
the cassette is mounted on or removed from the camera. If the
cassette were mounted as is usually done at room temperature, it
would make the removal of the blow pipe unavoidable with the consequent sudden cooling of the crystal to room temperature, whenever
a new X-ray photograph is to be taken. This has been overcome in
the following manner. The cylindrical film-cassette is split into
two halves, which are hinged on one side as shown in Fig. 2 (the
side facing the camera microscope). After loading the X-ray film,
the cassette is opened slightly, slipped over the screen tube F.1
and the two halves locked in position by locking screws shown in
Fig. 2. It is then mounted on the camera base in the usual manner,

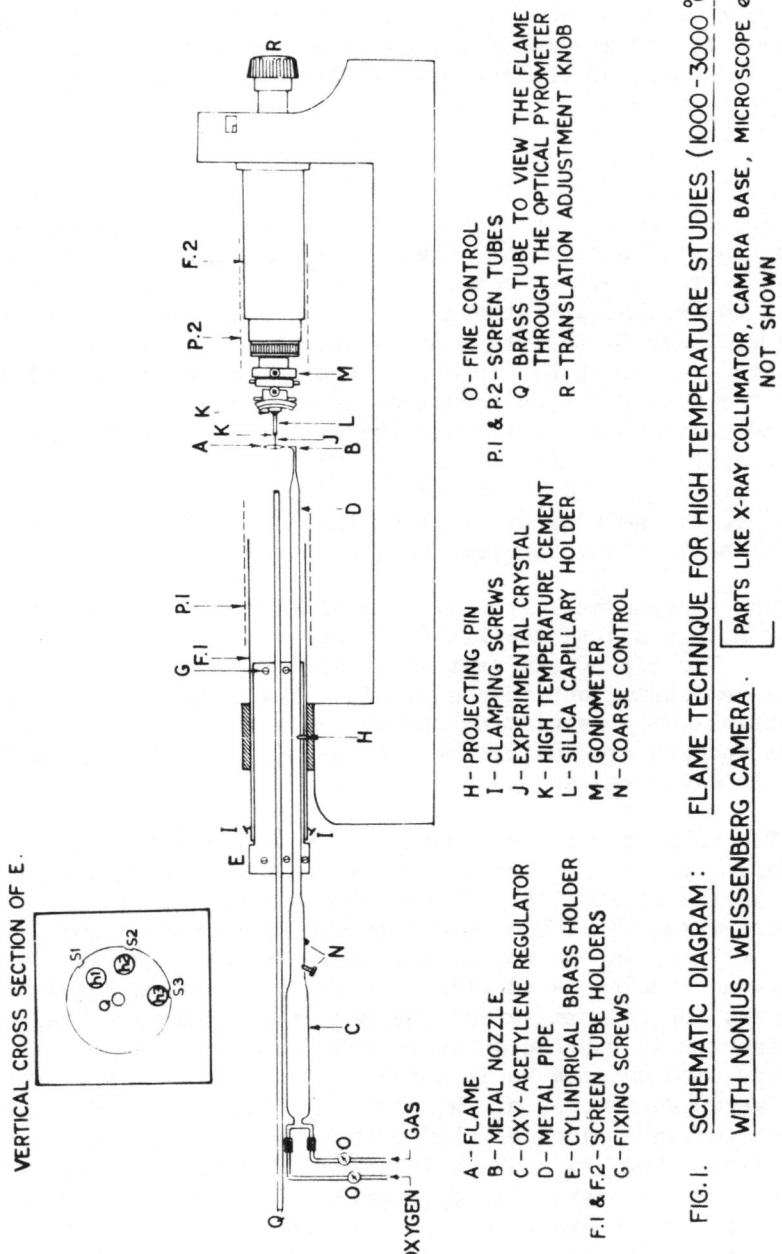

VERTICAL CROSS SECTION OF E.

OXYGEN ⌐ GAS

A - FLAME
B - METAL NOZZLE
C - OXY-ACETYLENE REGULATOR
D - METAL PIPE
E - CYLINDRICAL BRASS HOLDER
F.1 & F.2 - SCREEN TUBE HOLDERS
G - FIXING SCREWS

H - PROJECTING PIN
I - CLAMPING SCREWS
J - EXPERIMENTAL CRYSTAL
K - HIGH TEMPERATURE CEMENT
L - SILICA CAPILLARY HOLDER
M - GONIOMETER
N - COARSE CONTROL

O - FINE CONTROL
P.1 & P.2 - SCREEN TUBES
Q - BRASS TUBE TO VIEW THE FLAME
    THROUGH THE OPTICAL PYROMETER
R - TRANSLATION ADJUSTMENT KNOB

FIG. I. SCHEMATIC DIAGRAM :   FLAME TECHNIQUE FOR HIGH TEMPERATURE STUDIES (1000-3000 °C)

WITH NONIUS WEISSENBERG CAMERA.   [PARTS LIKE X-RAY COLLIMATOR, CAMERA BASE, MICROSCOPE etc
                                    NOT SHOWN]

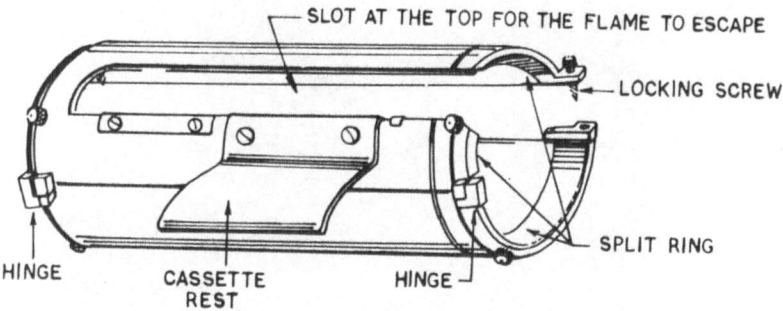

FIG. 2. HINGED SPLIT CASSETTE FOR TAKING HIGH TEMPERATURE X-RAY PHOTOGRAPHS ON THE NONIUS WEISSENBERG CAMERA USING FLAME TECHNIQUE.

the base having first been brought to the extreme left of the camera track. This procedure prevents the film from coming in contact with the flame during cassette-mounting and removal. Contact with the flame during the Weissenberg traverse is avoided since there is a slot in the cassette top as mentioned before.

The screen tubes P.1 and P.2 are permanently placed on the holders F.1 and F.2. They are much shorter than normal so that rotation photographs can be taken by just slipping them back on the holders. P.2 is provided with a 6.5 cm x 1.5 cm slot at the top through which the goniometer arcs can be reached for crystal centering and alignment at any temperature. After the crystal alignment, the slot is closed by a thin outer tube (not shown in figure). The temperature of the crystal is measured by a Ribaud optical pyrometer by observing the crystal through the metal tube Q fixed along the axis of the cylinder E.

The device has been used for taking single crystal photographs of MgO (at 2100°C) in the following manner. A mixture of oxygen and coal gas was used instead of acetylene as the latter was not available at that time. The high temperature arrangement, except for the flame, was first set up. A view of this arrangement is shown in Fig. 3. The experimental crystal, 1 cm long and 0.5 mm x 0.3 mm cross section, was aligned in the usual manner. A rotation photograph at room temperature was taken for one hour, using copper radiation. The crystal was then withdrawn from the

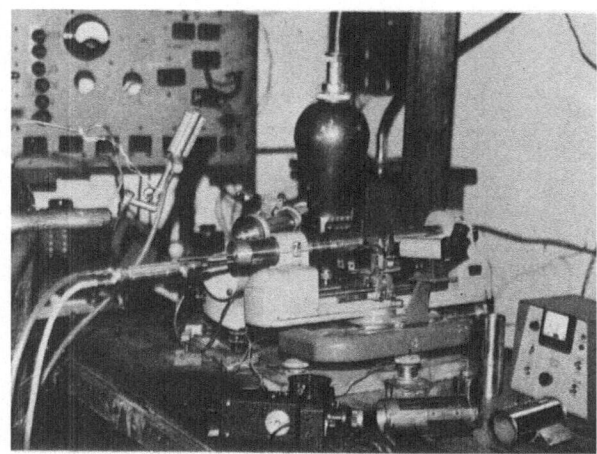

FIG·3· A VIEW OF THE FLAME HEATING DEVICE
ON THE NONIUS WEISSENBERG CAMERA

X-ray beam by turning the translation adjustment knob R. The
flame was lighted up and the gas flow adjusted till the visible
outline of the flame was about 2 cm long and 3 mm wide. The crystal was then brought into the flame by turning R as slowly as
possible in order to minimize any undesirable effects like crystal-
cracking on sudden heating. The high temperature rotation picture
was taken only for twenty minutes as the crystal vapourised rapidly
by dissociation beyond 1800°C itself in spite of its very high
melting point (2800°C).

The high temperature Weissenberg picture was taken by reducing
the circumferential gap between the two screen tubes P.1 and P.2 to
about 1 cm, which just prevented the first layer spots from being
recorded. This gap width was not reduced further to avoid the
heating of the screen tube edges. During the high temperature
exposure, a fresh region of the crystal was fed into the flame once
in ten minutes, by turning the knob R. This was necessary since
the crystal evaporated for the reason mentioned before. Fig. 4
shows the high temperature (2100°C) Weissenberg pattern along with
the room temperature picture, taken after the crystal was cooled
to room temperature.

ADVANTAGES AND DISADVANTAGES

The main feature of this device is its simplicity and the ease

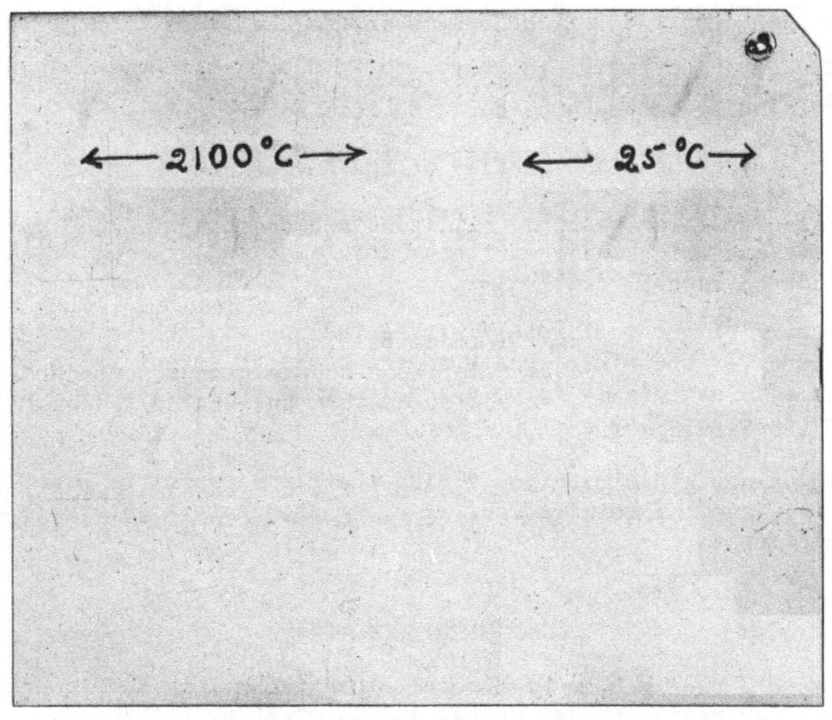

FIG. 4. <u>ZERO</u> <u>LEVEL</u> <u>WEISSENBERG</u> <u>PHOTOGRAPH</u> <u>OF</u> <u>MgO</u> <u>AT</u>

<u>2100°C</u> <u>AND</u> <u>AT</u> 25°C

with which single crystal data at very high temperatures can be collected. Among its advantages are:

1. No water-cooling is needed even at the highest temperature. The goniometer remains cold since the lateral temperature drop from the flame is extremely high.

2. Crystal-mounting and alignment at high temperatures are made with the same ease as at room temperature.

3. The cassette-mounting and removal does not disturb the heater assembly. The movement of the cassette during Weissenberg photography is not hindered.

4. No increased X-ray absorption is produced by the flame.

5. The diffraction spots at 2100°C are just as sharp and straight as those at room temperature, indicating that there are no significant temperature gradients or fluctuations in the X-ray-

irradiated portion of the sample inside the flame.

6. The device permits a straight forward application to single-
    crystal diffractometers, based on Weissenberg geometry.

    Among the disadvantages of the device are:

1. Substances which react in oxidizing and reducing flames, and
    also those which react with the water liberated in the flame
    cannot be investigated.

2. A part of the reciprocal lattice falling within the slot pro-
    vided for the flame-escape in the cassette cannot be recorded.
    This is not likely to be a problem when the device is adapted
    to diffractometers.

3. Temperature distribution inside the flame cannot be exactly
    evaluated. Investigations at close temperature intervals are
    difficult.

CONCLUDING REMARKS

Modern research in high temperature technology is producing
an ever increasing number of new and interesting refractory
materials. Detailed single crystal analysis at high temperatures,
which has not been done so far, can contribute significantly to a
better understanding of these materials--coordination in crystal
structure, thermal vibrations, and electron densities in atoms and
hence the ionization state of the atoms and the nature of chemical
binding at elevated temperatures. There is, therefore, a particular need for high temperature devices which permit these investigations above the melting point of platinum. Admittedly, the
flame technique does not provide for experimentation under highly
controlled conditions of temperature and atmosphere. The paucity
of single crystal X-ray data at high temperatures and the simplicity and inexpensiveness of the technique, however, make the flame
approach for getting these temperatures profitable, particularly in
ordinary X-ray laboratories, even though the information obtained
may be of a limited nature.

ACKNOWLEDGEMENTS

We thank Prof. P.S. Narayanan for his kind interest.

REFERENCES

1. R.A. Gubser, W. Hoffmann and H.U. Nissen, "Röntgenaufnahmen mit der Buergerschen Präzessionskamera bei Temperaturen zwischen 1000°C und 2000°C," Zeit. Krist. 119, S.264-272 (1963).

2. M.A. Viswamitra, K. Jayalakshmi and V. Kalyani, "A miniature furnace suitable for X-ray Weissenberg Photography up to 1000°C", J. Appl. Cryst. 3, 227-229 (1970).

3. M.A. Viswamitra and K. Jayalakshmi, "A simple miniature furnace for routine collection of intensity data on the Hilger & Watts Y190 linear diffractometer", J. Phys. E : Sci. Instrum. 3, 656-657, (1970).

AUTHOR INDEX

Underlined numbers refer to papers in this volume

SUBJECT INDEX

567